IN THE SHADOW OF THE BRIDGE

IN THE SHADOW OF THE BRIDGE

BIRDS OF THE BAY AREA

PHOTOGRAPHS BY DICK EVANS

TEXT BY HANNAH HINDLEY

FOREWORD BY MARY ELLEN HANNIBAL

HEYDAY, BERKELEY, CALIFORNIA

Library of Congress Cataloging-in-Publication Data is available.

Jacket Art: Dick Evans

Jacket Design: Archie Ferguson

Interior Design/Typesetting: Archie Ferguson

Published by Heyday

P.O. Box 9145, Berkeley, California 94709

(510) 549-3564

heydaybooks.com

Printed in China by Regent Publishing Services, Hong Kong

10 9 8 7 6 5 4 3 2 1

CONTENTS

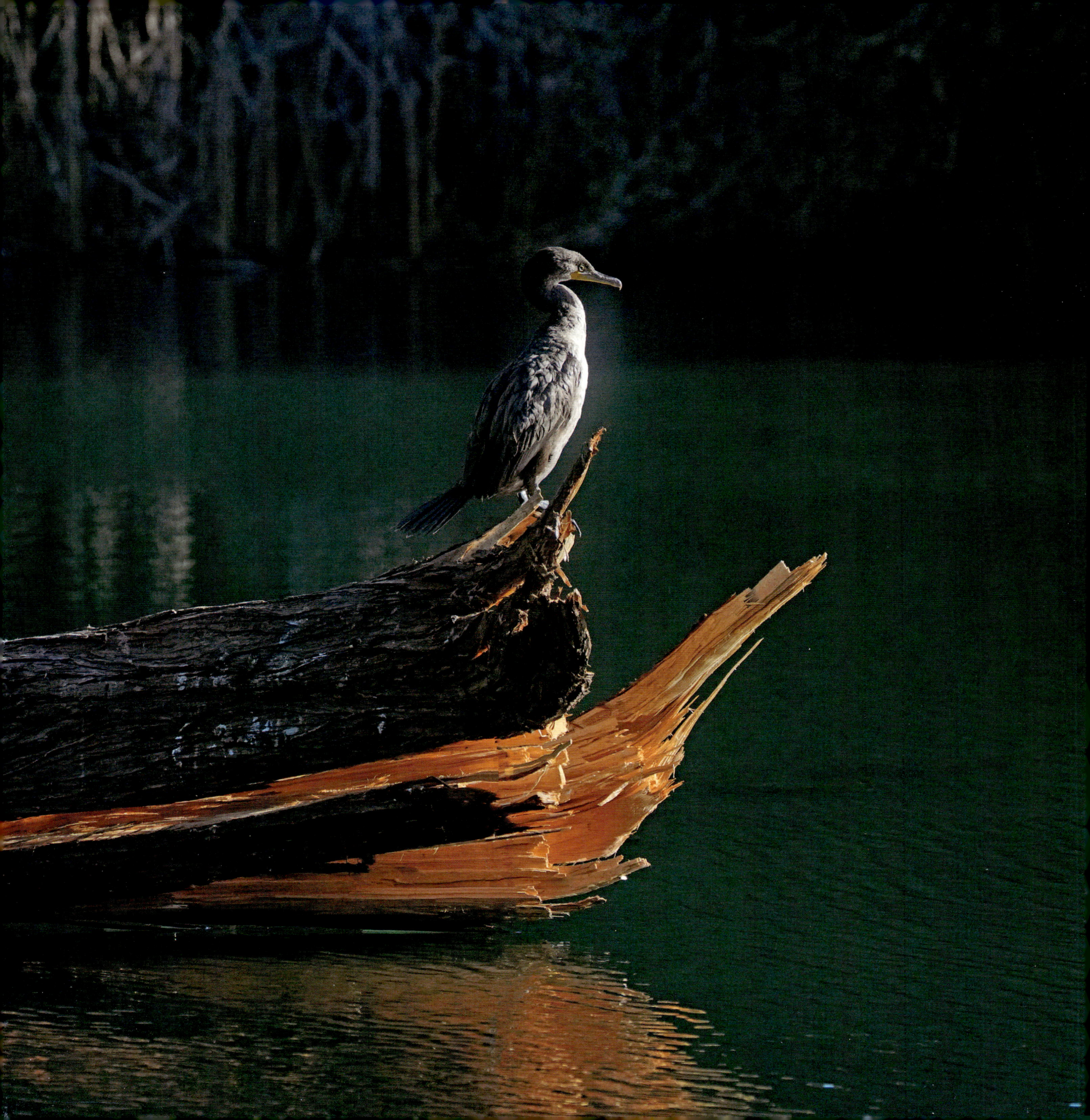

FOREWORD: LOOKING AT YOU

I tried to behave while tagging along with a Point Blue Conservation Science (Point Blue) birding group in fall 2024. Dick Evans was along with his camera, and avian ecologist Mark Dettling hauled a telescope and kept a list. Evans called us the "Fault Line Feathers Group" in honor of our San Andreas–defined zigzag path in Bolinas.

Frankly, I'm the person on a bird walk that the "real" birders would like to jettison. I'm chatty and I have questions. I'm hungry and waiting for lunch and trying to be quiet about the wrapper on my energy bar. I am pretty good at noticing a bird but hopeless at knowing what it is. I'm not particularly bent on improving my birding—I just am glad to be out looking for birds.

I do love learning from the real bird-watchers, who observe patterns and anomalies, get to know the behavior of different species, and stay highly tuned to seasonal changes and the birds they bring. Following Evans's photographer's gaze as it arches upward, querying the sky, then attempting to discern what Dettling is hearing in the trees, I'm guided to perceive more deeply. Birders are also superb citizen scientists. Every observation donated to eBird contributes to the pointillist portrait of birds as they stitch up the tapestry of life.

I look at the birds, I try to "see" them. I want to be involved with the birds and their lives, even as I am thrilled at their otherness—simultaneously outside me, indifferent, and yet mysteriously entwined with my own being. We frequently say that scenes of aesthetic overflow such as the natural beauty Bolinas provides are beyond words. But this unity and diversity are *prior* to words in human development and drive us to express what we see, feel, and attempt to know.

In the Shadow of the Bridge brings new and deep information to this quest. Hannah Hindley's storytelling about birds and the Bay is warm and accessible to all, while managing to capture a progressive view of ecology and its many entanglements. How does the deeply fissured Bay Area geology provide structure to the view and habitat for such a variety of birds? (Dettling counted more than 80 species that day.) How do Indigenous people here integrate avian doings with their own, helping to create what we call nature? Hindley touches down on human history, Earth-system science, land use change, and species' evolution, and how these impact our birdy experience all around the Bay.

A Double-crested Cormorant gazes across Mallard Lake in San Francisco's Golden Gate Park, perched atop a tree toppled in a recent winter storm.

Of course, I find myself staring at Dick Evans's photographs. Wildlife photographers "get the shot," never a mean feat with a subject that is often cryptic, mobile, and far away. But there are subtle and important differences in how bird photographs are indeed captured. Evans gives us many spectacular portraits of birds in situ, living their lives, mating, eating, adrift on the water, or gliding in the sky, all the while following patterns in time that are millions of years old.

But there's another surprise in these photographs as well. Many of Evans's portraits have a focus on the bird's eye. Evans is looking at the bird and taking the picture, but the bird is not precisely looking back at him. I find myself entranced with these persistently perceiving birds. Evans has fixed his eye on their eyes in a direct encounter. A biologist will tell you the bird is looking for potential prey, a mate, or a predator. Fine, we're all doing some version of that, right? But it doesn't quite explain the charismatic presence these birds have. Looking at them informs us as well about the moment we live in, its history in the making. The birds are seeking, recognizing, knowing. We're doing that, too. And as Wallace Stevens put it, the bird is "involved in what I know." We seek to know this place of land and water—this wonderful book helps us know it better.

—MARY ELLEN HANNIBAL

Ancestors of this Common Raven have lived alongside human communities in the Bay Area for thousands of years. Their lives and routines continue to run in parallel with ours as we share urban space with these ever-present companions.

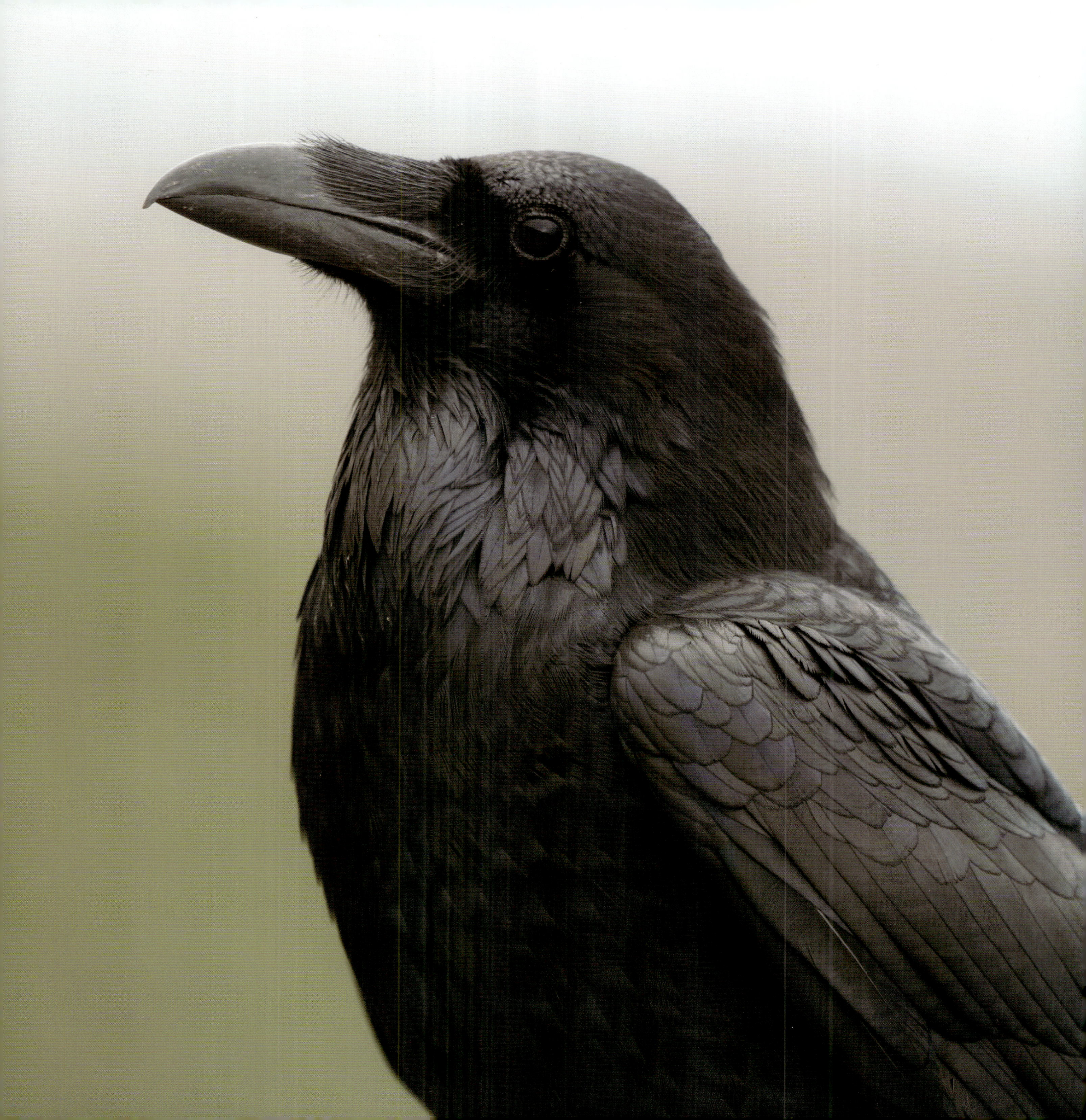

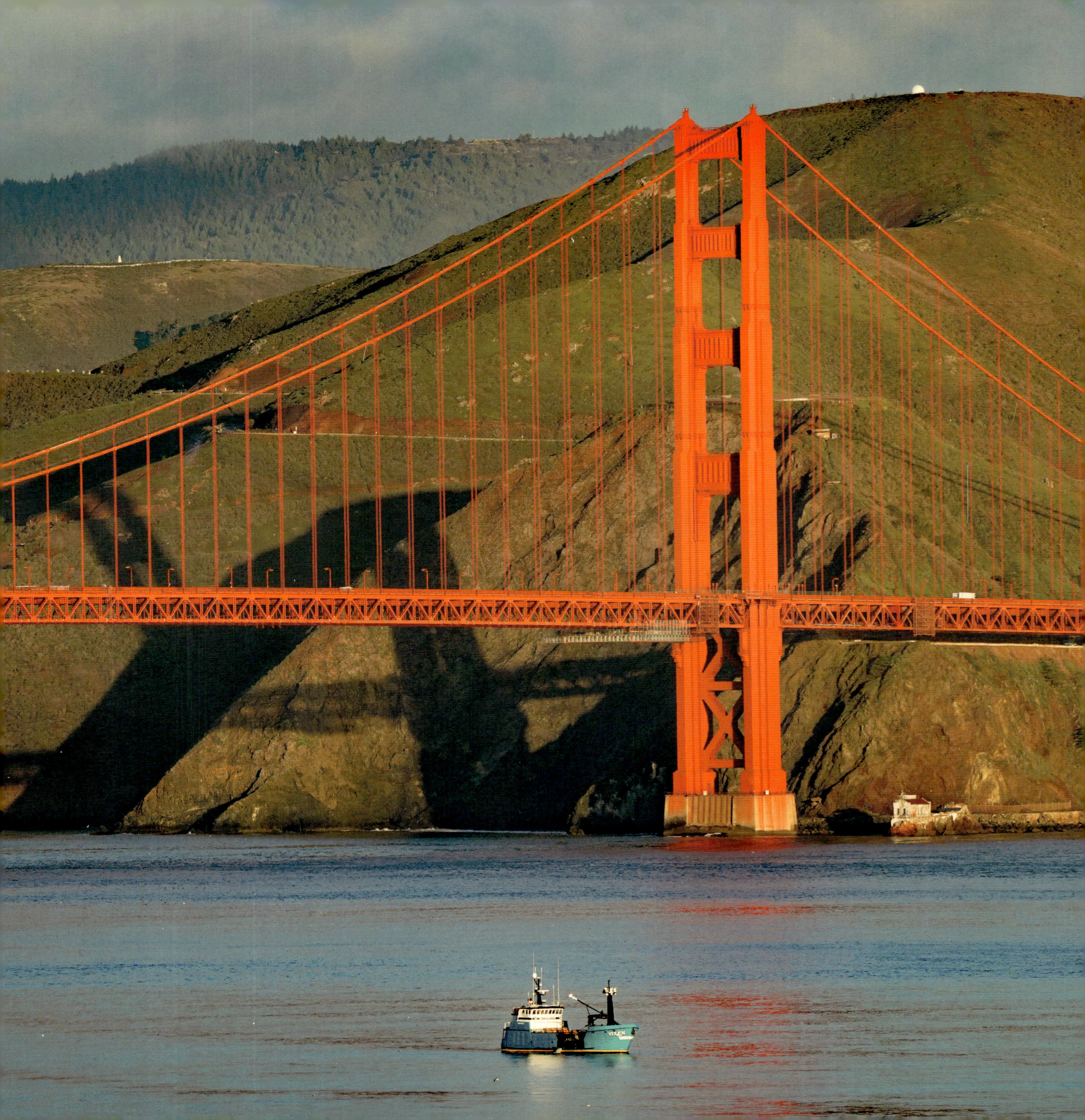

INTRODUCTION: A WALK AMONG BIRDS

Step outside your door on an early autumn morning, and look up. If you're standing within sight of the Golden Gate Bridge, you might see raptors migrating from the north, rising on the morning thermals above the Marin Headlands. Perhaps you can spot the yellow flash of a Townsend's Warbler in a neighborhood tree, coming from the forests of Alaska's Inside Passage. The greater Bay Area is an avian convergence zone. Arriving from far corners of the globe, birds connect this region across space. A Tundra Swan landing in the Cosumnes River Preserve south of Sacramento has traveled from the edge of the Bering Sea—over 4,000 miles. A tiny Wilson's Warbler pausing in the Presidio along its migration is like a living needle, stitching a continuous path between Costa Rica's cloud forests and the Canadian Rockies.

Every winter, Sandhill Cranes arrive by the thousands to feed in the winter fields and waterways of the California Delta. The species has made this same migration for more than 2 million years—or, by some estimates, for as long as 10 million years. They are descendants of an unchanged lineage dating perhaps as far back as the Miocene, when the Andes first began to rise and when the Eurasian and African continents first collided. Stand in a flooded rice field at the edge of San Francisco Bay, where the cranes arrive en masse at dusk in a churn of wings and water, and you can feel the tug of deep time.

In 2023, a flock of critically endangered California Condors flew over the East Bay, their slow shadows sliding over the docks of Alameda on their way toward Mount Diablo. It was the first time in a hundred years. But in the Pleistocene, these vultures and their kin soared across North America, floating hundreds of miles a day on wings as wide as U-Haul trucks. Now, the California Condor is the last surviving member of its genus, a holdover from days when mammoths and dire wolves ranged across the region. Cranes and condors, swans and warblers—to watch birds is to meet space and time, wrapped in feathers.

This book project began with the simple act of taking a walk. During the constrained early days of the 2020 pandemic, Dick Evans slung his camera over his shoulder and stepped out his door in San Francisco to stretch his legs when the world felt as though it was closing in. And what a place to go for walks: More than half of all the bird species in the United States can be found in the long

Tides push hundreds of billions of gallons of water under the Golden Gate Bridge daily, feeding the largest estuary on the California coast and all the birds that call it home.

shadow of the Golden Gate Bridge. Look up, and you might find Sharp-shinned Hawks circling the Marin Headlands, Tree Swallows swooping low over the lakes in Golden Gate Park, migratory songbirds thronging through city greenbelts in all the colors of the rainbow, rare vagrants taking refuge on our coast, Northern Pintails arcing between their breeding grounds in Siberia and their winter refuges in the California Delta.

In parallel, I too had been taking long walks—exploring my homeplace with fresh eyes after my career as a wilderness guide in Alaska was cut short by the pandemic. Each step I took deeper into the oak savannas and Doug-fir forests of home brought me closer to appreciating the profound wildness of my own Bay Area backyard.

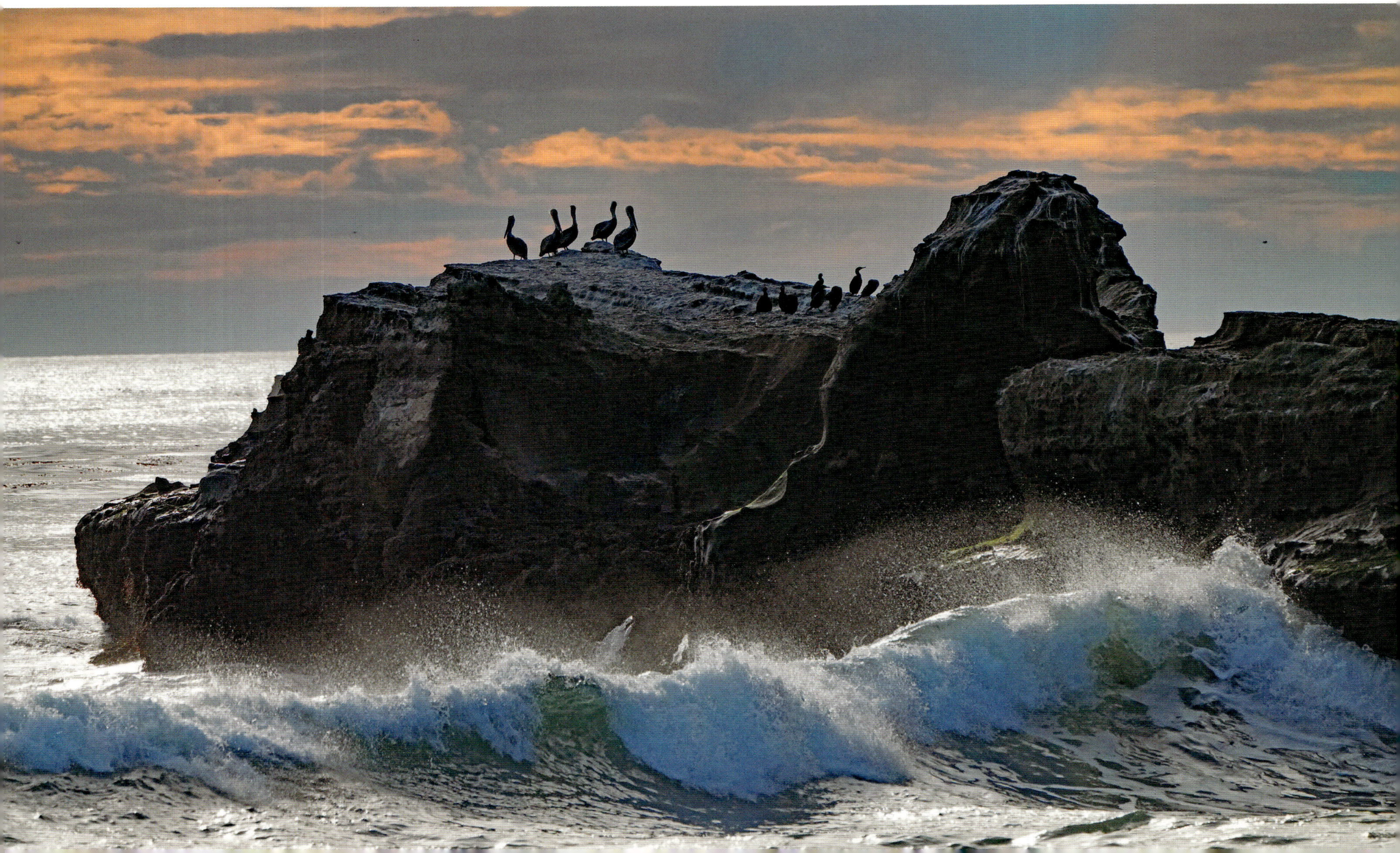

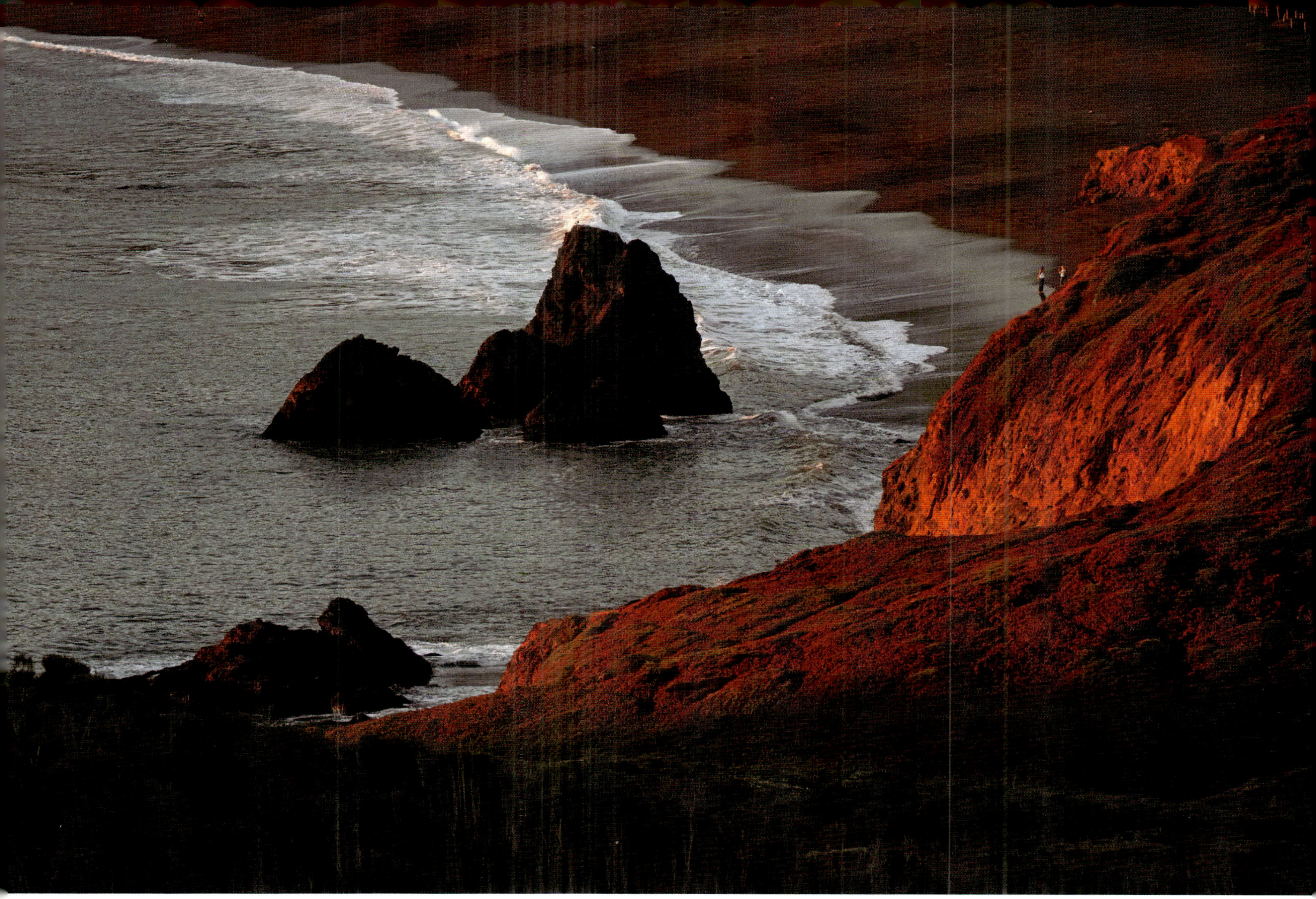

From Año Nuevo State Park (left) to the Marin Headlands (right), anywhere the Pacific meets the California coast, birds throng to take advantage of cliffy roosts, sandy beaches, tide pools, mudflats, estuaries, and coastal scrub. Where sea meets land, birds of all feathers hunt, nest, preen, and forage.

When Dick approached me about putting together a story about the birds of the region—told in pictures and in words—I knew this was a special project. We recruited Christine Walker from Point Blue Conservation Science to be our deeply attuned guide to the ecological and conservation stories across the region. The three of us began going on walks together.

In the story that follows, we invite you to join us on those walks. Let your eyes lead you among the flocks of birds that call the Bay Area home, from the shores of the Laguna de Santa Rosa to the

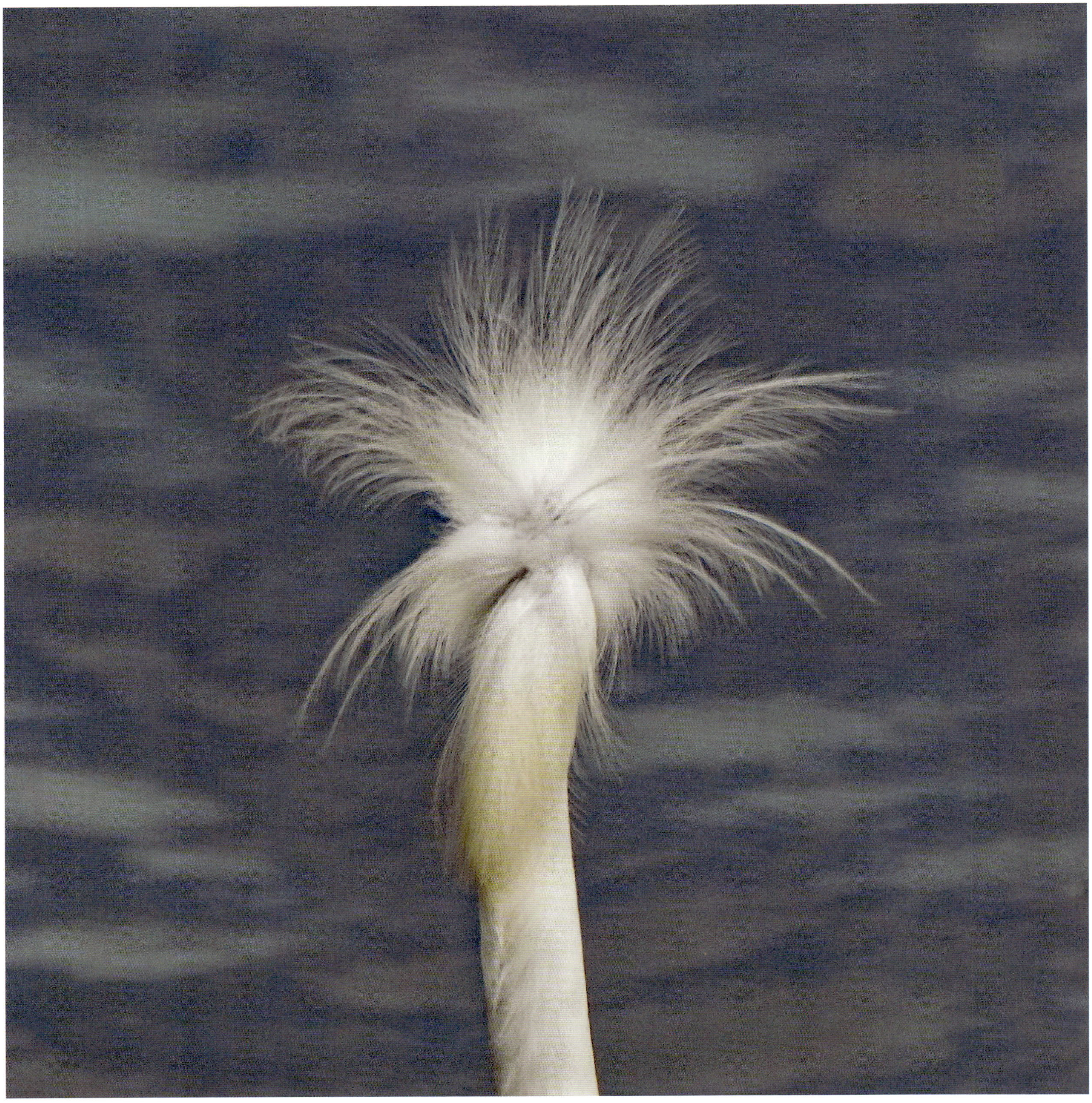

tidal marshes of Oakland The book is divided into three sections that explore the histories, the communities, and the complex conservation questions that weave together to make this region remarkable for bird diversity and for the study of birds.

First, we uncover the natural forces that have shaped the greater Bay Area into such a bird-beloved place. Step back in time to watch the slow movement of tectonic plates, the rise of mountains, and the merging of great waters as the tides from San Francisco Bay rise to meet a dozen Sierra-fed rivers to form the sprawling California Delta. This confluence is the largest estuary on the Pacific Coast. The forces that have sculpted this region—from glaciers and fog to subduction and oceanic upwelling—bring birds from distant latitudes right to our doorsteps as they converge along North America's great westernmost migratory highway—the Pacific Flyway.

At its heart, this book is about relationships. In the middle section, we look at the human histories that have developed in response to the birds of the Bay. Strike out to sea with us to look for nesting seabirds among the foggy crags of the Farallon Islands—the same remote archipelago where hungry nineteenth-century "eggers" from San Francisco collected so many Common Murre eggs that the birds were nearly wiped out forever. Visit the Grace Hudson Museum & Sun House to sit with master basket weaver Susan Billy and learn how feathers are woven into Pomo baskets, in a process that treats each bird not as a resource but as a connective thread to a whole living ecosystem. Follow us at dawn to the Palomarin Field Station to catch birds in gauzy mist nets on a quest to understand where these migrants come from—and what they might tell us about our changing world. Join us in conversation with local artists who are building a movement to interweave science and art in search of a richer understanding of birds and their habitats. Peek behind the doors of the specimen collections at the California Academy of Sciences to see how the bodies of birds—a plucked feather, a snipped hummingbird claw, the invisible double-helix of an owl's DNA—can tell stories about how things were, and what the future may hold.

A sudden gust of wind rearranges the head feathers of a Snowy Egret (left, also pictured below).

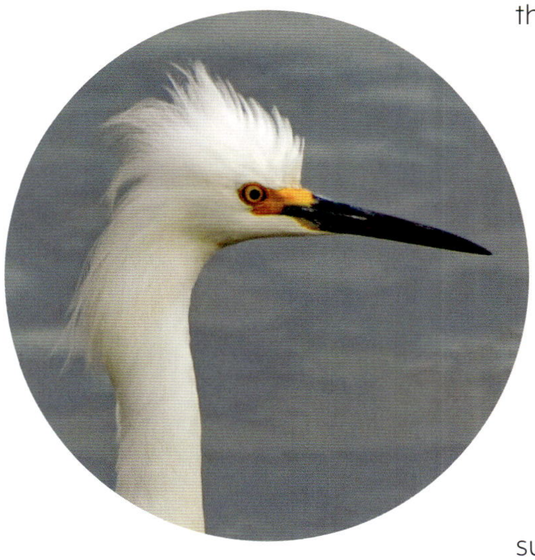

In the final section, we explore the evolving challenges that Bay Area conservationists face. How do wildlife managers weigh the choice between letting endangered Spotted Owls disappear, or systematically killing off the hundreds of thousands of invasive Barred Owls that have supplanted them? How can city planners mediate sea-level rise in a way that sustains the dwindling tidal marshes that Ridgway's Rails and Salt March Common Yellowthroats rely on? What do Central

Valley rice farmers and Snow Geese have in common? There aren't tidy answers. But what's clear is that now is a time for talking across borders. Some of the most hopeful stories of bird conservation in the Bay Area are grounded in unlikely partnerships: among ranchers, schoolchildren, the military. "Birds help us connect across difference," says David Hage, cofounder of the ecological nonprofit Weaving Earth.

In this final section, we also turn our attention to the educators who show us the entry points into bird-watching and bird conservation. These are the guides, the guardians, and the mentors who help us look ever more closely and celebrate the beauty that's worth protecting.

Just as the text in the book weaves together these three broad themes, the photos aren't tightly organized around topic or species so much as around theme. We encourage you to view them through the lenses of *place* and *relationship*, and—when you linger to appreciate the sharp silhouette of a Steller's Jay against the sky, or heavy beads of water gleaming on the back of a Ring-necked Duck—we hope you might linger with a deepening sense for the conservation efforts that are ongoing in the region to protect the diversity on display in these pages.

The truth is, it's a complicated time to be a bird. In the last 50 years, three billion birds have disappeared from North America—about 30 percent of the population that existed in the 1970s. Birds are vulnerable. Their respiratory systems can't handle the smoke from supersized wildfires fueled by climate change. City windows knock them out of the sky as they fly through urban mazes. Pesticides and environmental pollutants weaken their eggs and dull their feathers. For the swirling rivers of birds that migrate through the Bay Area every spring and fall, any disruption along the paths they travel—a late snowpack, a haywire tropical storm edging farther north than usual, the warming of a cold ocean current that once carried rich snacks for far-traveling birds—might topple their success. A Townsend's Warbler on the shore of Bolinas Lagoon also depends on the health of ecosystems along the Columbia River Gorge, in Alaska's Alexander Archipelago, in Haida Gwaii.

Beginning in the early 1900s, miners would bring caged canaries with them under the earth. If a canary died, it was an indicator that odorless carbon monoxide gas was entering the tunnel. The birds—more sensitive to the poison than humans—were early warning signs of danger. Today, too, birds are often the first to know when something is wrong. If we pay attention, we can make out the shape of the coming peril. "Birds are the nervous system of an ecosystem. They're also the Twitter feed," says David Hage. "They're like a tapestry, connecting everything. They're letting us see a huge story."

Hooded Mergansers (male pictured here) embody a bridge between the land and waters of coastal California. They hunt underwater for fish and crustaceans, and they rely on forests for the tree cavities where they nest.

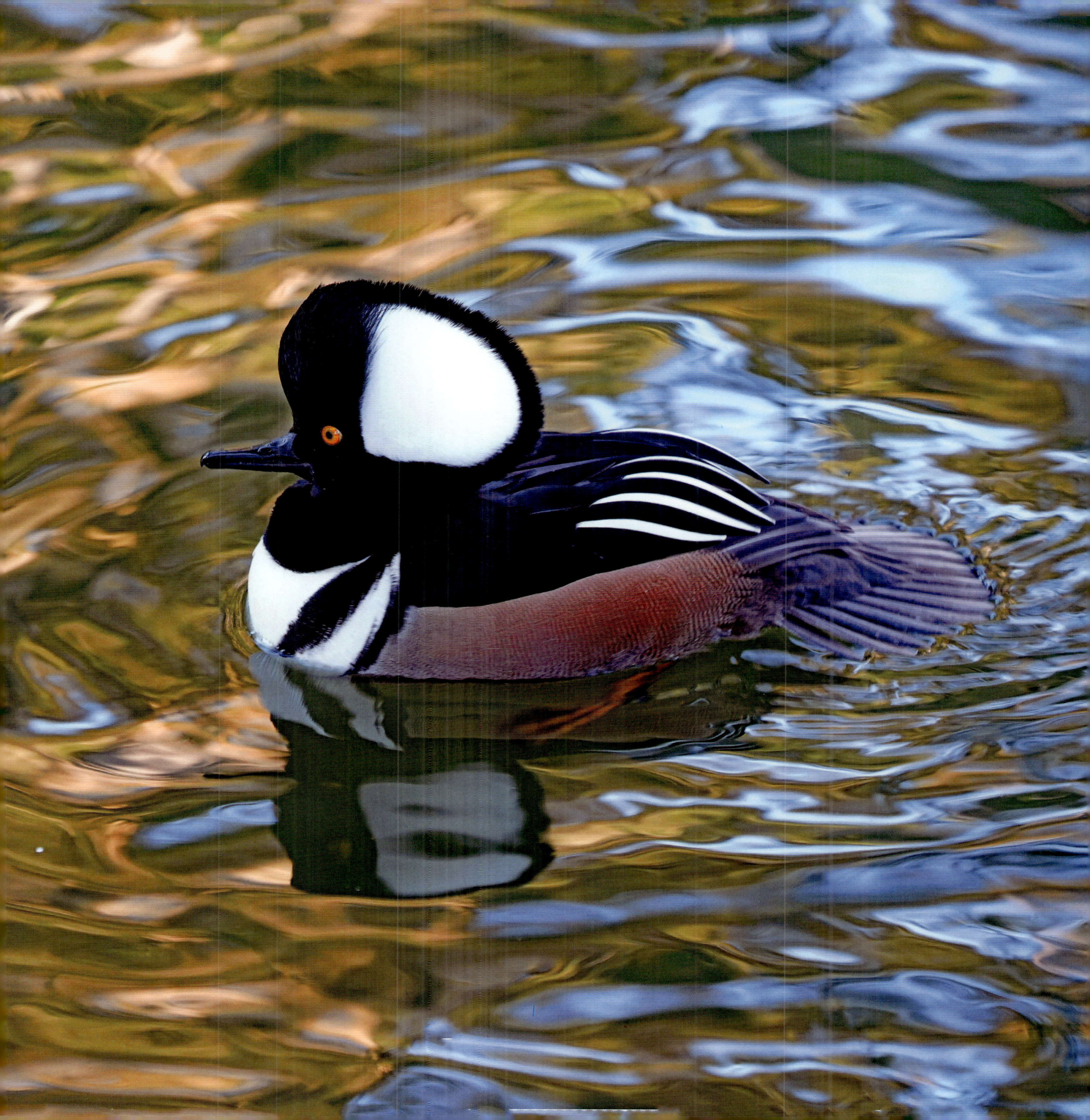

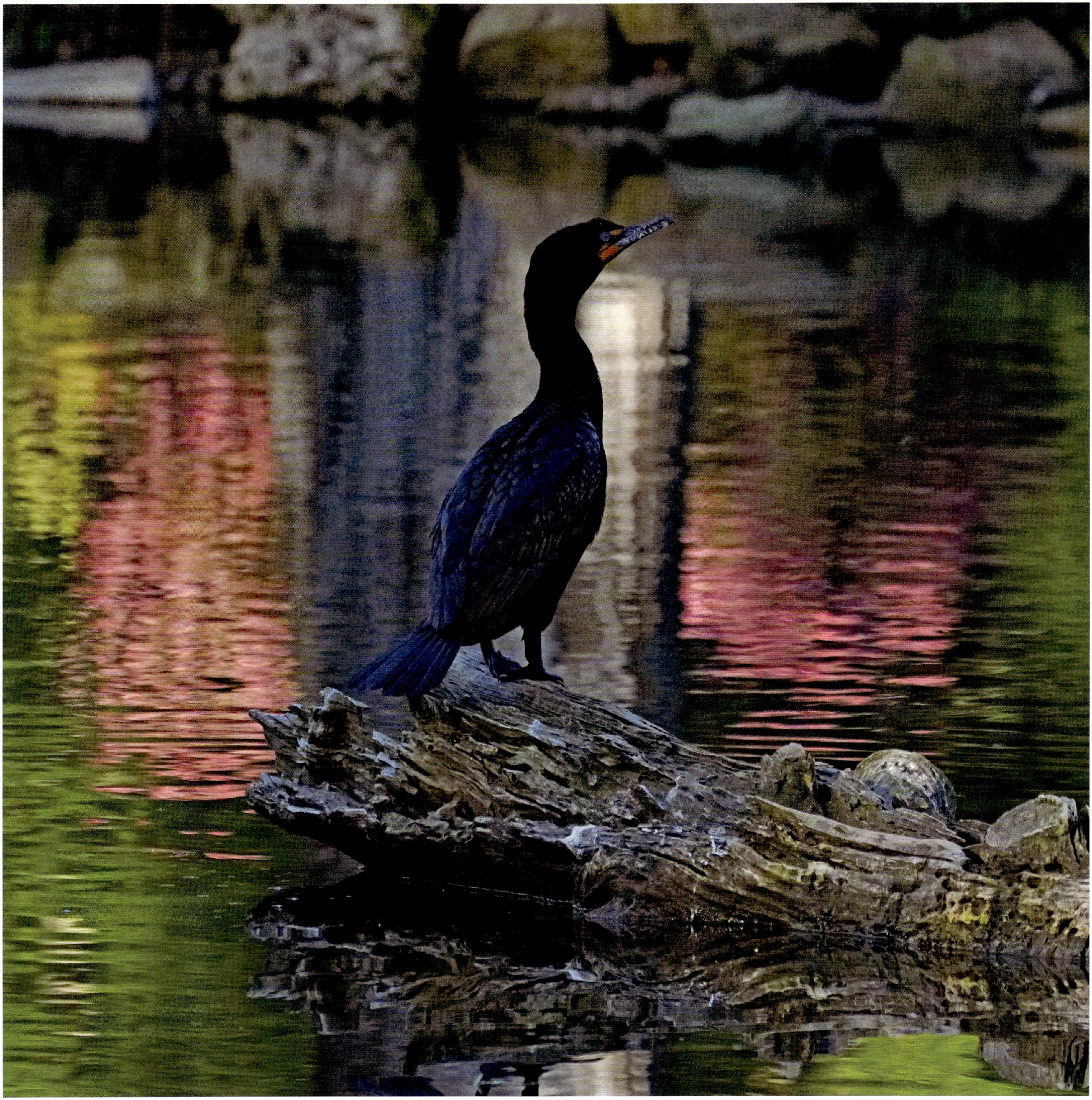

For 60 years, Point Blue Conservation Science (formerly the Point Reyes Bird Observatory) has monitored the birds of the Bay Area. The organization's long-term data archives enable us to see the shape of the tapestry the birds are weaving. There's value in consistent monitoring—after all, it's hard to know what new information birds are telling us if we haven't been recording their earlier stories too. Because seabirds are quick to respond to changes in their environment—sometimes within a week—they are especially strong indicators of larger forces at work, whether the changes are physical (such as storms and upwelling), chemical (acidification, for example), or biological (such as food availability). "If you want to learn about Central Valley salmon, look at Cassin's Auklets two years beforehand," says Pete Warzybok, program leader for Point Blue's Palomarin Field Station on the Farallon Islands. "They're experiencing the same ocean conditions that the juvenile salmon are experiencing." Across years, geography, and generations, birds are the keepers of histories and the harbingers of change.

For all the humans who love birds, it is an exciting time to be in the Bay Area. The region has a century-long record of conservation battles and victories and is home to environmental organizations, artists, clubs, storytellers, educators, and societies dedicated to celebrating and sustaining local ecosystems. In 2020, Governor Gavin Newsom signed an executive order for 30 percent of all of California's land and water to be permanently protected by the year 2030. The 30x30 goal aims to restore biodiversity and build climate resilience across the state. Since 2010, there has been a 40 percent increase in land protected by California land trusts. These protections include land trusts with Indigenous leadership, such as the Ramaytush Ohlone Land Trust of the San Francisco Peninsula. Of all the times to be looking closely at Bay Area birds, this moment is alive with challenges and possibilities, both.

The pictures in this book were all taken on publicly accessible lands. You can explore each of these places: the beaches of Monterey, the windswept cellblock on Alcatraz Island, the lakeside paths in Golden Gate Park. The Farallon Islands National Wildlife Refuge, though closed to public landings, can be seen from the decks of visiting boats. This book is an invitation. Come walk these same trails; come seek these same bright flashes in the trees.

This book is also an invitation to see yourself in its pages. After all, this story is about relationships. We hope you might consider your place in this region not just as a resident but as an engaged neighbor. We all live along the Pacific Flyway, birds and people alike. In this greater-than-human community, there are numberless ways to participate. On our next walk, perhaps we'll find you out in the woods too: your head tilted, your ears tuned to the conversations unfolding in the branches above.

Unlike many waterbirds, Double-crested Cormorants lack fully waterproof feathers—this helps them dive more effectively, and also means shoreside visitors are likely to see cormorants perched on stumps and rocks while these birds wait for their plumage to dry.

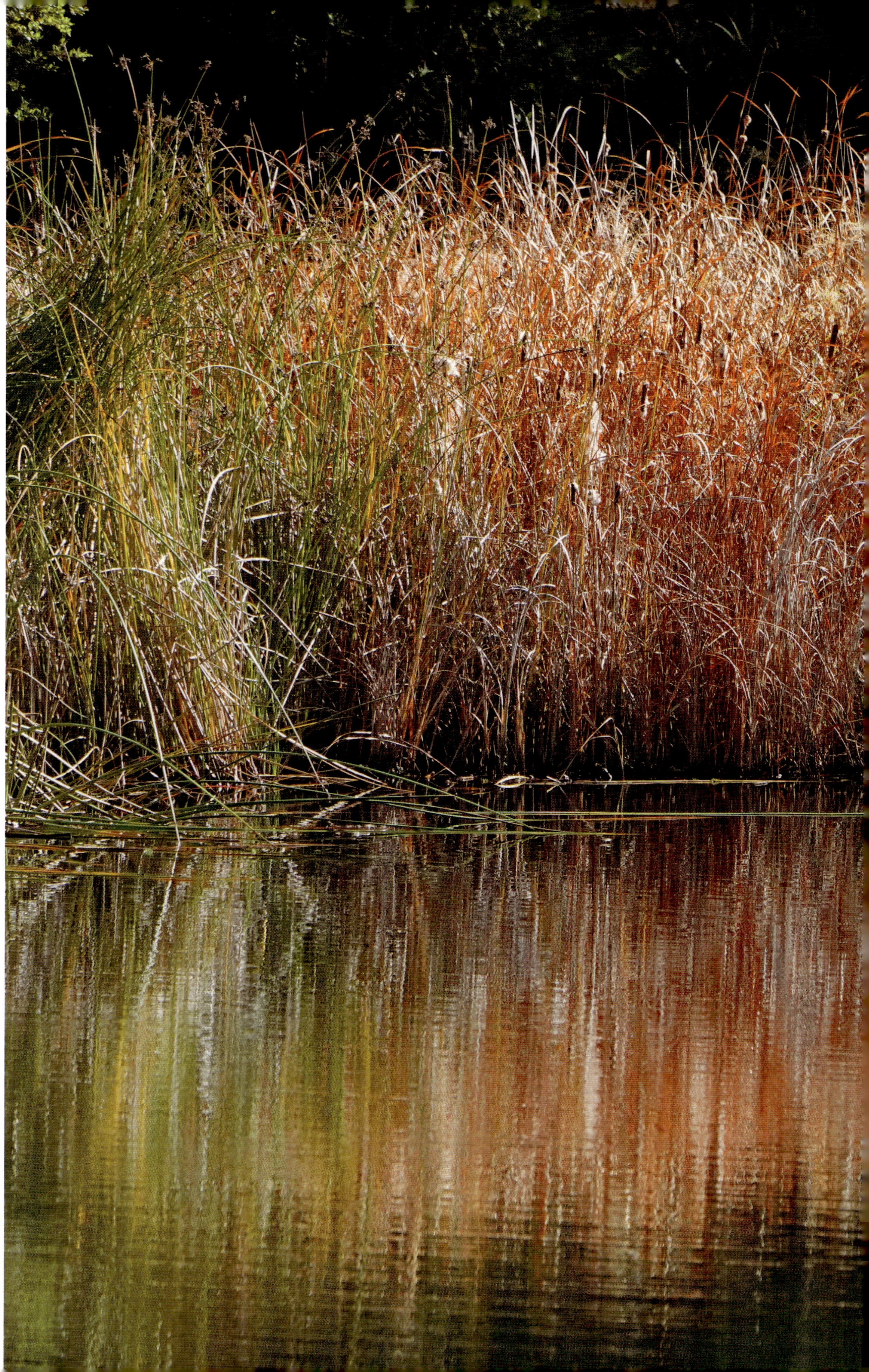

Reedy lake margins like these on Lake Lagunitas are a perfect place to listen for Red-winged Blackbirds, Marsh Wrens, and countless other hidden singers.

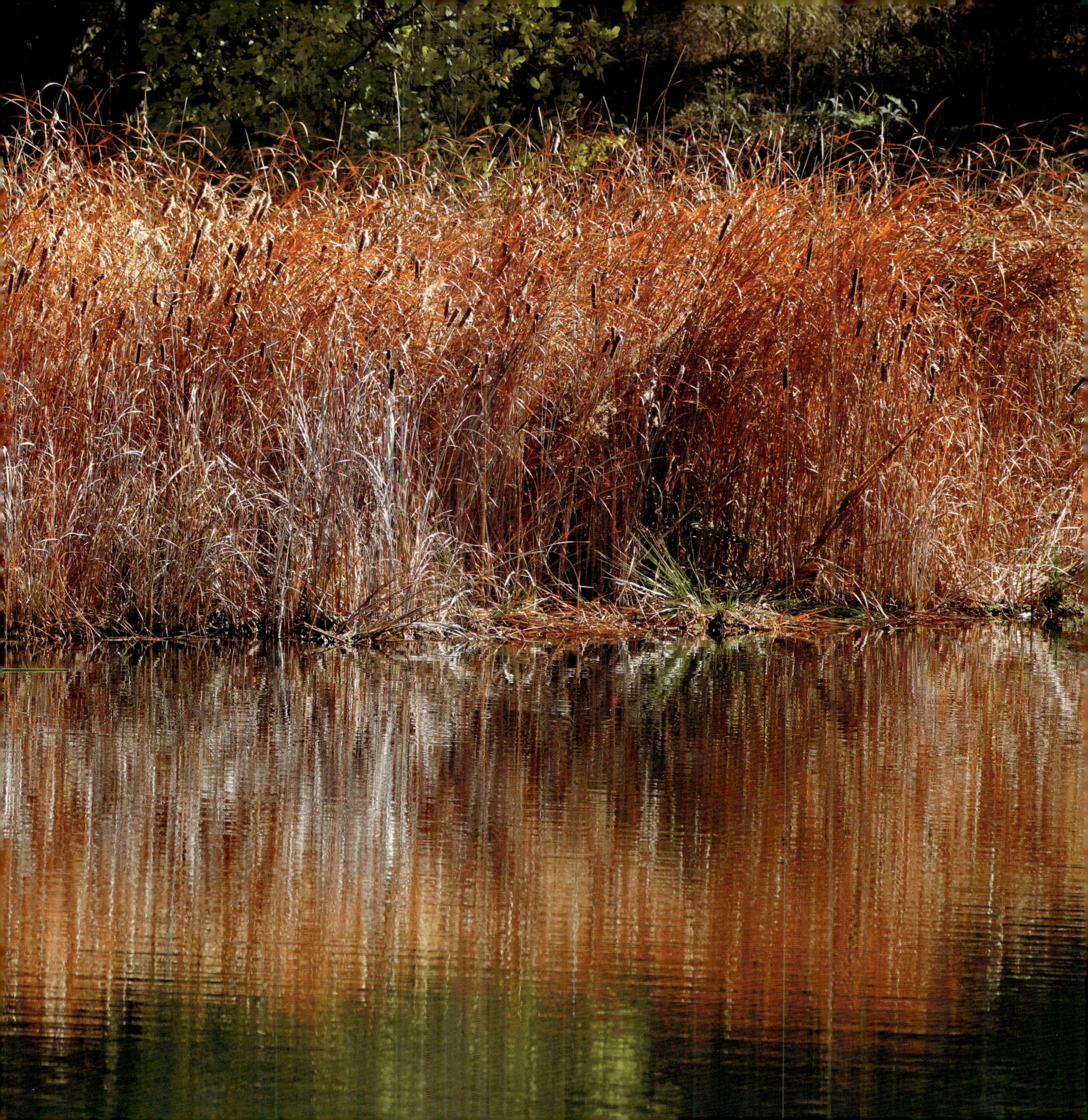

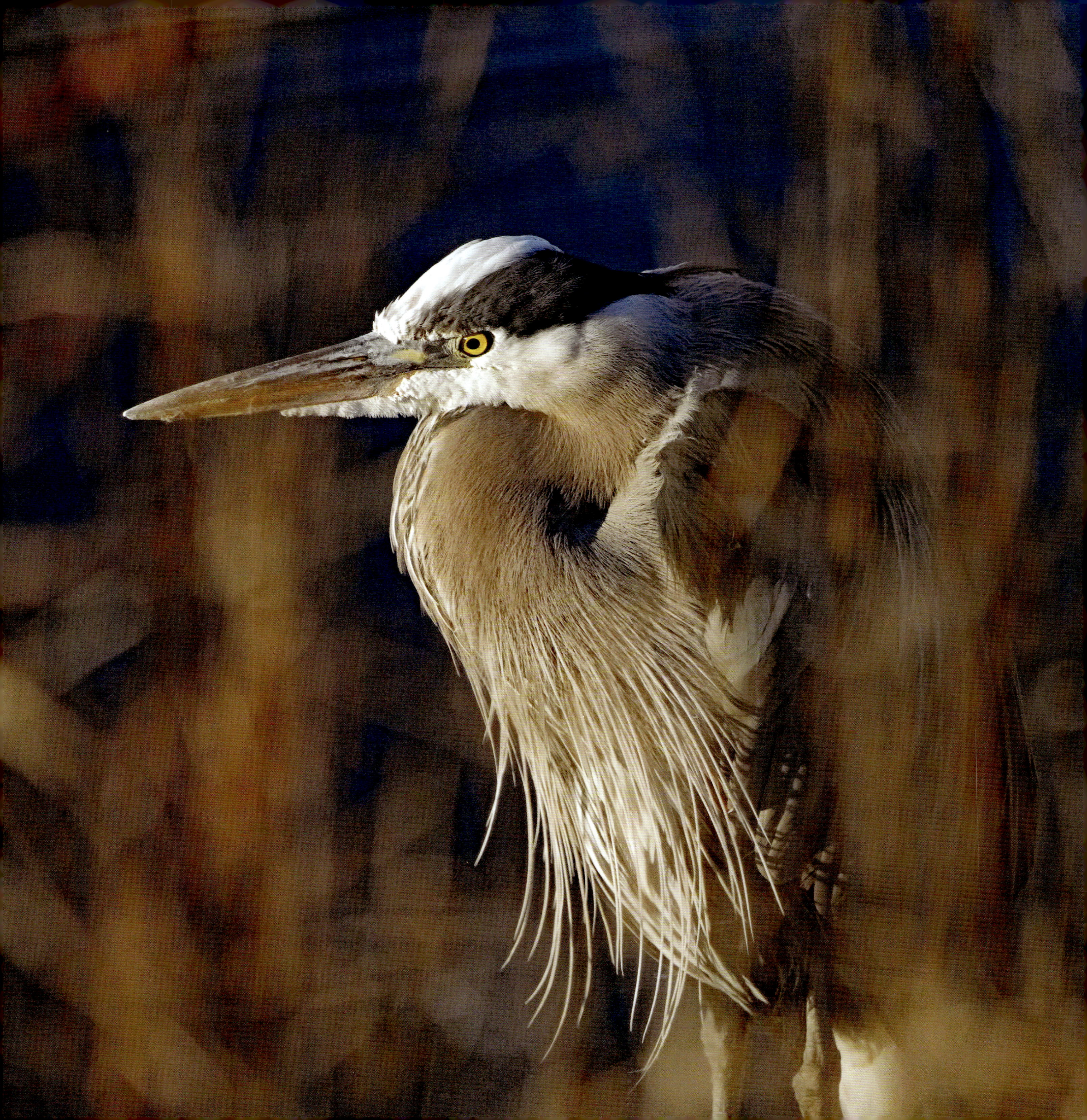

PHOTOGRAPHER'S STATEMENT: IN PLAIN SIGHT

In March 2020, I was driving with my son-in-law Praveen Madan from San Francisco to the Cosumnes River Preserve near Sacramento in the Central Valley. We were taking the scenic shortcut through Antioch and the Sacramento–San Joaquin Delta when the car radio played the news: the jinxed cruise ship with several verified cases of COVID-19 had been cleared to enter San Francisco Bay that morning, not to dock in San Francisco or Oakland but rather to dock and be quarantined at a military base in Solano County—which by chance we were entering that minute and planned to be hiking later in the day. We briefly considered turning around but concluded that if COVID could travel 30 miles in the cold, wet Delta in a few hours, we were probably doomed anyway.

At the time, I was also busy coauthoring my second documentary photography book with Heyday—this one on San Francisco's Chinatown, following a book three years earlier on San Francisco's Mission District. The Mission and Chinatown books were built on the foundation of hundreds of hours of walking the neighborhood streets and talking with (and photographing) the residents. However, the pandemic mostly shut down my urban walking and talking. Instead, my wife, Gretchen, and I redirected our hiking activity to the parks, preserves, and trails anywhere within a two-hour drive of our home in San Francisco. That translates roughly to the Russian River to the north, Monterey Peninsula to the south, and the Cosumnes River Preserve to the east, all in the shadow of the Golden Gate's influence. We explored the Pacific coastline, the giant San Francisco Estuary, and the encompassed wetlands, forests, and chaparral-carpeted hills.

I took my camera on most of these outings and quickly realized that this region held a trove of local and migratory birdlife. In total I have captured approximately fifteen thousand images of 165 species. Because of the region's unique abundance and variety of birds, it has also nourished a long-standing, dedicated, and knowledgeable cadre of birding clubs, scientists, and bird conservancy organizations. Some such as the California Academy of Sciences date back to the mid-1800s, and others, including my book partner, Point Blue Conservation Science, started in the mid-1900s.

Dry winter grasses on the shores of Muzzi Marsh frame the piercing eye of a Great Blue Heron.

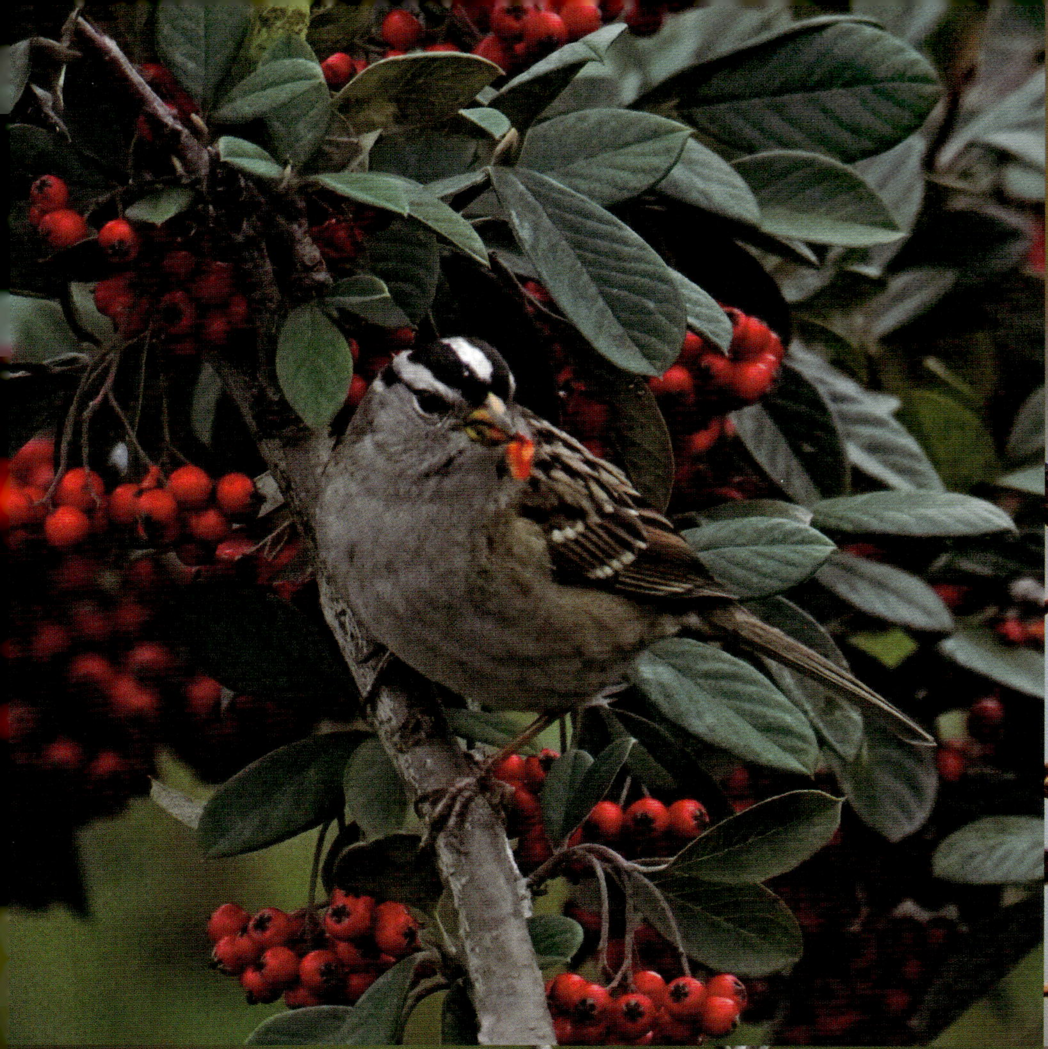
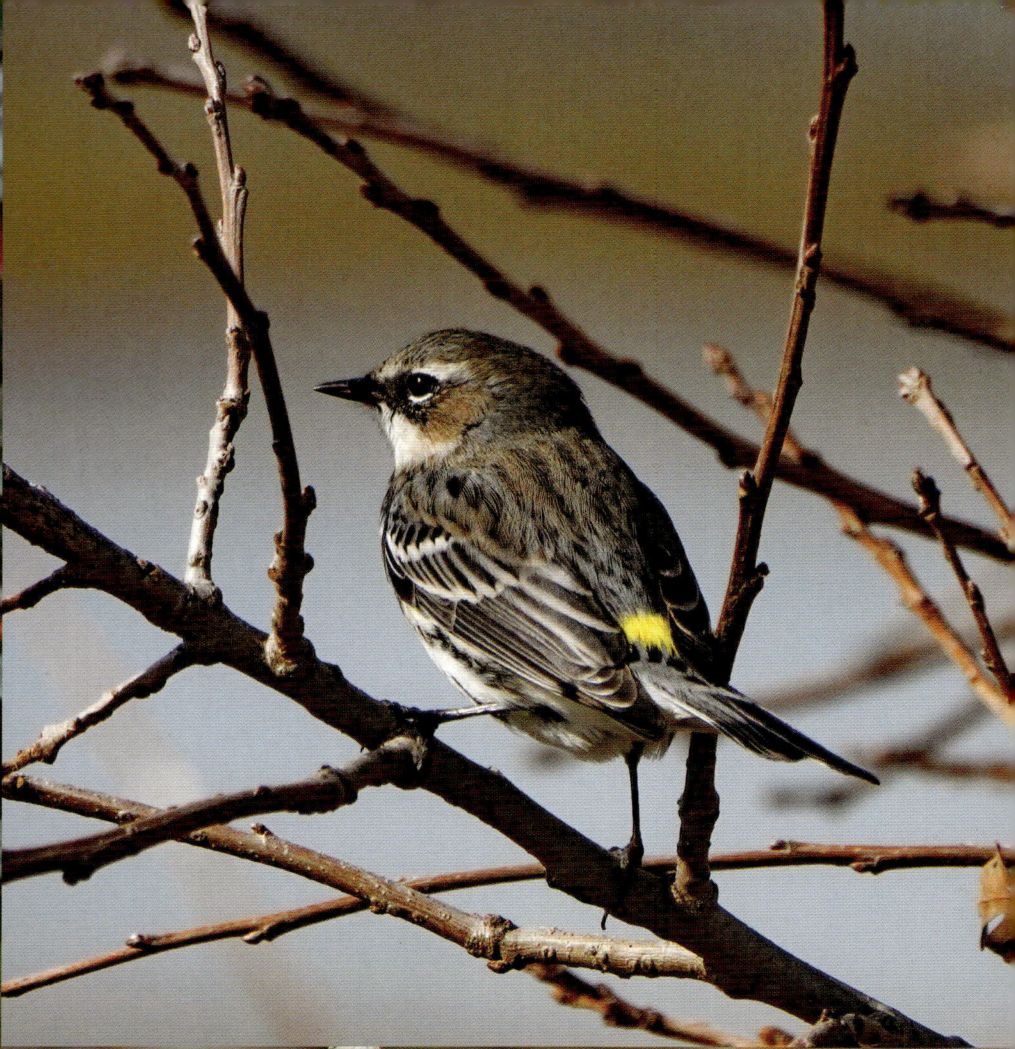
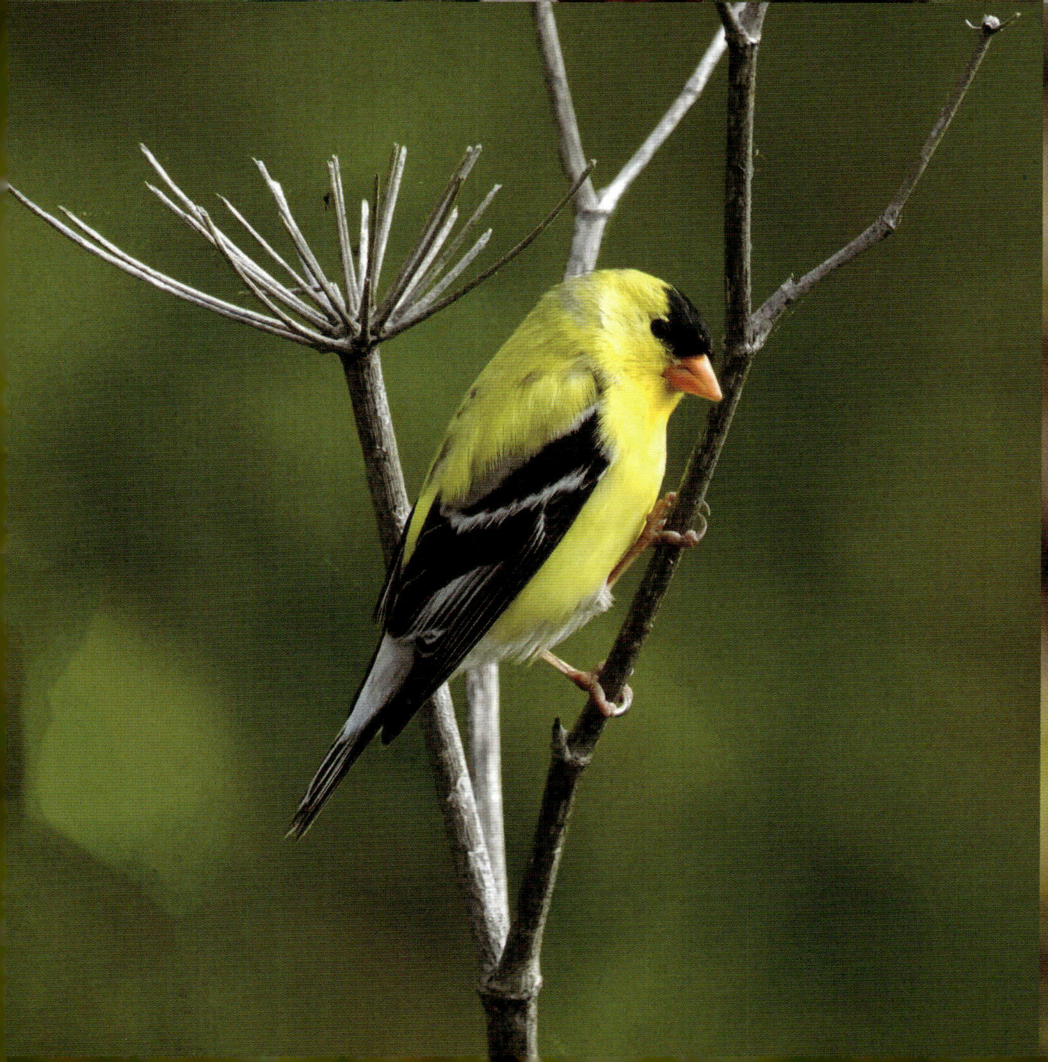
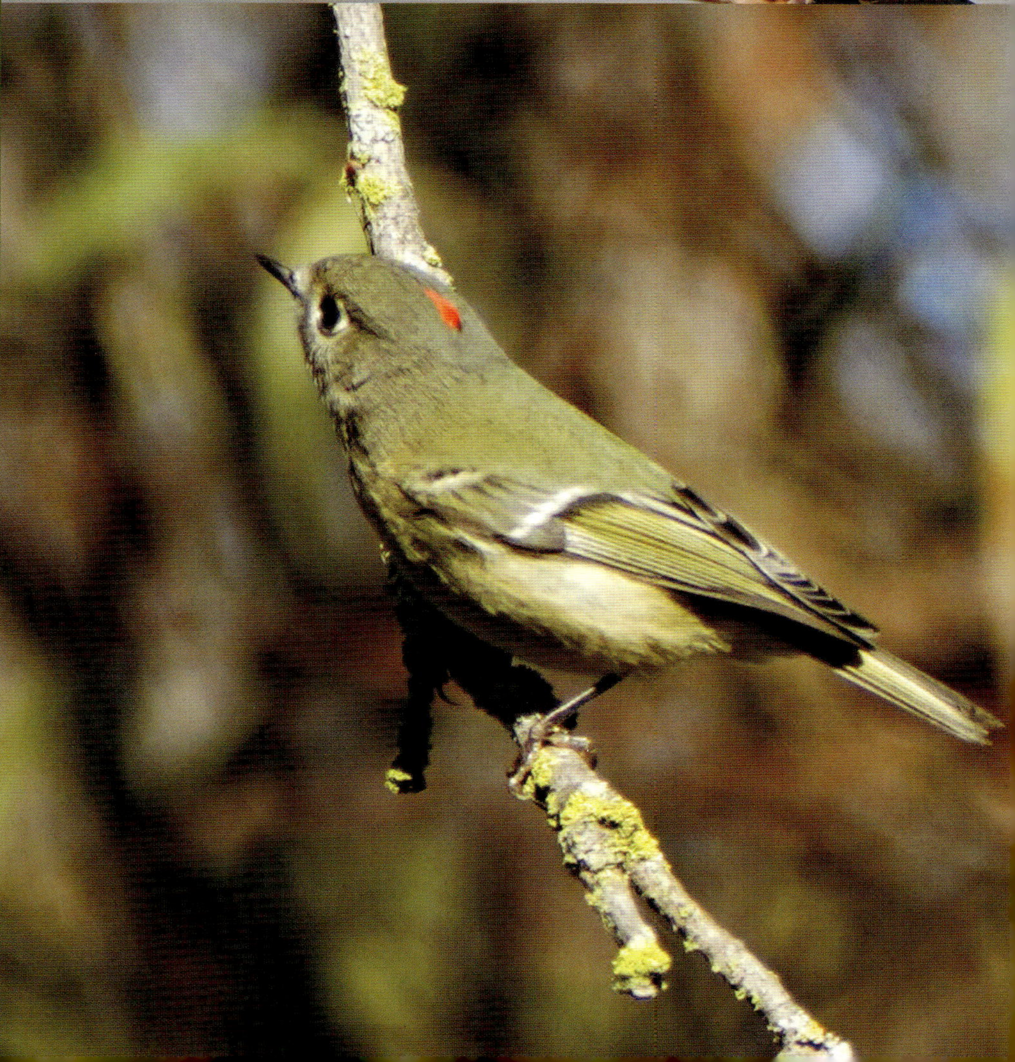

I have never been a believer in predeterminism, but neither does it seem plausible that our lives unfurl totally randomly. I grew up on a hilly cattle ranch in western Oregon, where I spent hundreds of hours with birds in nature. I also realized at an early age that I wanted to see a good deal of the world, and studied engineering as a way to make that happen. Marrying someone (55 years ago) who also wanted to see the world undoubtedly helped. We've been lucky enough to live in Washington, California, Ohio, and West Virginia as well as Ghana, Switzerland, Paris, and Montreal—and each new location has afforded opportunities to appreciate birds. Making my home now in San Francisco has fostered enduring relationships with the people, birds, and organizations of the Bay Area, including with Heyday and my coauthor Hannah Hindley, bringing this book of Bay Area birds to life.

When I was fifteen in 1962, my mother, a lifelong lover of literature and teacher of languages, gave me a copy of the then-controversial best seller by Rachel Carson, *Silent Spring*. The title came from the fear that the pesticide DDT would wipe out swaths of avian species and silence our springs. It nearly did. This book is my small contribution to support those who are working hard to protect our future birds of spring.

As seasons change, a walk along any path in the Bay Area might carry us among local beloved birds like the White-crowned Sparrow and the American Goldfinch (top and bottom left, respectively) or treasured migratory visitors like the Yellow-rumped Warbler and the Ruby-crowned Kinglet (top and bottom right).

A song sparrow (below) gluts on late-summer blackberries.

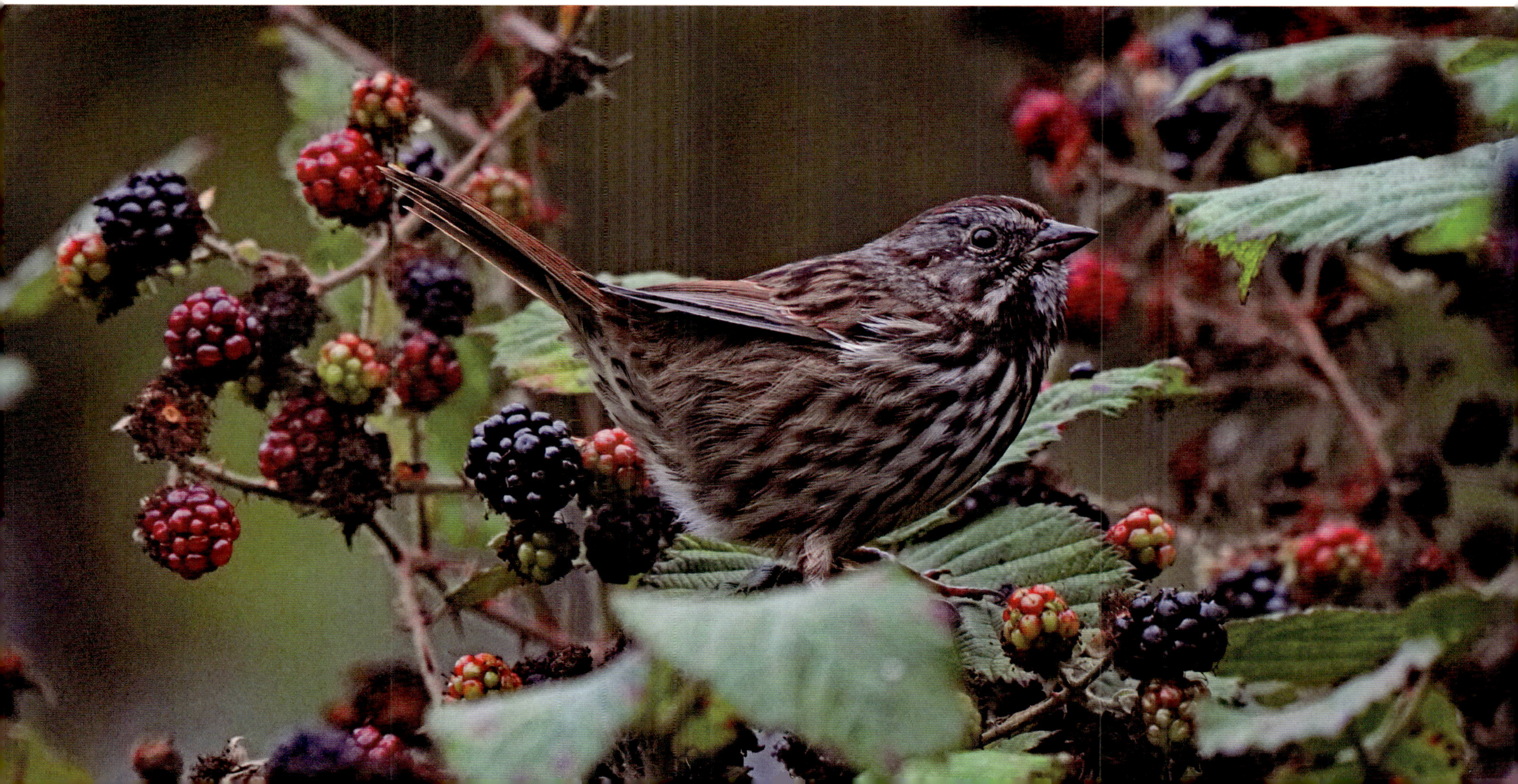

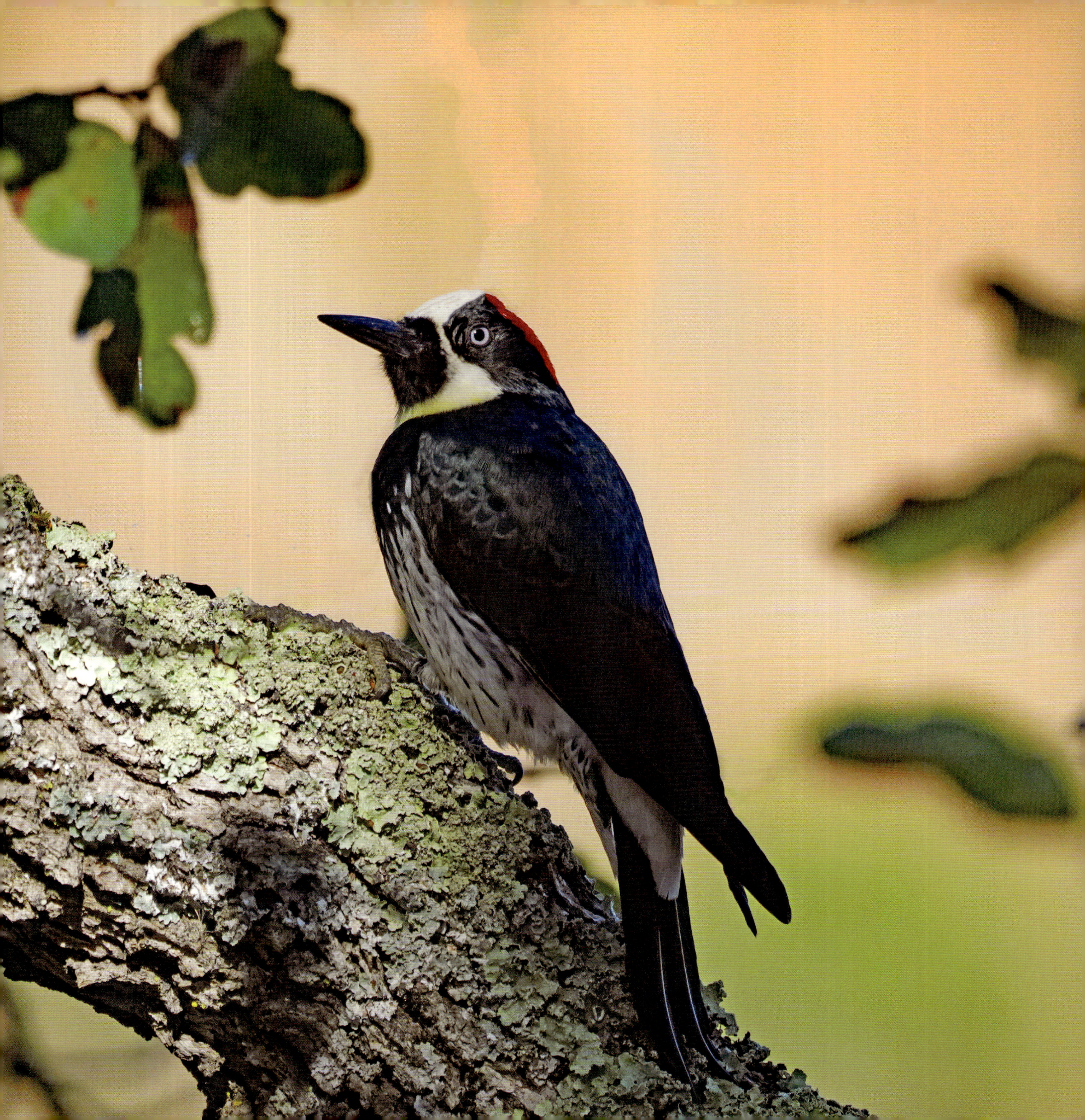

WRITER'S STATEMENT: HOMECOMING

In the late spring of 2020, I returned to my family home in the North Bay, driving through the night, my Subaru rocking in the highway wind. It was the height of the pandemic. When I finally reached home, my mother and I went out for daily walks on the roads branching out from our hometown: Henno Road, Nun's Canyon, Adobe Canyon. Acorn Woodpeckers drummed above us in the roadside oaks. At the same time, Dick Evans was walking the public lands of San Francisco, eyes open to the patterns birds were weaving.

Mine had been a reluctant homecoming. I was supposed to go to Alaska that season, guiding wilderness trips as I had done every summer for a decade. I missed Alaska's ankle-deep carpet of moss, wolf tracks pressed into soil under towering spruce trees, the summer songs of forest birds.

Then, one day in this difficult time, as I sat under a tree on the Bear Valley Trail at Point Reyes National Seashore, I heard a familiar Alaskan sound: the clear trill of a Varied Thrush, like a referee's whistle blown on some unseen playing field in the woods. As I lingered, another familiar song washed over me: the complex, cascading notes of a Pacific Wren, as bubbling and full of movement as running water. If I closed my eyes, I could have been 2,000 miles north in Alaska's temperate rainforest.

Walking the coast later that afternoon, I watched long skeins of Brown Pelicans moving south along the shore like fat pterodactyls. All through my twenties, working in the winters on a small boat in the Sea of Cortez, I had witnessed this same population arrive to breed and fish alongside bottlenose dolphins and blue whales.

My homeplace, I began to realize, was a global crossroads. I unclenched my jaw. I had found myself—inadvertently, obscurately—at the center of something extraordinary. Time and again, the birds continue to shake me awake to the rare beauty of the Bay. As Dick's photos put in plain sight, no place is wilder than our own backyard. No place feels more precarious, either.

Artist and birder Jenny Odell writes, "Our citizenship in a bioregion means not only familiarity with the local ecology, but a commitment to stewarding it together." To be a true citizen in this state of change, I know that I will be called not only to love this place and its dwellers—mother, husband, neighbors, Wrentits, and raptors alike—but to care for it too. When the first Varied Thrushes return in the autumn to fill the coastal forest with song; when the last cranes of spring rise from the shallow delta waters and, with voices like rattling bones, ascend—how will I show up for them?

A familiar denizen of western oak woodlands, the Acorn Woodpecker is a highly social bird, living in large "clans" which cooperatively help raise the next generation.

FOLLOWING: Butts up! Northern Pintails, a Mute Swan, Canada Geese, and a male Mallard (left, clockwise from top left) all show off their best angles. These species are dabblers, meaning they tip their bodies and duck their heads underwater to forage for food. In contrast, Hooded Mergansers (right) dive for their food, seizing small fish with their serrated bills.

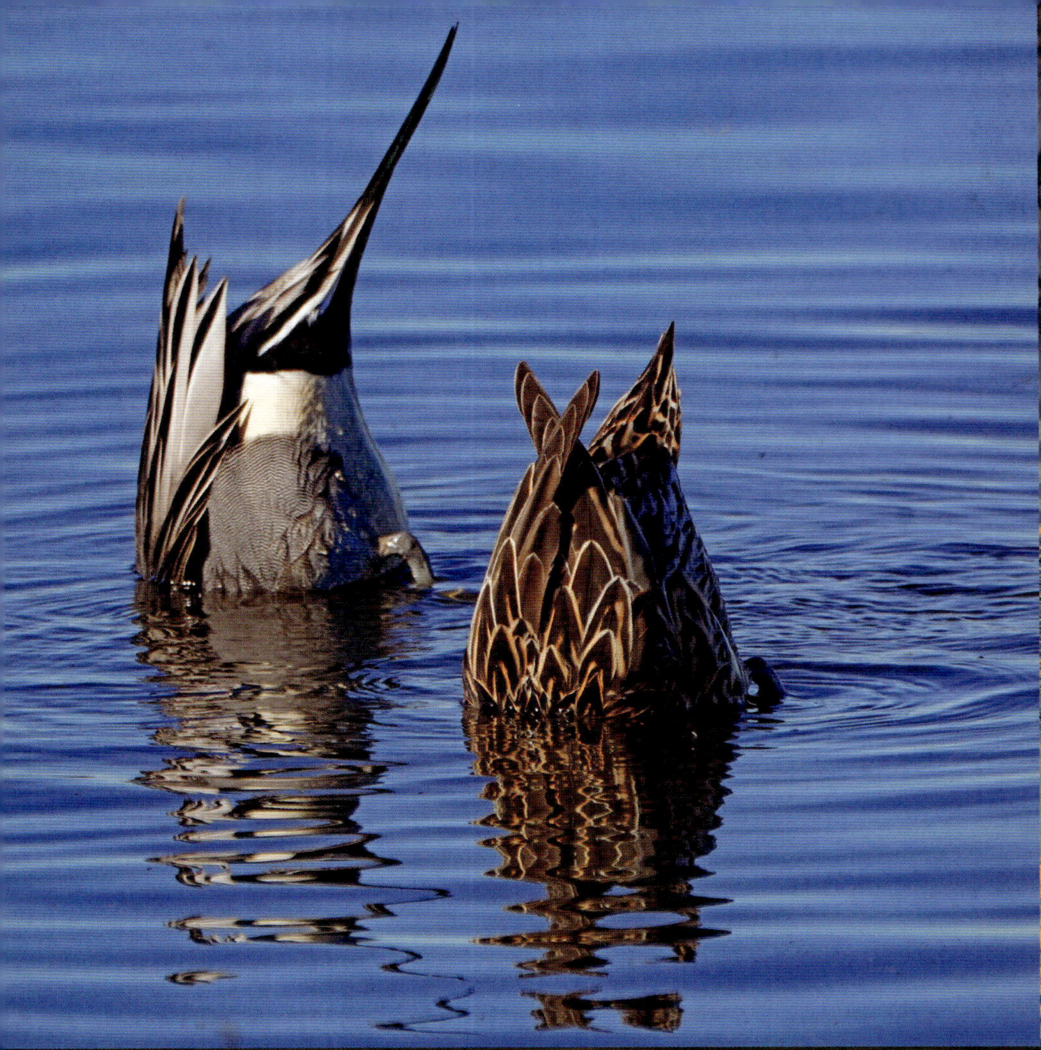
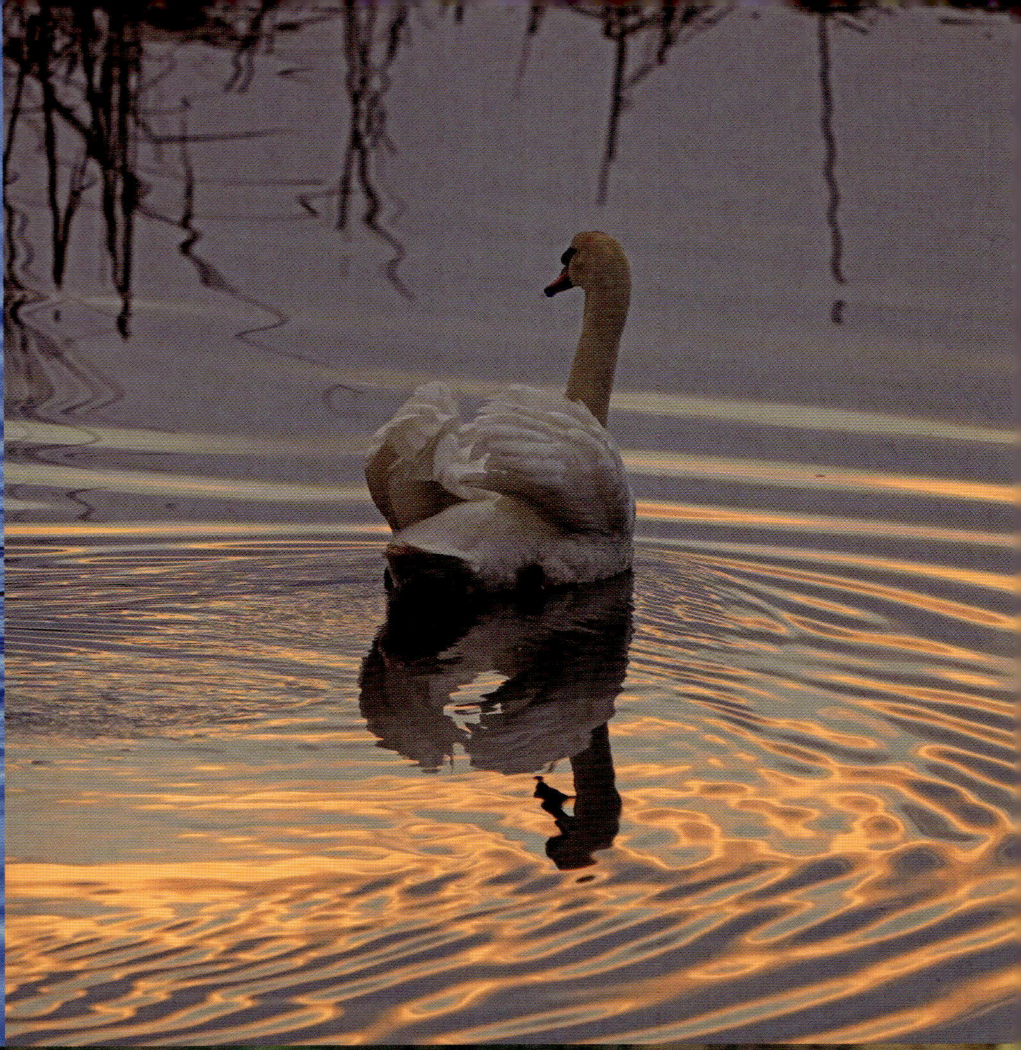
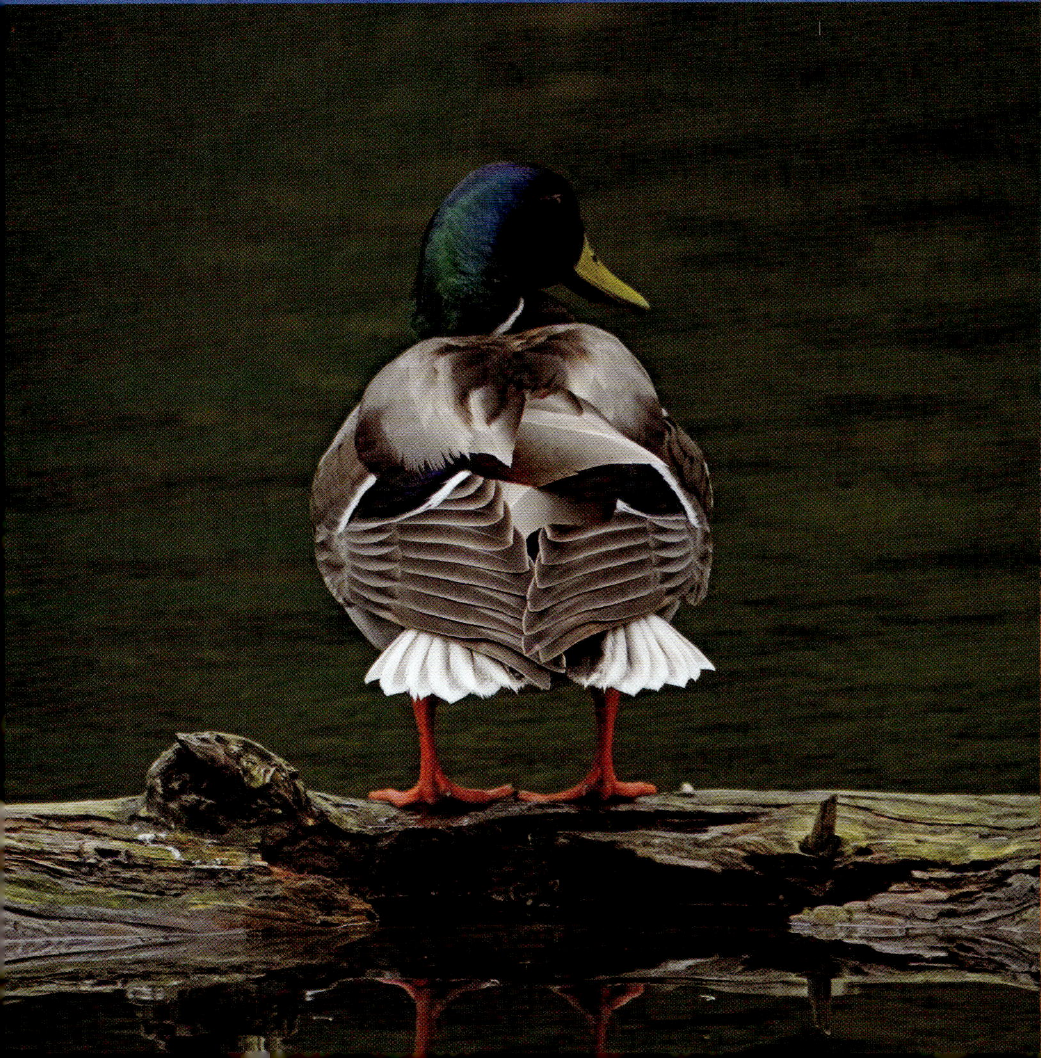
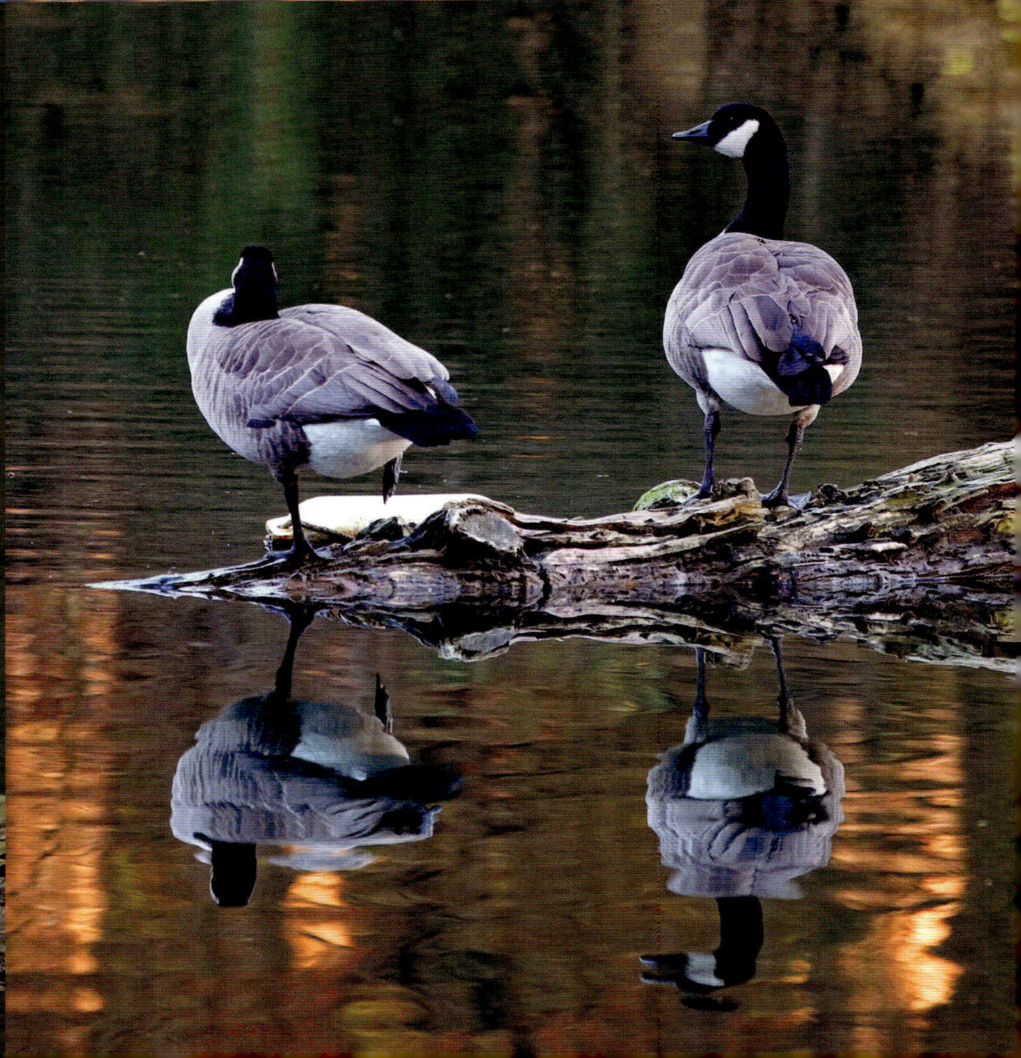

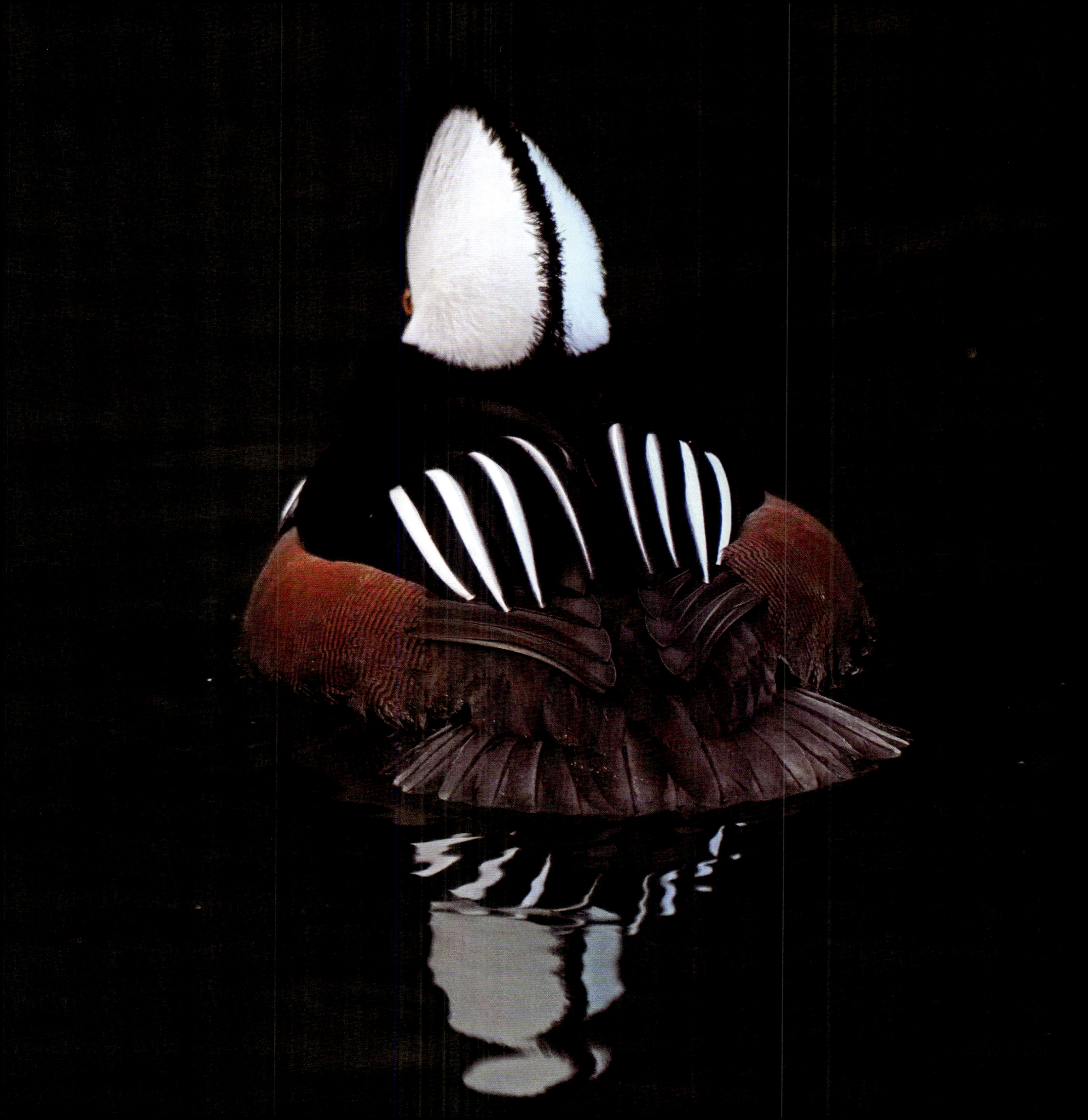

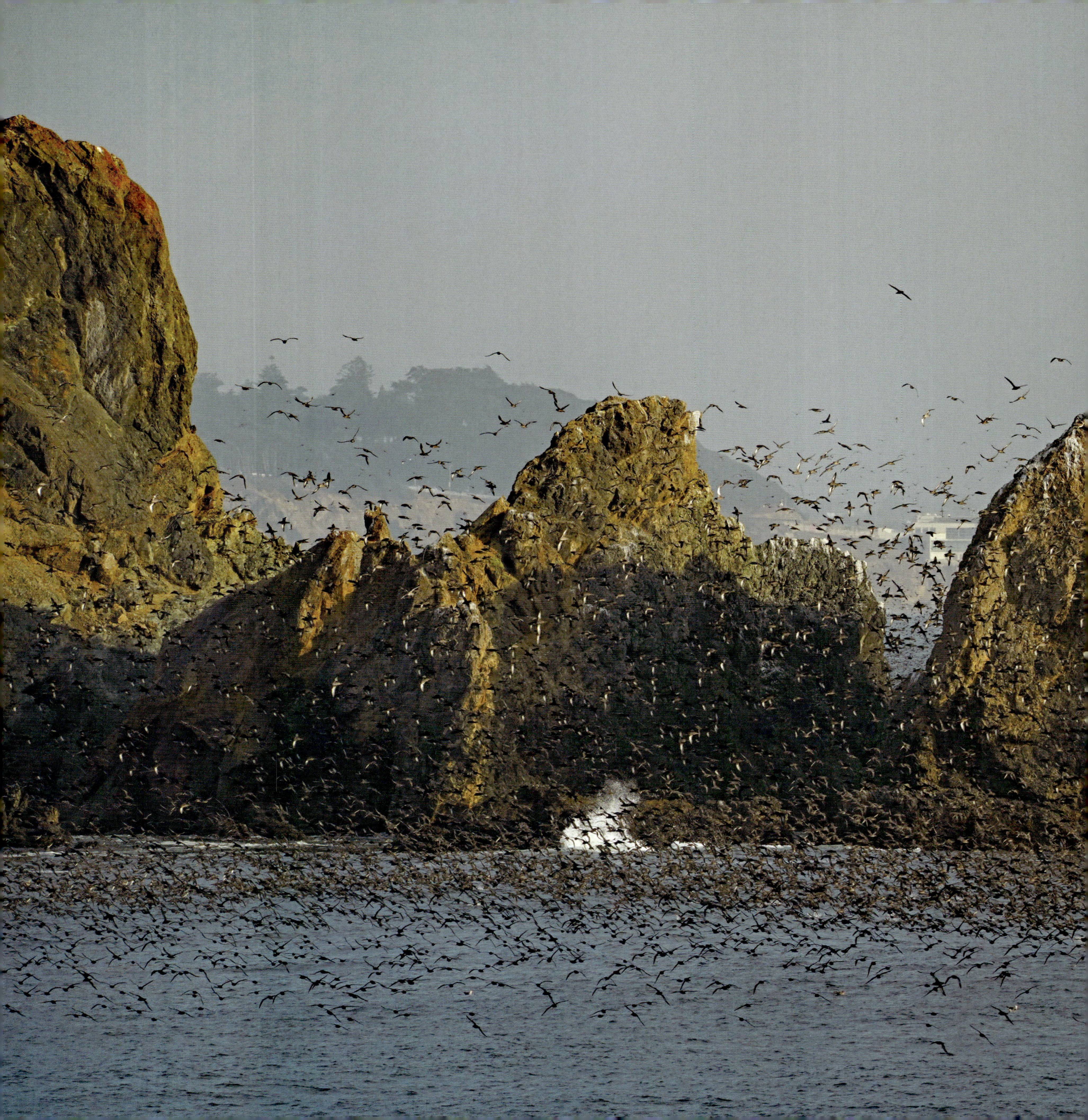

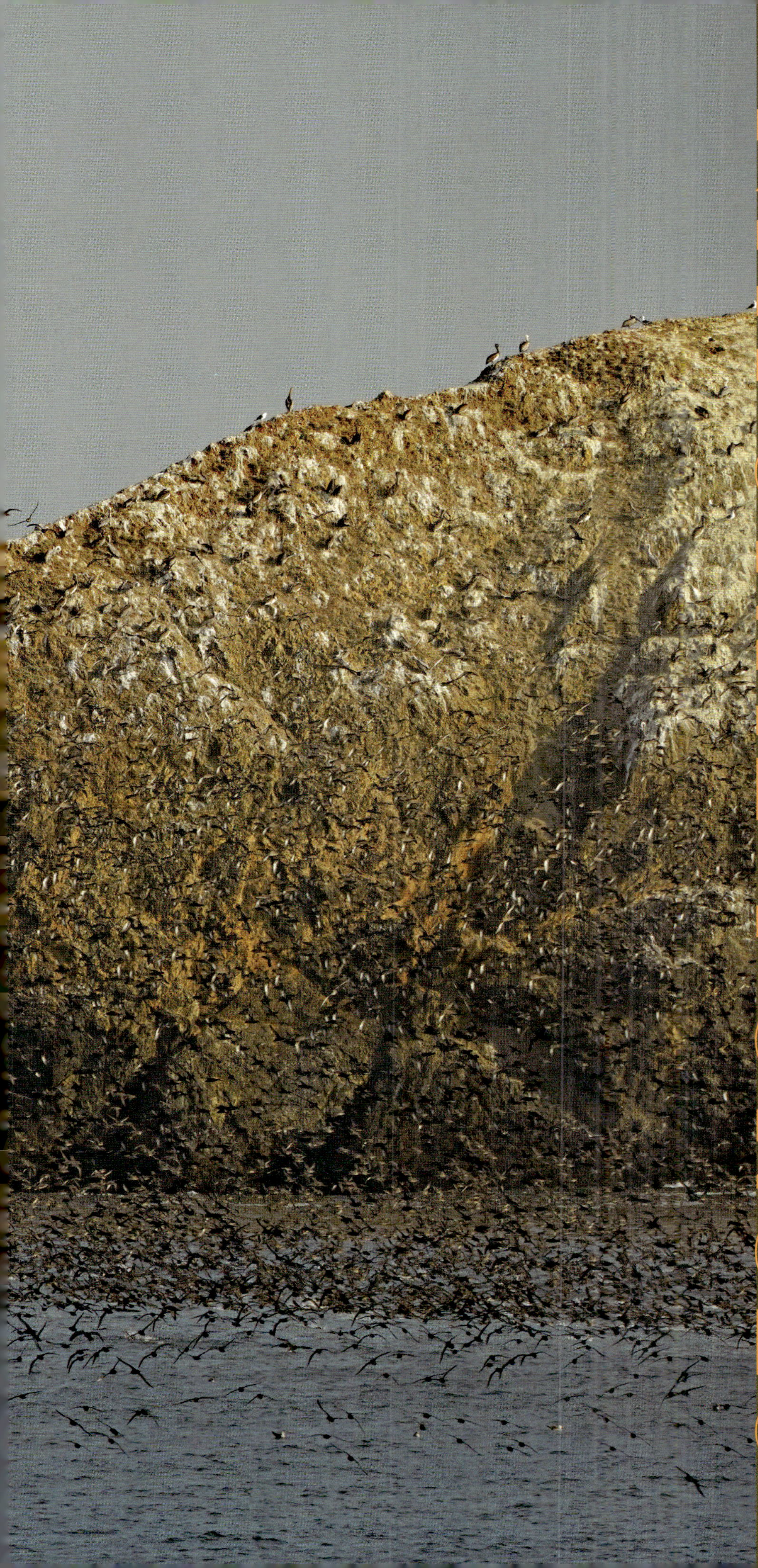

NATURAL FORCES

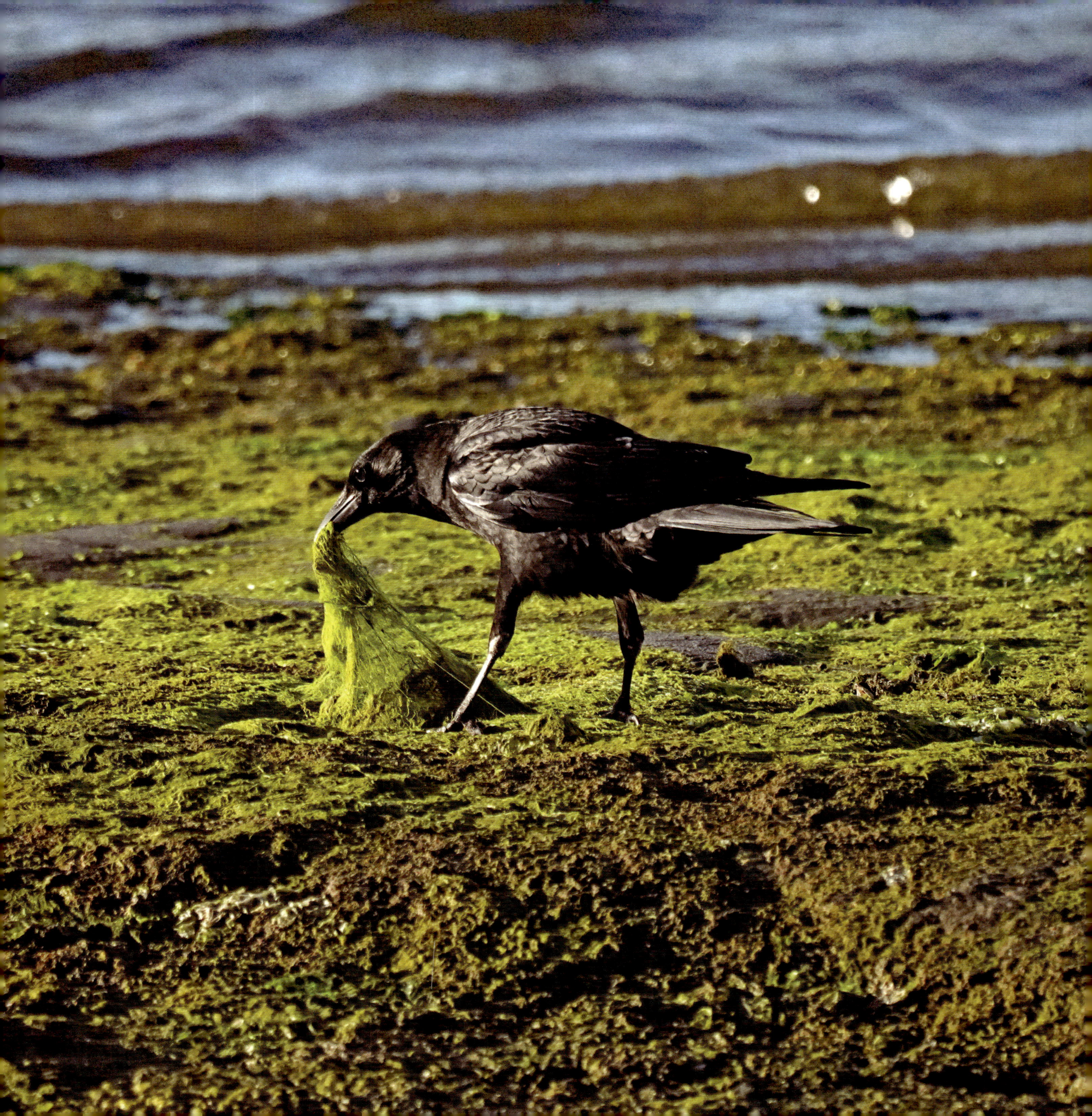

The greater Bay Area is a mecca for bird life. The mighty Pacific Ocean steers our seasons and moderates our climate, and as it pushes deep inland from San Francisco Bay, it feeds a sprawling estuary bigger than any on the West Coast. Slow tectonics, swift sea-level changes, and the inexorable parade of rising mountains and melting glaciers have all set the stage for the kaleidoscopic communities of birds that call this area home today.

Early one morning on the shore of the Bolinas Lagoon, a young biologist holds up a Wrentit, writing down the number printed on a small aluminum band on its ankle. The band shows that the bird has been captured here before. In fact, it likely has never flown more than a few hundred yards from where we stand amid salt grass and willow. Wrentits are the least migratory bird in North America—they rarely venture farther than 1,300 feet from where they were born. With a pale, penetrating eye, the bird turns its head and evaluates the humans surrounding it. If birds could tell histories, this one would carry the stories of ancestors that have lived right here without straying: generations of small brown birds that have watched the region change around them, like still stones in the river of time.

The California that long-ago Wrentits witnessed was a land of fresh water and valleys. Twenty thousand years ago, the Bolinas Lagoon was a low valley, with a creek cutting through its heart along the San Andreas Fault. A bird would have needed to flap more than 20 miles west, across a wide, grassy prairie, to find the edge of the Pacific Ocean. On that ancient coastline, sharp hills rose up from the grasses—hills that would someday submerge to become the Farallon Islands. The basin that San Francisco Bay would one day fill was an undulating landscape of valleys with rivers draining out of the Santa Clara and Central Valleys. Bison and mammoths followed these watercourses on their seasonal migrations, and ravens winged above the sprawling herds. Where saber-toothed cats and short-faced bears left behind carcasses, enormous condor species arrived to glut on the kills.

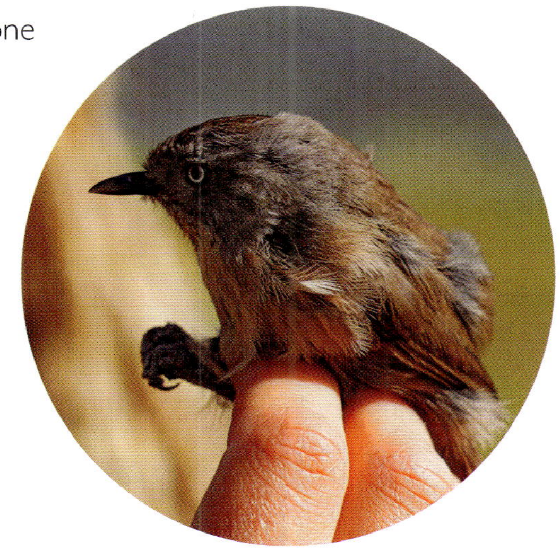

PRECEDING: Throngs of Sooty Shearwaters darken the sky offshore from the Marin Headlands, attracted to rich upwelling offshore. These birds take advantage of ocean winds to cover vast distances along their migrations—upwards of 500 miles in a single day.

LEFT: A clever Common Raven rummages under mats of algae for snacks at low tide.

RIGHT: A Point Blue bird-banding apprentice gently holds a Wrentit at the Palomarin Field Station in Bolinas, California.

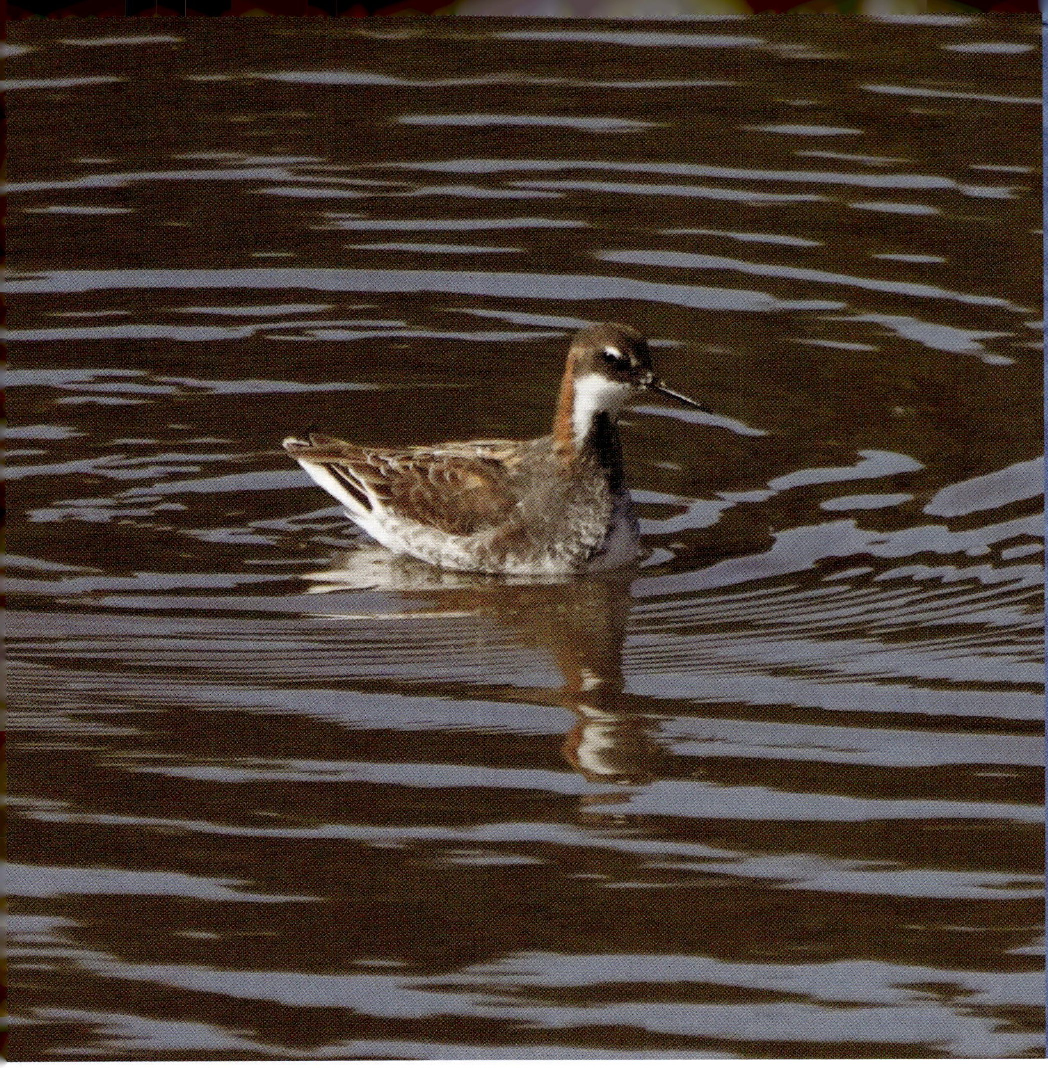
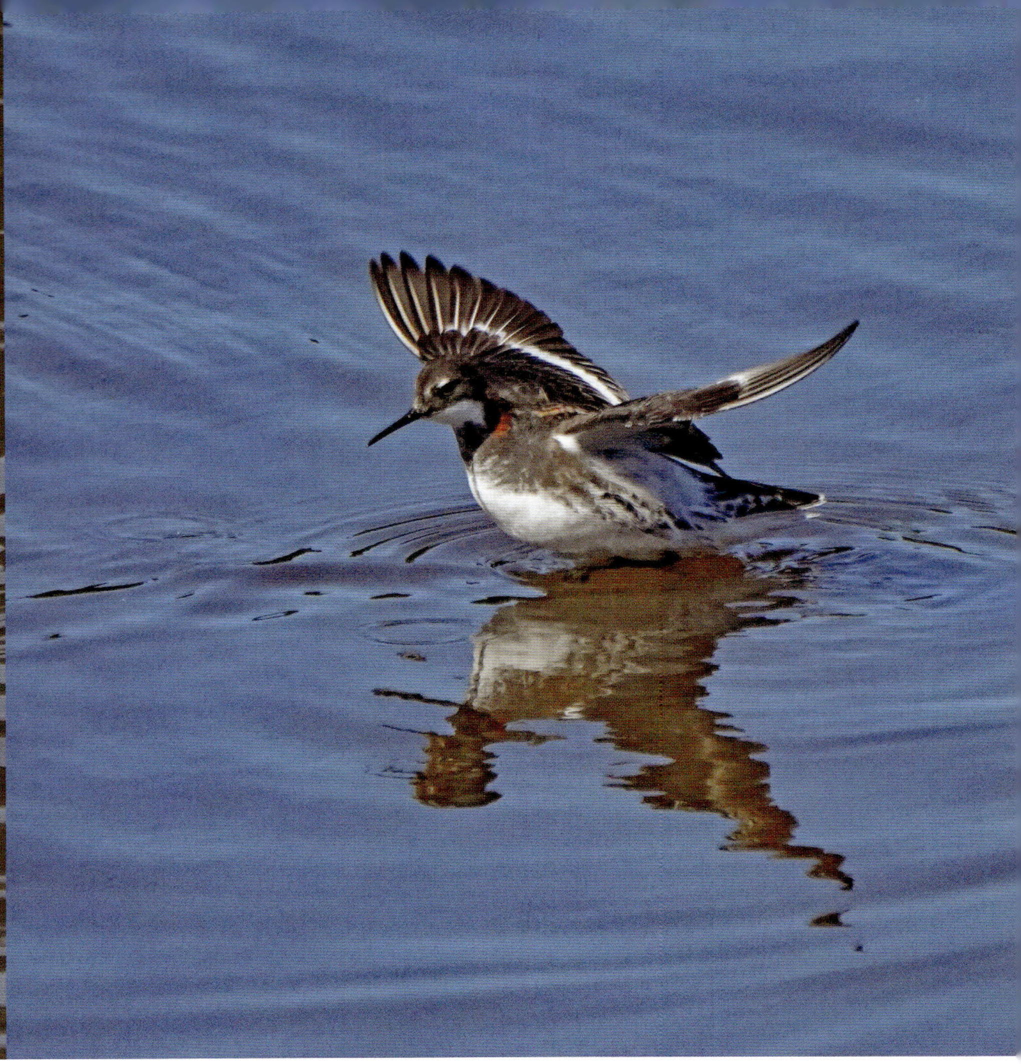

Delicate Red-necked Phalaropes (above) travel one of the longest North American migrations of any bird, breeding in the high Arctic tundra and passing briefly through Bay Area waters on their way to winter homes off the Pacific coast of South America. Phalaropes are among the few shorebirds considered pelagic: They spend much of their lives flying over the open ocean.

This was the tail end of the Pleistocene—the last great ice age—but the Bay Area's climate was mild, and the coastal prairies and forests provided refuge for bison and birds alike. Winters were cool and wet; summers were warm and dry. Much of the world's ice was locked up to the north and south in continental ice sheets. Two hundred miles to the east, glaciers dragged their slow bodies across the Sierra Nevada. Then, 10,000 years ago, the climate began to warm, and melting ice raised sea levels. All around the globe, the continental shelves that exist today are the drowned shoulders of continents that the rising ocean subsumed at the end of the Pleistocene.

These shallow waters, bathed in sunlight and fed by nutrient-rich runoff from coastal rivers, have sustained humans for thousands of years; globally, 90 percent of the fish in people's diets come from the continental shelf. Off the coast of San Francisco, this ecosystem is also fueled by the California Current, a strong nearshore flow that runs north to south. Boundary currents like this run along the east edge of all of Earth's ocean basins, and the planet's rotation pulls these currents toward the equator—a phenomenon known as the Coriolis effect. Where the California Current is pulled seaward, deep, cold water fills into the void left behind. Summertime winds from the north amplify this effect, pushing surface waters away and allowing cold water to rise up to the continental shelf from the abyssal plain.

Today, these forces in combination bring richness to the Bay Area's offshore waters that attracts birds by the millions. Upwelling carries nutrients to the surface and fuels an intricate web of life, from anchovies to colossal whales to a dozen different species of breeding seabirds: dense colonies of Common Murres; Pigeon Guillemots with legs as red as race cars; Tufted Puffins flapping on fast, whirring wings. Another thirty-five seabird species regularly come to the Farallon Islands to feed along their long migration routes: throngs of scoters, Pink-footed Shearwaters, and Black-footed

The Farallon Islands are home to California's largest breeding colony of Common Murres (below). Every year, more than 300,000 murres nest on the Farallon Islands' granitic cliffs—that's more than double the human population of Berkeley! The murres share this craggy landscape with a dozen other nesting seabird species. As murres establish themselves, they tend to displace cormorants. Can you spot the few remaining nesting Brandt's Cormorants among the flock?

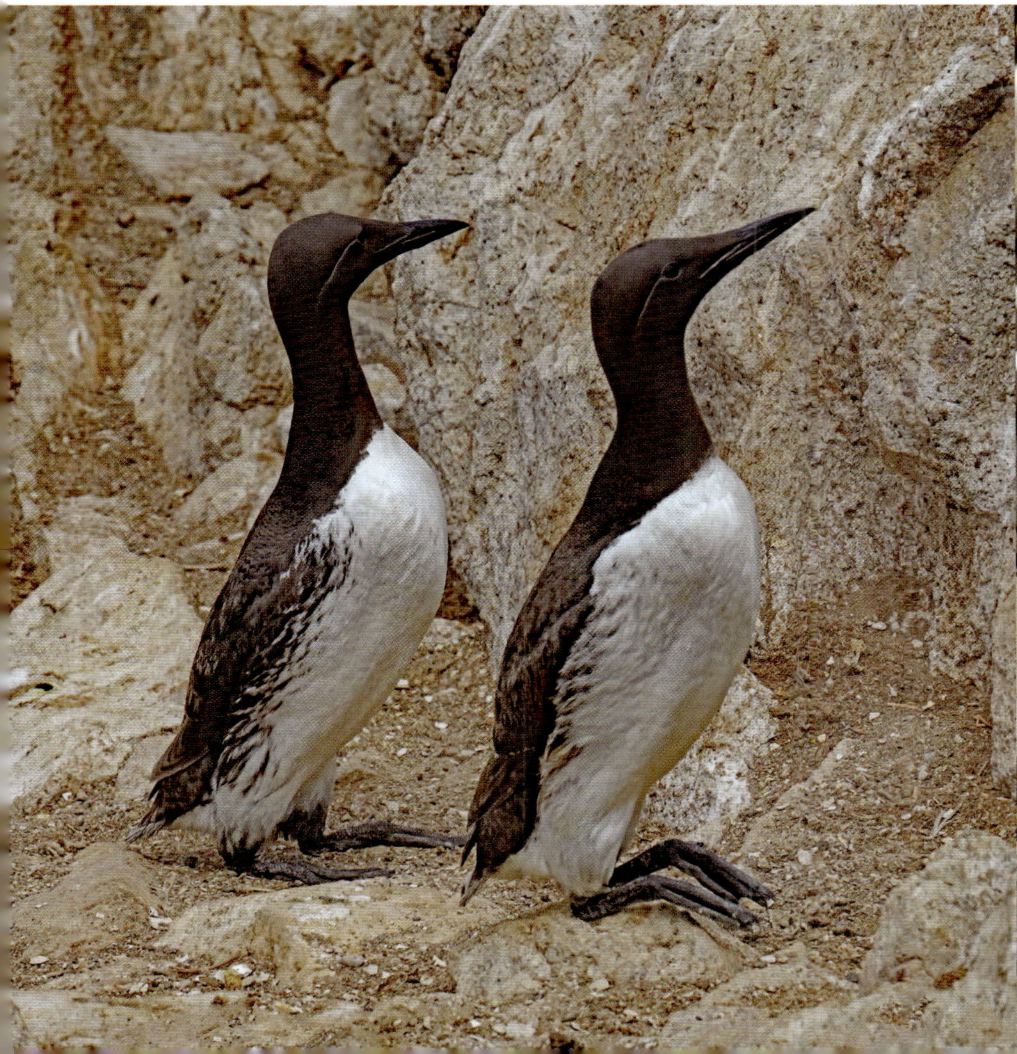

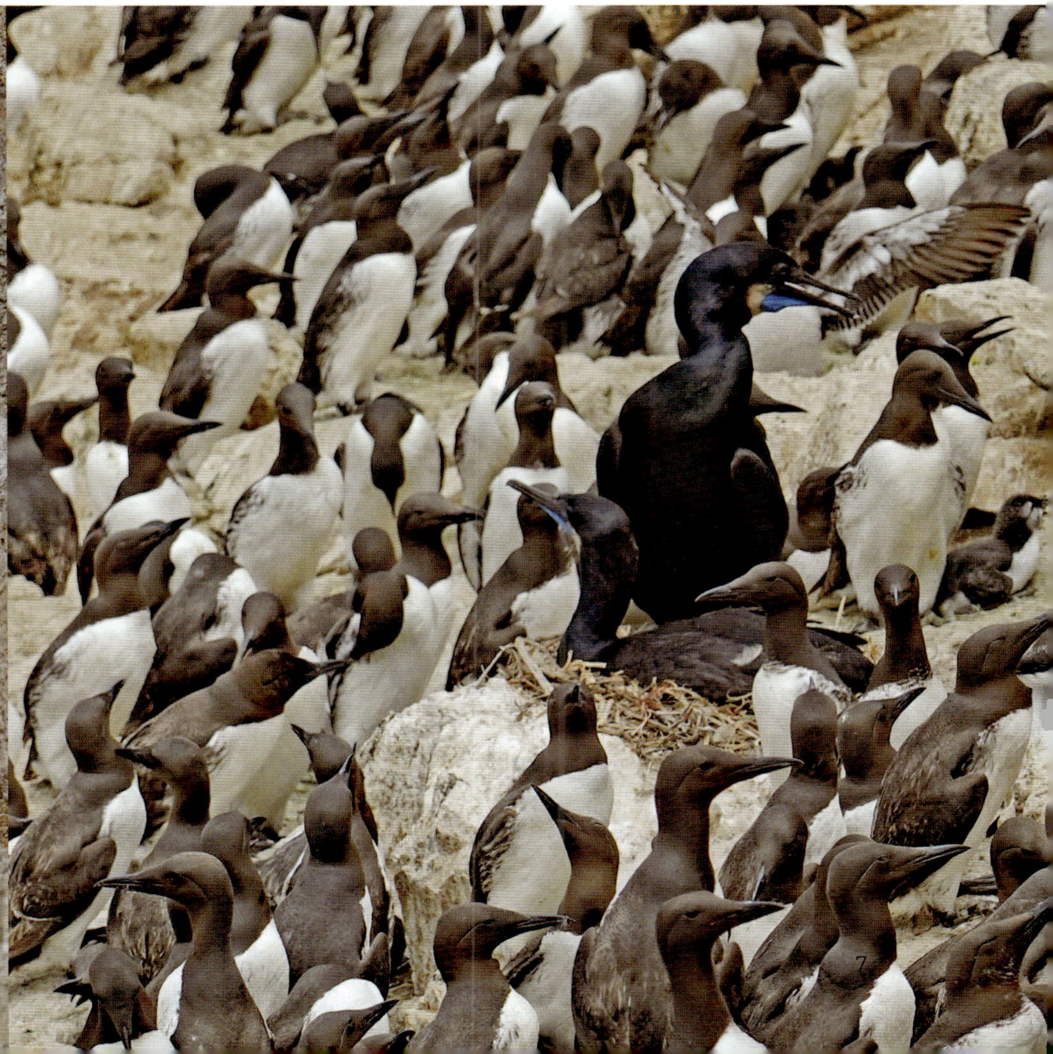

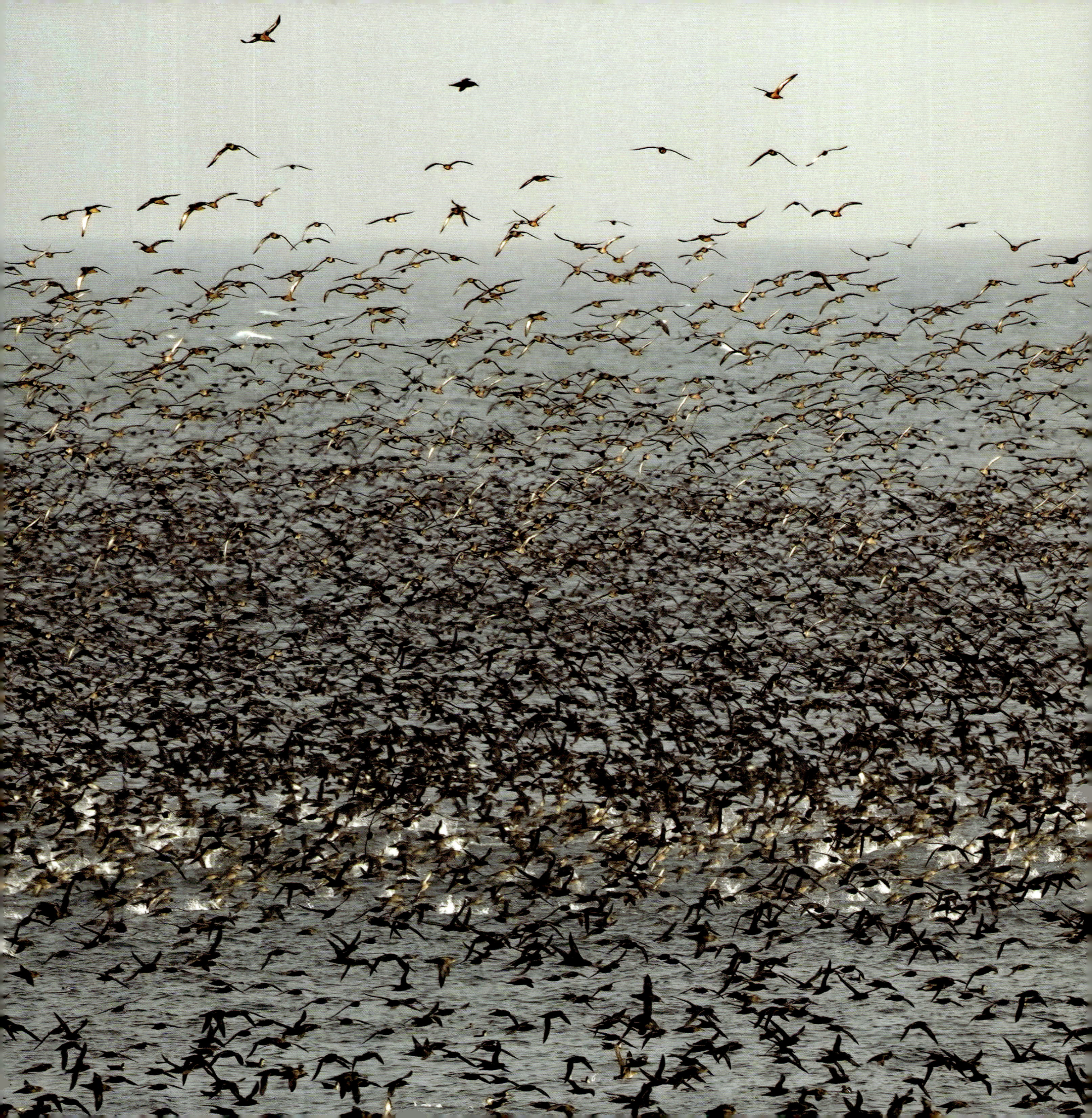

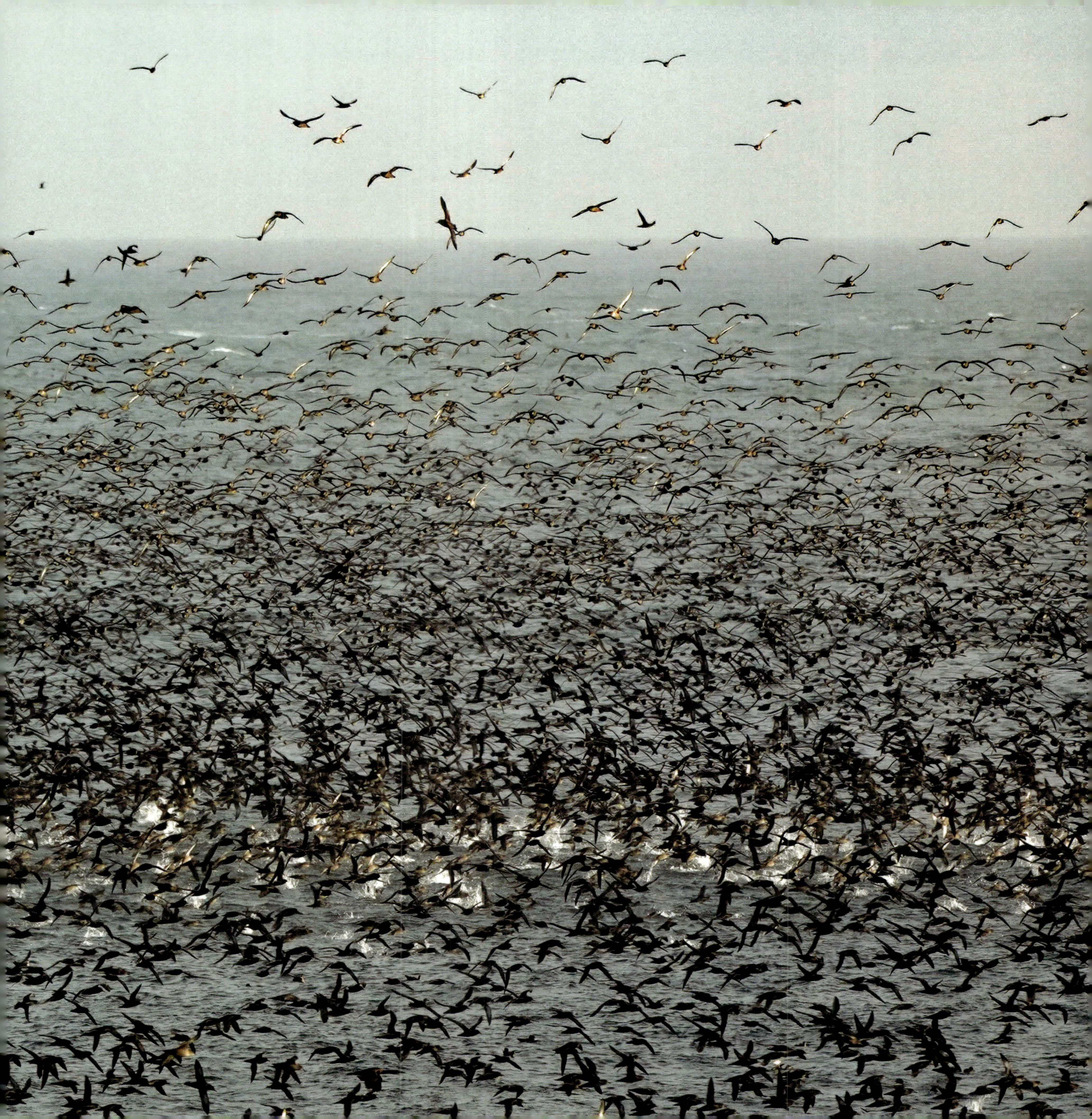

Albatross with wingspans as long as a kayak. And over two dozen other rogue seabirds visit the Gulf of the Farallones well outside their usual range, including the Manx Shearwater, a far-straying species from the Atlantic Ocean. The broader region teems with so much avian life that the Audubon Society has designated it a high-priority Global Important Bird Area, stretching from Mendocino in the north to Monterey Bay in the south.

Late-Pleistocene sea levels rose past the continental shelf, filling the bay and creating hundreds of thousands of acres of tidal marshes sprawling inland toward Suisun Bay and beyond. Meltwater and rain from the Sierras braided its way downslope into the Central Valley. Narrow riparian forests followed the water courses. Some of that water fanned out and sank into willow groves—some as big as 200 acres, alive with birdsong when Spanish colonists arrived 10,000 years later. They called these isolated forests *sausals*. And where the rivers' force was strong enough to flow into incoming tides,

PRECEDING: Like a roiling wall of smoke, hundreds of thousands of Sooty Shearwaters fill the air above Rodeo Beach at Marin Headlands. In the summer, vast flocks of Sooty Shearwaters follow the California Current toward their Southern Hemisphere nesting grounds, pursuing squid and anchovy en masse.

BELOW: Offshore from Alcatraz Island, Pigeon Guillemots use their powerful wings to "fly" underwater in search of small fish, worms, and crustaceans.

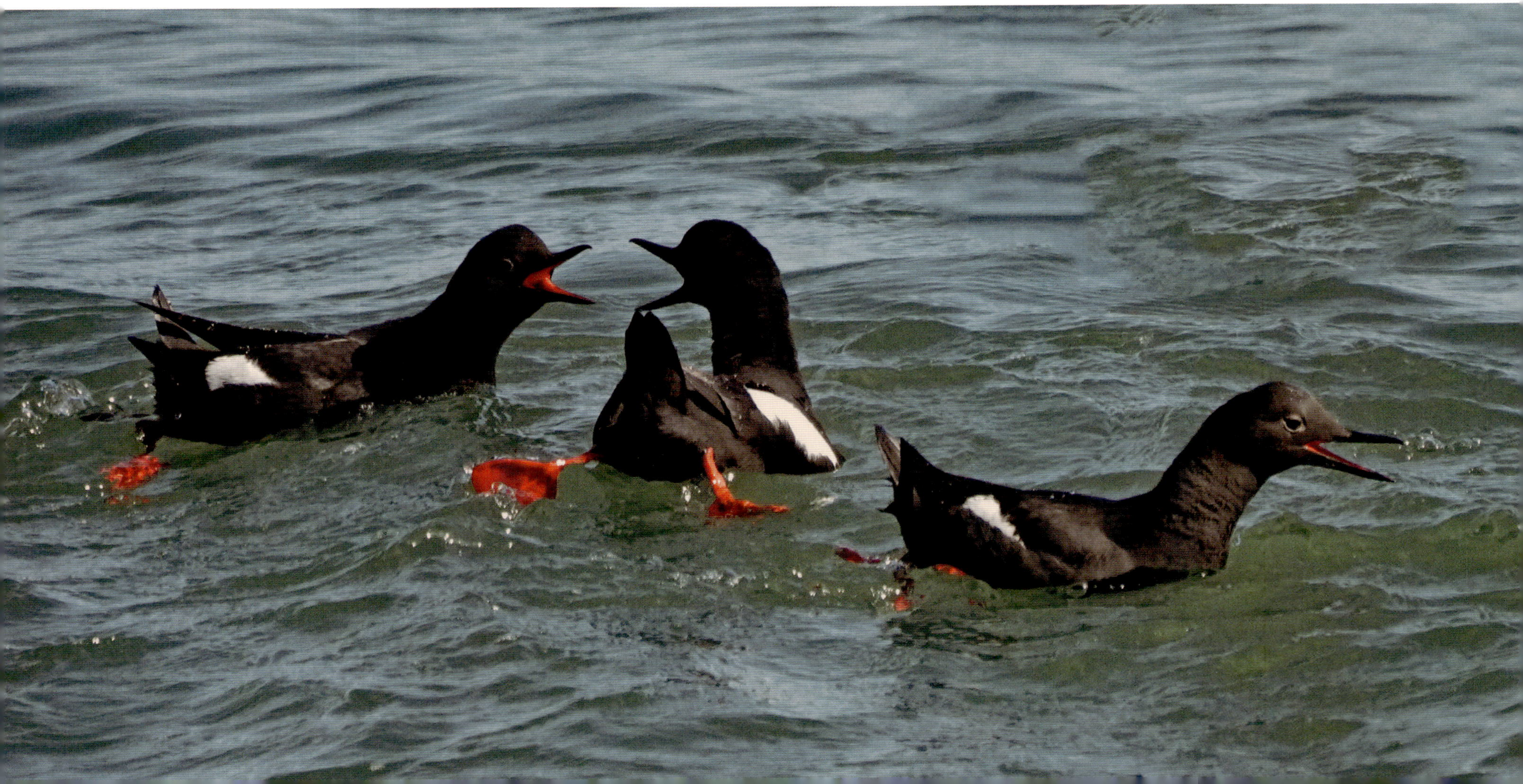

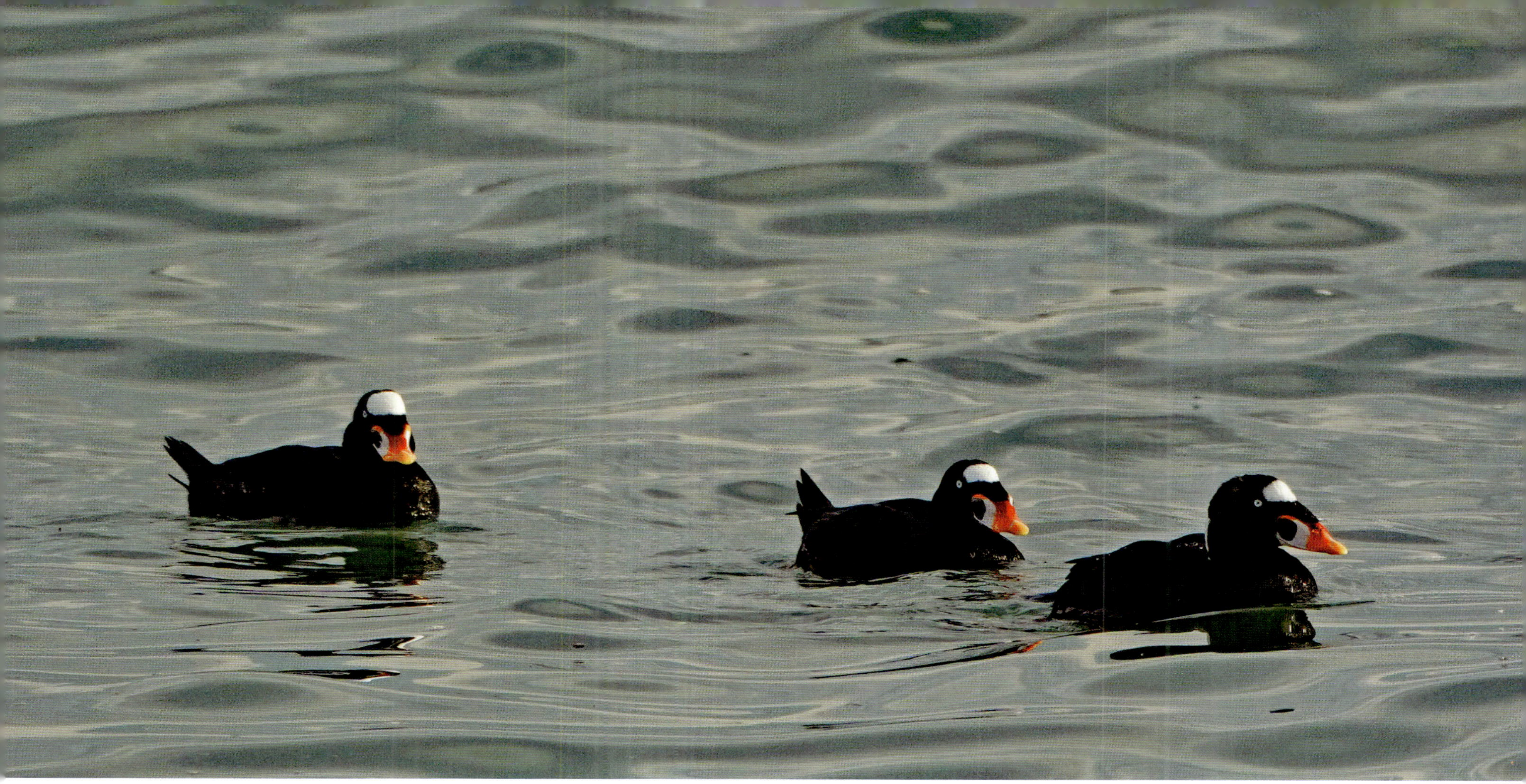

ABOVE: Of all birds to thrive in California's confluence of natural forces, Surf Scoters are among the most impressive: These muscular sea ducks effortlessly dive for food in strong ocean currents and powerful surf.

intricate channels wove our Delta, dendritic and meandering. Our Bay's regional influence stretches more than 100 miles inland from where the tide first passes under the shadow of the Golden Gate Bridge.

With these massive forces at work—ice melt, ocean fill, the braiding of vast waters—the Bay Area became the West Coast's largest estuary. And where water flowed, life followed. Today, the grand avian migrations that Californians witness across the Bay Area result from the somewhat literal tides of change.

As all of us who practiced earthquake drills under elementary school desks know, the tides shift on land as well, in long periods of slow-sliding tectonic plate activity and in sudden jumps that topple buildings and open gashes in the rock and soil. Calaveras, Hayward, Greenville, San Gregorio—a half dozen interlocking fault lines cut through the Bay Area. The queen of all of these is the San Andreas:

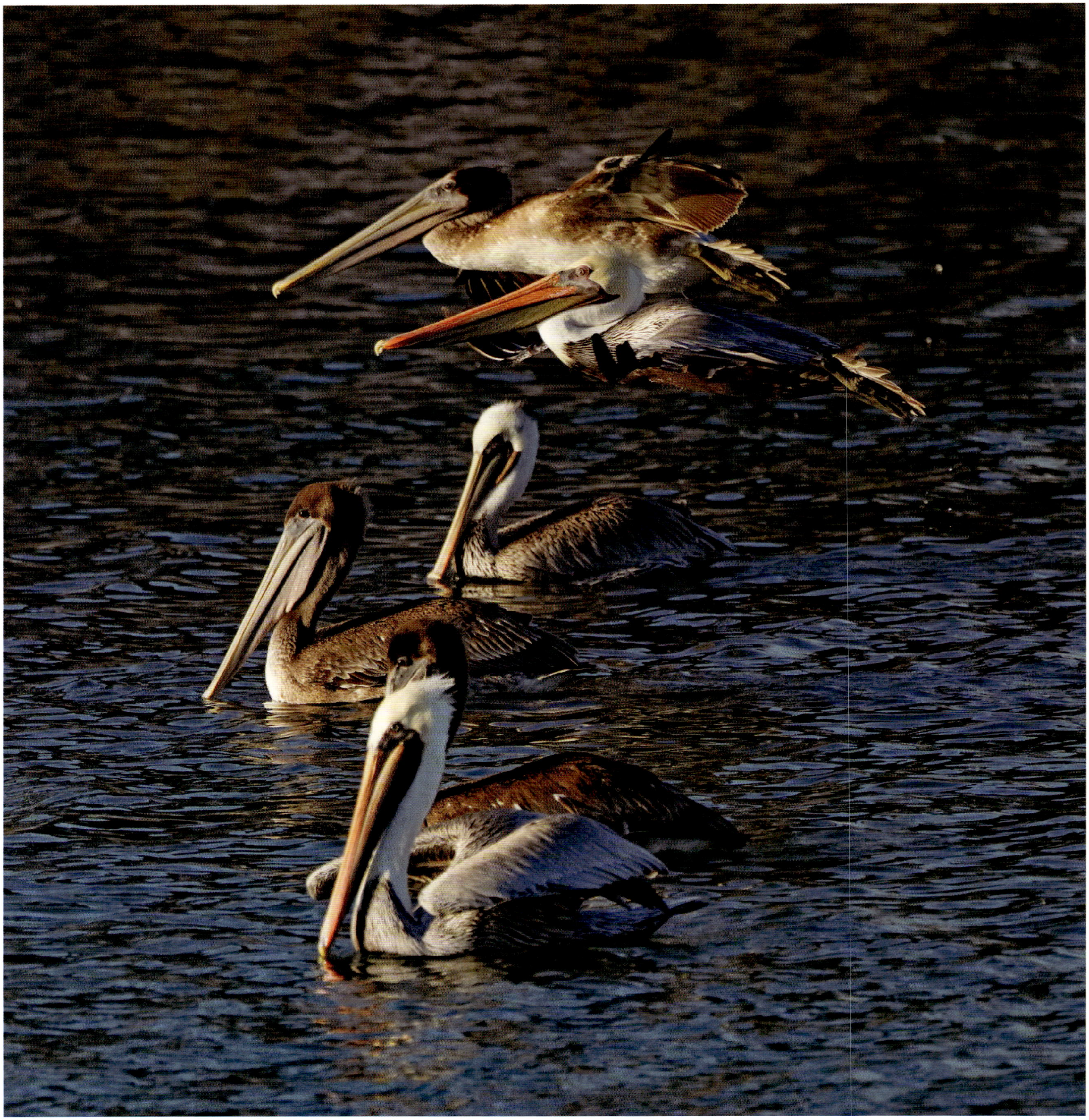

Enormous pelicans are drawn to the California coast for seasonal feeding. In the summertime, Brown Pelicans (left) take plunging dives in search of fish, while winter-visiting White Pelicans (right and below) scoop and dabble in wetlands, bays, and lakes

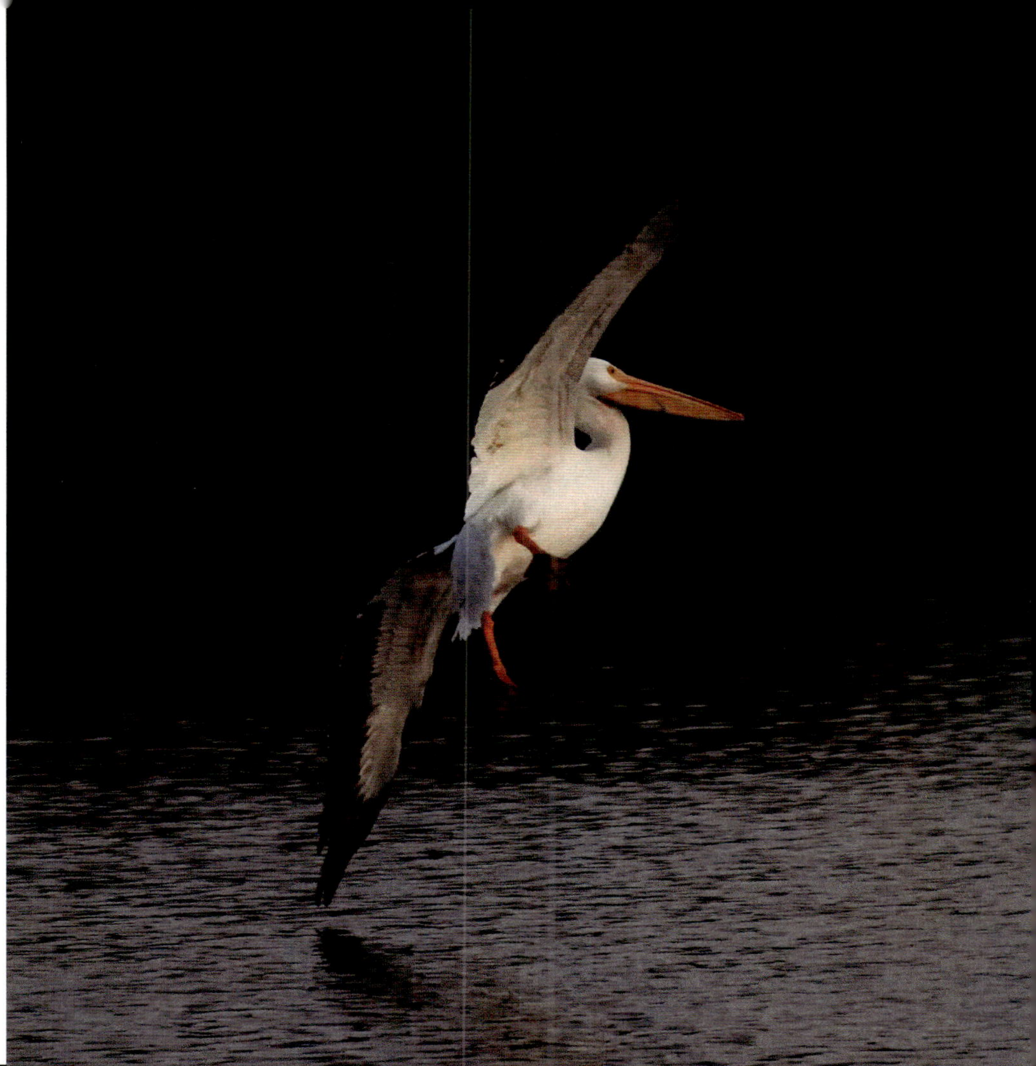

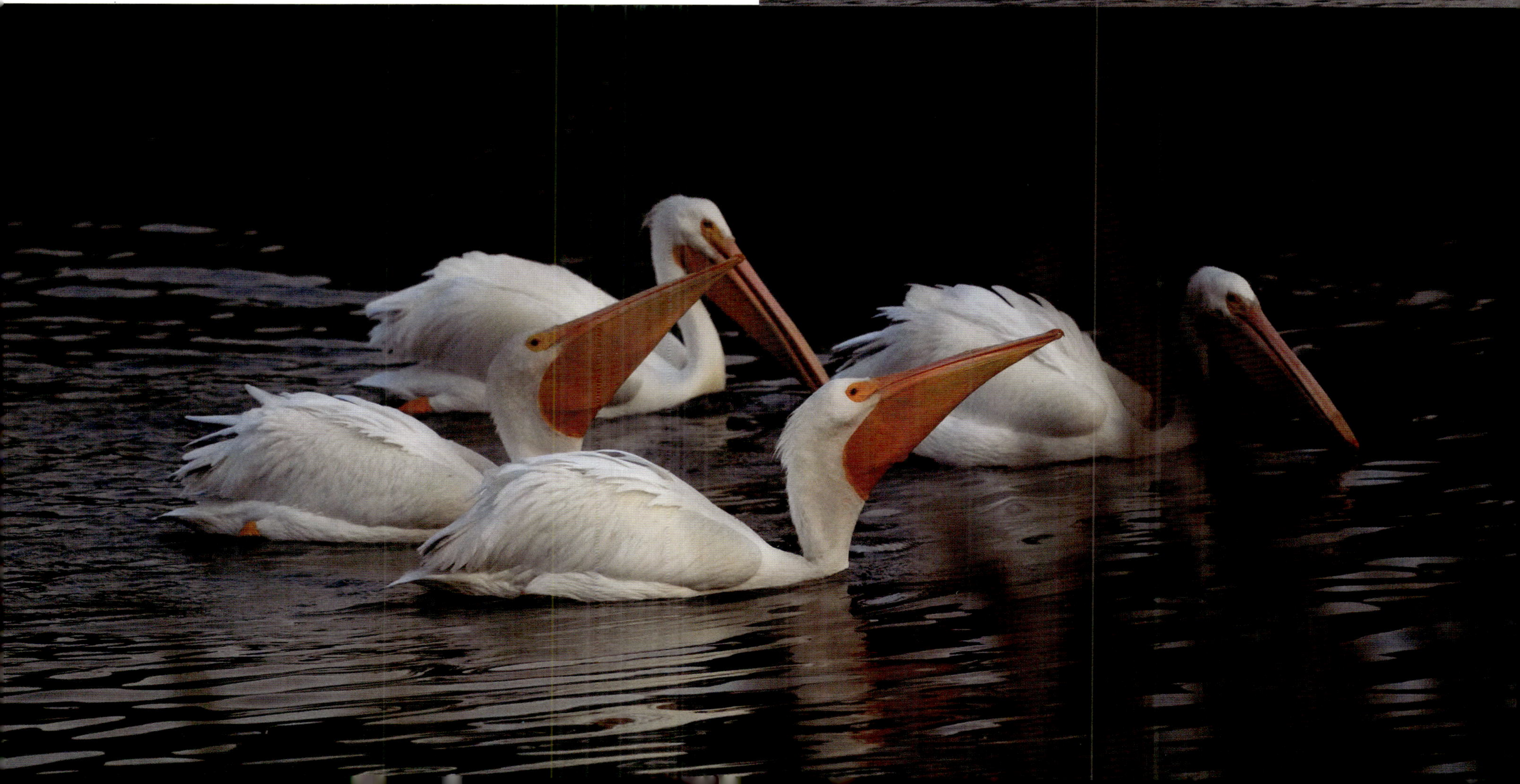

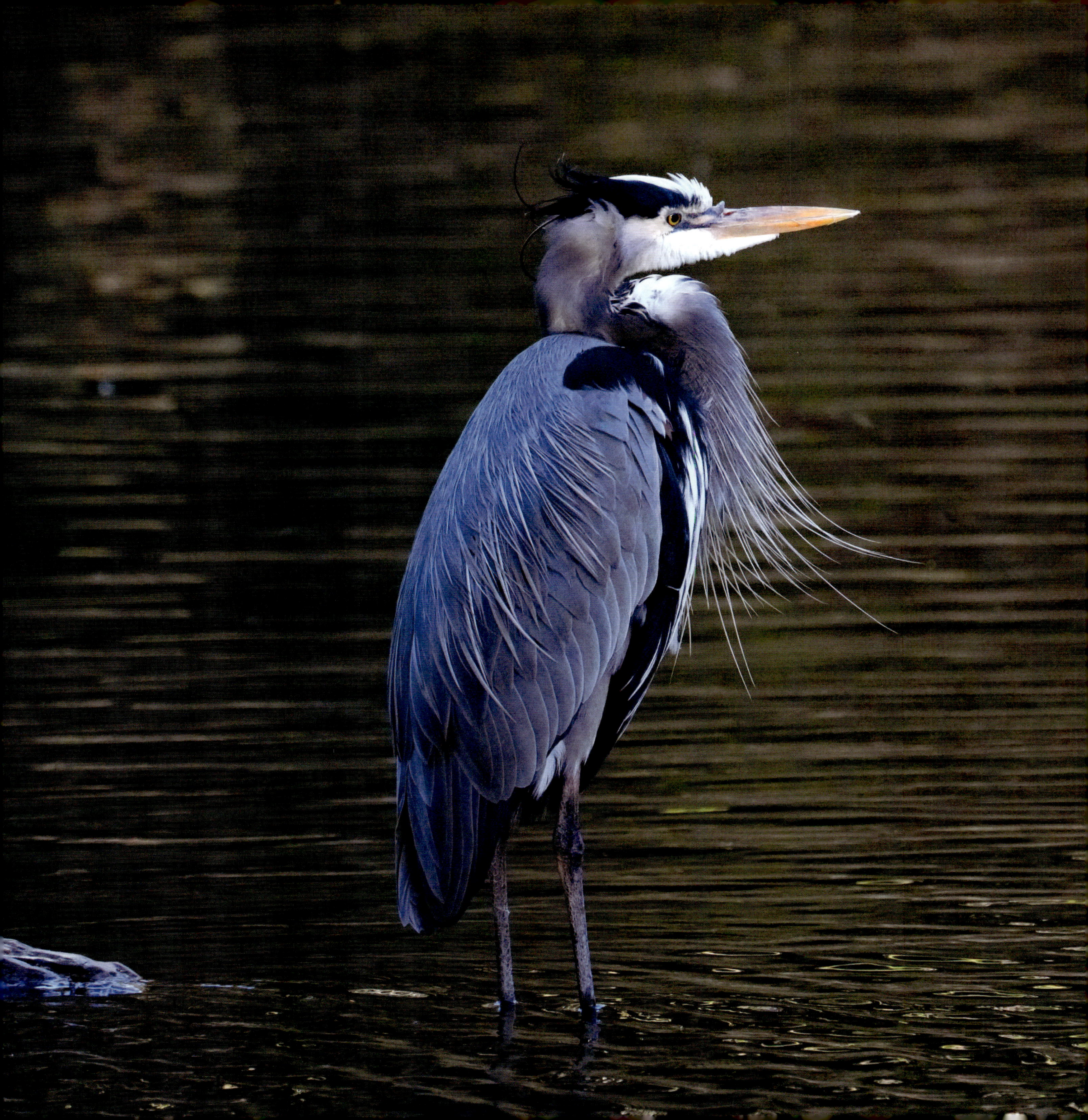

When the bird is in repose, a Great Blue Heron's (opposite) unique, long vertebrae double back on themselves, allowing the heron to fold its neck into an S-shape and then swiftly extend it to catch its prey. These primeval birds can be found nesting on Alcatraz Island, at Don Edwards National Wildlife Refuge in the South Bay, and (aptly) at Golden Gate Park's recently renamed Blue Heron Lake.

The forces that have shaped the Bay Area—sea-level rise, sediment runoff from the Sierras, the intermingling of tides—have sculpted vast networks of tidal channels and marshy sanctuaries (below) ideal for wading birds.

the 800-mile-long slip-strike fault responsible for San Francisco's most cataclysmic earthquakes. On the west side of the San Andreas, the Pacific Plate moves northwest, an inch per year. As it grinds and catches on the edge of the North American Plate, the land under us ripples and shakes.

But 200 million years ago, a third tectonic plate—the Farallon Plate—occupied the space where the North American Plate and the Pacific Plate touch today. As the Pacific Ocean widened, the heavier Farallon Plate pushed up against the side of the continent and, heavier than the rock it met, slid under the lip of the west coast. When one plate subducts beneath another, the heat of the earth's mantle melts the heavy rock, and the fresh magma seeks a path back toward the surface, just like bubbles rising slowly in a lava lamp. This is how the Sierra Nevada range formed. The curved shoulders of Half Dome, the granite prominence of El Capitan—these are the rounded tops of mountain-sized lava lamp bubbles that cooled slowly under the earth's crust and, under the scouring of many ice ages, emerged at the surface like polished marbles.

Today, the high spine of the Sierra partners with the Pacific Ocean to the west to create a vast geographic funnel that forms part of one of North America's four major migratory bird routes: the Pacific Flyway. Seen from the air, the Sierras help to orient birds in the same way that signposts and median strips steer us on a drive.

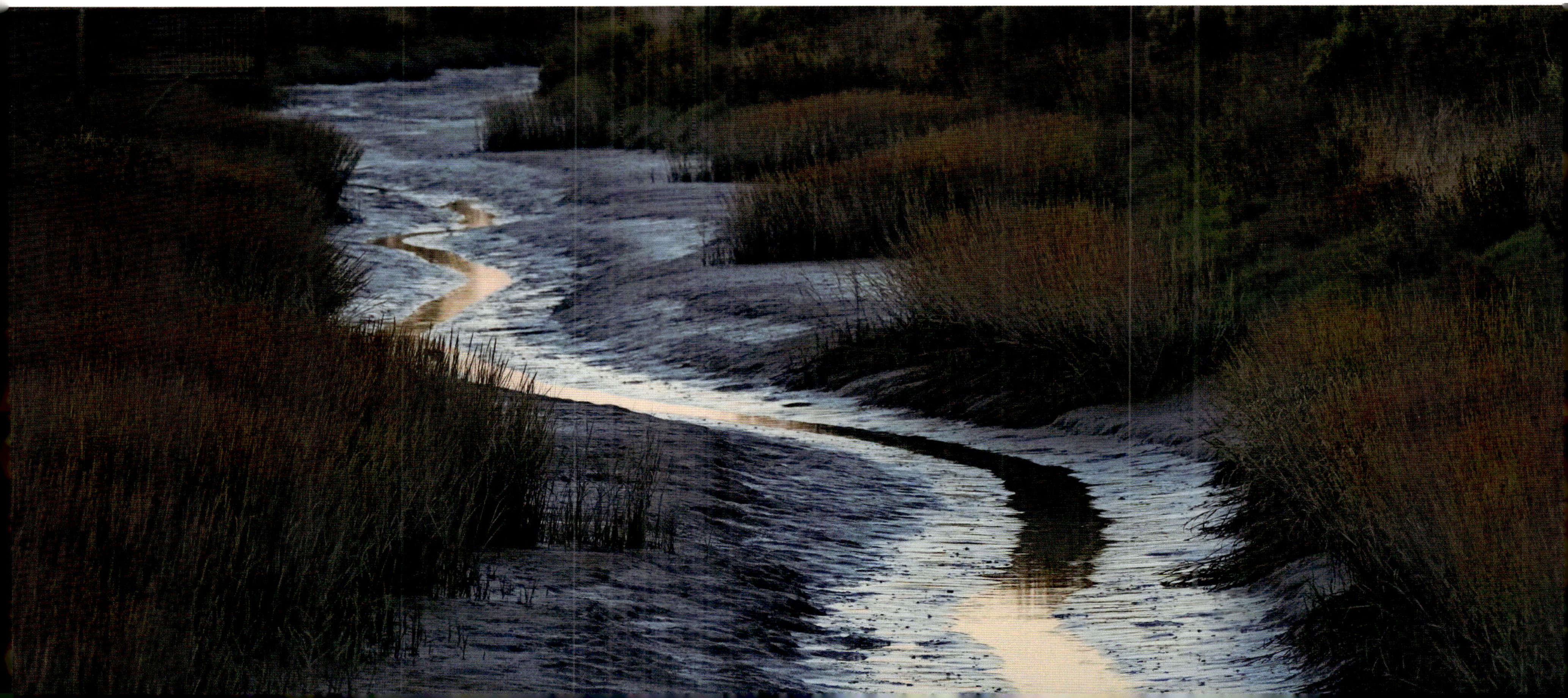

The mountains feed the Bay Area's weather systems too. During the Miocene period, about 25 million years ago, California was a temperate rainforest bordering a tropical sea. Summer rainstorms meandered across forests of figs and palms. But in the Pliocene Era, around 5 million years ago, the Sierra Nevada tilted and became several thousand meters higher. The uplift blocked off summer moisture and seeded the conditions for our Mediterranean climate and its accompanying species.

Of course, today's climate is also a story of change. In 2017, my hometown near Santa Rosa burned in what was, at the time, the most destructive wildfire in California history. Three years later, it burned again. Around the bay, sea levels have risen eight inches in the last 100 years; they're projected to rise nearly ten inches more by 2050. Alongside the deep natural forces that have shaped California—shifting plates, glacial ebbs and flows, a natural fire regime that enriches and seeds life across the dry summer hills—human influences are now also playing out across the landscape. And if climate change is the smoking gun, birds are often the first to feel the recoil. And bird-watchers have often been the first to notice.

Today, as I look into the searing eyes of the Wrentit, it seems clear that the bird not only holds histories of what this region once looked like but also will be a front-row witness to the ways the state will continue to change. For a hyperlocal species like the Wrentit, climate disruption could spell disaster. In West Marin, scientists are still unraveling the effects of the 2020 Woodward Fire, which took nearly a year to fully extinguish. While other bird species could relocate to escape the charred wake of the fire, the Wrentit, with its stubborn quarter-mile range, could not. The species is recovering slowly. But in Bolinas, where a muddy mix of land use and climate change has caused a Doug-fir forest to fill into an area that was once dominated by coastal scrub, Wrentits' homebody nature has proven beneficial. Where other birds have been displaced, Wrentits stoically persist. Unwilling—or unable—to disperse, they now nest in the trees. For Bay Area residents for whom the region is a deep-rooted home, perhaps the Wrentit is a guide for how to navigate the California to come. In this state of change, how do we adapt? And how do we defend our feathered neighbors who cannot?

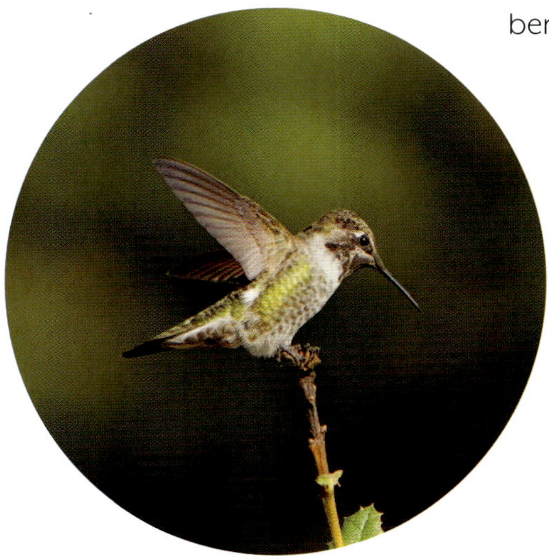

Tales of fire and birds run deep in the Bay Area: In Ohlone and Coast Miwok creation stories, the Anna's Hummingbird (female pictured here) is the original bearer of fire.

Notoriously secretive and hard to see, Ridgway's Rails are best observed during King Tides, when the rising water pushes them to higher ground.

FOLLOWING: Turkey Vultures like this one, and other New World vultures such as California Condors, have circled and cleansed the Bay Area landscape for millennia. As apex scavengers, they maintain vital ecological balance by preventing the spread of disease and recycling nutrients.

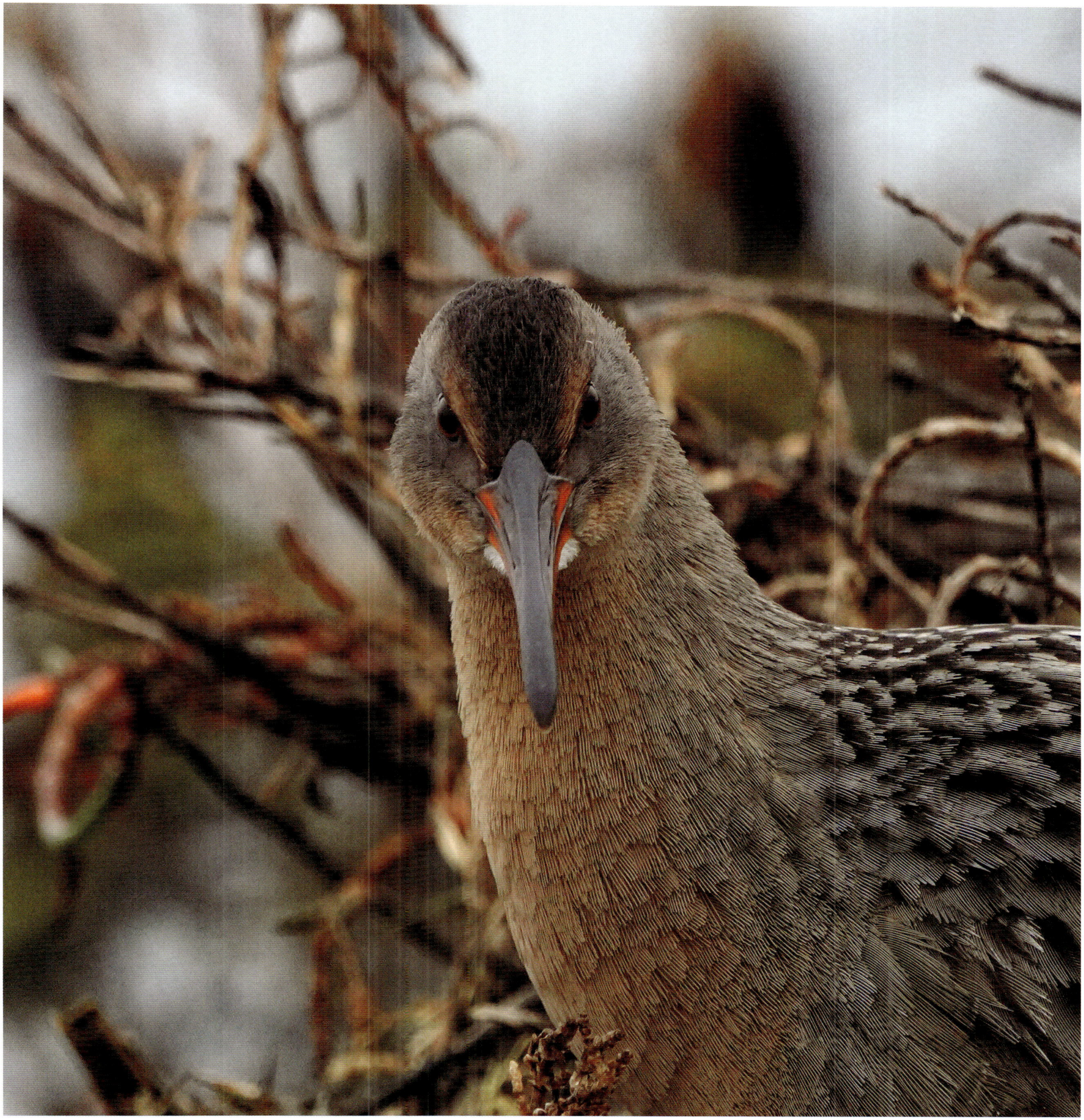

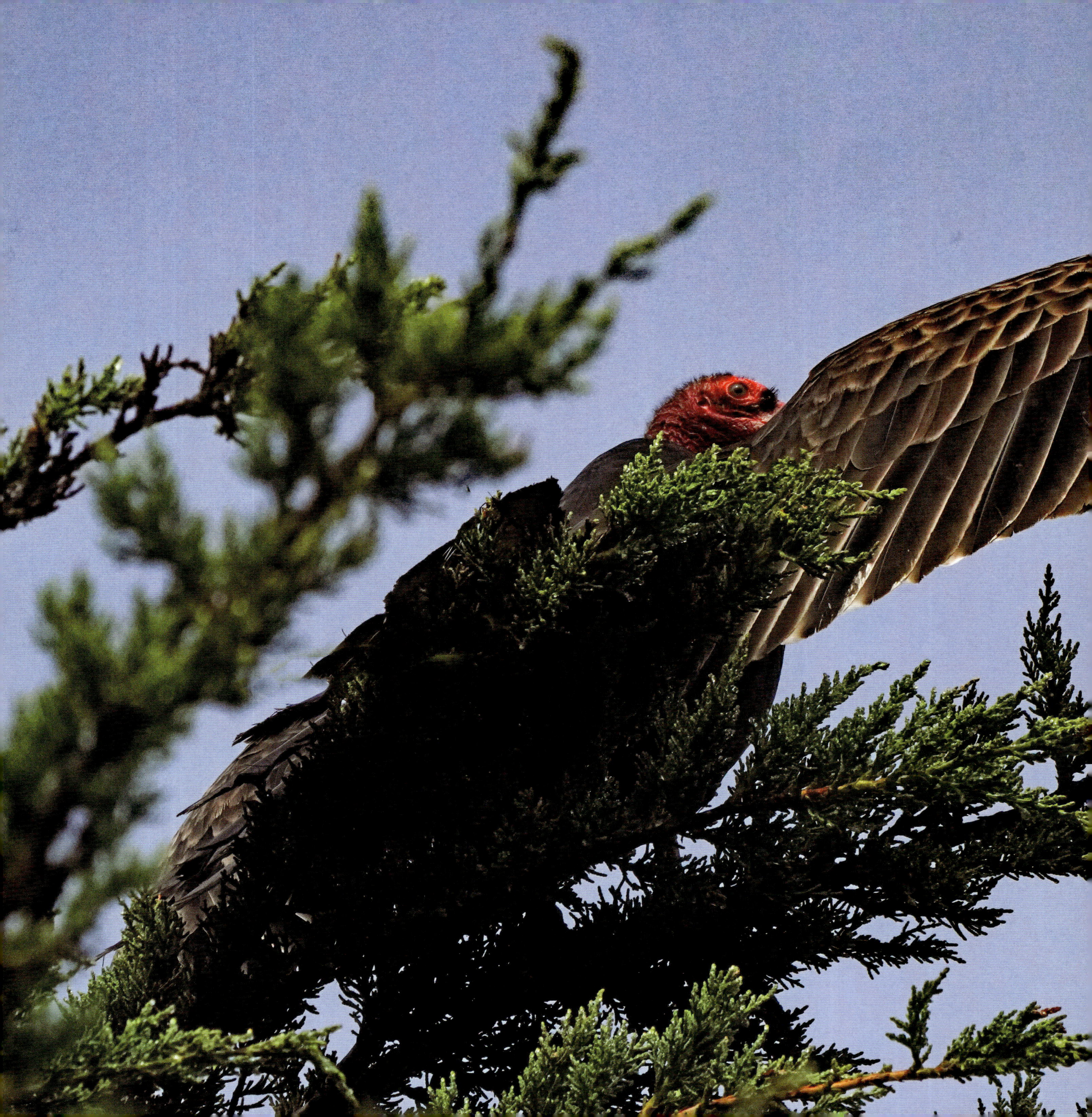

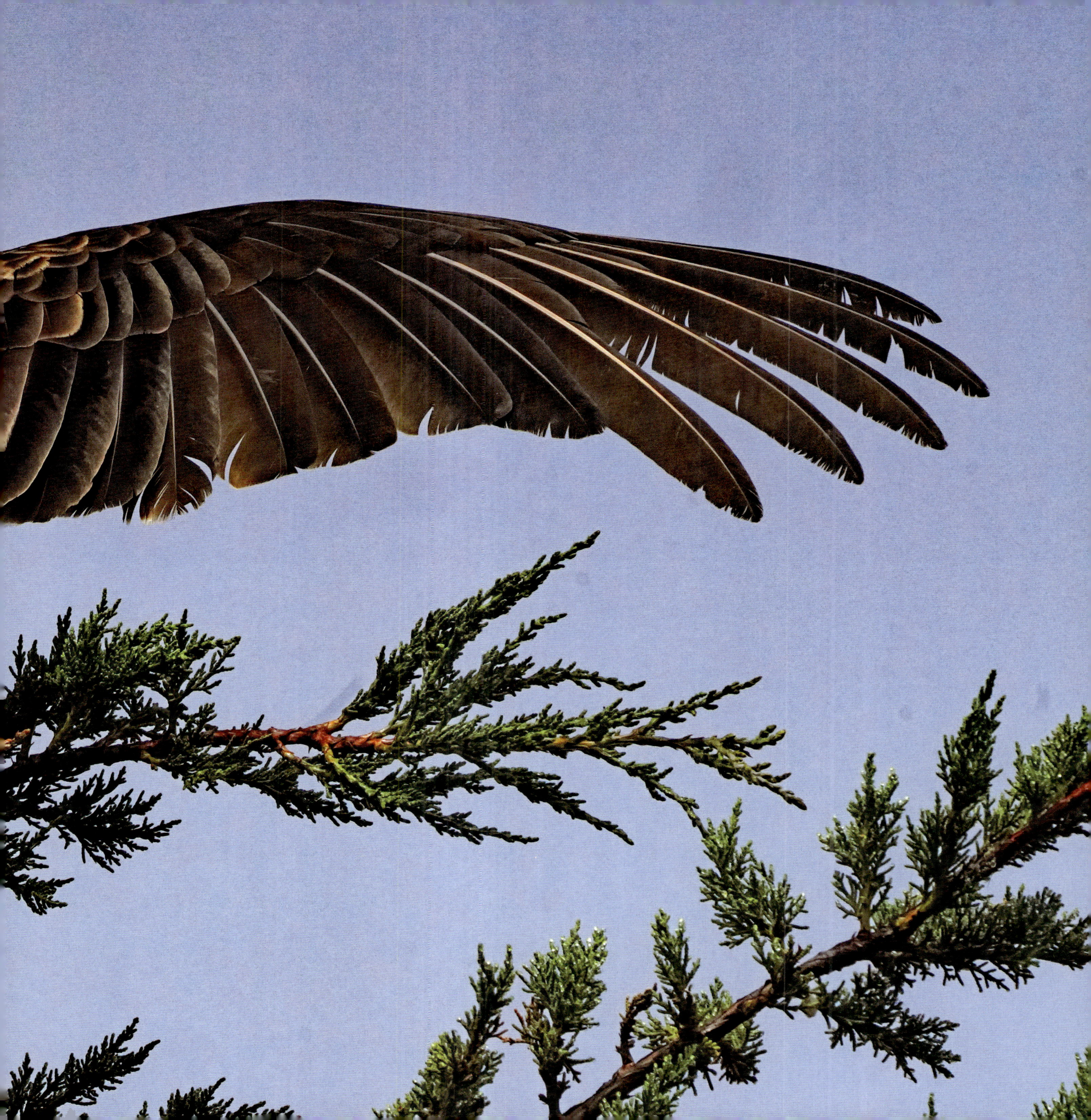

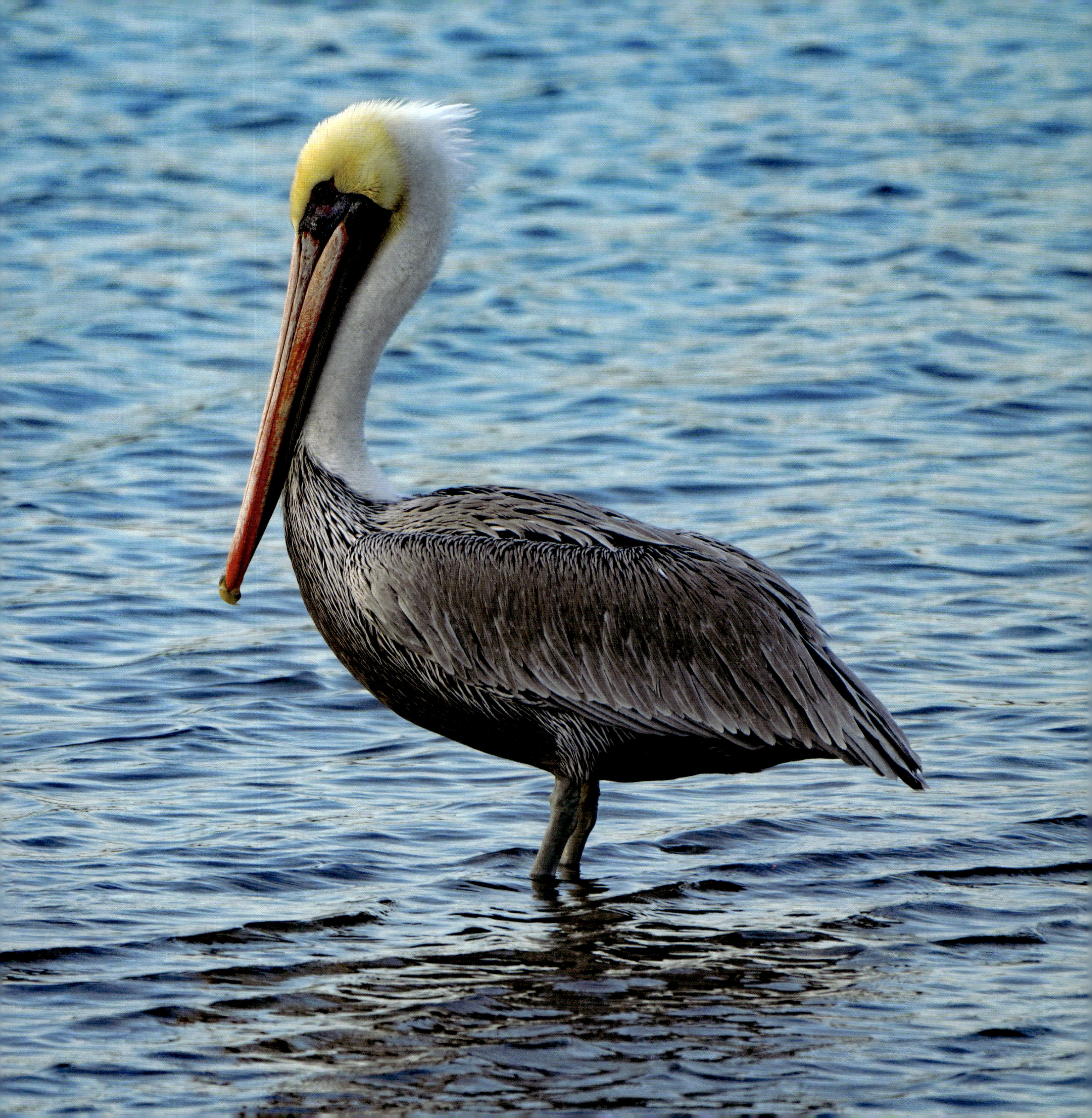

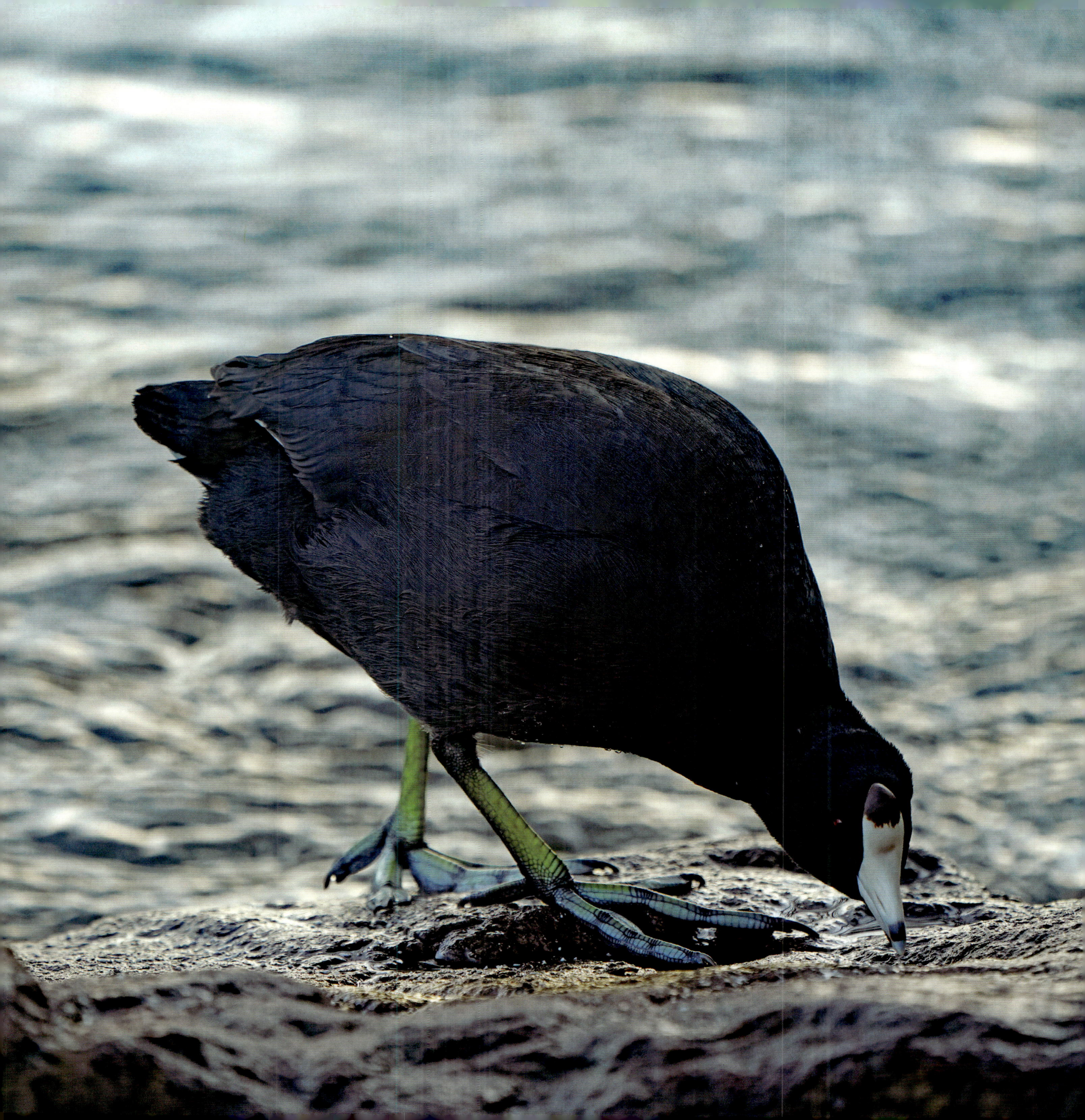

PRECEDING: A Brown Pelican (left) with striking adult plumage dips its feet in coastal shallows. These impressive birds were listed as an endangered species in 1970, at a time when the pesticide DDT was interfering with the calcium metabolism of birds, resulting in thinner eggshells that easily broke during incubation. Now, in a conservation victory, recovered Brown Pelicans are a common and celebrated sight along the Bay Area's coasts, and they were removed from the Endangered Species list in 2009. The Bay Area's diversity of coastal and freshwater ecosystems attracts seabirds and shorebirds alike, but the American Coot (right) is remarkable in its ability to navigate both land and water with equal ease on its elongated lobed toes.

RIGHT: A female Cinnamon Teal flings bright beads of water as she splashes in the Las Gallinas Wildlife Ponds at the edge of San Rafael. The male can be recognized by his striking rusty orange plumage and fiery red eye.

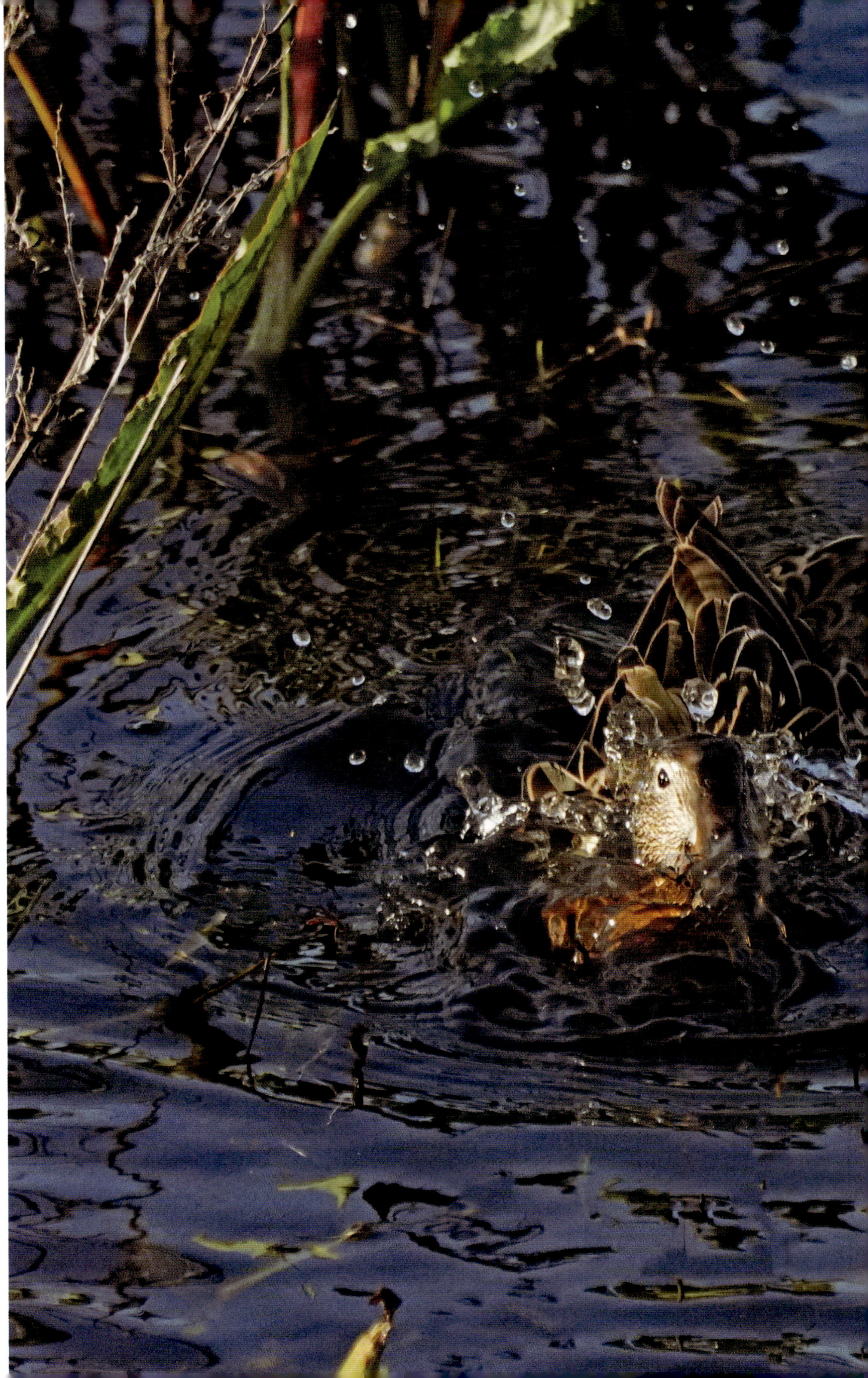

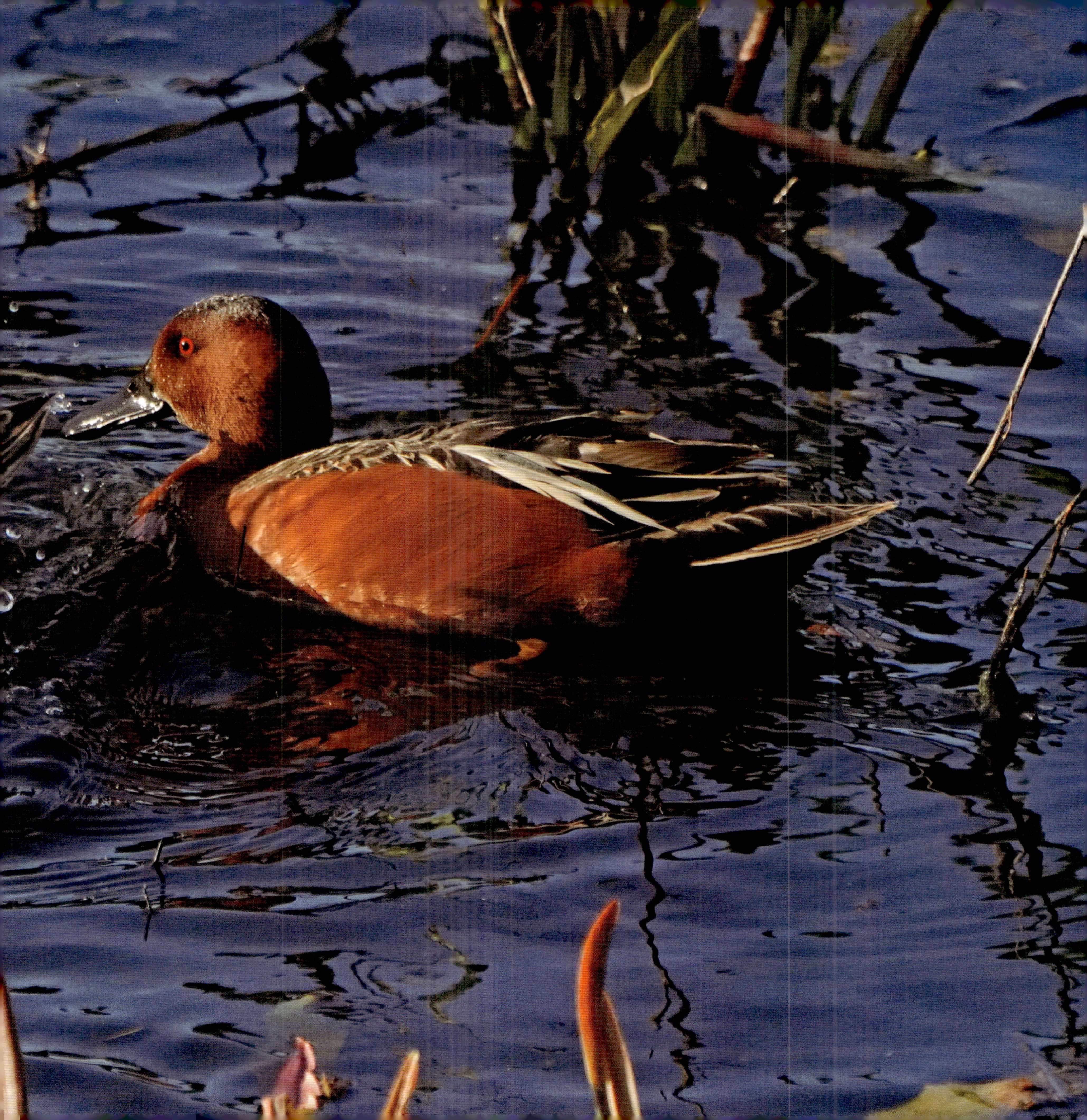

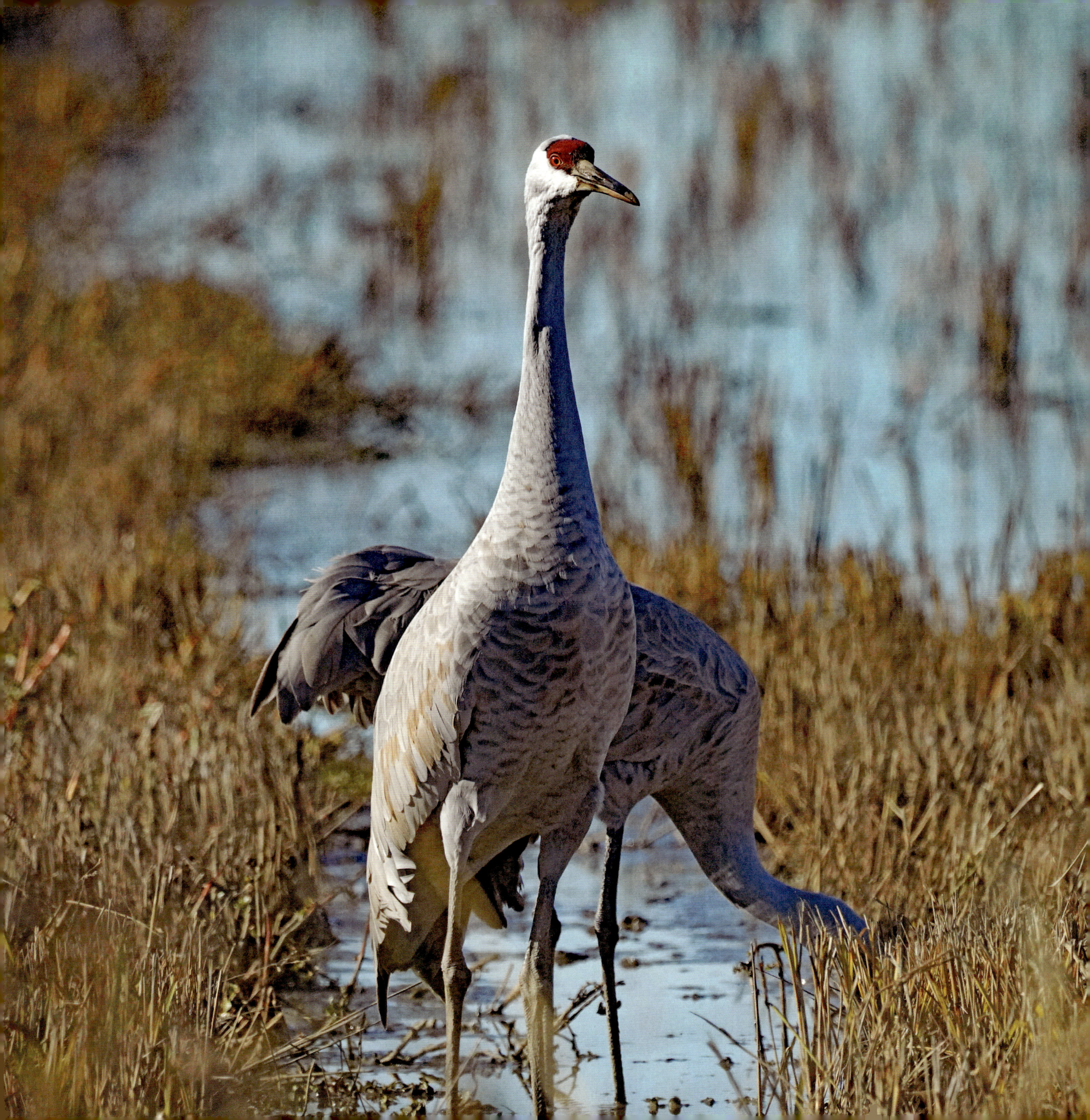

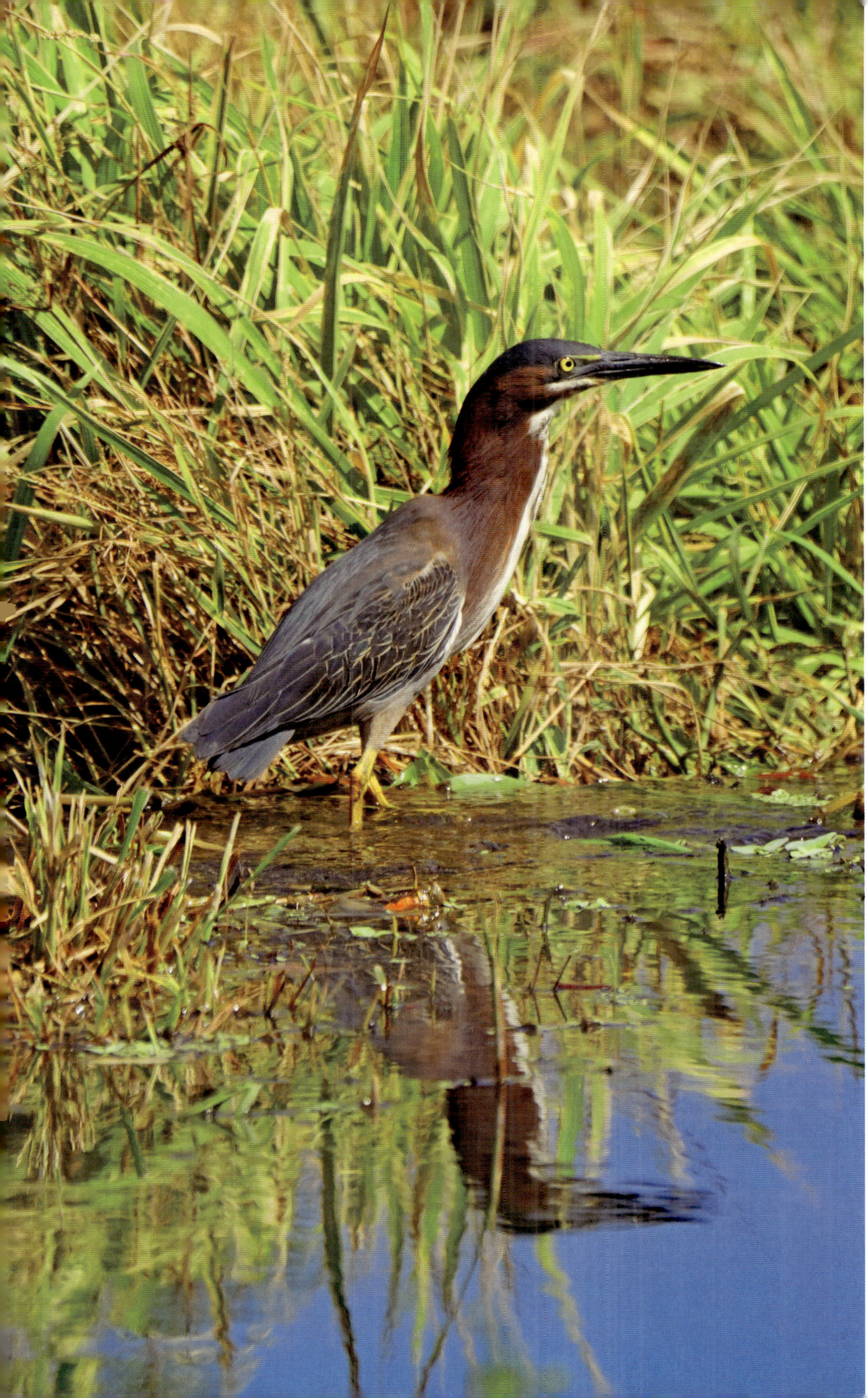

Two Greater Sandhill Cranes (left) forage in the winter waters of the California Delta. In the summer, thousands of these prehistoric birds head to the mountain meadows of the northern Sierra Nevada and the Cascades and the high desert of northeastern California to nest.

Often hidden from sight behind dense fresh water foliage, a secretive Green Heron (right) steps into the open in the South Bay Wetlands, revealing its rich chestnut neck plumage and radiant yellow eyes. The South Bay Salt Pond Restoration Project is a 50-year effort to restore 15,000 acres of tidal marshland in south San Francisco Bay—the largest wetland restoration West of the Mississippi!

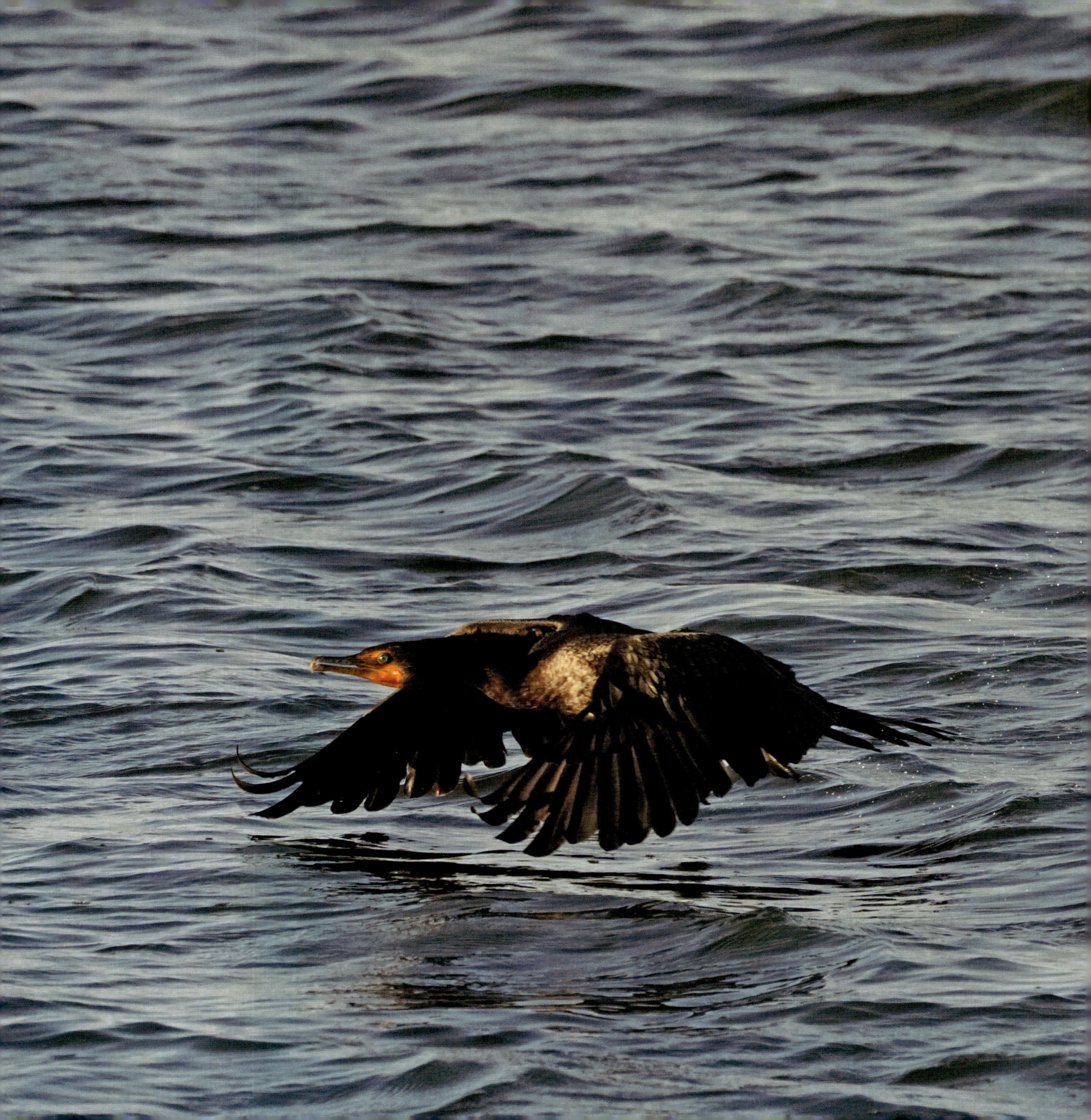

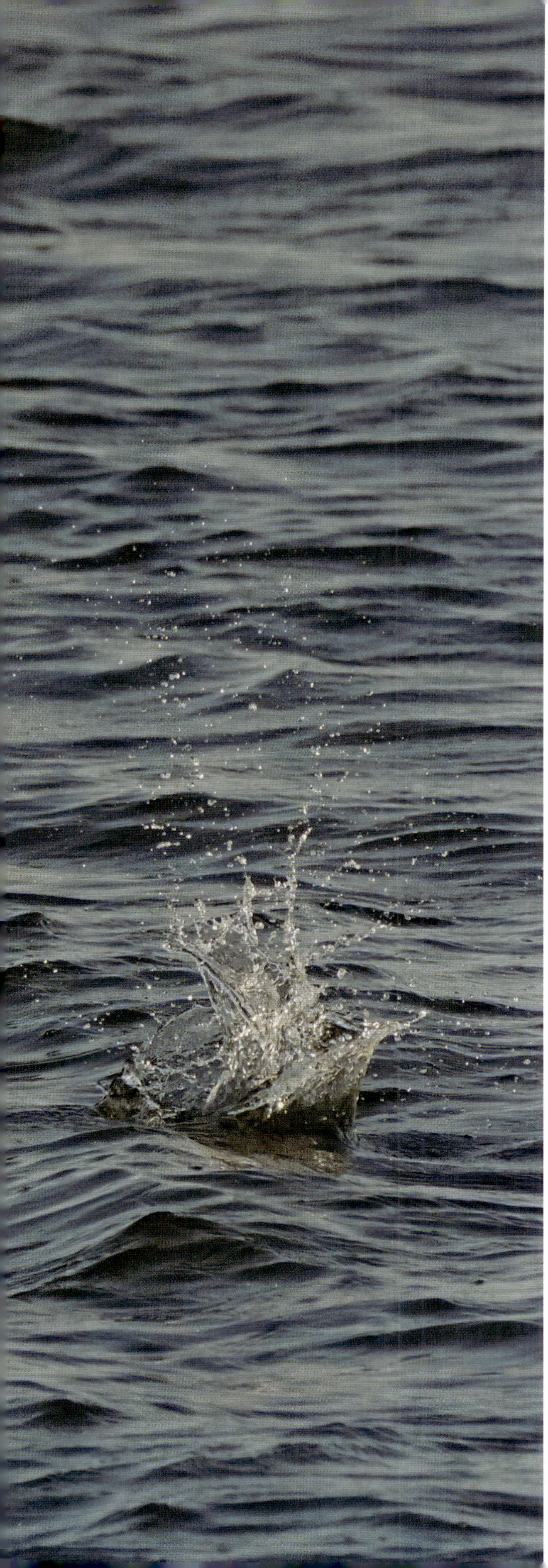

MOVING MOUNTAINS

In Tomales Bay, 30 miles north of San Francisco, the wind has picked up, and choppy waves jostle our kayaks. Across from us, the Double-crested Cormorants on Hog Island are unfazed. Tightly packed on the guano-whitened rock, they stretch their wings to dry in the stiff breeze. Liz Wilhelm and Dallas Smith, owners of Point Reyes Adventure Company, lift their paddles and let their boats sway in the swell. Liz encourages us to pause and look around from our vantage point in the middle of the mile-wide bay. “That’s the North American Plate”—with her paddle, she indicates the eastern shore that we launched from, with golden, undulating hills stretching out toward the interior of Marin County. “But over there on the Point Reyes Peninsula”—she points to our west, where pinkish-gray skeins of fog are rising above a forested ridgeline—“that’s an entirely different continental plate.” Where we drift in our kayaks, Liz tells us, we’re floating a couple of meters above the most notorious crack in California: the San Andreas Fault.

This strike-slip fault—responsible for the 1906 earthquake that destroyed 80 percent of the city of San Francisco—is also responsible for the slow northward migration of the hills to our west. The Point Reyes Peninsula has been slowly detaching from the North American Plate for 20 million years. Someday, it will become an island. At our particular moment in geologic time, the 10-mile land mass jutting out into the Pacific Ocean acts as a hook for lost birds flying over the ocean in search of refuge. The geology of the place—and the diverse, rich ecosystems it supports—makes it an avian hotspot. Four hundred ninety bird species have been recorded here—over 50 percent of all species of birds in North America.

Tectonic activity shaped the Sierra Nevada 250 miles to our east, too, in a grind of subduction and slow-cooling magma. Millions of years

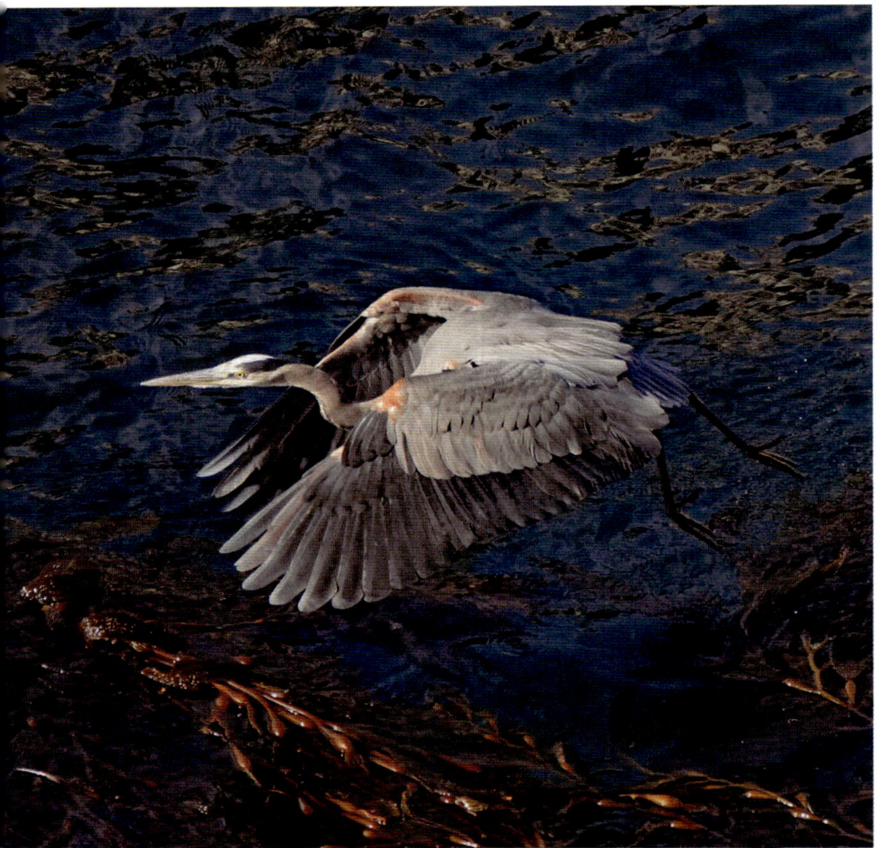

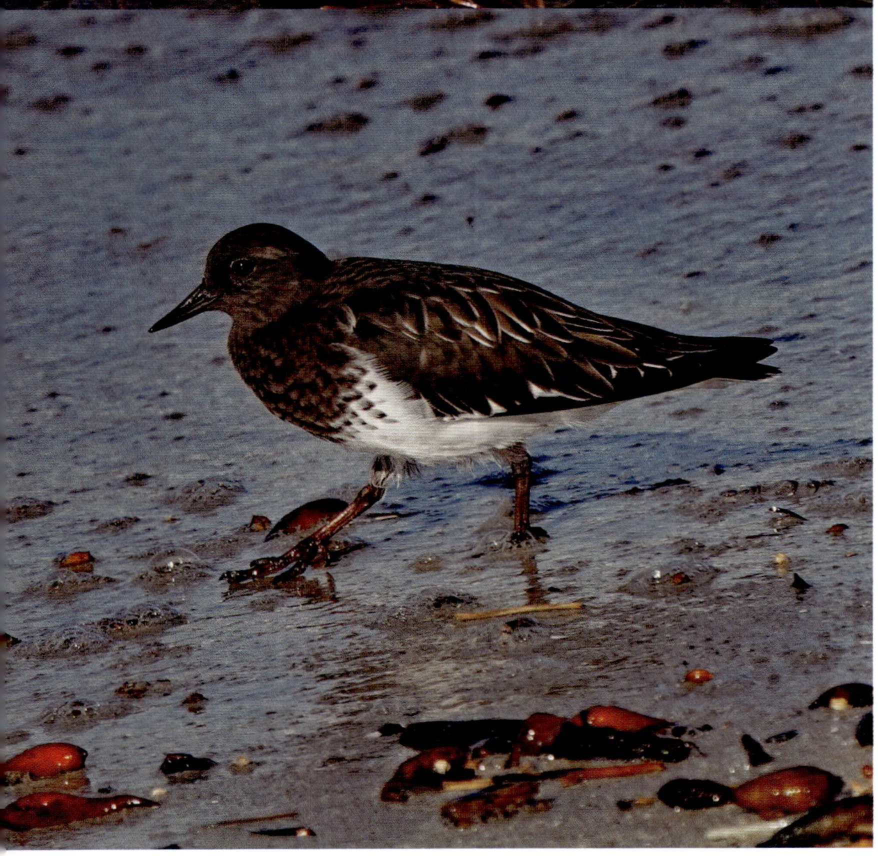

of glacial carving and rain washed much of the sediment off the granitic domes westward to form the flat basin that is now the Central Valley. Several hundred thousand years ago, the whole valley was a freshwater lake, one that drained into a river valley as deep as the Grand Canyon. That valley is now submerged offshore from Monterey, but it still supports the ecosystems of Monterey Bay. The mile-deep Monterey Canyon pulls cold water in from the abyssal plain, feeding Monterey Bay with upwelling nutrients that attract throngs of birds and other wildlife. The region is an Audubon-designated Global Important Bird Area, alongside the California Delta and the waters offshore from San Francisco. California's geologic history cuts a powerful throughline in the story of the region's birds.

Uplift, continental drift, erosion, flood—these set the stage for all the wonders a hiker might find within two hours of San Francisco: marshes in the South Bay loud with the clacking cries of Ridgway's Rails; foggy mornings on the Point Reyes Peninsula in search of vagrant birds blown in from other continents; Delta sloughs stirred up by the wingbeats of ten thousand ducks alighting in unison. Geology is not just rocks but also the forces of water and wind that wear on them and are shaped by those same bones of the earth in turn. These determine which plant communities thrive where, and therefore where animals will find their ideal niche, including the place of birds in the food webs of forest, shoreline, and wide river valley. Raptors ride the wind currents in pursuit of prey on the wing and spiral with vultures on uplifting thermals shaped by the mountains and passes that surround the Bay.

It's no wonder that locals take pride in plate tectonics around here; our kayaking guide Liz has a tattoo of the Point Reyes Peninsula on her left calf, a celebration of the slow and unstoppable northbound slide of the Pacific Plate. It's an acknowledgment of how beautifully small we are—humans and birds alike—in the face of the big forces that continue to shape our Bay Area backyard.

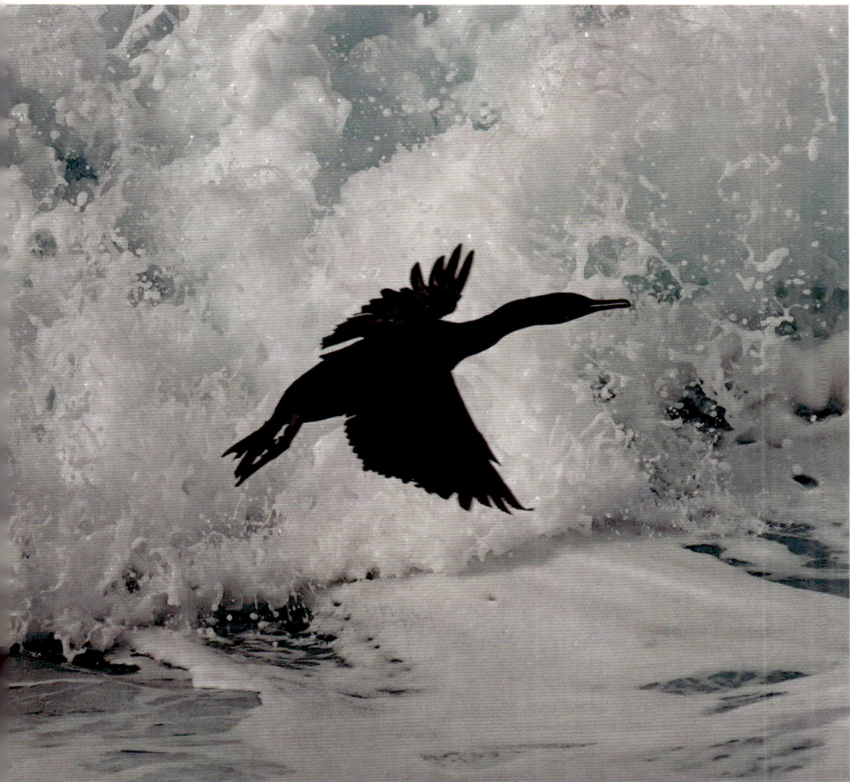

PREVIOUS: A Double-crested Cormorant, with heavy bones for masterful diving, has a hard time getting airborne. These waterbirds take off horizontally, using their feet to push off from the surface of the water, leaving a trail of tell-tale splashes behind them.

OPPOSITE: In Monterey, where unique underwater geology supports teeming bird populations, a Great Blue Heron skims over a bull kelp forest (above) and a Black Turnstone patrols the shore (below) in search of tasty invertebrates under rocks.

LEFT: A Pelagic Cormorant flies low across the surf at the dynamic edge where sea meets land near Monterey.

BELOW: Sanderlings scamper in and out with every wave on the outer coast of the Point Reyes Peninsula. With more than 80 percent of its perimeter hemmed by water, the fault line–sculpted peninsula may someday become an island as it continues to drift northward along the San Andreas Fault.

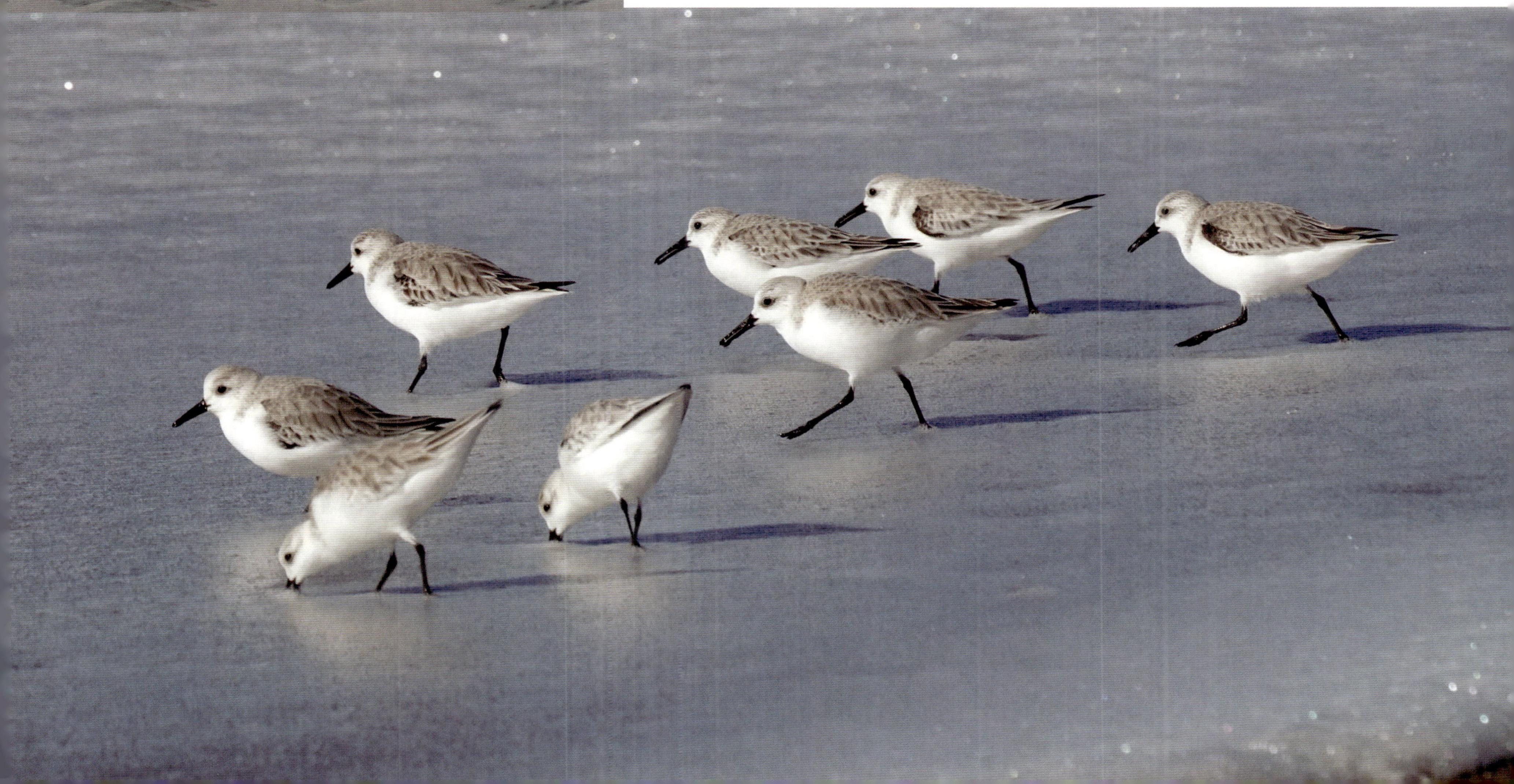

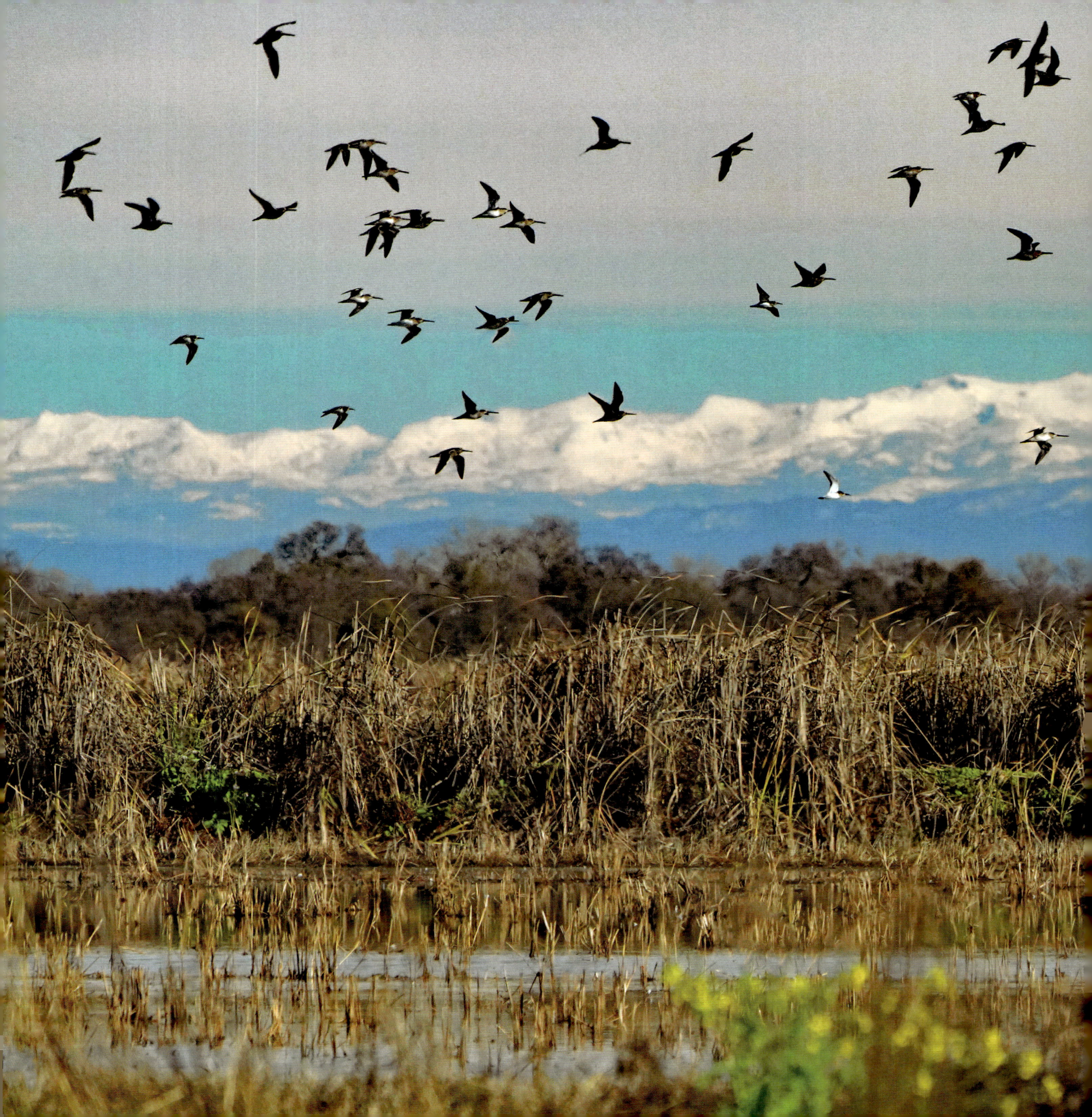

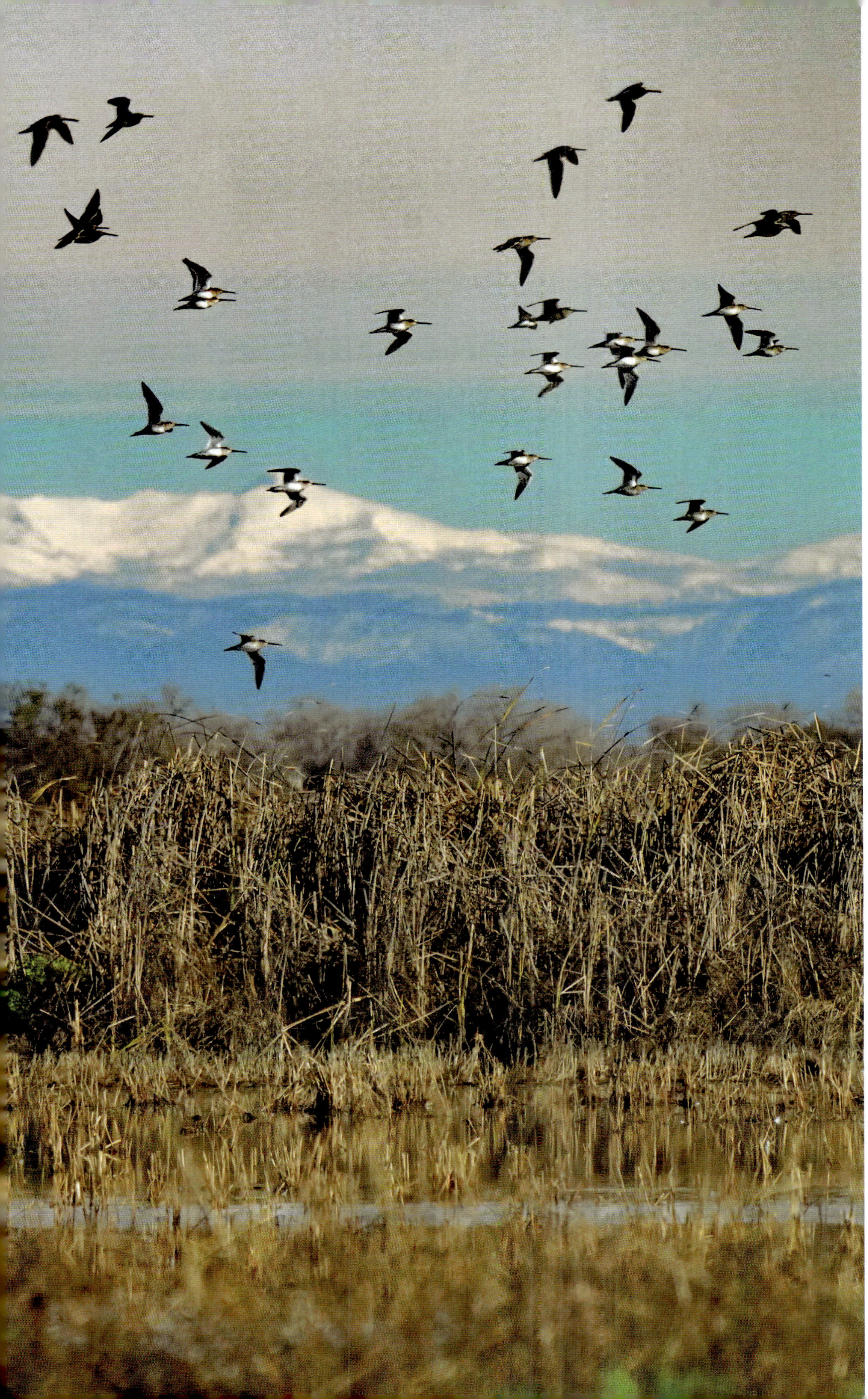

Where the distant snowy Sierra Nevada mountains rise above the rich delta flats—feeding the Central Valley with groundwater and eroded nutrients—a mixed-species flock of wetland birds takes flight.

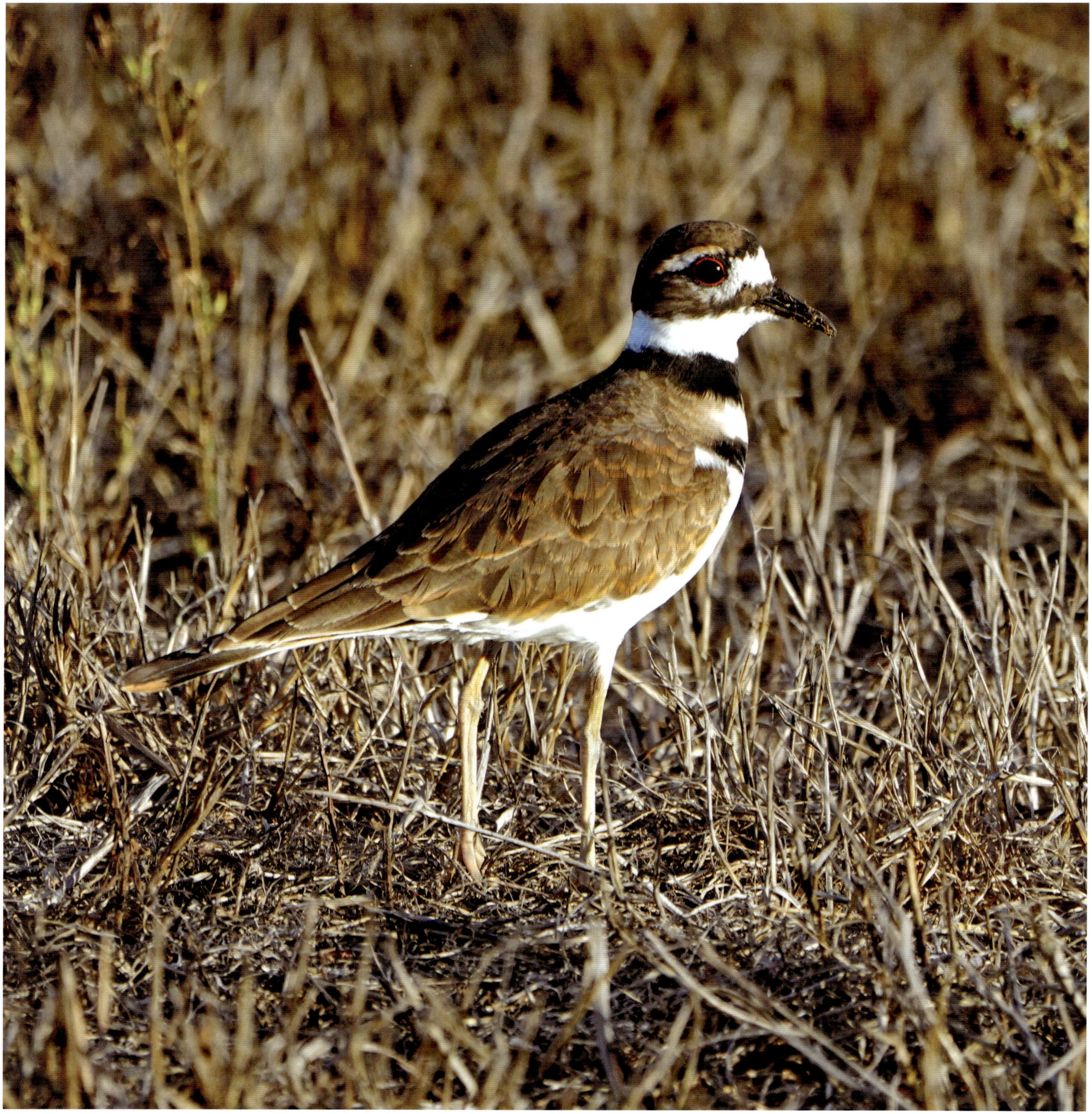

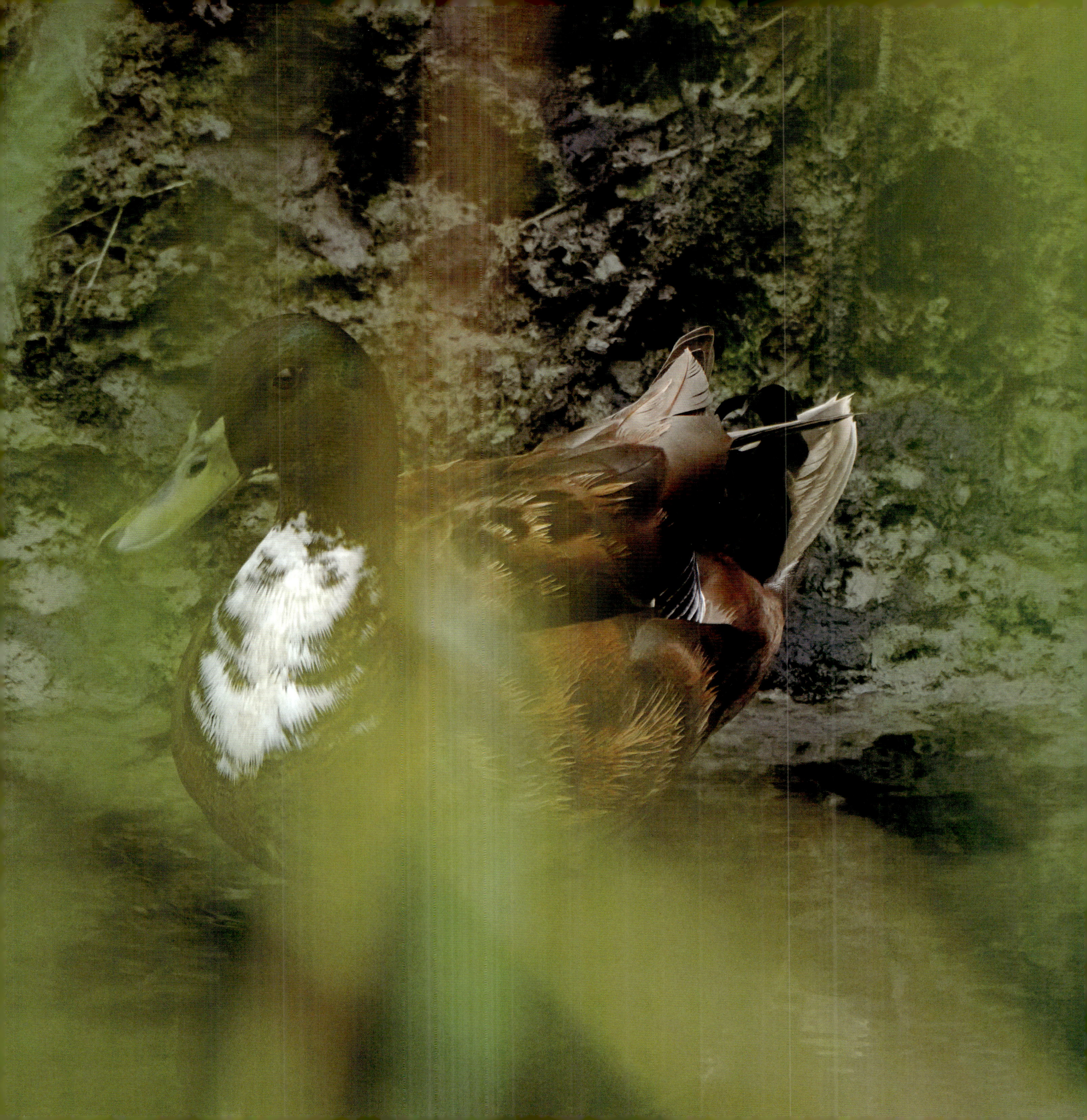

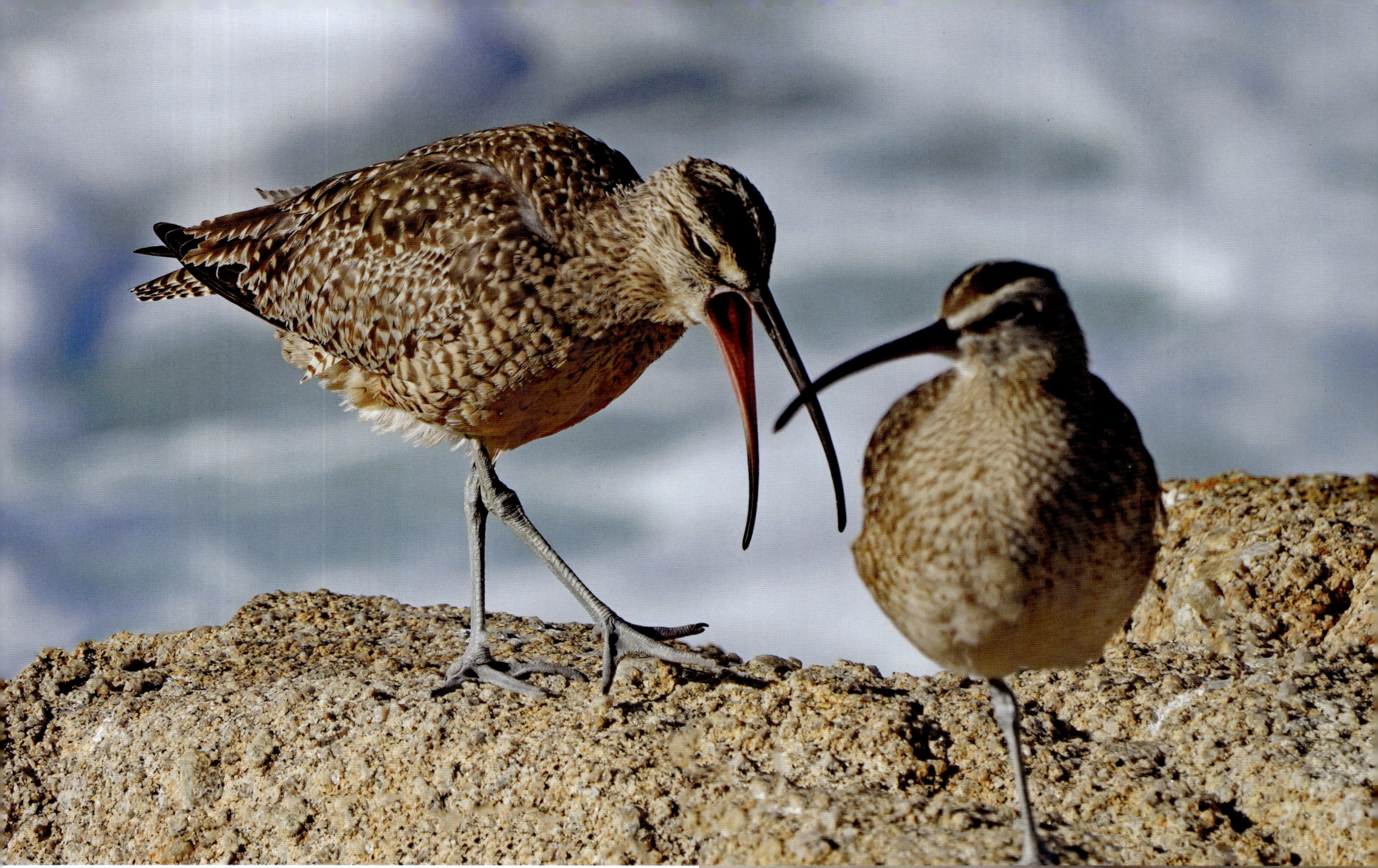

PRECEDING: Killdeer (left) are one of the Bay Area's most widespread shorebirds, particularly during the spring and summer breeding season. They can adapt to the many human-influenced landscapes of the region, and can be spotted on beaches, ponds, lawns, golf courses, parks, and parking lots—wherever they can scrape out a shallow nest on the ground. This unusual Mallard hybrid (right) is likely the product of a wild bird mating with an escaped domestic variety. As various waterfowl throng together in the Bay Area's wetlands and city ponds, species lines can grow blurry, and we are reminded that evolution is a complex ongoing process, especially at the interface between urban spaces and wild migratory flyways.

CLOCKWISE FROM TOP: Whimbrels, Black Oystercatchers, and a Spotted Sandpiper all strut and stalk among the rocks where the Pacific meets the shore. Of these species, Black Oystercatchers are almost exclusively rock-dwellers, choosing to spend most of their lives on sea stacks, stony reefs, and cliffs. Even oystercatcher eggs are hardy, capable of being submerged by high tides and storm surges. Parents simply roll the eggs back where they belong if waves knock them around on the rocky shore!

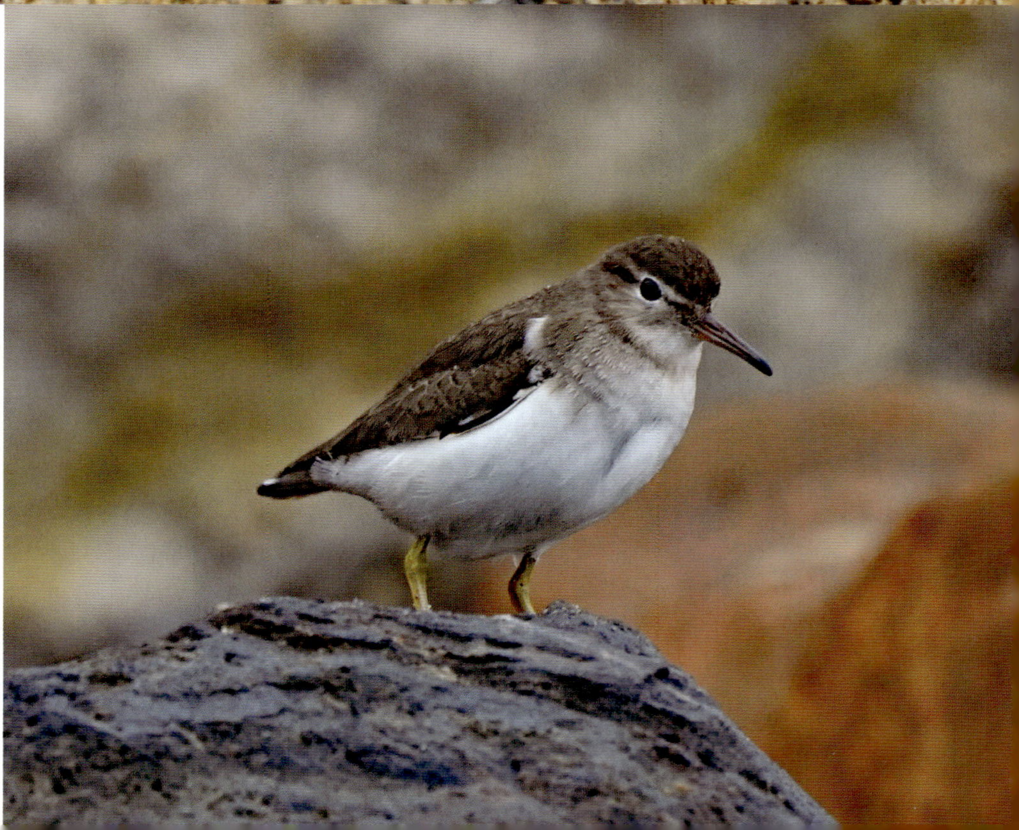

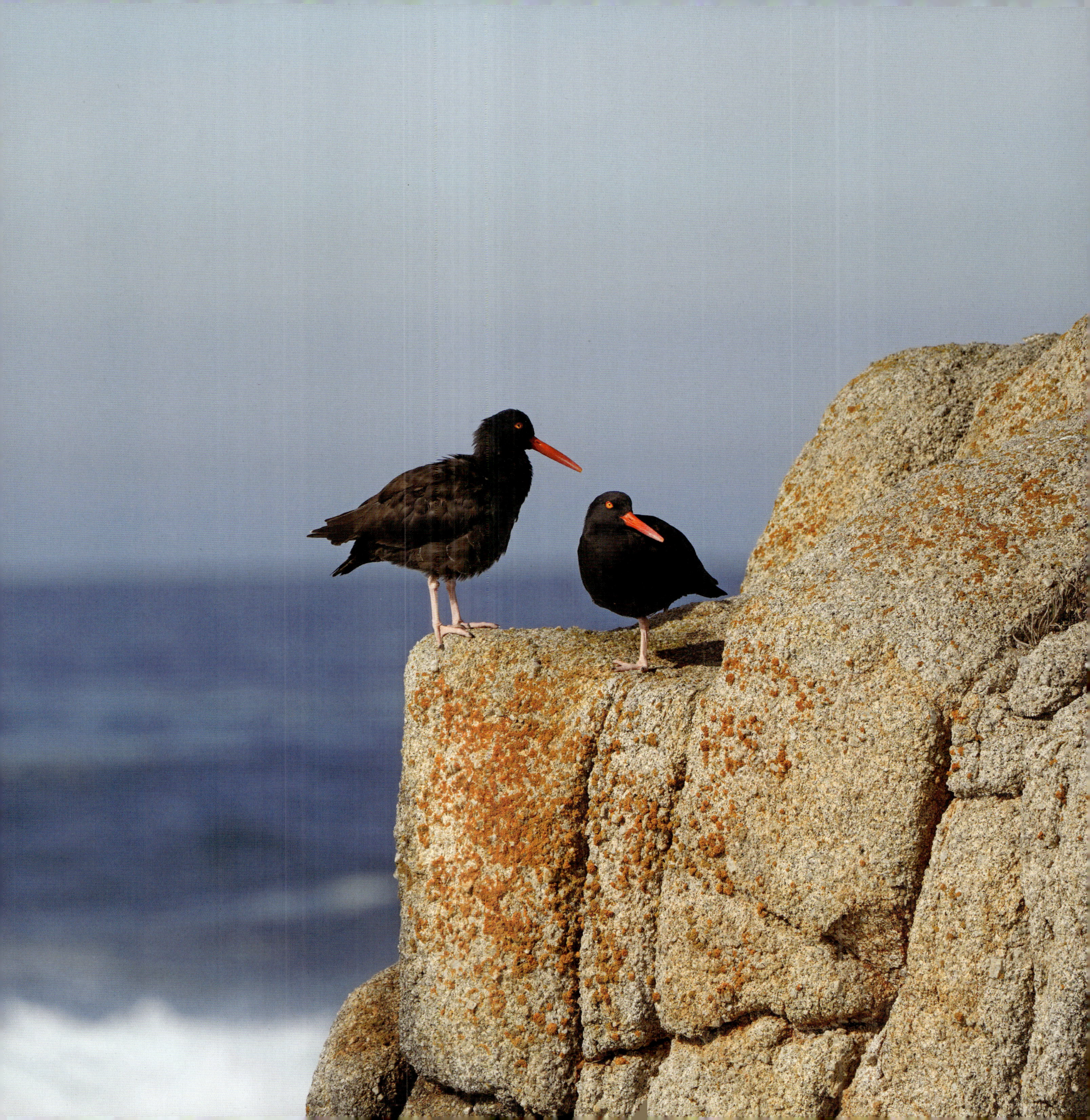

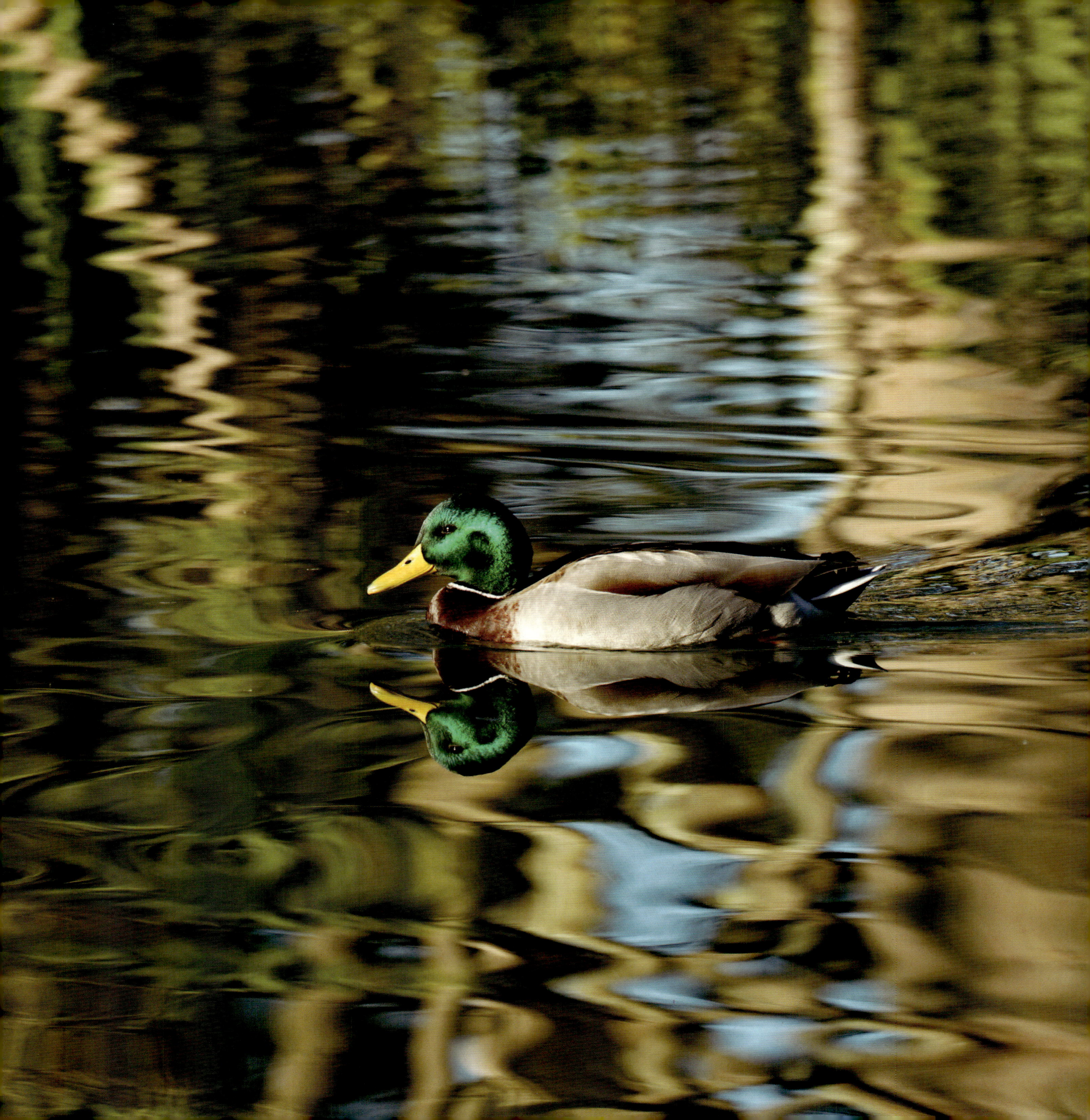

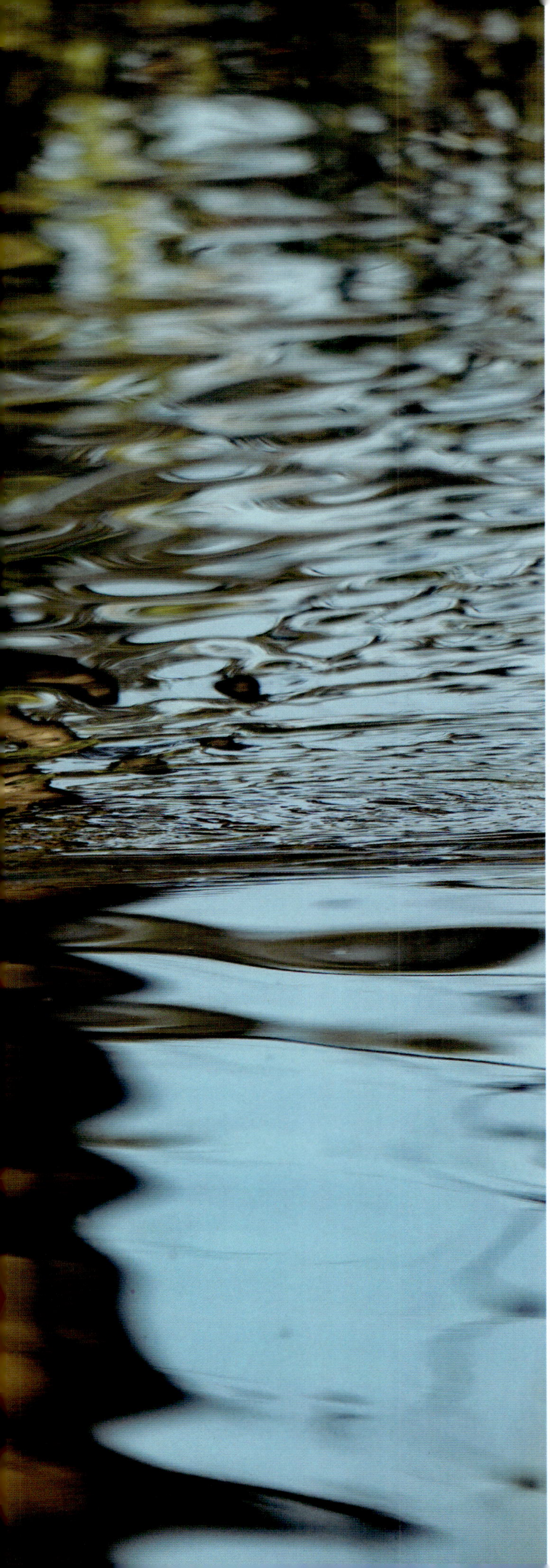

COOL FOG, GOLDEN SUMMERS

It's a cool spring morning in Golden Gate Park, and the water in Blue Heron Lake is a mirror. A Mallard Duck sits at the edge; the reflections that surround him are a dazzlement of color and branching patterns. This garden-like park was built 150 years ago on top of a rolling landscape of sand dunes and still ponds. Yet the ocean's influence abides. This morning, the Pacific unfurls a soft blanket of fog across the park and the surrounding city. From within the cloudy blanket, unseen birds call out: the buzzy song of a Townsend's Warbler passing through on its journey north; the dry, scratching chirps of a Chestnut-backed Chickadee; the loud *tseee* of a Black Phoebe. It's easier to hear them than to see them through the encroaching fog, but the birds have no trouble finding their way: UV vision and specialized color-sensing cones in their retinas help them orient in the fog.

It's a good thing that the Bay Area's birds know how to navigate with low visibility: San Francisco sees nearly 200 days of fog a year. Every spring, a high-pressure weather pattern called the Pacific High settles over the region. Winds from the Pacific High blow surface water away from the shore, and cold deep-sea water rises to take its place. At the frigid surface, water vapor condenses. Heavy, dense layers of fog are born. As warm air rises inland in the spring and summer, the fog gets sucked into coastal cities and forests all along the coast.

The Pacific High is responsible not just for the fog but for the greater Bay Area's Mediterranean climate throughout the year. During the spring and summer months, the pressure system sits over the region and deflects incoming storms, preventing rainfall. In the autumn, the Pacific High weakens and moves offshore, allowing low-pressure winter rains to roll across the

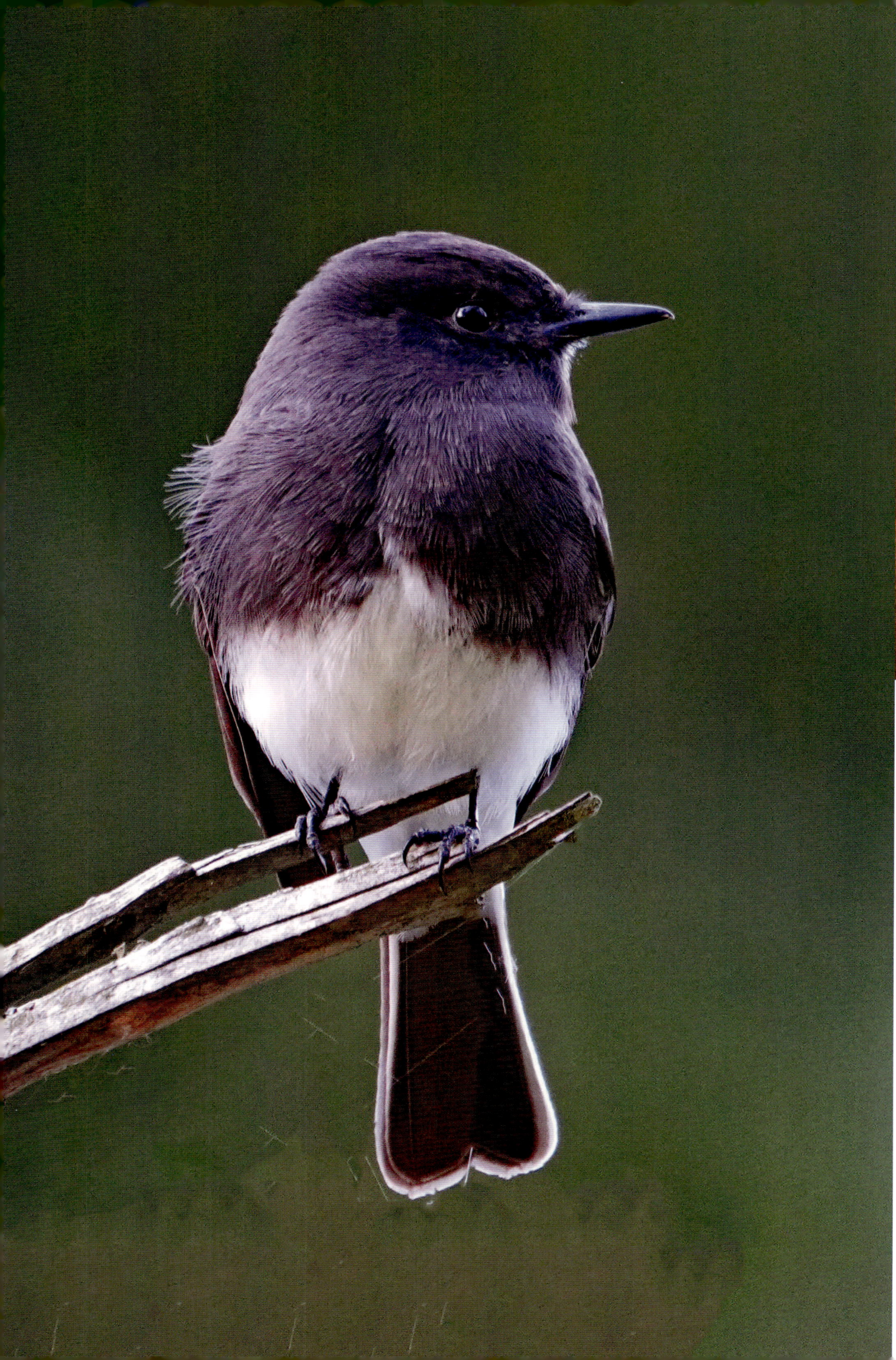

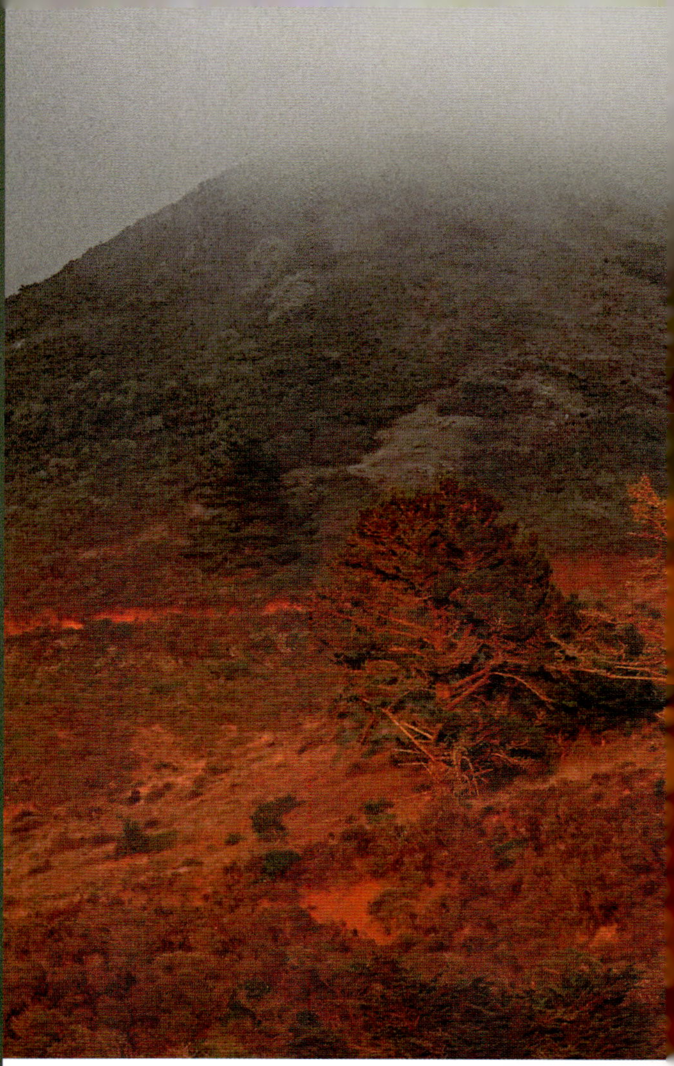

PRECEDING: A male Mallard glides across Blue Heron Lake in Golden Gate Park. Over 270 species of birds have been counted in this beloved San Francisco park, which is often bathed in summer fog.

LEFT: A Black Phoebe perches on a branch in Golden Gate Park, waiting to dart into the open to catch insects on the wing.

TOP: Fog rolls in over dry coastal hills at sunset in the Marin Headlands.

RIGHT: A Chestnut-backed Chickadee (pictured here in Golden Gate Park) is adaptively colored to blend well with the damp coastal forests where it lives.

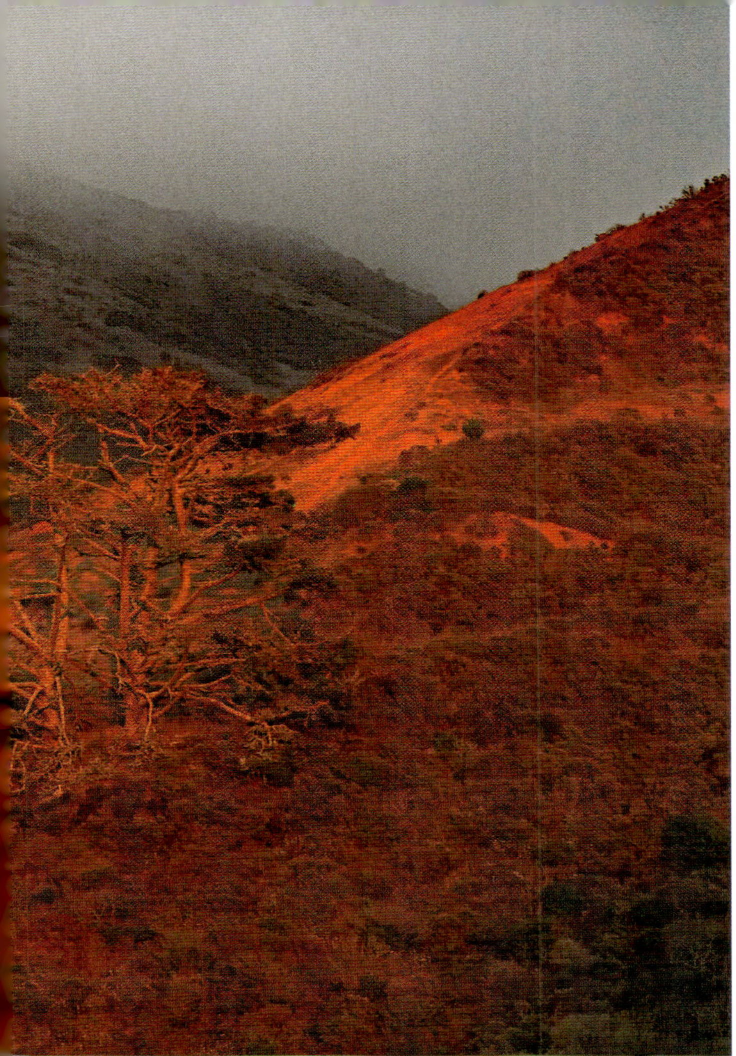

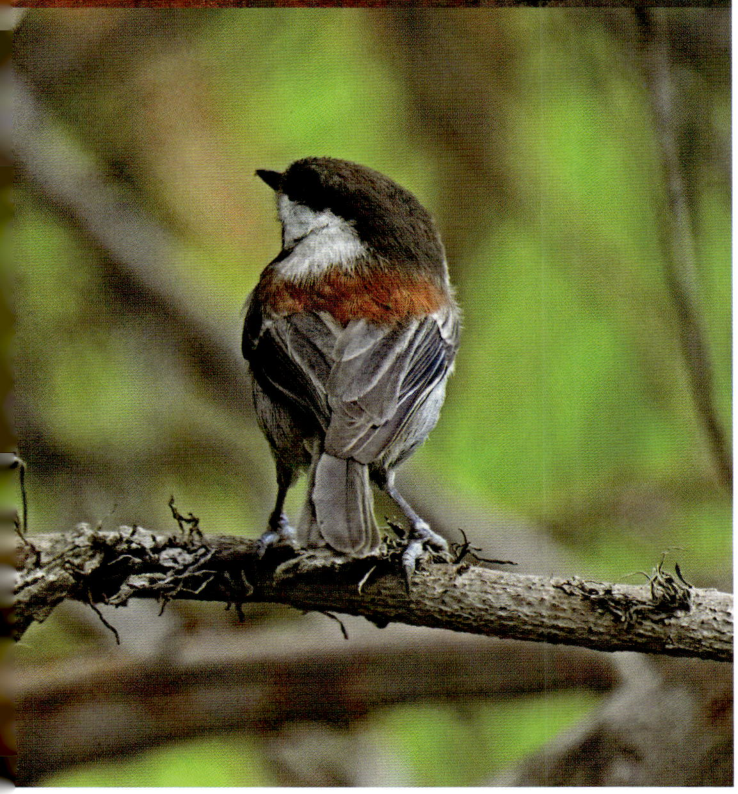

state. Our region is one of only six places on earth with a Mediterranean pattern of warm, dry summers and wet, temperate winters. The climate feeds the ecoregions around the Bay: tall redwood forests dripping with summer fog and the waterfall-like songs of Pacific Wrens; dry oak savannas pulsing with the nasal *wha wha wha* of the White-breasted Nuthatch.

But California's climatic sweet spot is shifting. Atmospheric scientists project longer droughts, more powerful superstorms, mounting wildfires, rising seas. Bird communities are at the front lines of anthropogenic climate change. In 2015, after four consecutive years of record-breaking drought, California's population of breeding ducks fell 30 percent compared to 2014. In the spring of 2021, young Barn Owls sheltering in backyard nest boxes died of heat stress all across the state. Birds at the edge of colossal wildfires, like those that have swept through Sonoma and Santa Cruz, lose body mass—the smoke keeps them from foraging.

Climate disruption also affects phylogeny, the timing of cyclical patterns in the natural world. When does the first coast live oak bud in the spring? When does the first Golden-crowned Sparrow return in the fall? These phenomena are connected, and birds are the weavers that stitch them together. Across the foggy strait from Golden Gate Park, hawks in the Marin Headlands depend on seed dispersion to feed insect populations, which in turn feed songbirds, which the raptors eat. As climate change throws individual species off-kilter—causing early blooms or late hatches—the clockwork of the seasons gets thrown off. The birds at the center of these changes may not have time to adapt before the web is unwoven.

Our mild climate has drawn many humans to live close to the Bay just as it draws thousands upon thousands of birds. Whatever the future may bring, our bird neighbors will be the heralds of change, and learning to understand the stories they tell may guide us toward choices we can make to fortify our shared homeplace—and to adapt, together, to what may come next. Just as we do on a foggy morning walk through Golden Gate Park, we'll need to practice keener ways of listening.

Red-tailed Hawks are one of the most successful raptors in the New World, equally at home nesting in the coastal forests of Half Moon Bay (pictured here), and on skyscraper ledges and in urban trees in San Francisco.

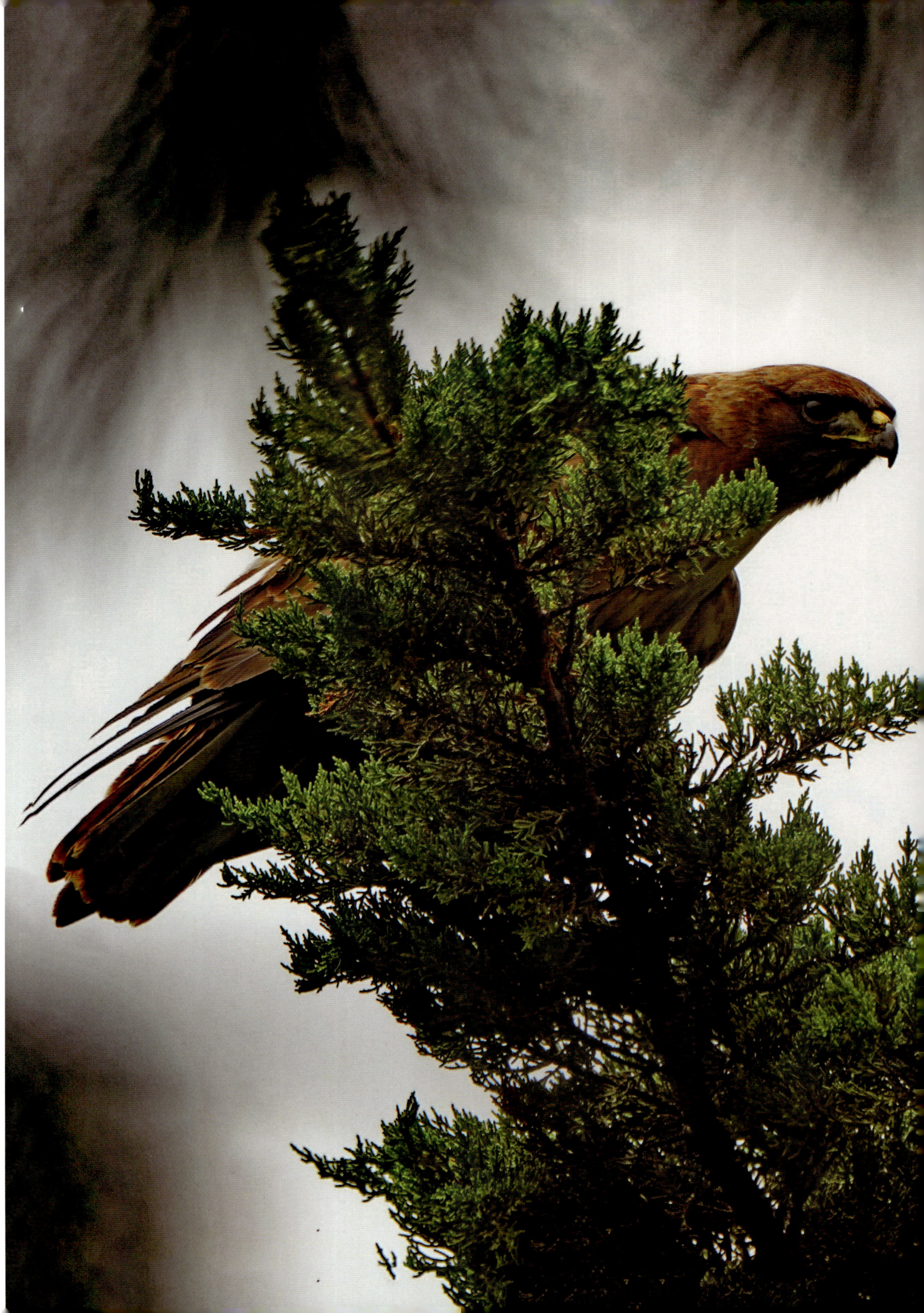

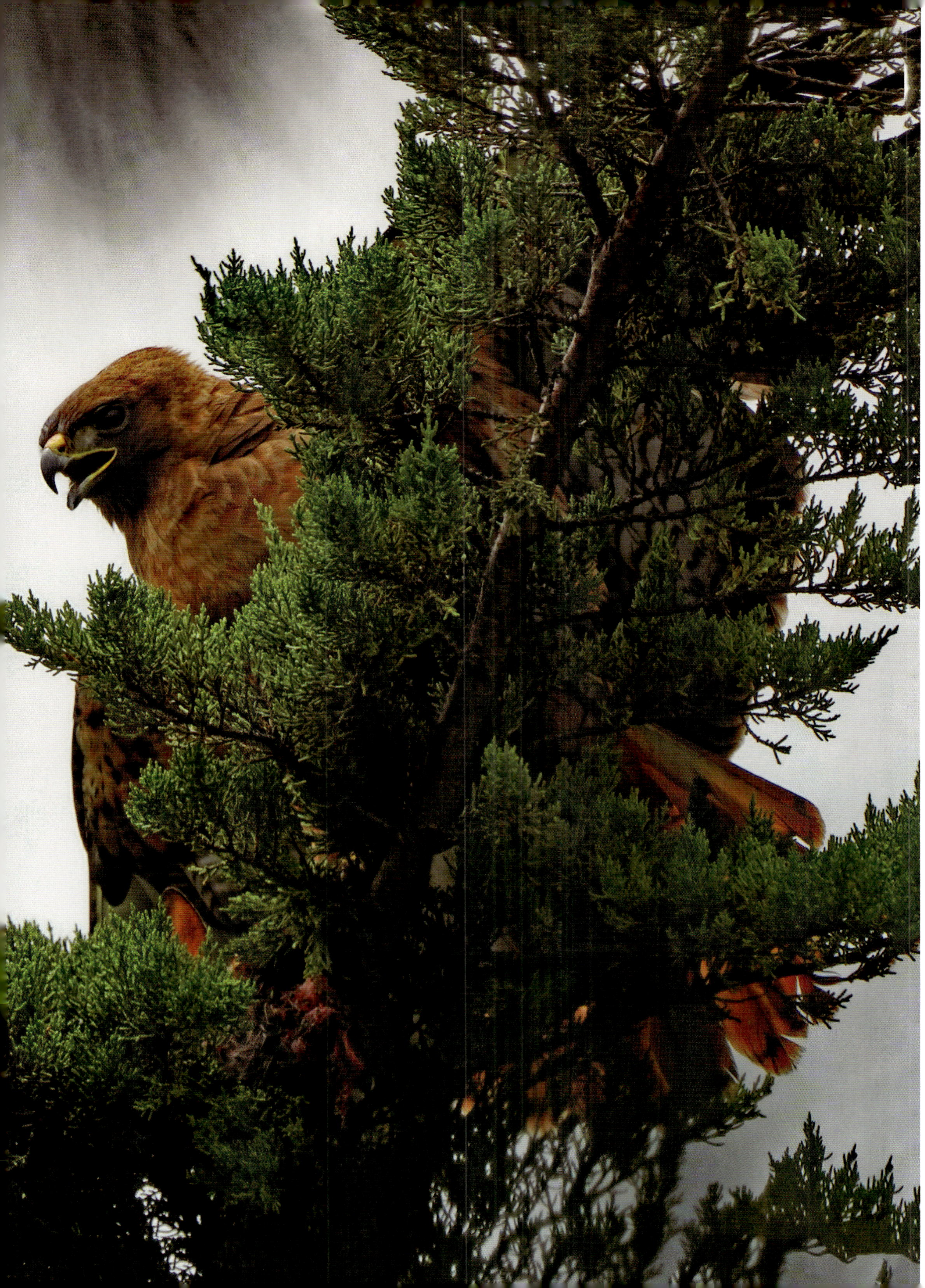

FOLLOWING: The Lesser Goldfinch (top left) and the Bushtit (bottom left) occupy different niches in the Bay Area ecosystem. Seed-eating Lesser Goldfinches frequent open, brushy areas and are familiar denizens of urban and suburban habitats; insectivorous Bushtits are more often found in mixed woodlands, acrobatically foraging for bugs. Both are likely to make a stop in your garden! Acorn Woodpeckers and Scrub Jays (top right, bottom right) are extraordinary "ecosystem engineers" in the oak woodlands of the Bay Area. Their respective behaviors of caching and scatter-hoarding acorns significantly contribute to regenerating oaks and revitalizing their associated ecosystems. A Barn Owl (opposite) gazes out over a familiar Bay Area palette of golden marshlands and pink, rolling hills near Pacheco Pond in Novato.

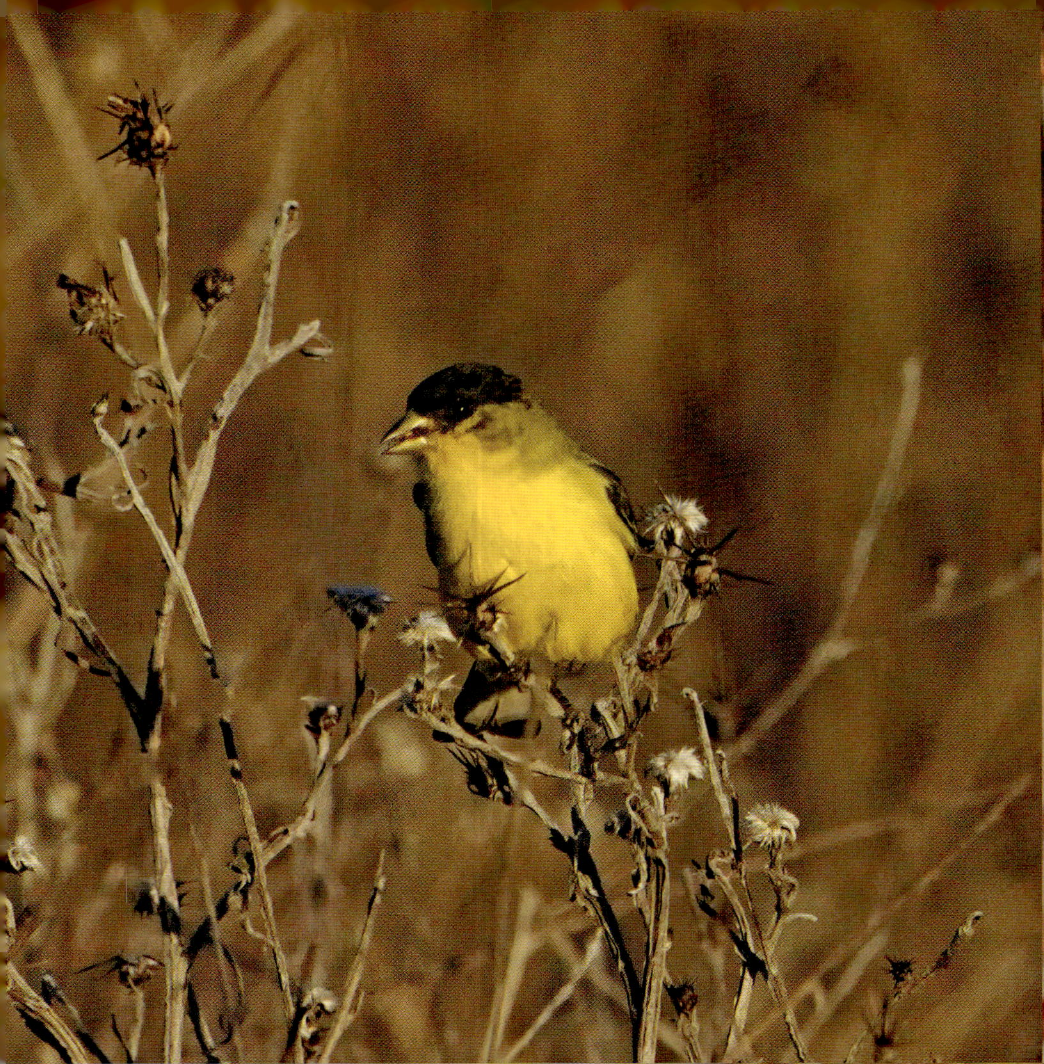
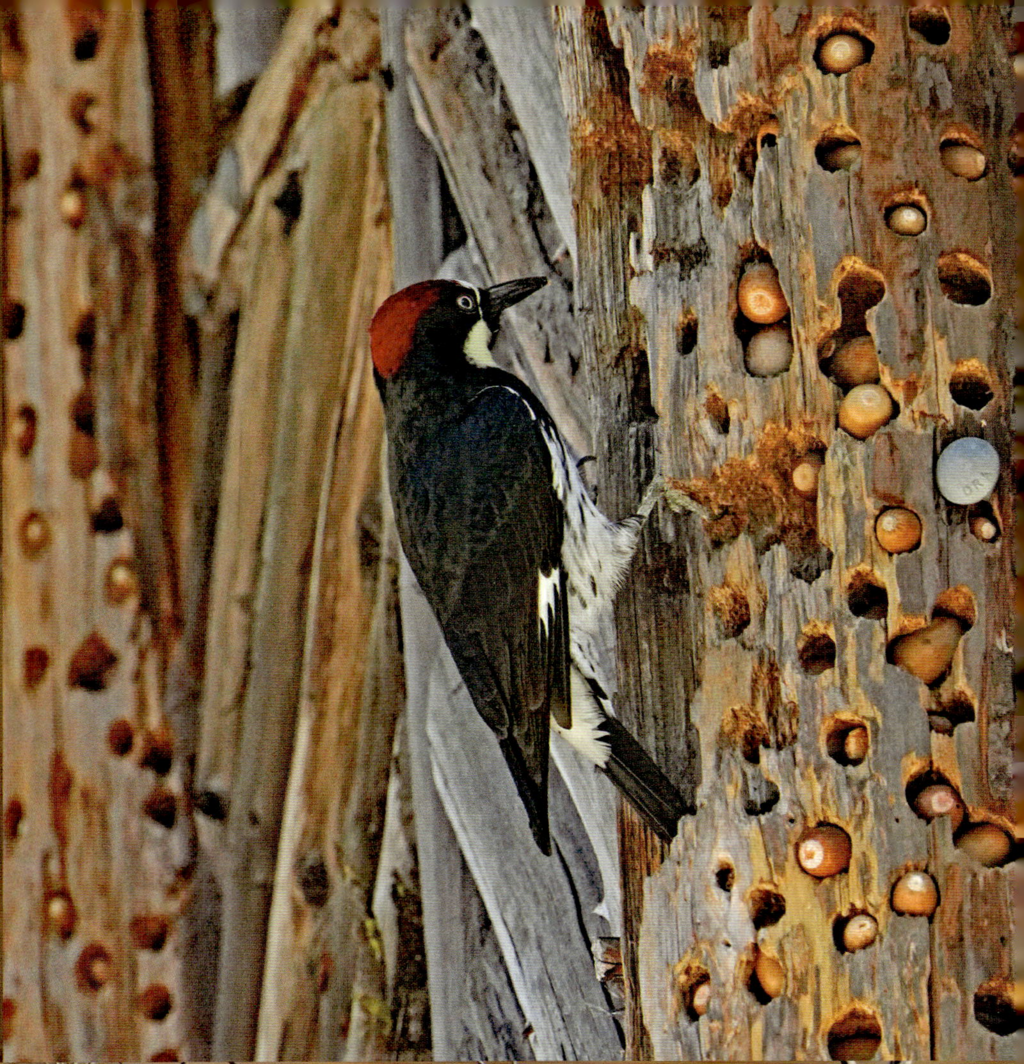
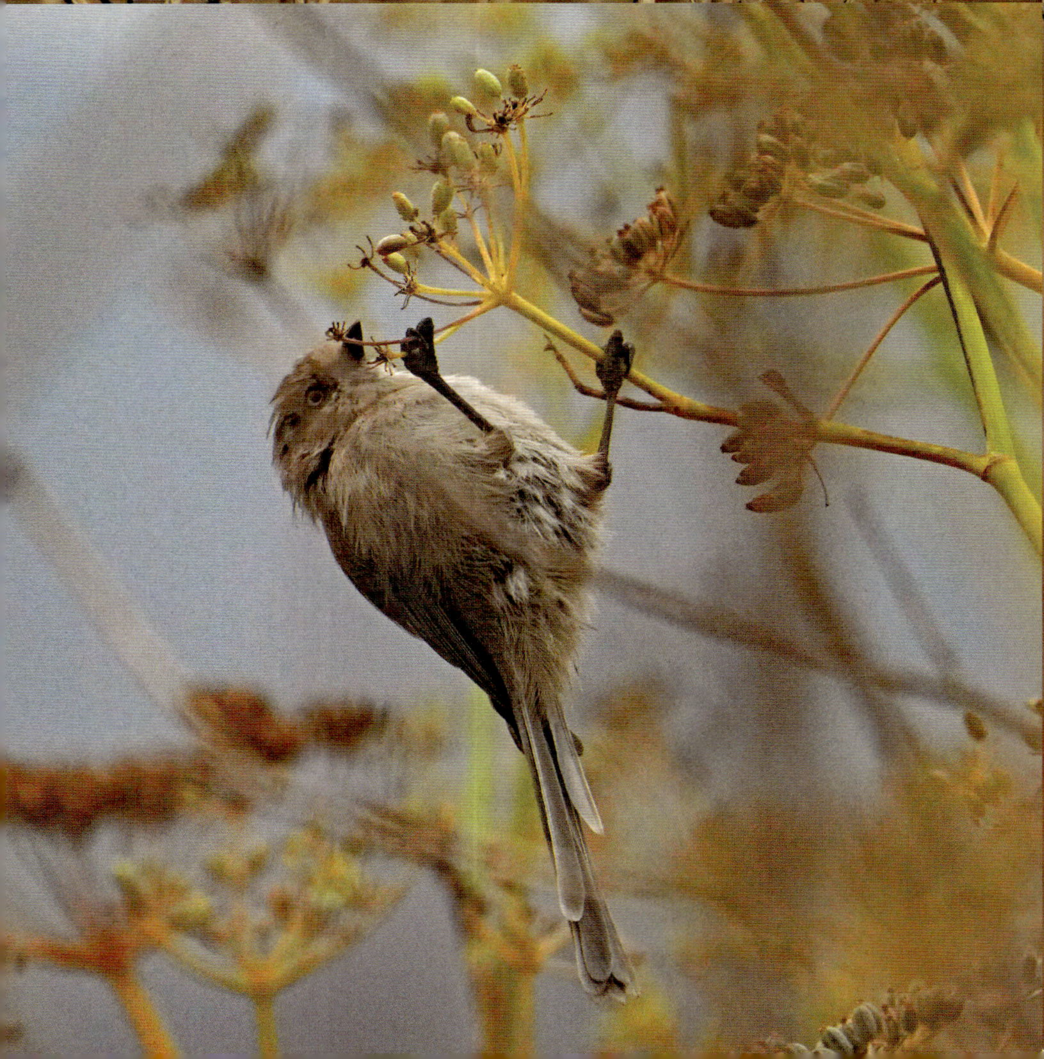
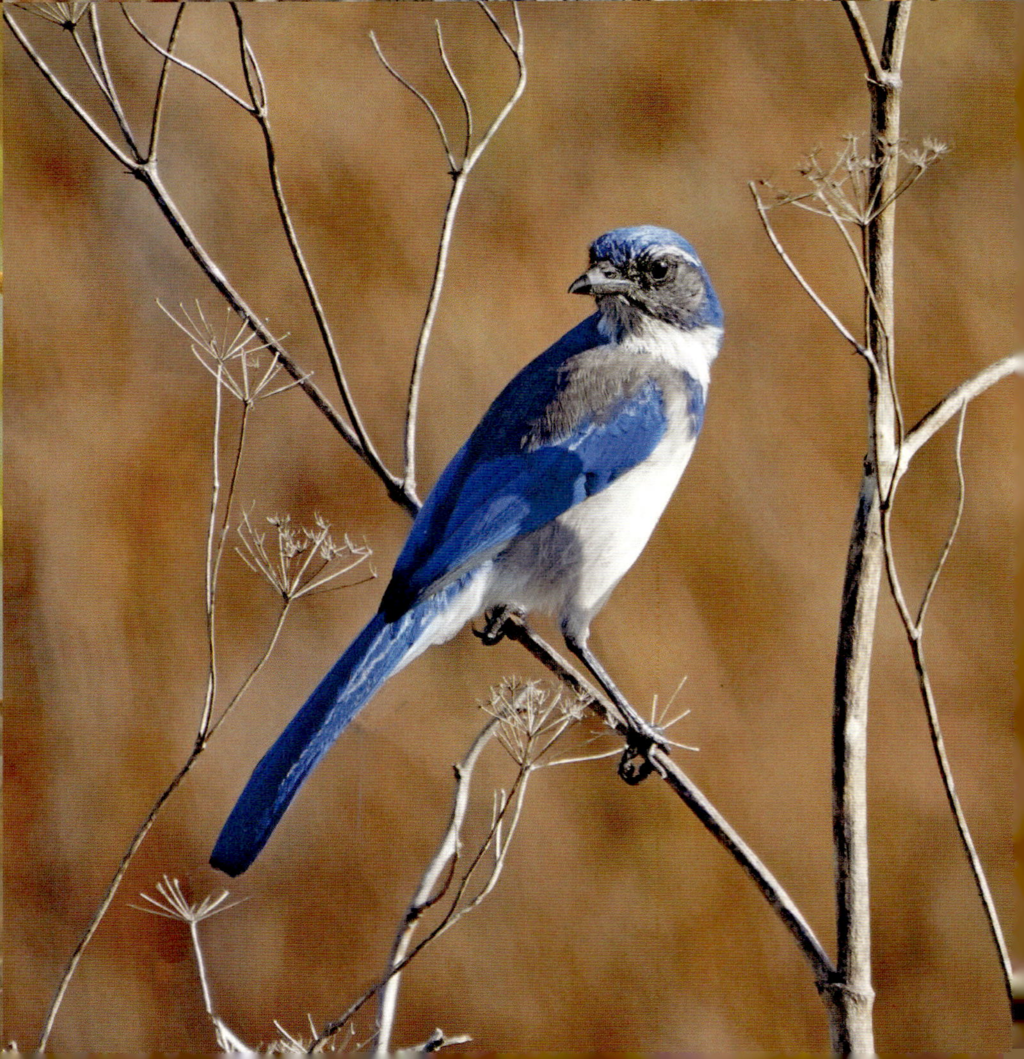

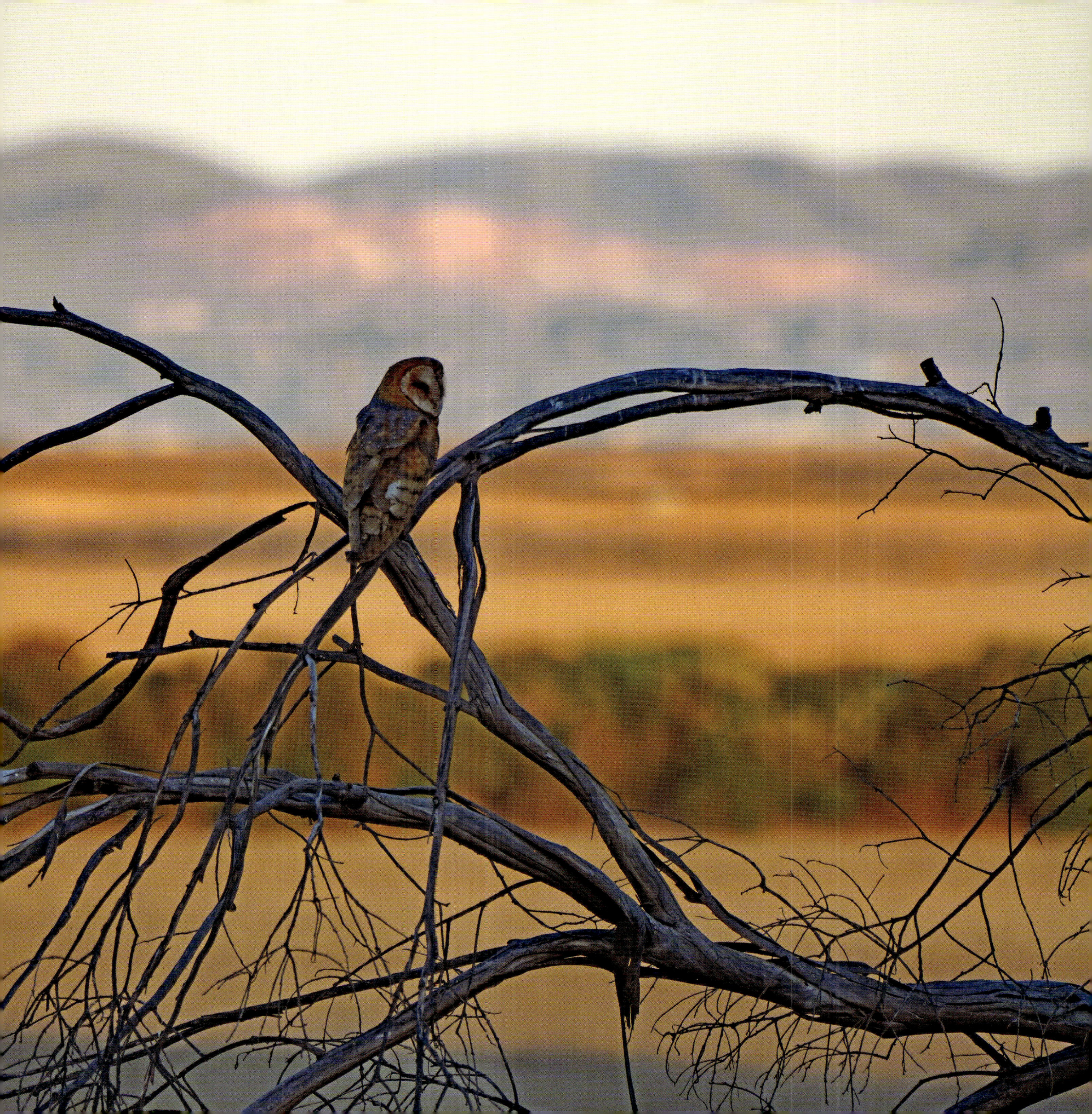

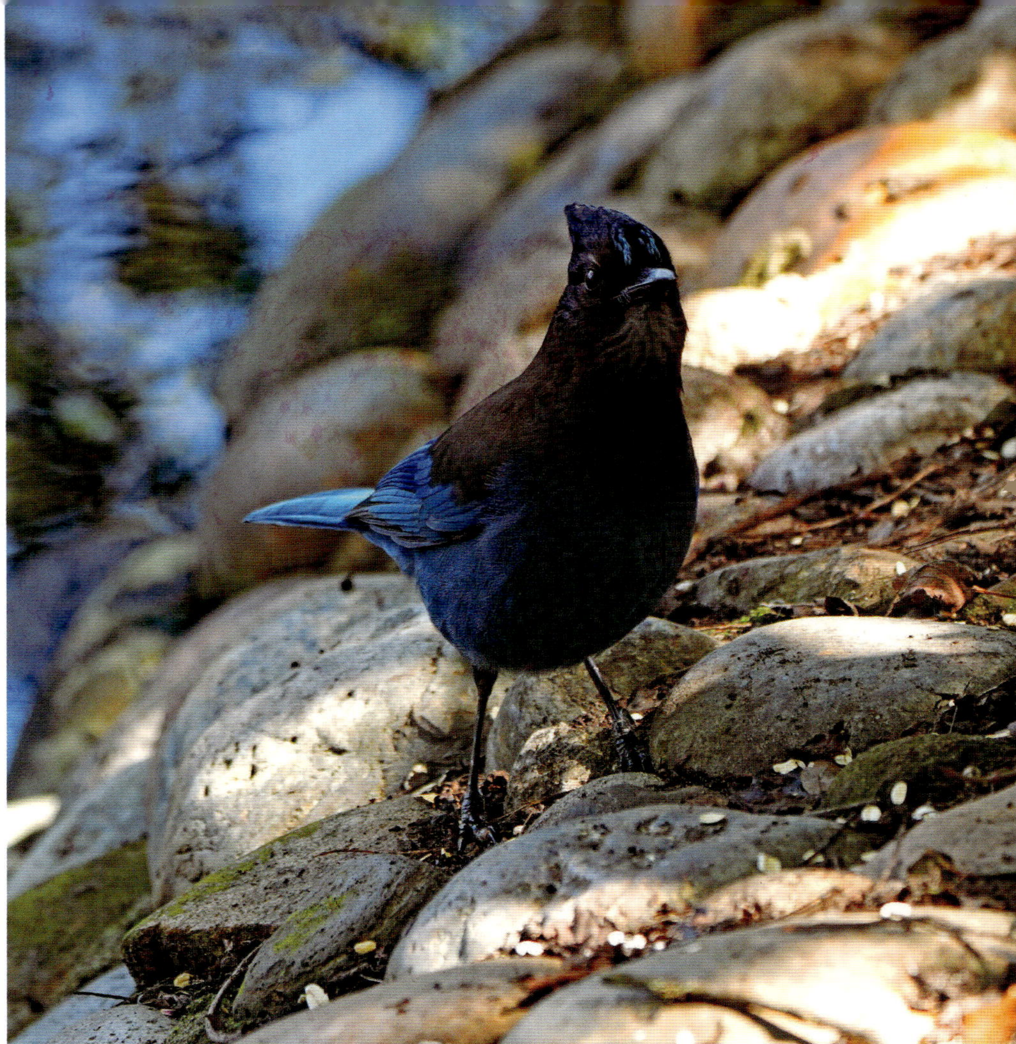

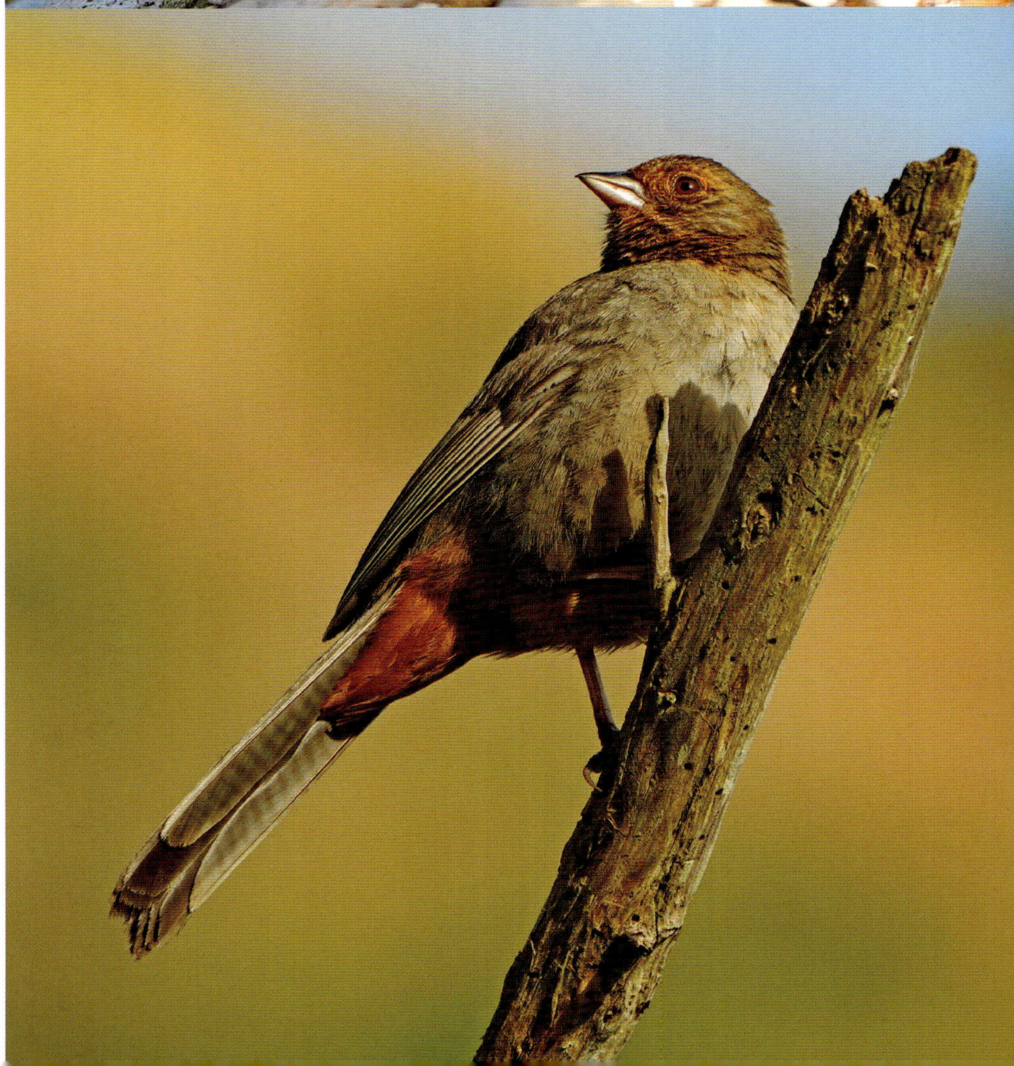

CLOCKWISE FROM TOP LEFT:
The Steller's Jay, American Robin, Band-tailed Pigeon, Marsh Wren, White-breasted Nuthatch, and California Towhee each claim their places among the Bay Area's diverse climatic microregions, from sunbathed oak woodland and foggy coastal forests to tangled pond margins and city parks.

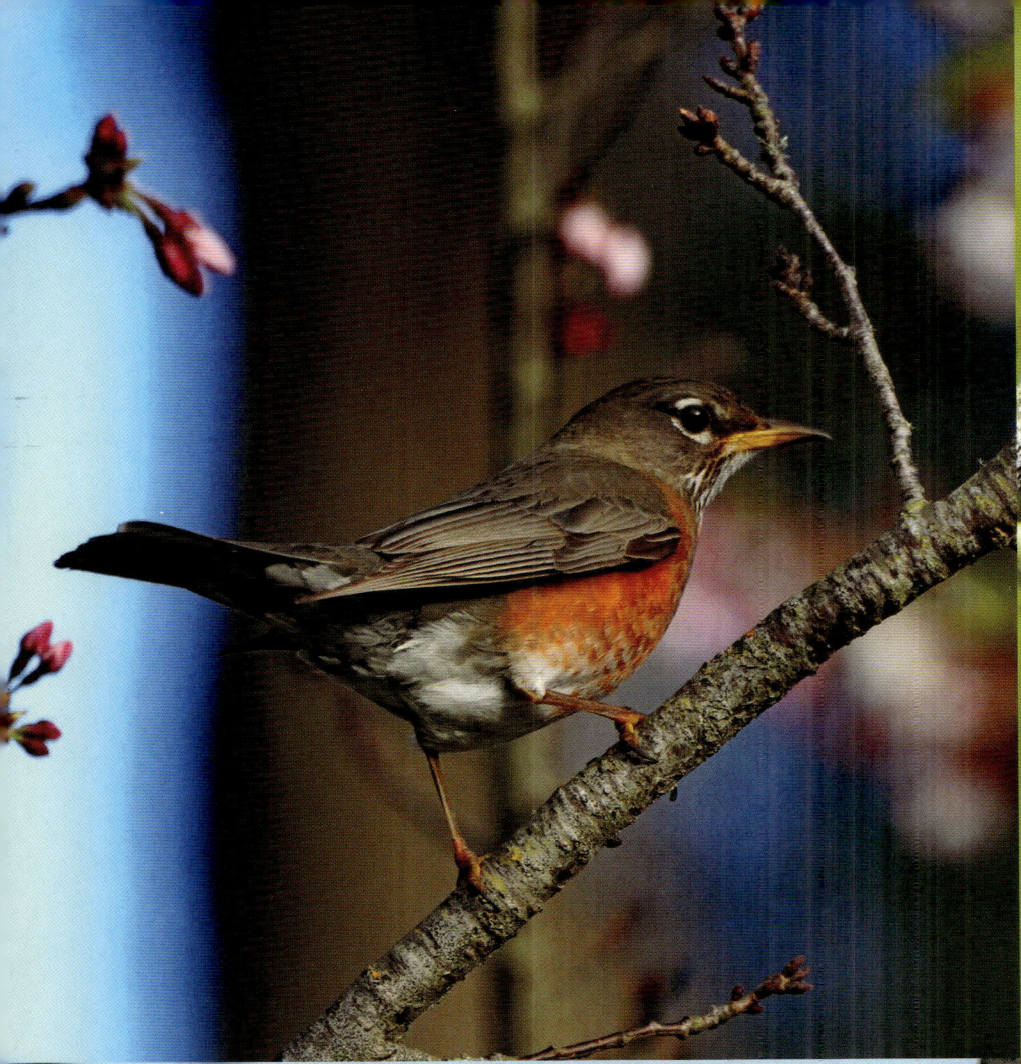
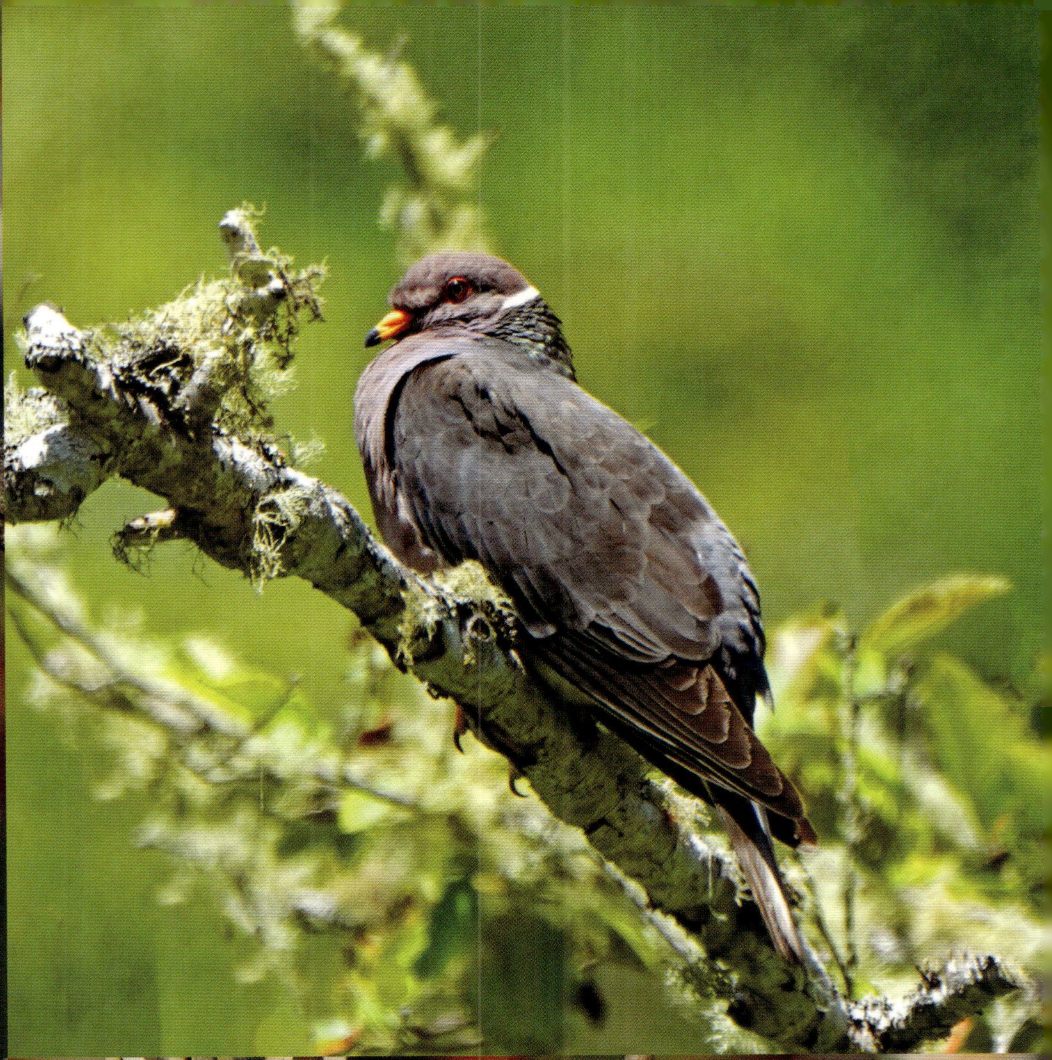
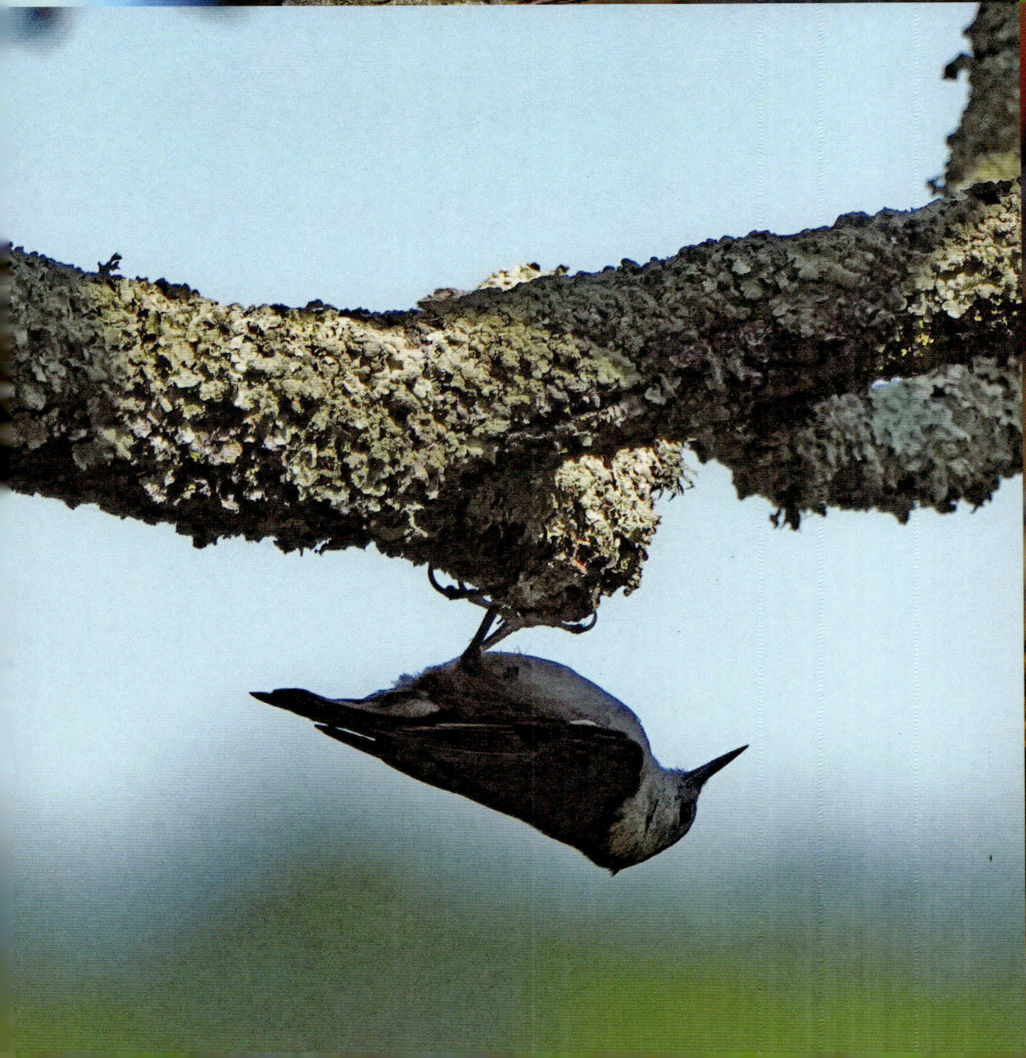
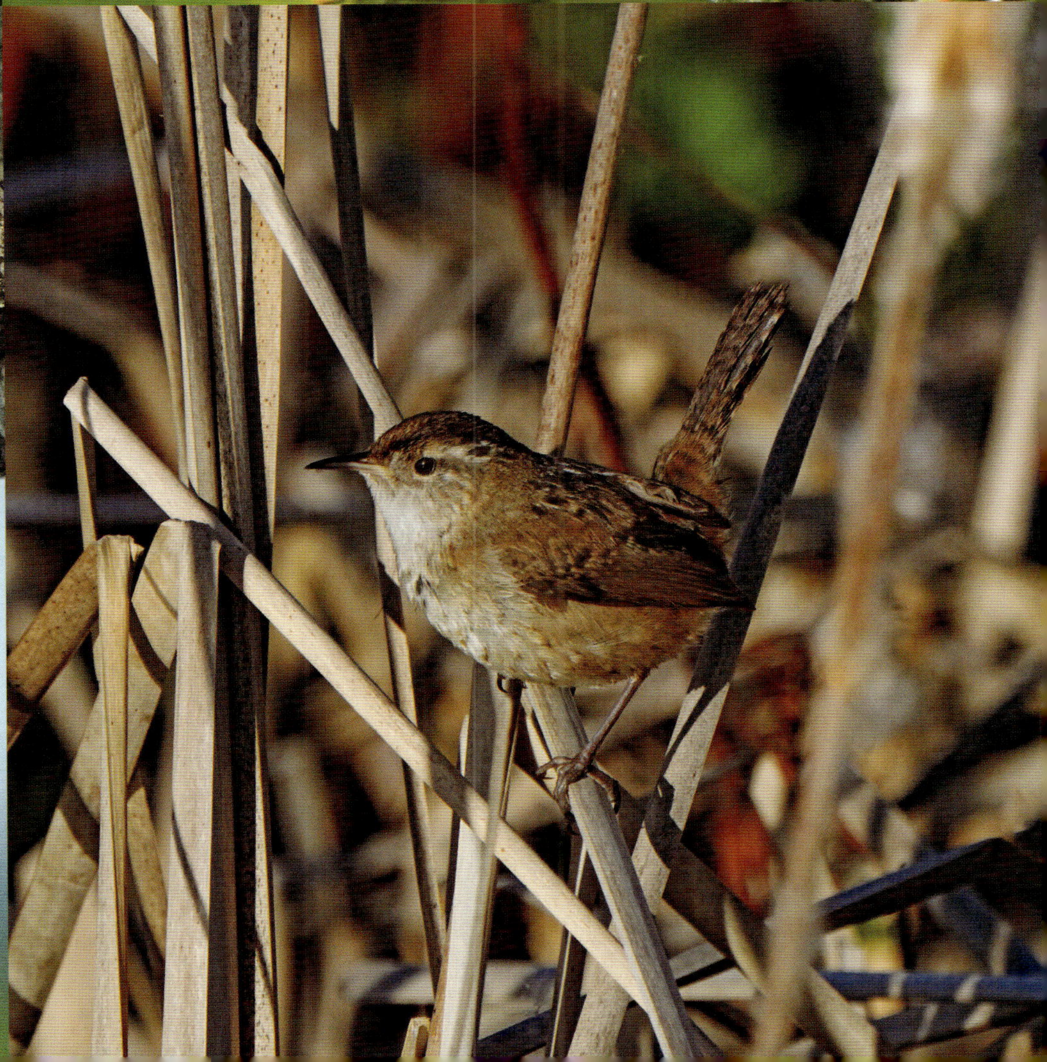

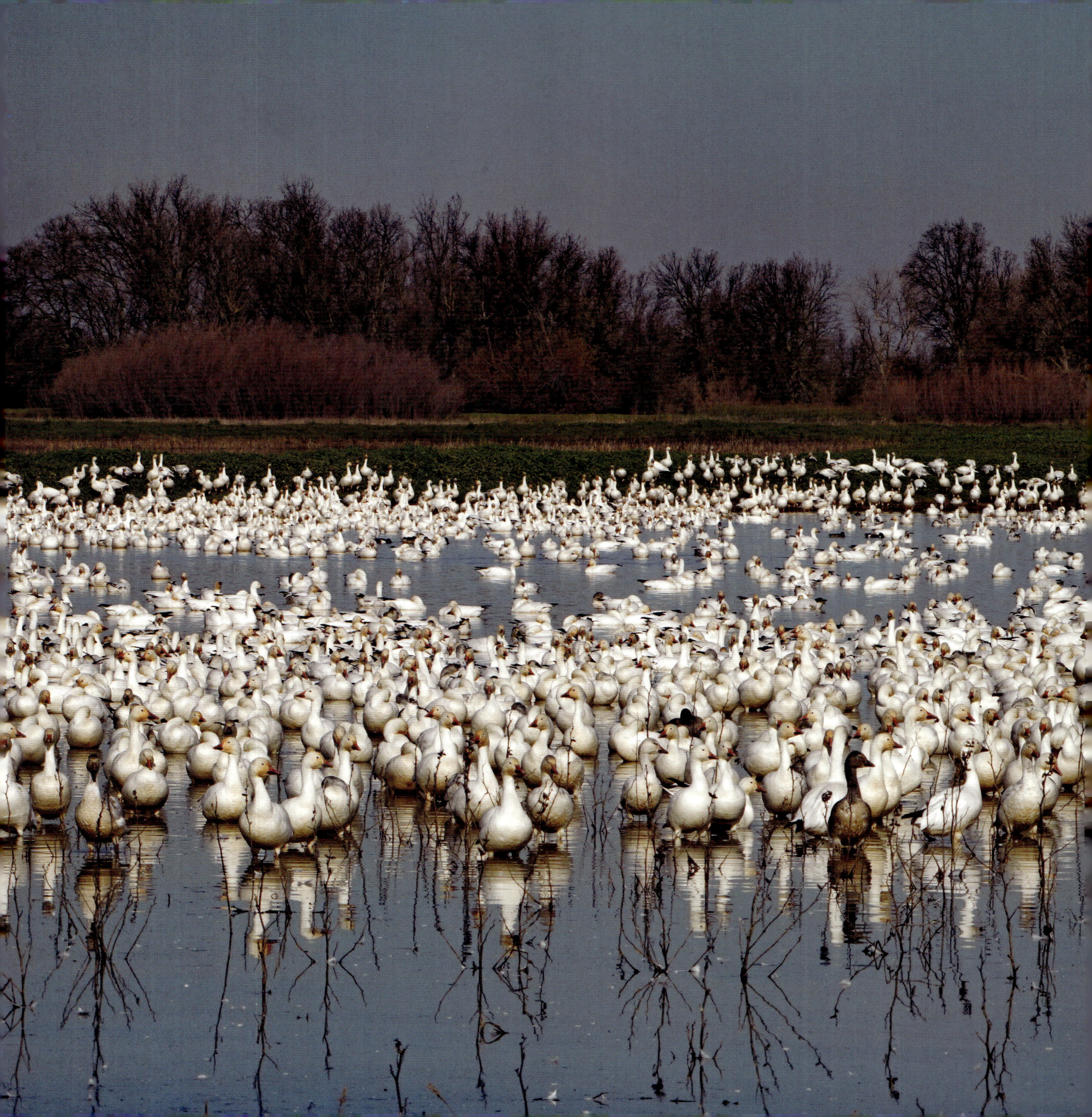

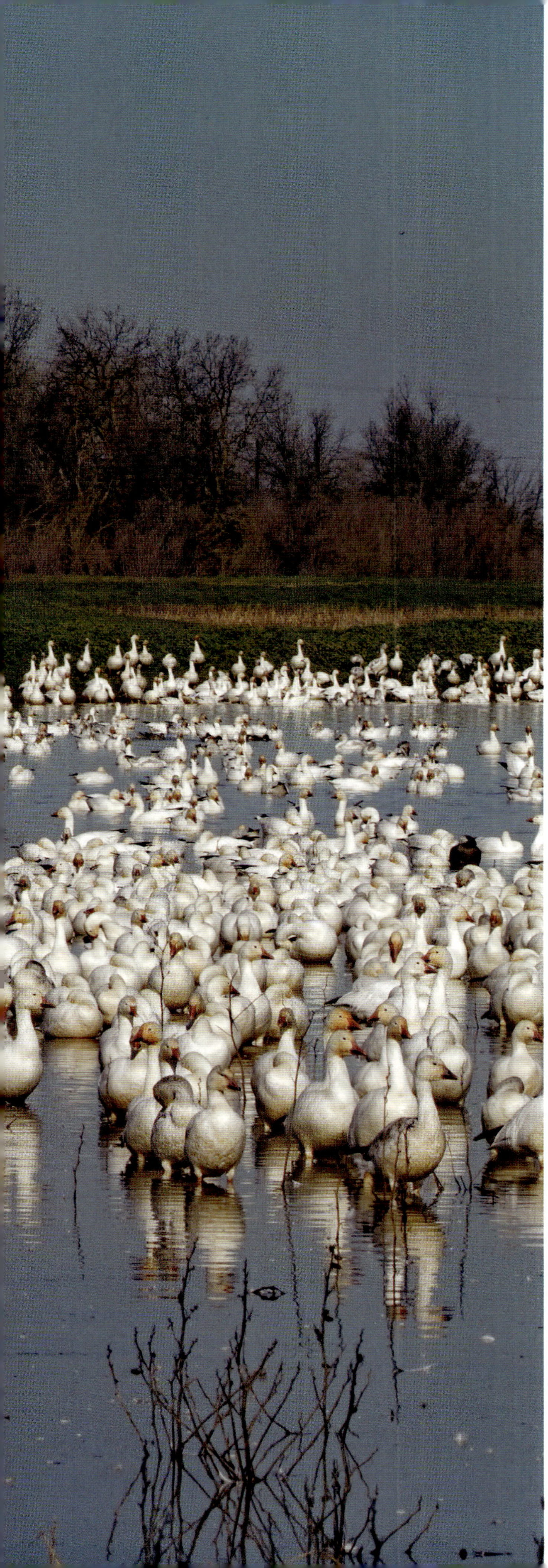

THE FAR REACH OF THE BAY

When you enter the Cosumnes River Preserve, a boardwalk leads through the woods to a high bridge. In midsummer, the ground below the bridge still holds moisture. Willow branches interlace, and raccoon tracks crisscross in the mud. A Scrub Jay scolds from an oak, and the call of a lone Wrentit resounds from the brush like a dropped Ping-Pong ball. It is a quiet evening; cattails are in silhouette against a pink sky. The occasional Spotted Towhee sings out, high and thin.

In the winter, though, the 50,000-acre preserve is anything but quiet. The Cosumnes River swells into sloughs and side channels as its waters seek the Delta and, eventually, the Pacific. The region is over 60 miles inland from San Francisco Bay, but only five feet above sea level, and the long fingers of high tide reach east, backing up the flow of the rainlogged river and pushing water deeper into the Delta channels. Waterfowl arrive with wings like thunder. Vast flocks of Snow Geese fill the sky like a shaken snow globe. Northern Shovelers drift across still water. At sunrise, the air is a cacophony as birds take flight—as many as fifty-eight thousand of them in a single day. At dusk, thousands of Sandhill Cranes arrive en masse to roost in shallow water. Their percussive calls penetrate to the bone. Their breath forms clouds in the winter air. Here, the seasonal dance between water and birds follows much the same choreography that it has for millennia.

Snowmelt runs into the Cosumnes River from snowy pine forests, and rain fills seasonal streambeds that weave through oak woodland and pastureland. The Cosumnes River watershed—encompassing all the land that channels water toward a common low point—is 940 square miles, twice the size of the city of Los Angeles. A dozen other rivers flow in parallel with the Cosumnes from the Sierra Nevada, but in this story of colossal water

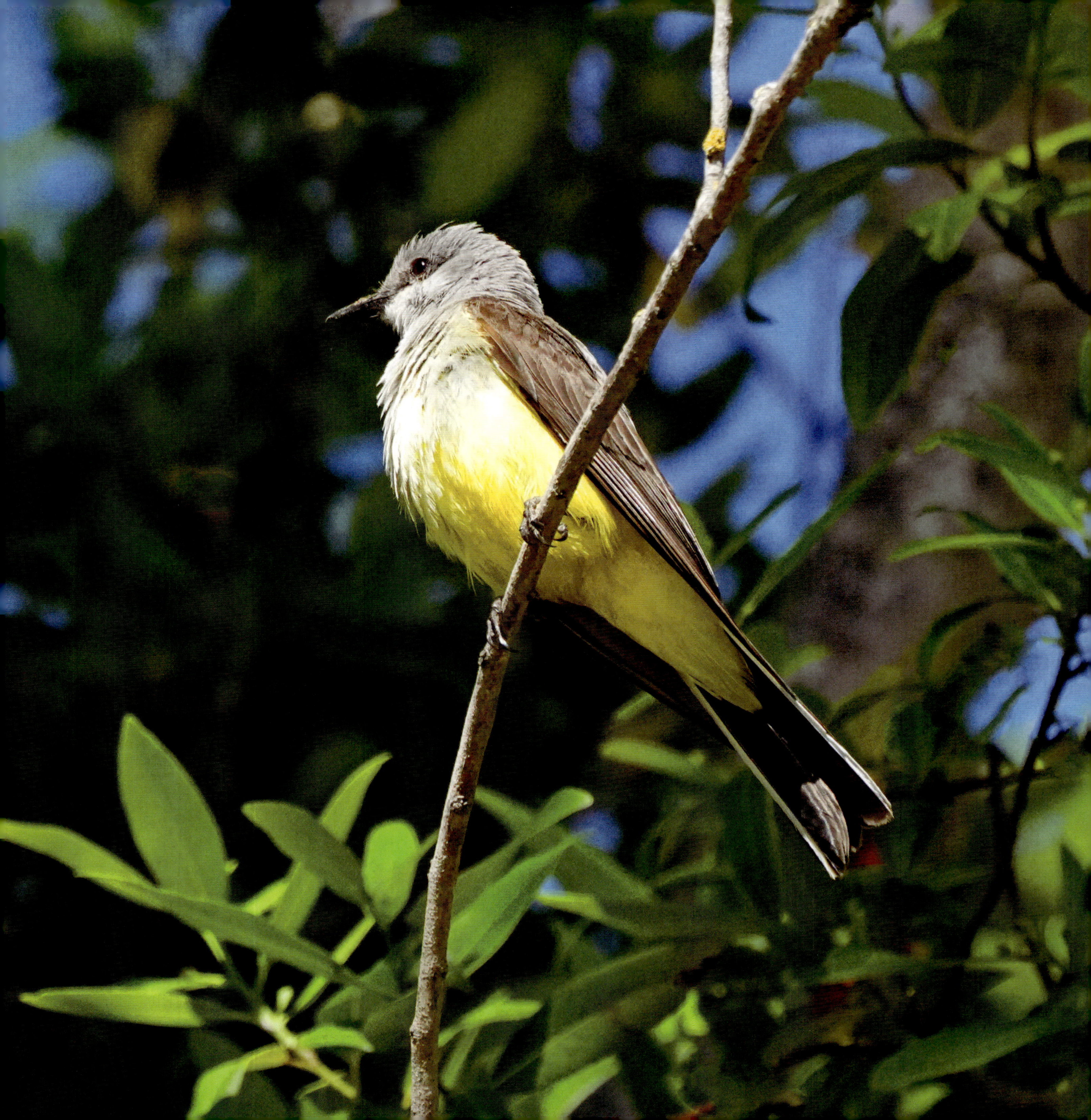

PRECEDING: In the wintry California Delta, enormous flocks of Snow Geese gather to search for food in the shallow waters. When spring comes, they will begin a northward migration to their nesting grounds in the Arctic tundra.

LEFT: The California Delta sees seasonal waves of winter migrants in the wet season and spring-and-summer-nesting birds like this Western Kingbird as the waters subside and riparian trees leaf out.

BELOW: A male Northern Shoveler peers out at delta waters through a sea of dark feathers.

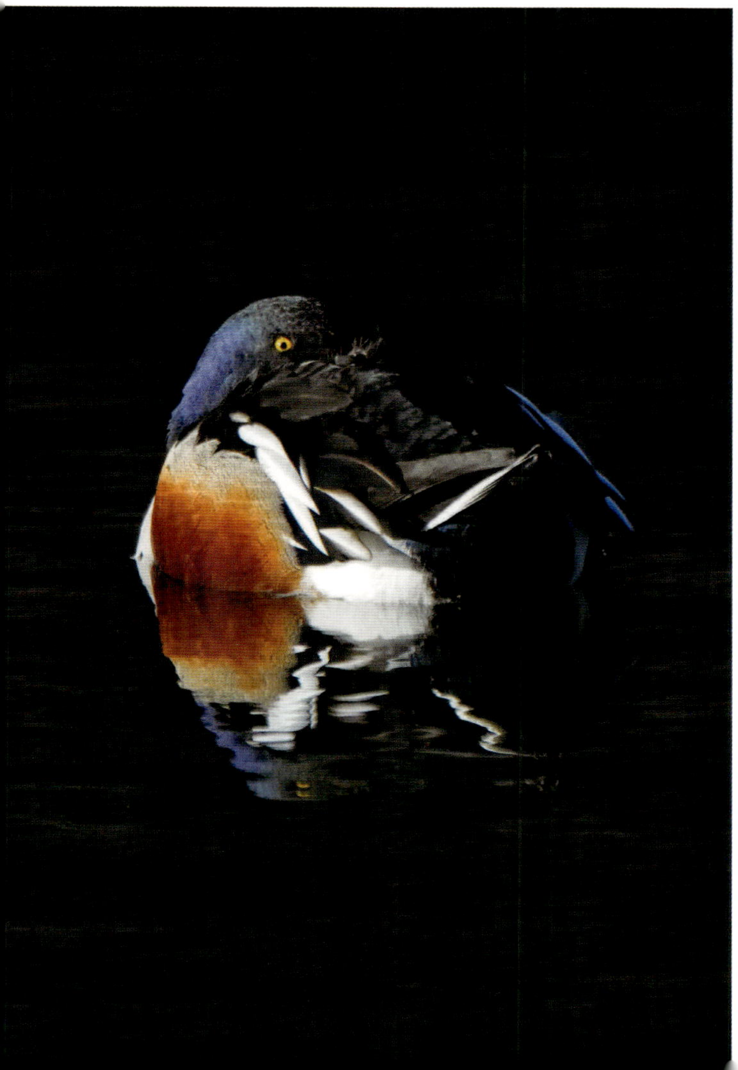

movement, the Cosumnes River is unique: It is the last remaining undammed river in the Central Valley.

Two hundred years ago, huge floods would descend from the Sierras, overflowing riverbanks and feeding floodplains and tidal marshes with rich sediment. Slow-moving currents nurtured a rich aquatic food web. But large-scale levee construction rearranged the geometry of the Delta. As finally 95 percent of the wetlands were drained, the sinuous channels, reeds, and mud that torrents of visiting birds depended on were narrowed to a patchwork of preserves.

The Consumnes, however, is regenerating—the river has run wild since 1987, allowing flooding to deposit sediment, riparian forests to regrow, and the bright songs of neotropical birds to fill the air. Kayak to the "Tall Forest" on the Cosumnes River in the summer and you might think you've traveled back in time: This is one of the only riparian corridors remaining in the Central Valley that looks much the way it did before European contact. As you drift under the canopy of 75-year-old oaks and the cool shadows of California grape, see how many bird calls you can tune in to: the high, sweet pulse of a Yellow-rumped Warbler on its way to the mountainous highlands of Honduras; raspy chirps from Tree Swallows with winter homes in Guatemala; Western Kingbirds en route to the open scrub of northern Mexico, calling out in explosive squeaks like a dog's squeeze toy. More than two hundred bird species have been documented in this 100-acre tract. The 2009 Delta Reform Act has mandated that managers look for solutions to heal the degraded Delta, and the Cosumnes River is their primary case study for guiding restoration efforts. The tendrils of the Bay Area extend far inland, our region connected by flowing water and the living things that travel in that current.

Here, as we pause by the muddy banks of the free-flowing river, it is hard to know whether we're standing 200 years in the past or in the future. Time and water alike seem to move differently here.

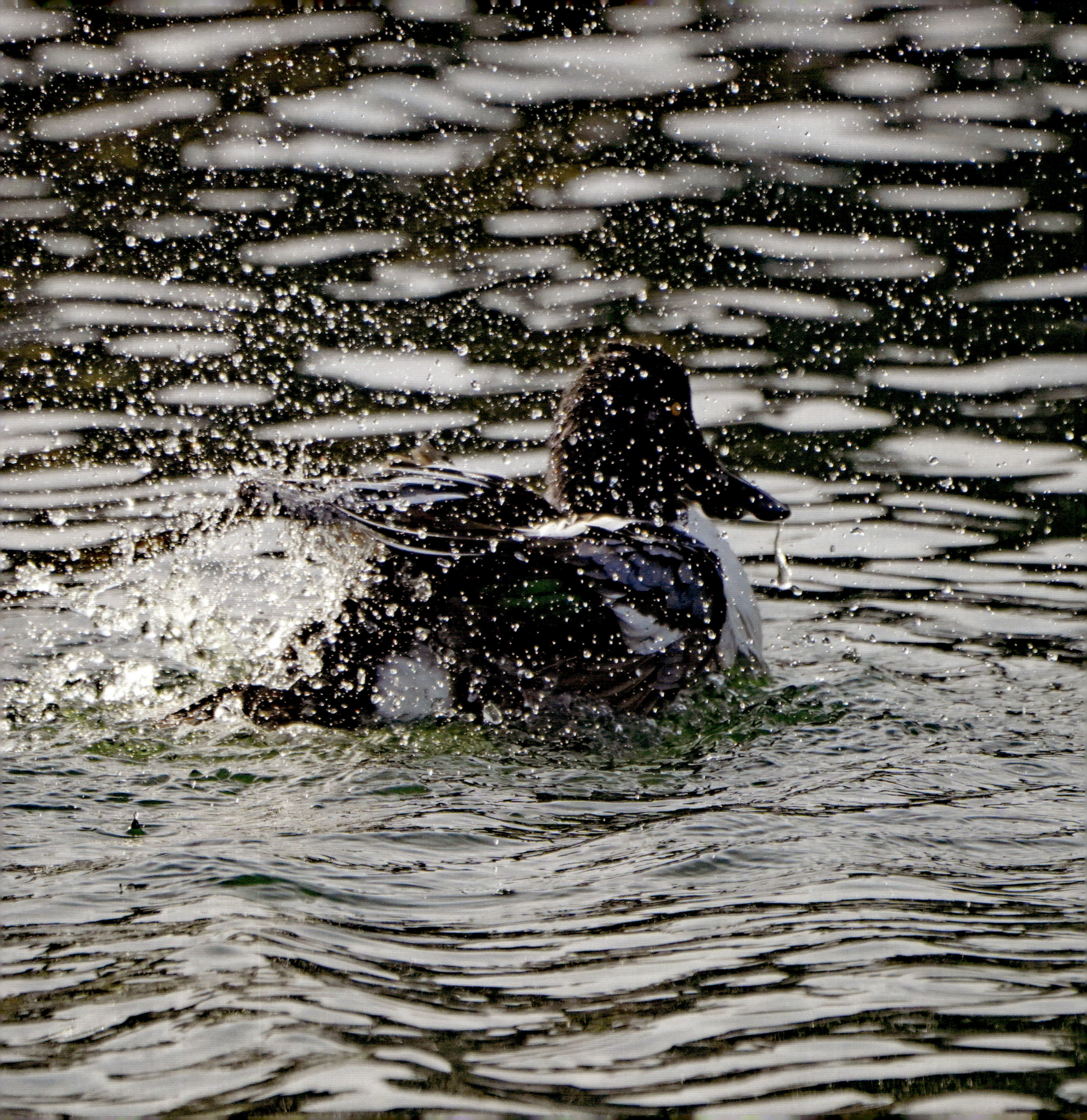

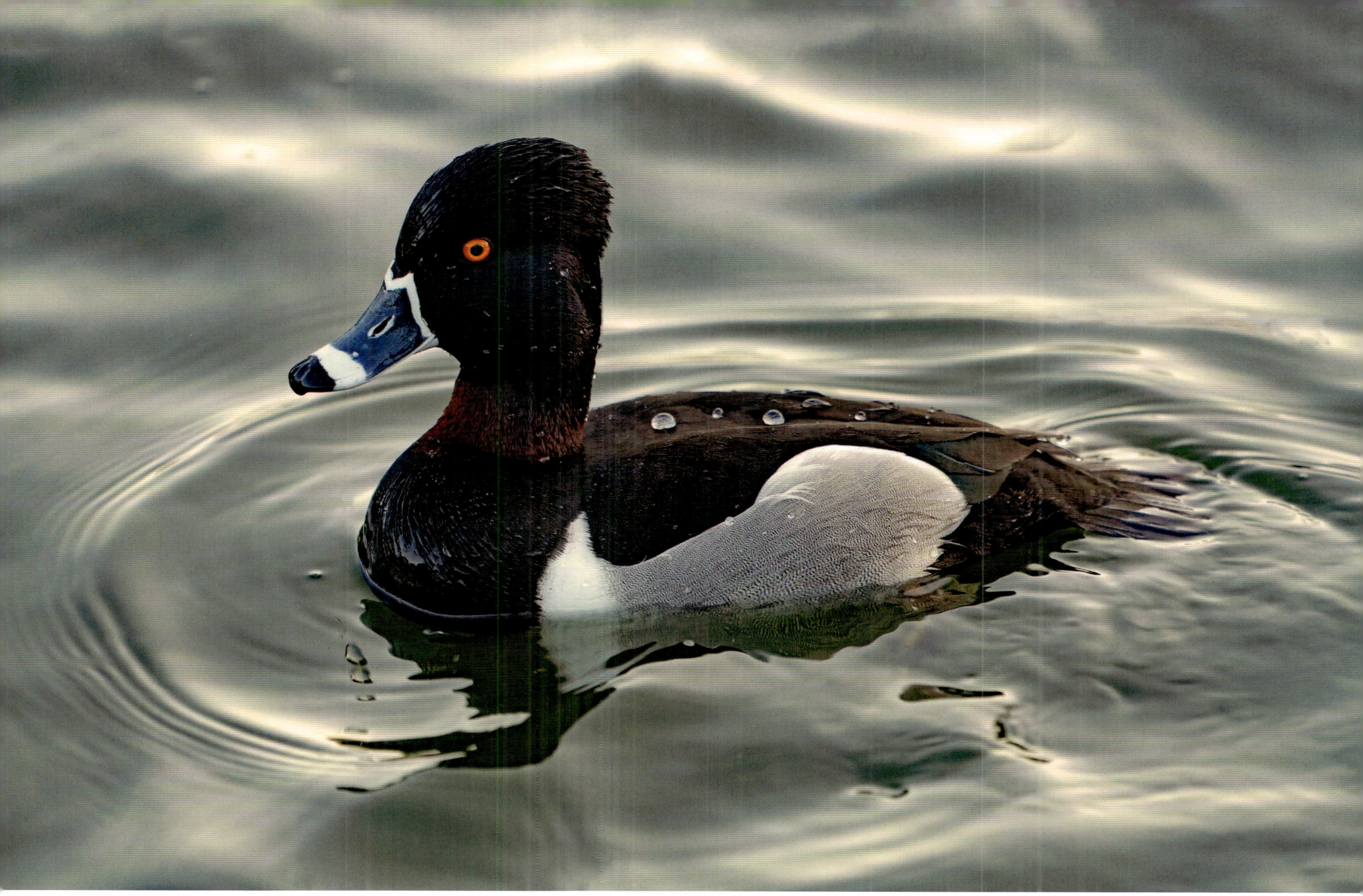

A duck's world is a world of water. A Northern Shoveler (left) drenches the air with a vigorous shake. Pearly water droplets condense atop a Ring-necked Duck's oily feathers (above). Ducks distribute oil from their preen gland to keep themselves waterproof and to float easily across lakes and wetlands.

FOLLOWING, CLOCKWISE FROM TOP LEFT: The Greater Scaup, Common Goldeneye, Bufflehead, Ring-necked Duck, Common Merganser, Blue-winged Teal, Wood Duck, and Hooded Merganser can all be found by sharp-eyed birders visiting the California Delta, especially during winter months. Ducks like these are beneficiaries of careful management, in large part thanks to hunters. The purchase of Federal Duck Stamps—required annually of duck hunters by the Fish and Wildlife Service—are among the most successful conservation tools ever created to protect habitat for birds and other wildlife.

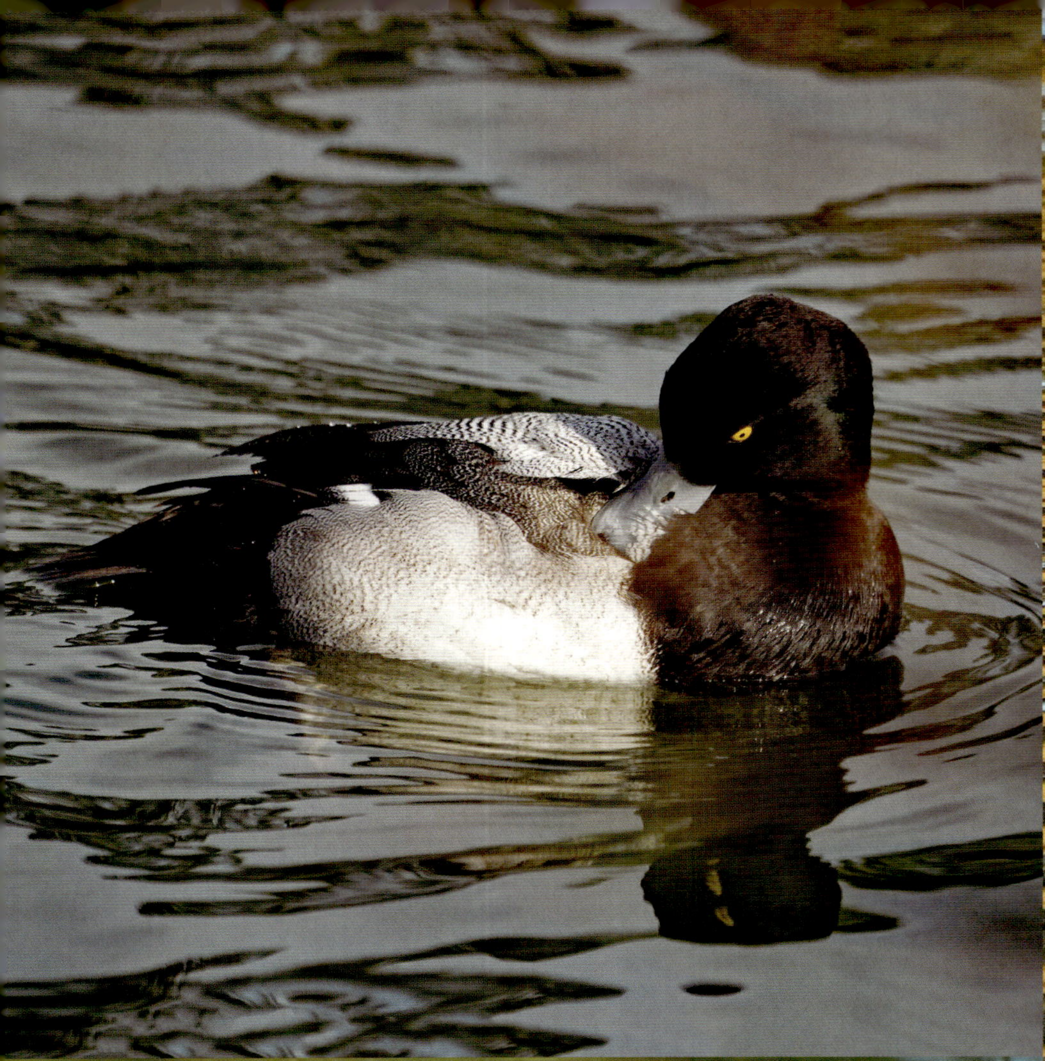

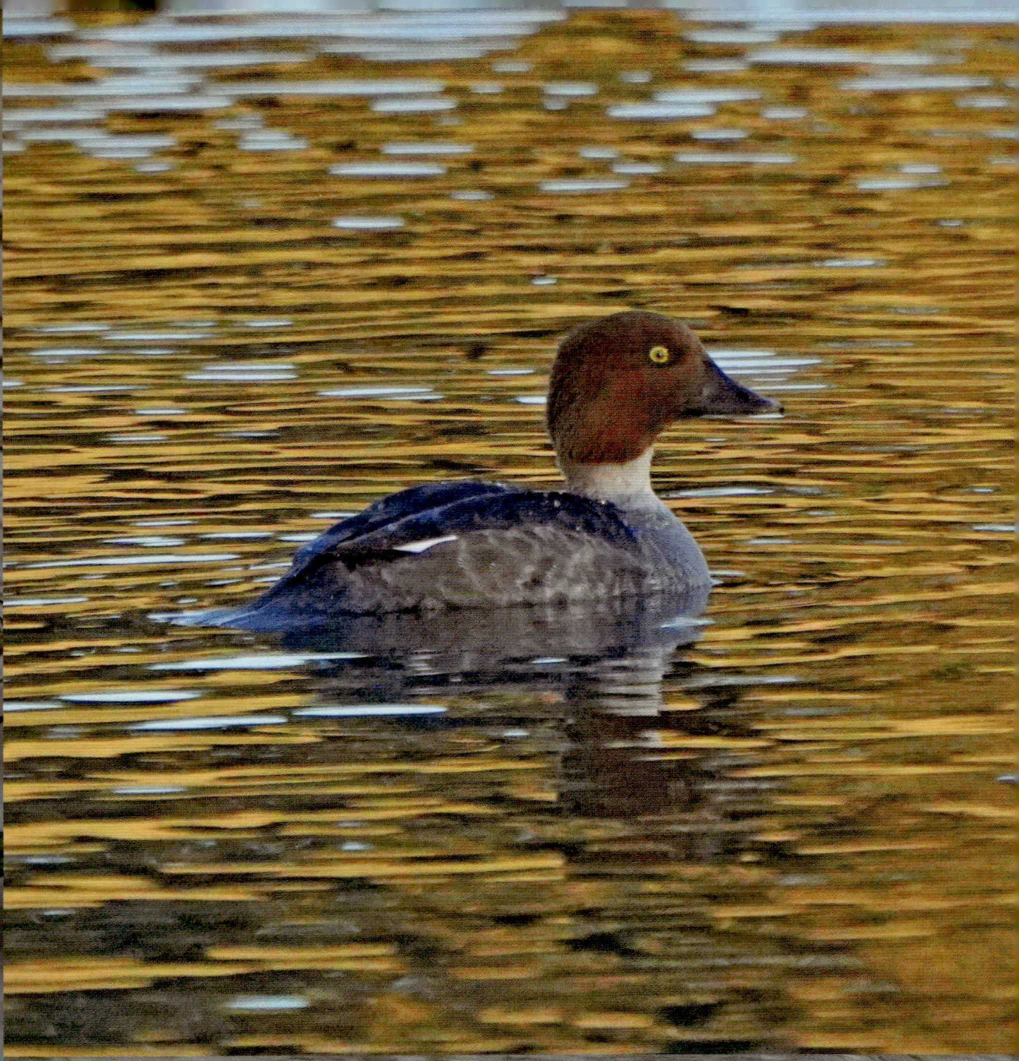

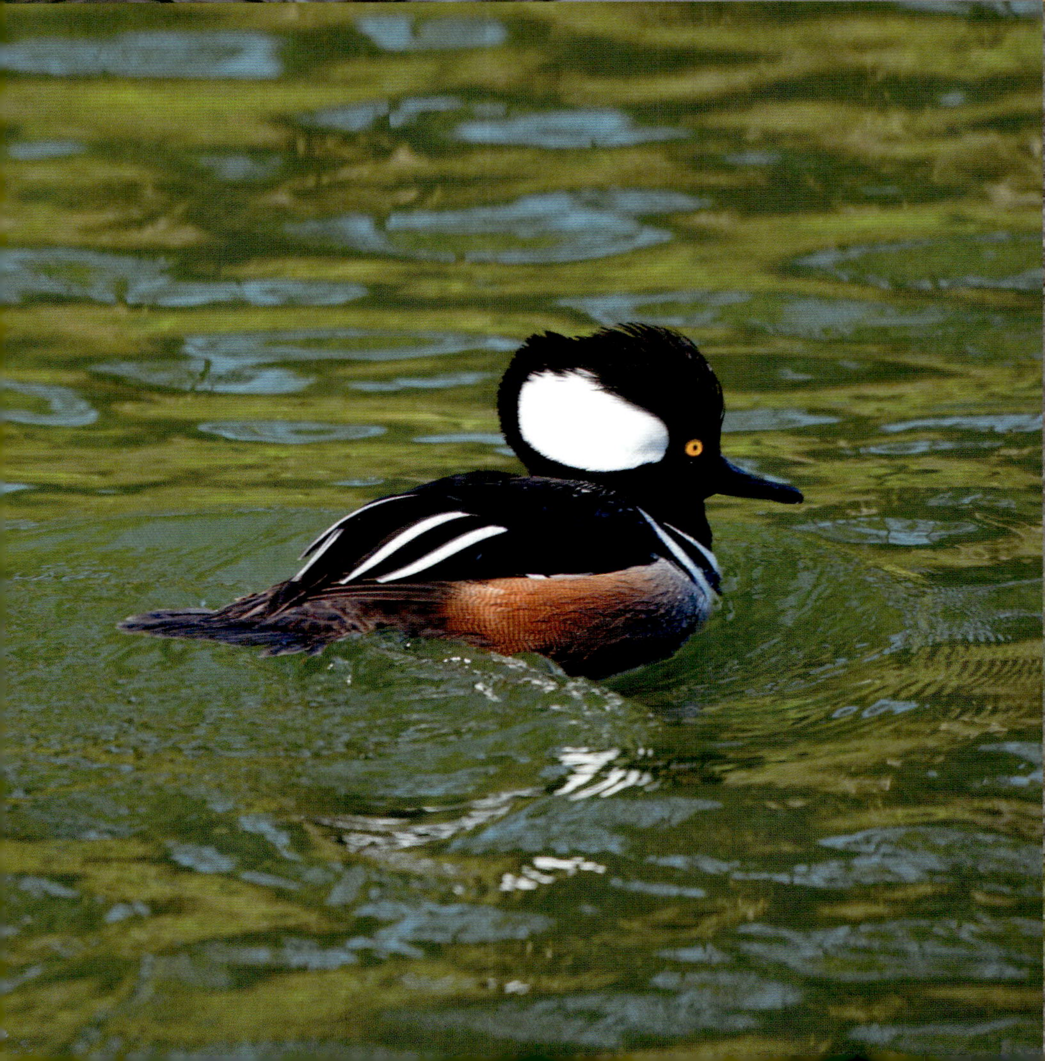

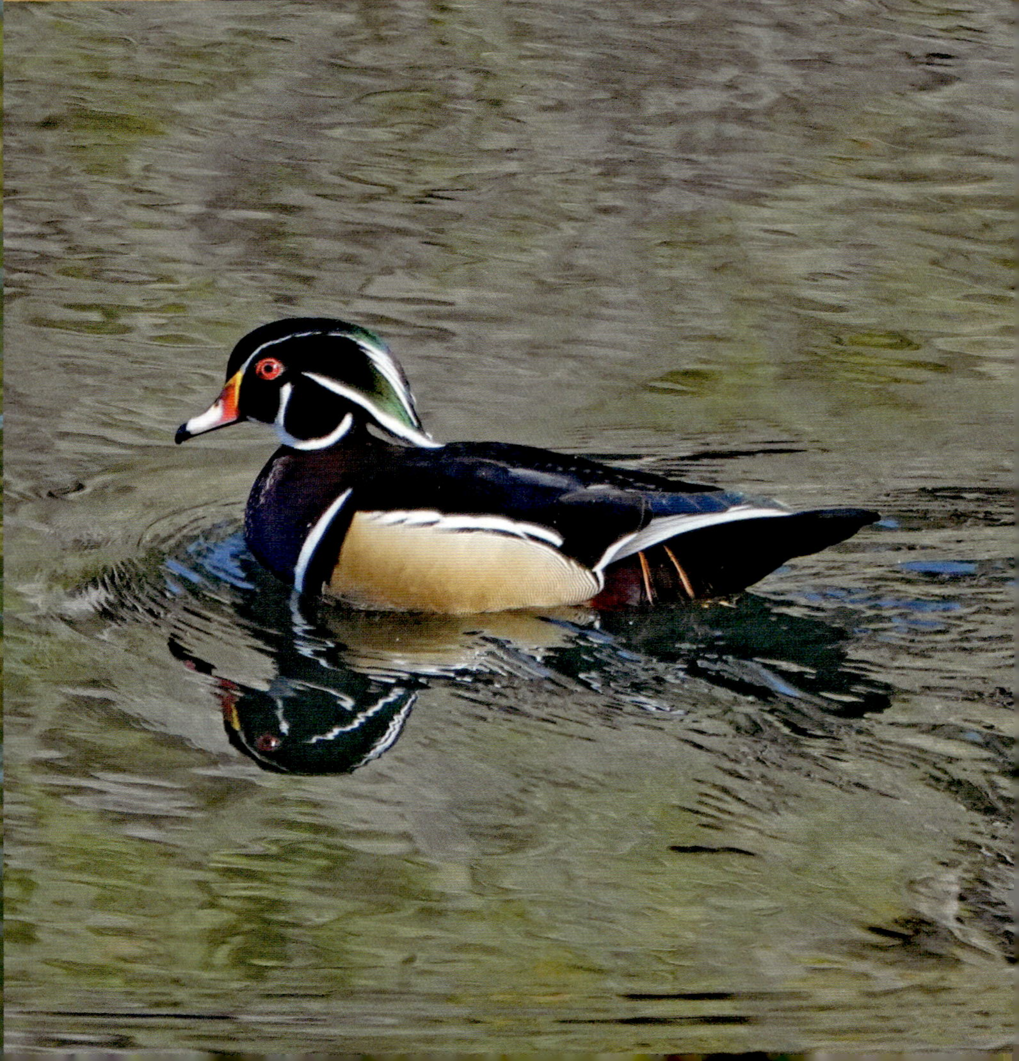

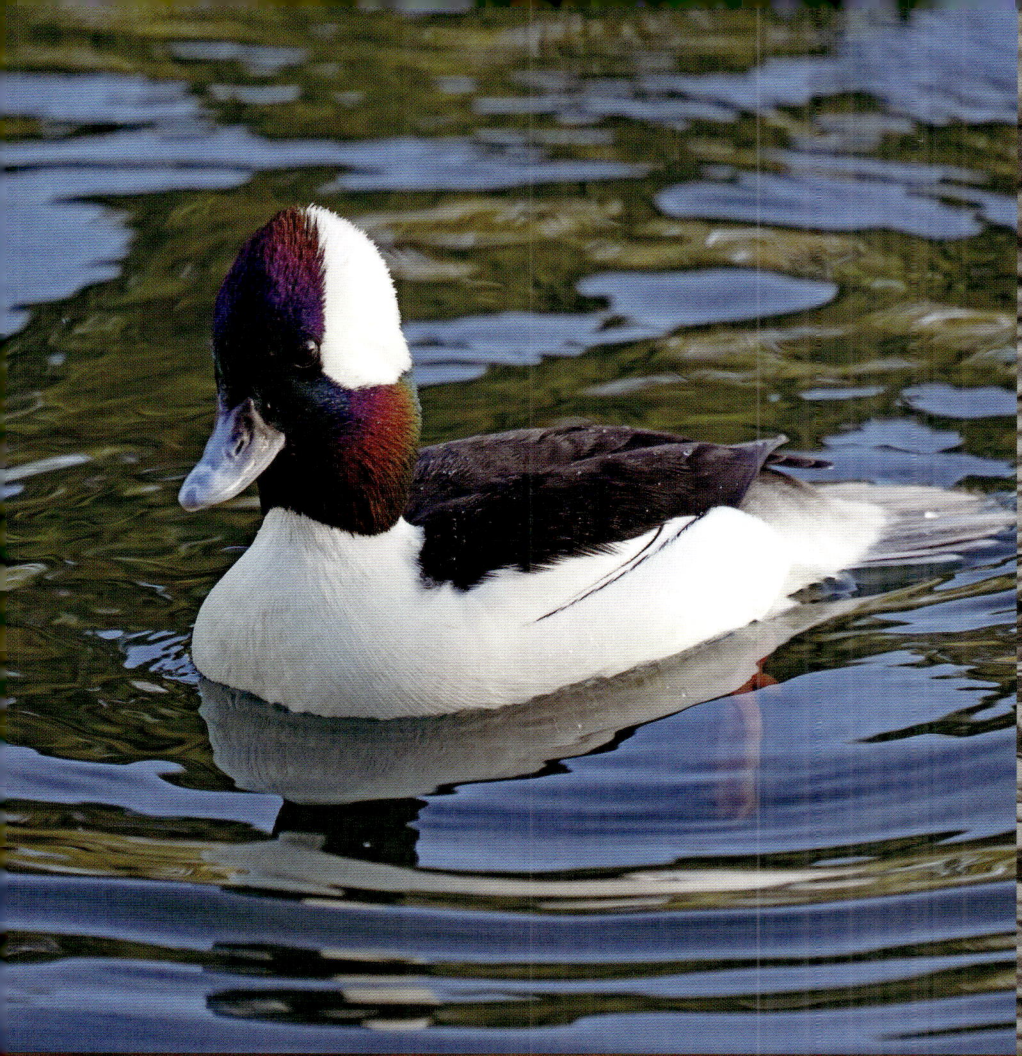

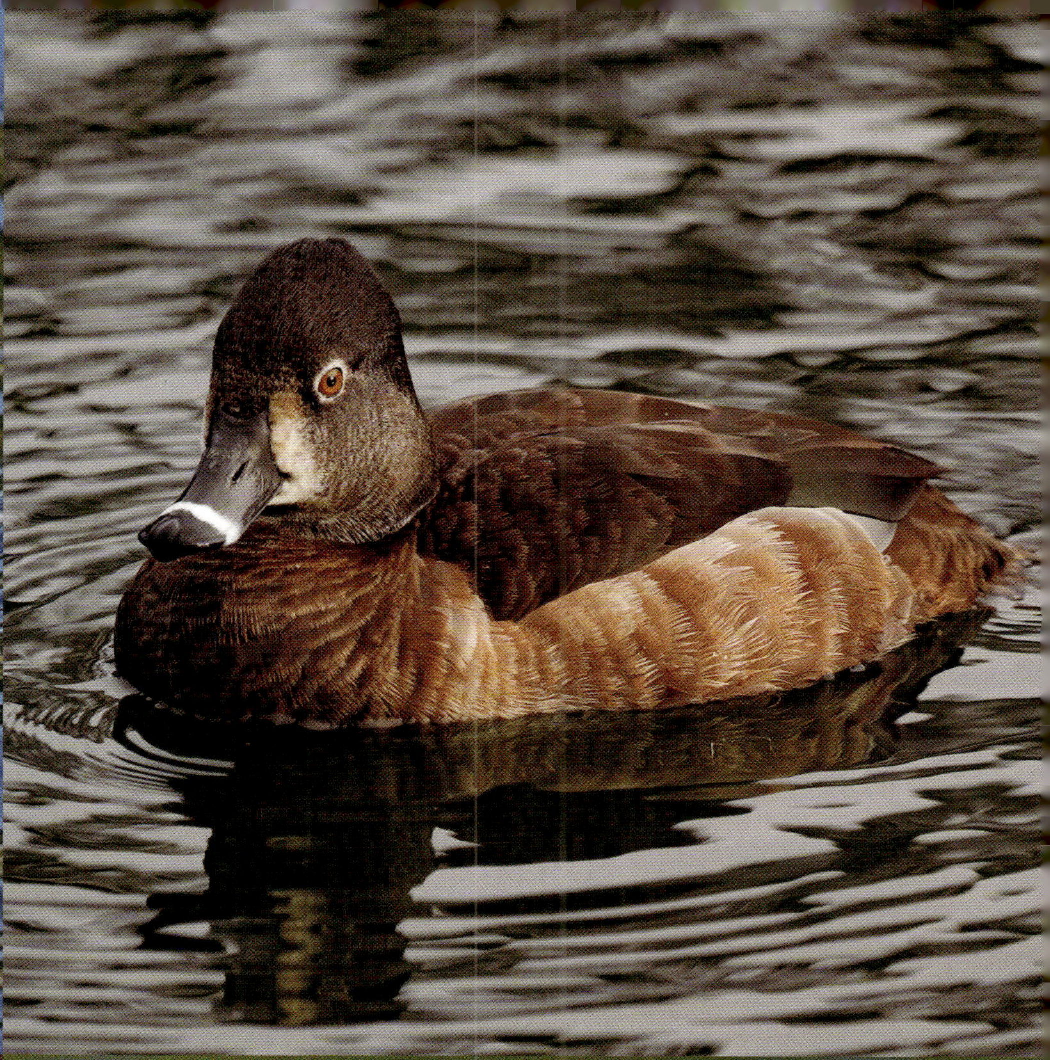

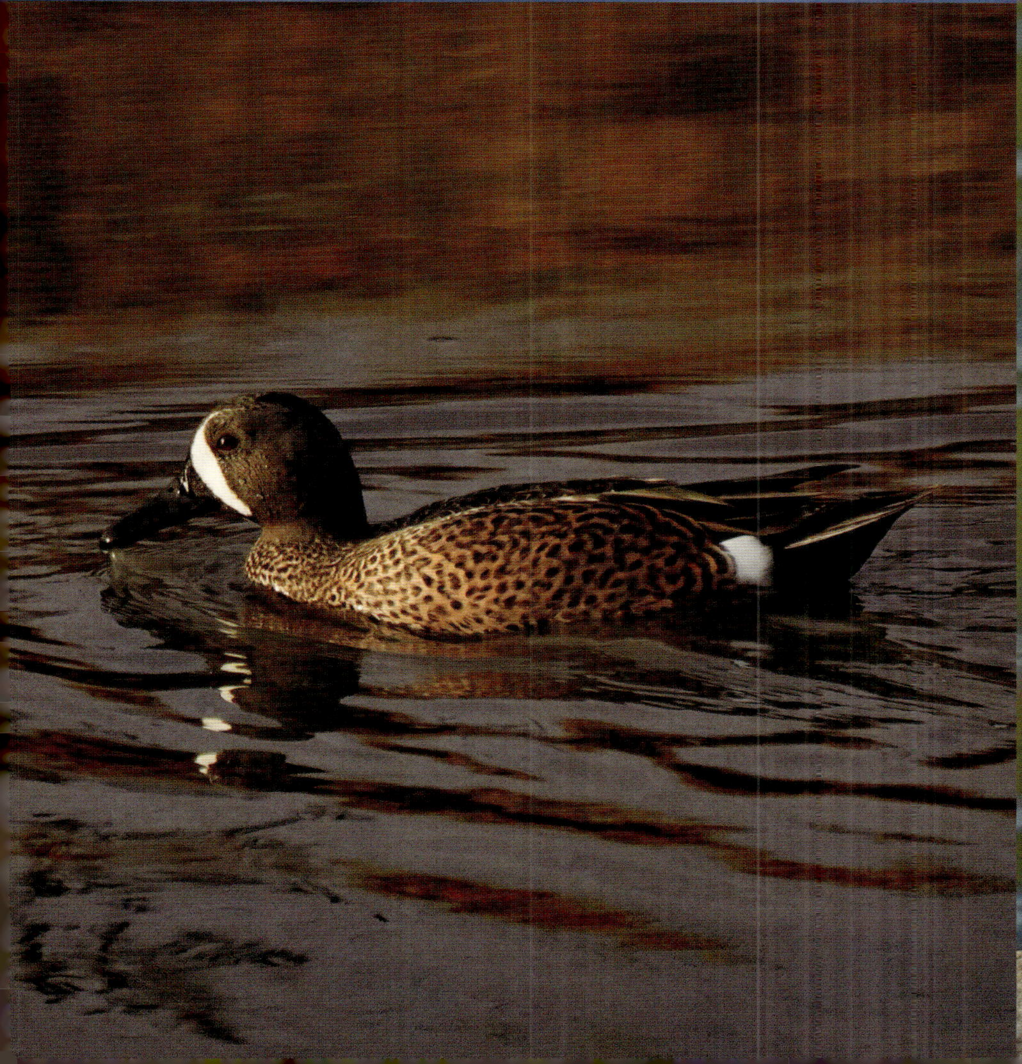

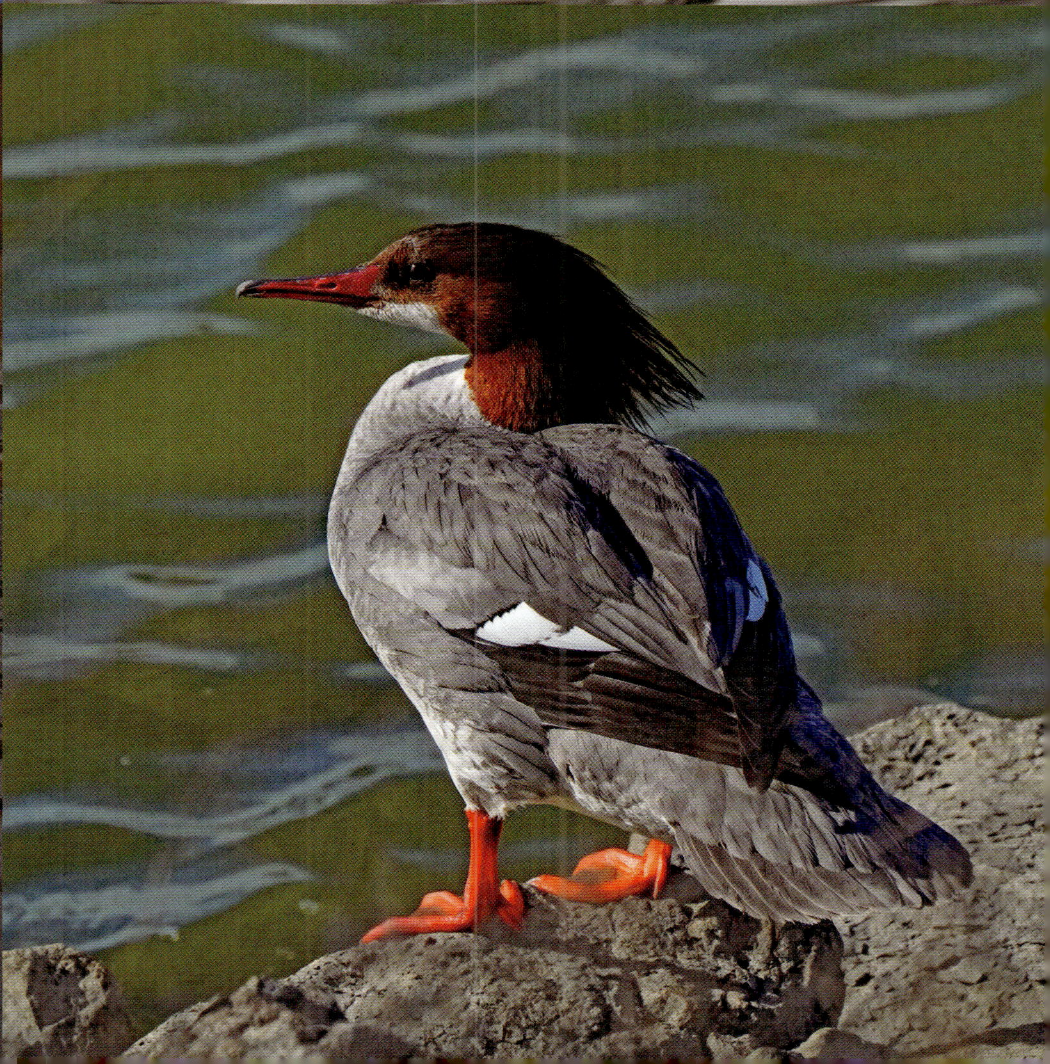

The Green-winged Teal is the smallest dabbling duck in North America, and one of the fastest—reaching speeds of up to 60 miles an hour and maneuvering in flocks with aerodynamic twists and turns. They are among the first spring migrants to depart the Bay Area for their northern nesting grounds.

FOLLOWING: The patterns in a Gadwall's (left) plumage mirror the refractive ripple patterns in the surrounding water. The Bay-Delta region is a crucial area for waterfowl, historically rich in wetland and riparian habitats but now heavily modified. It's a focus area for conservation efforts like the Central Valley Habitat Joint Venture, which aims to protect and restore habitat primarily for waterfowl. A Ring-necked Duck's (right) amber eye shines the same luminous orange as the reflections that encircle it.

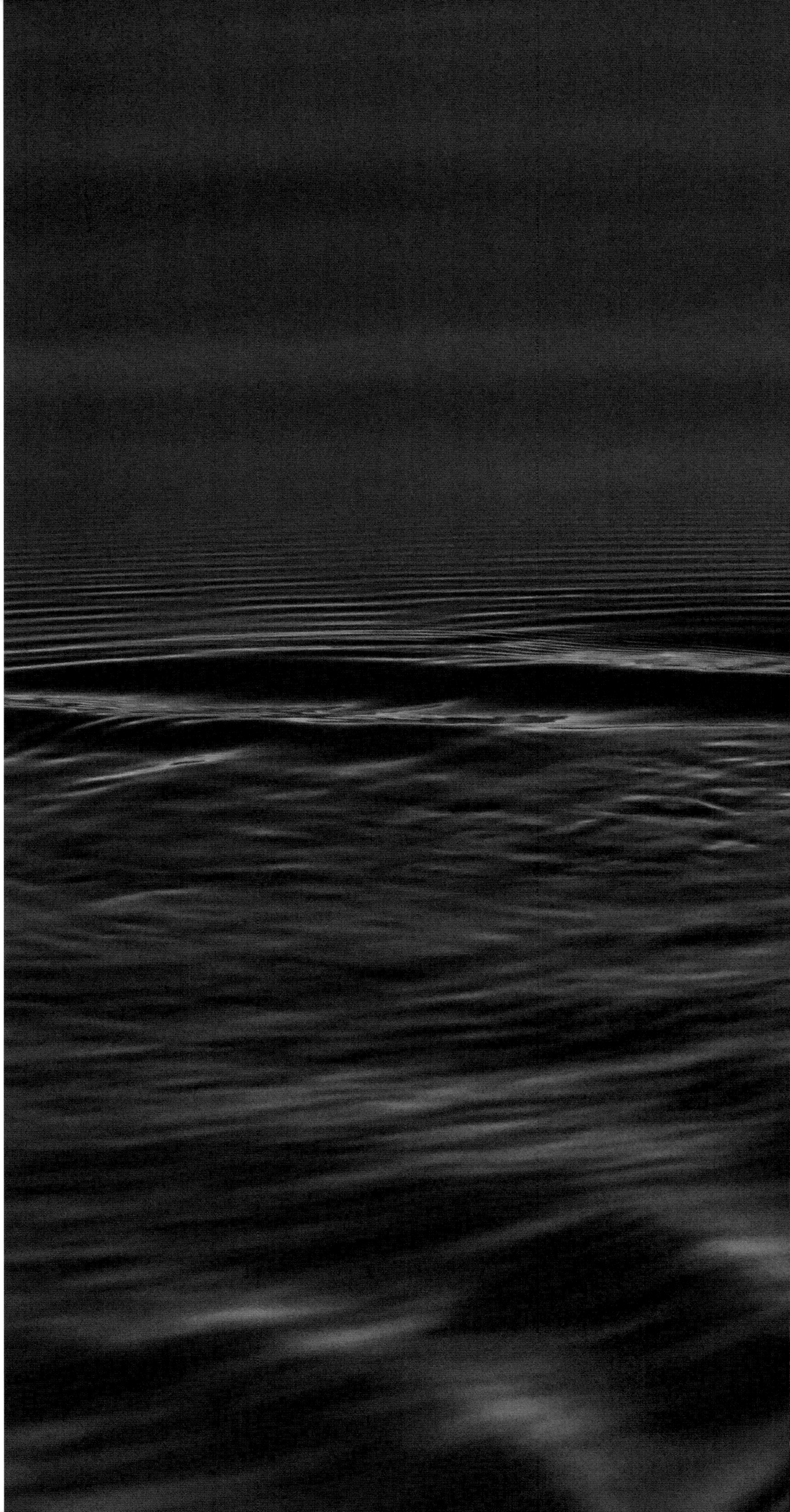

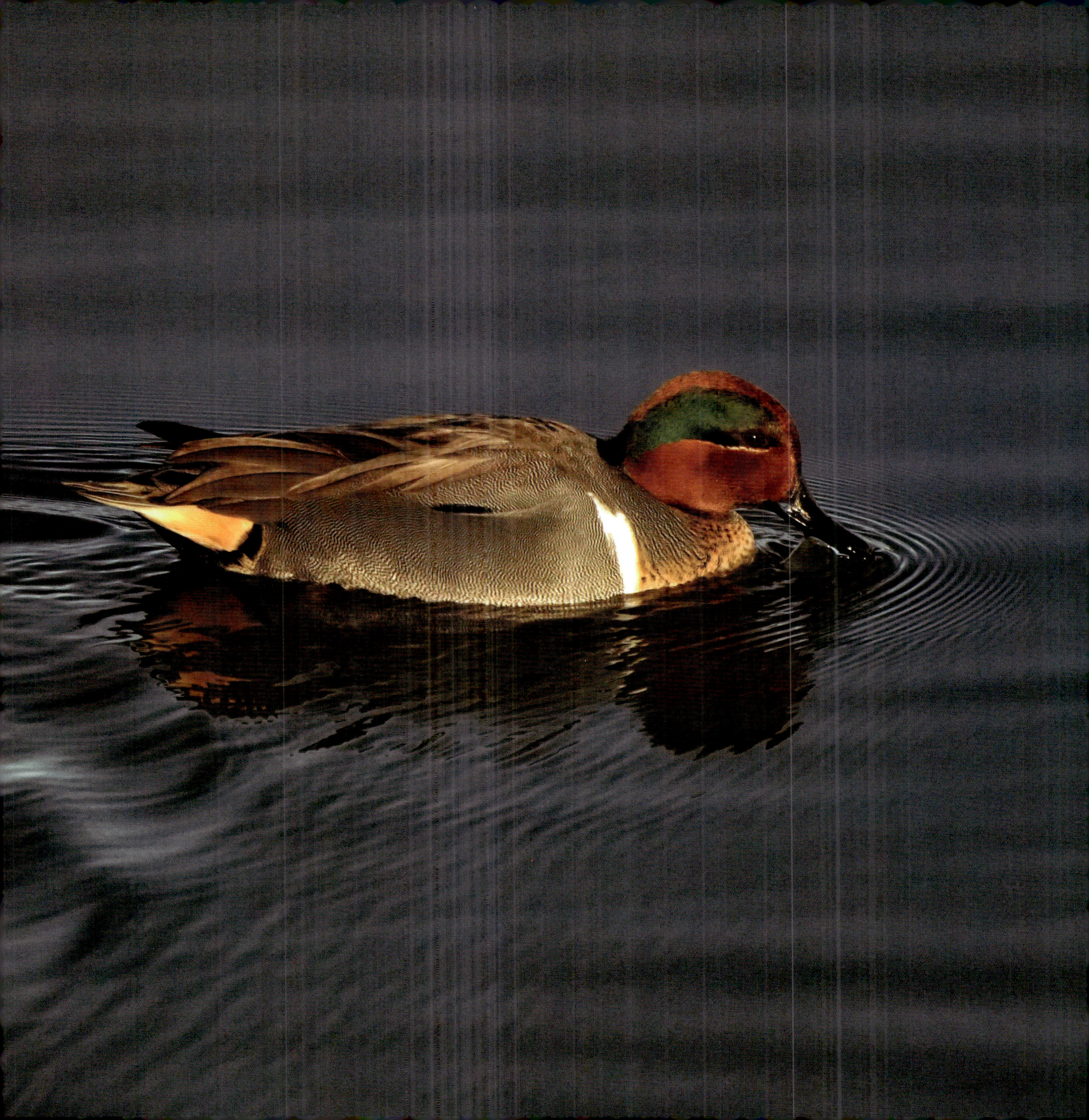

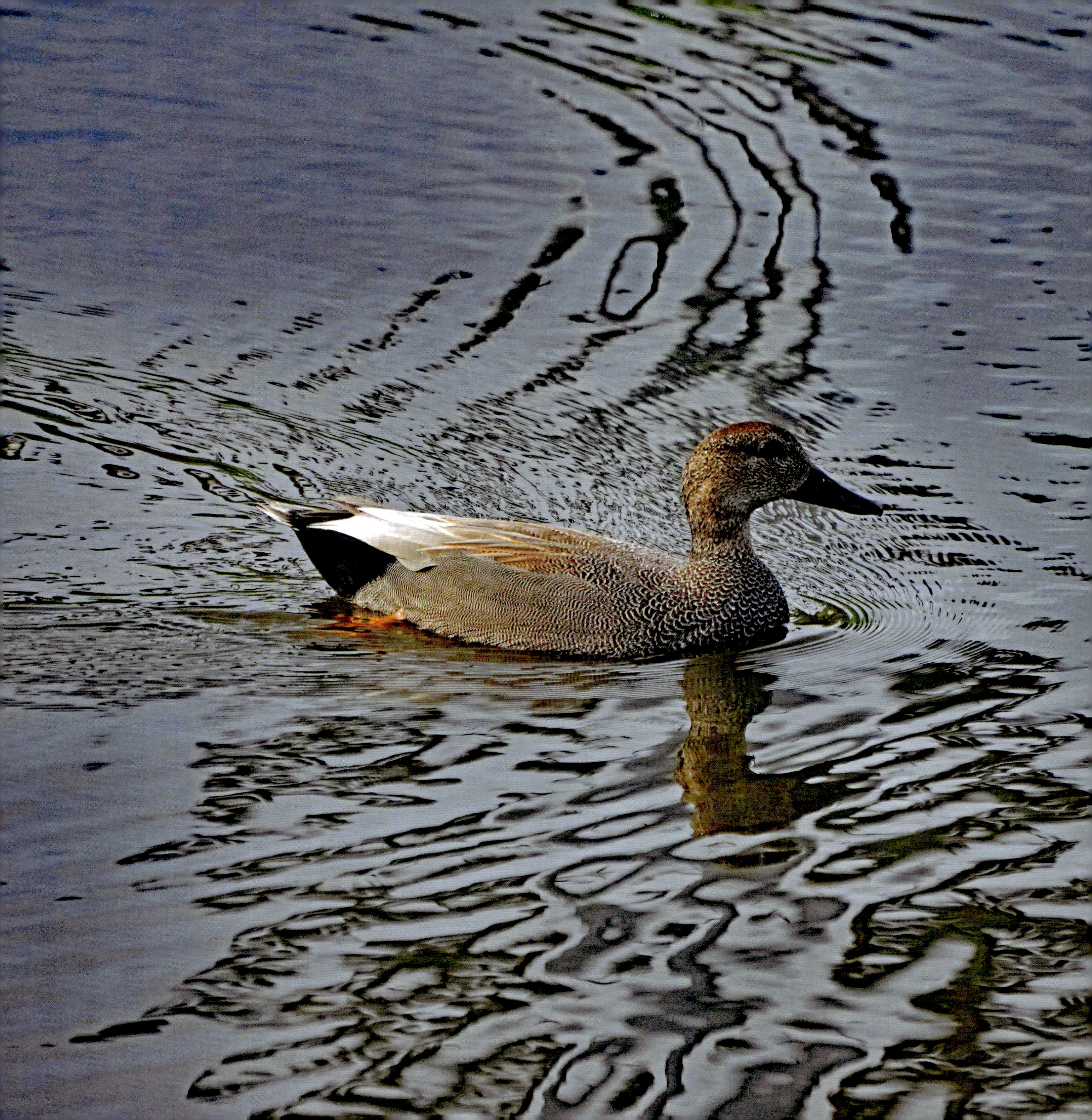

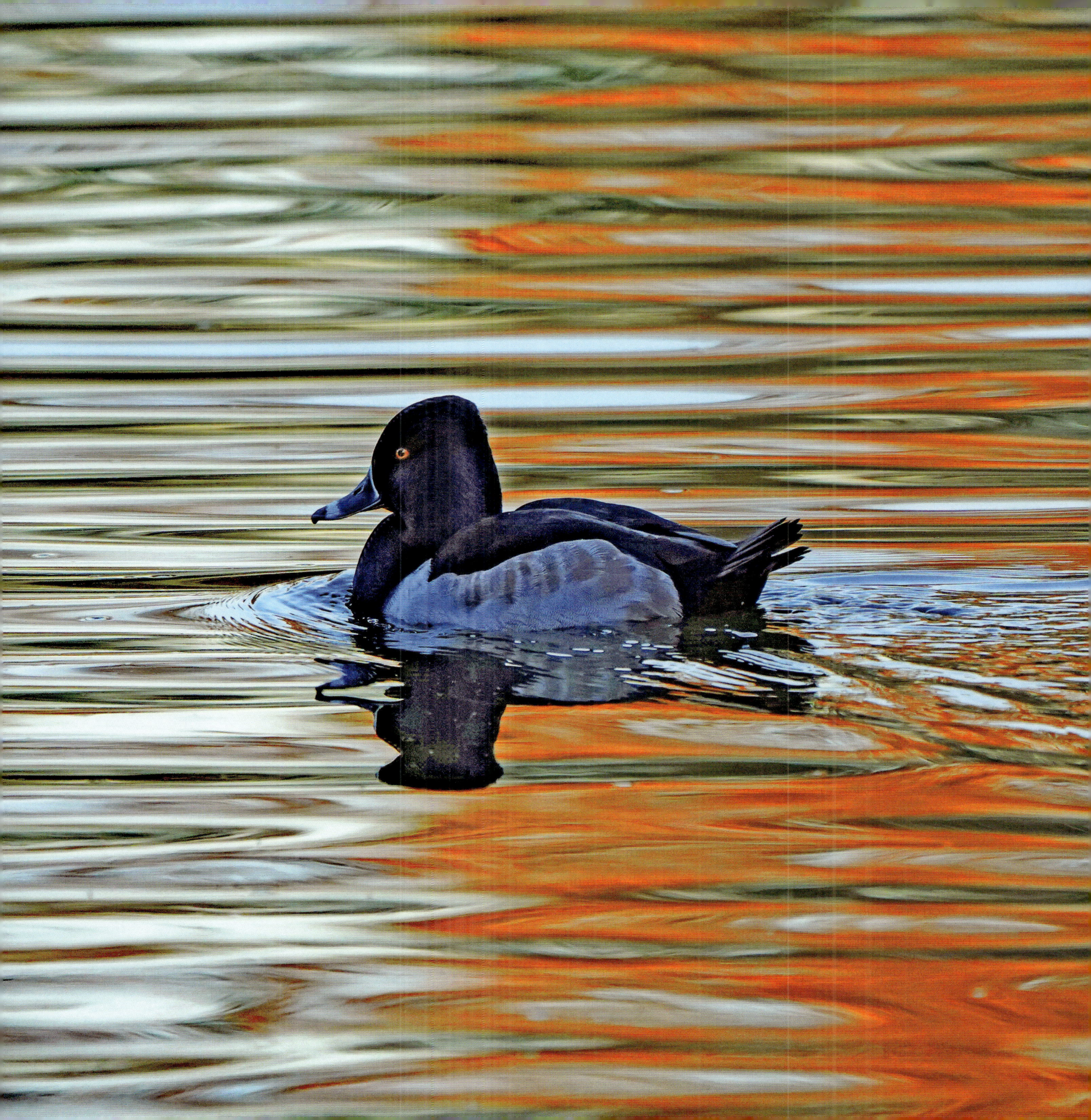

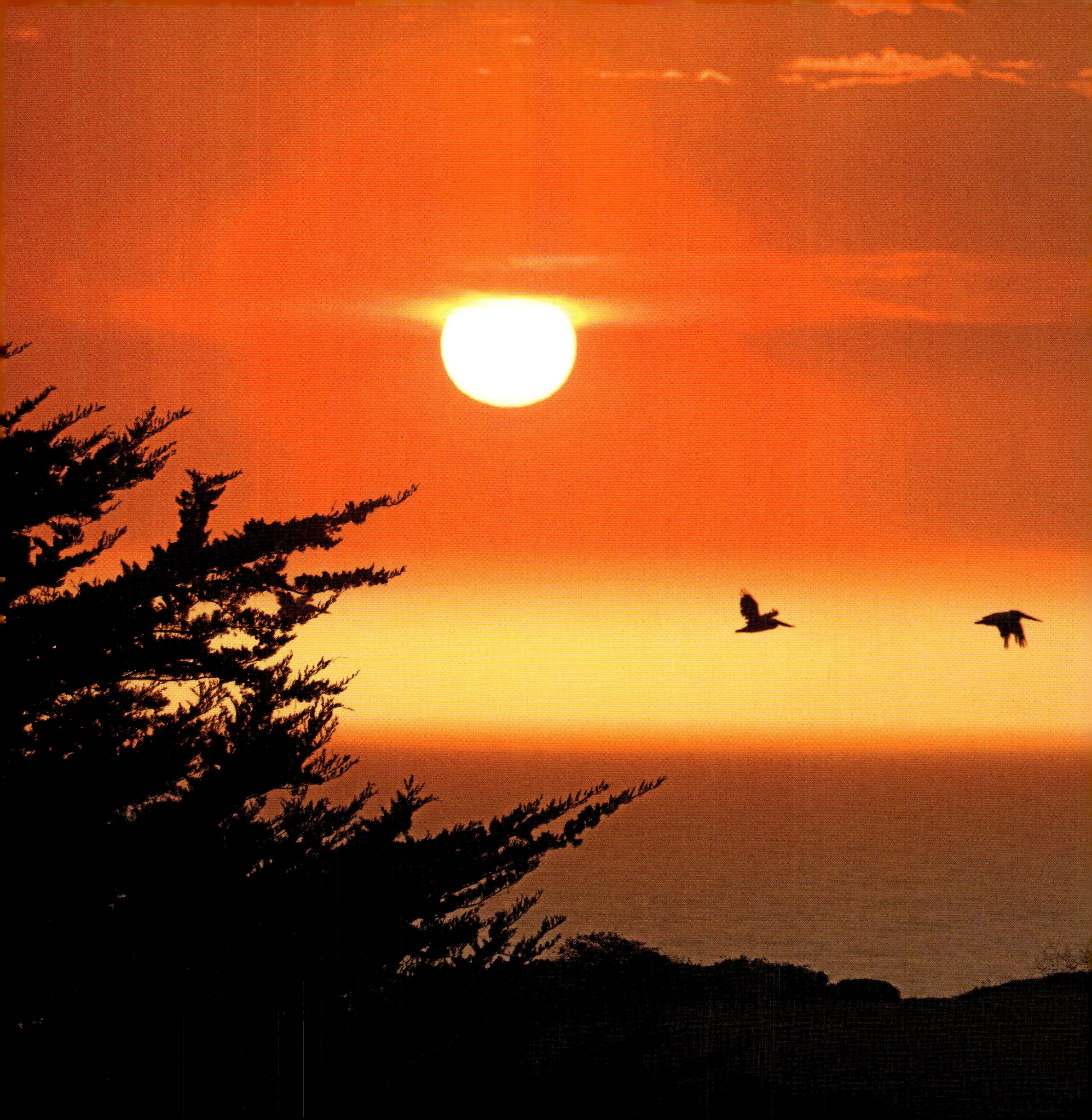

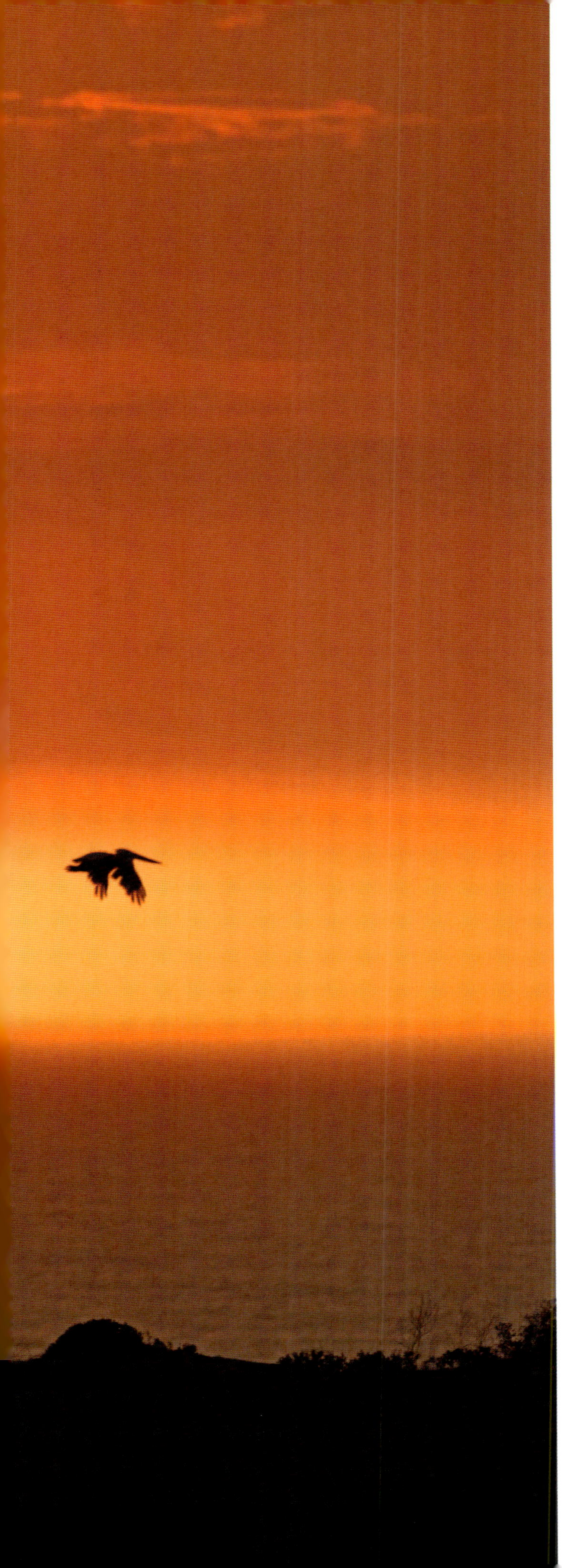

FLIGHT PATHS IN THE NIGHT

In early May, I sit on my patio in Berkeley and watch the stars come out. The sun set long ago, and the neighbor's brugmansia blossoms are opening in the night and unspooling their lemony fragrance over fences and into open windows all down the block. I've stayed outside long past dark, engrossed in a writing project, my laptop like a lamp in the night. And then, I hear it. At first I barely notice it, being so accustomed to the everyday background chatter of birds in the neighborhood. Then I sit up. I close my laptop and listen. Above me, in the darkness, birds are calling. As I tune my ears, it's clear that I haven't heard just an idle chirp from a confused bird out past its bedtime. The whole night is alive with soft, sharp little calls, like chimes in a dark breeze. I turn my face to the sky, but I can make out no movement. Unseen, a river of birds passes above me, and I am adrift in the current, listening.

On any given night during spring and fall migration seasons, as neighborhoods sleep, hundreds of thousands of birds will take wing on their seasonal journeys, navigating above increasingly complex landscapes illuminated by city lights and punctuated by skyscrapers. Over the course of the spring, more than a billion birds pass through California, and 80 percent of them do it under cover of darkness. Then they'll do it all in reverse come fall. Why and how remain a mystery even after more than a century of scientific study. In 1880, astronomers at Princeton University invited an ornithologist named William Earl Dodge Scott to view the full moon through one of their telescopes. What Scott saw dazzled him: silhouetted against the moon, the shapes of migrating birds in the night. In September 1896, a history professor named Orin Libby spent five hours on a Wisconsin hillside, tallying nearly four thousand flight calls. During World War II, radar operators began to detect "phantom signals" in the night. In one such instance, British biologists David Lack and George Varley used a telescope to visually confirm that a flock of gannets flying overhead was

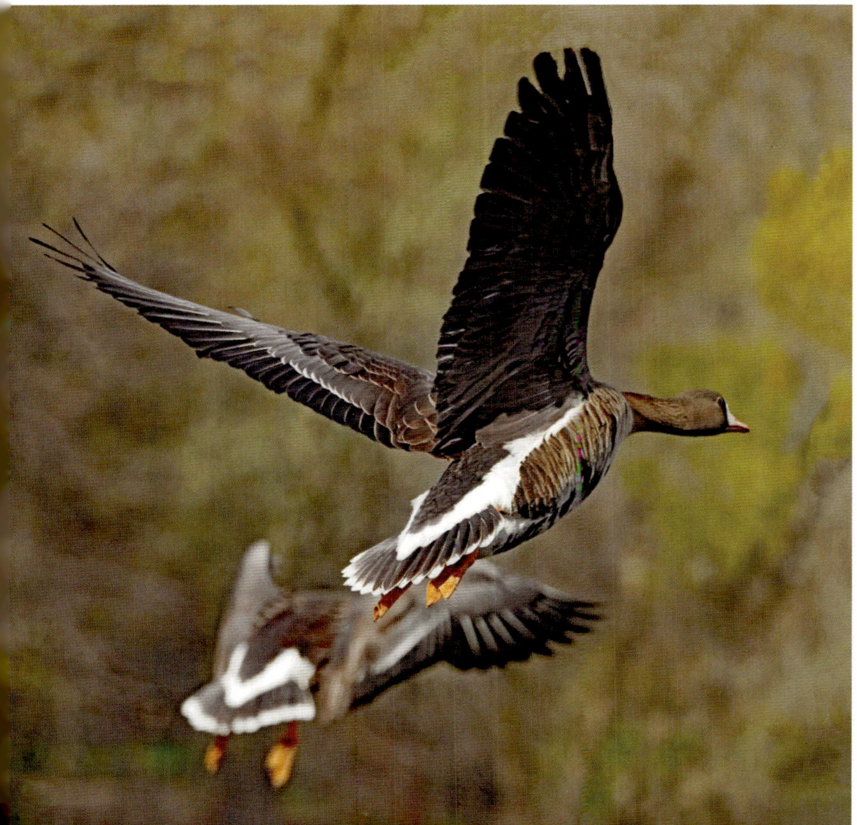

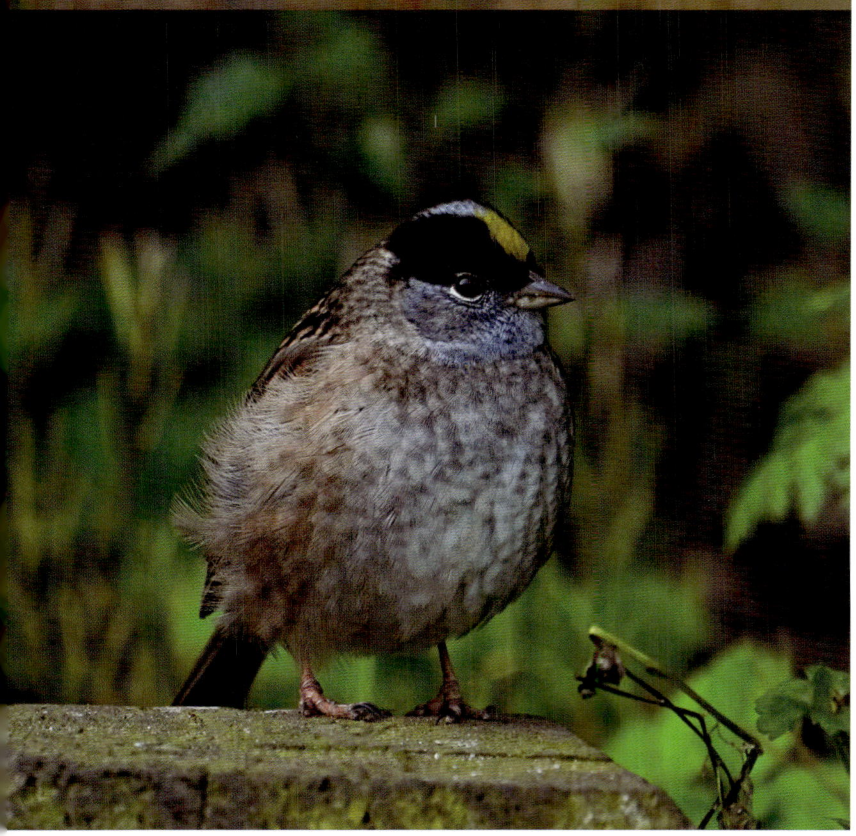

responsible for the mysterious military radar signal. When physicists eschewed the results, insisting that these large-scale signals must be caused by atmospheric ions, Lack wrote that he and Varley "accepted their view, provided that the ions were wrapped in feathers."

In the intervening years, the shape of these massive migrations has come into focus, even if the details remain fuzzy. The birds that pass through the Bay Area are traveling along a 4,000-mile avian superhighway that stretches from Alaska in the north to Patagonia in the south. On this globe-spanning journey, they are guided by an internal magnetic compass, celestial cues, and patterns of light polarization at sunset. A migration map shows the widespread regions that our birds arrive from during their fall migrations—Bristol Bay, the Alaskan Coastal Plain, the Tongass rainforest—and the routes our resident summer birds travel as they peel off and head to lower latitudes: the coasts of Mexico, the cloud forests of Central America. The far-flung flight paths all converge in the Bay Area like streams of sand passing through the neck of an hourglass.

Hemmed by the Sierra Nevada to the east and the open Pacific to the west, the region's geography funnels shorebirds, waterfowl, and songbirds toward San Francisco Bay. Here, on the banks of California's largest estuary, feathered travelers find food and respite from their long journey. The region's habitats, ranging from redwood forests to coastal prairie, make this a comfortable place not just to rest but to linger. For many migratory birds, the Bay Area will be their winter home. Every fall, the Palomarin Field Station in Bolinas hosts a Golden-crowned Sparrow party—complete with handmade golden crowns—to celebrate the homecoming of our familiar songbirds as they pour in from their northerly breeding grounds.

PRECEDING: A squadron of Brown Pelicans skims low along the California coast. These huge, stocky birds are often seen surfing updrafts along cliffs or the steep faces of ocean waves.

OPPOSITE: Greater White-fronted Geese (top) are the first migratory geese to arrive in the California Delta each fall, arriving as early as late August from the Yukon-Kuskokwim Delta in Alaska. By the time Golden-crowned Sparrows (bottom) arrive in the Bay Area on their annual migration from Western Canada and Alaska, they have flown nearly 3,000 miles—almost entirely under cover of darkness.

Weather radar technology can now be used to follow the movement of massive clouds of birds in the night, in real time. Cornell Lab of Ornithology and Colorado State University post a collaborative forecast map that predicts pulses in nocturnal migration around the country. When they are passing above you, step outside and listen.

Those tiny buzzes and whistles we hear in the night? Only recently have biologists identified them at a species level, using spectrograms to visualize the auditory signals. Each sound is like the fingerprint of an invisible bird species passing overhead. Stations around the continent are picking up these odd nighttime sounds and using them to pinpoint which birds are on the move. In the Bay Area alone, 250 species of birds migrate through the region every spring, each with a distinct call that sizzles in the night. The data is overwhelming—thousands of hours of calls collected during the crush of spring and fall migrations—and scientists are innovating ways to process the barrage of sound with algorithms and the help of AI. But you don't need special equipment to tune in to the nighttime show. Train your ears using the recordings on the Nocturnal Flight Calls of North America website, and you might make out the differences between these night calls as you listen from your porch.

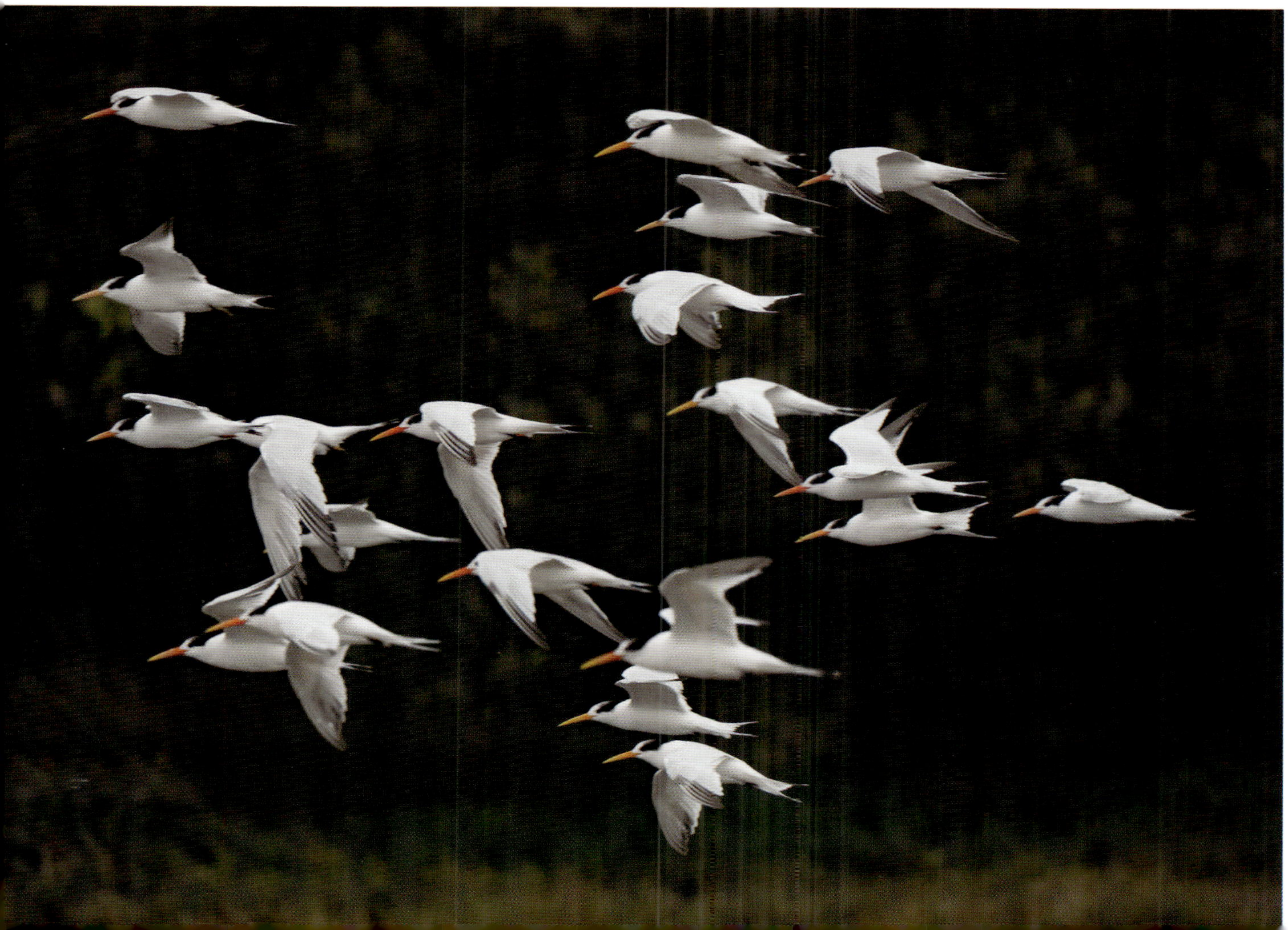

LEFT: More than 90 percent of all Elegant Terns nest on a single tiny island in Mexico's Gulf of California, but in the summer and fall, these sharp-winged aerialists can be found in familiar Bay Area sanctuaries like Crissy Field.

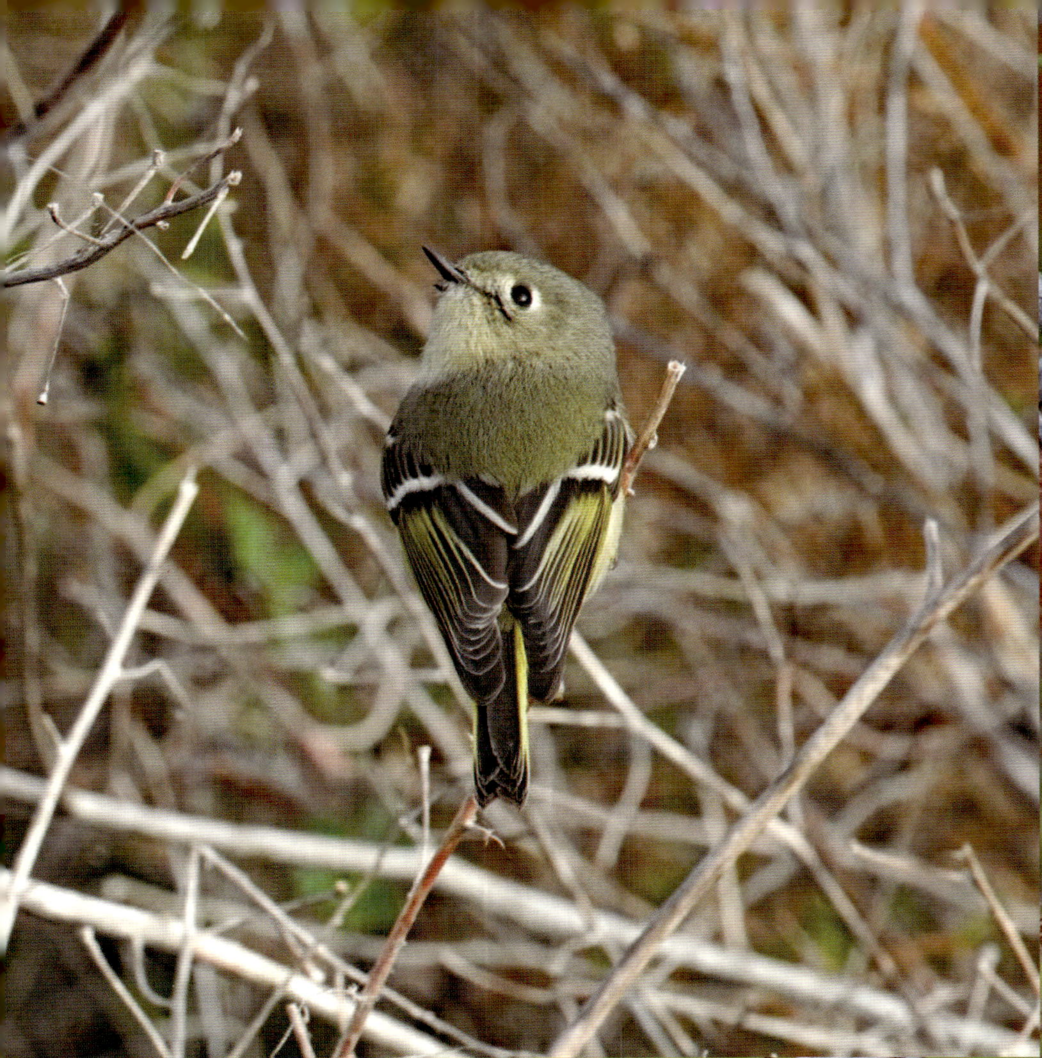
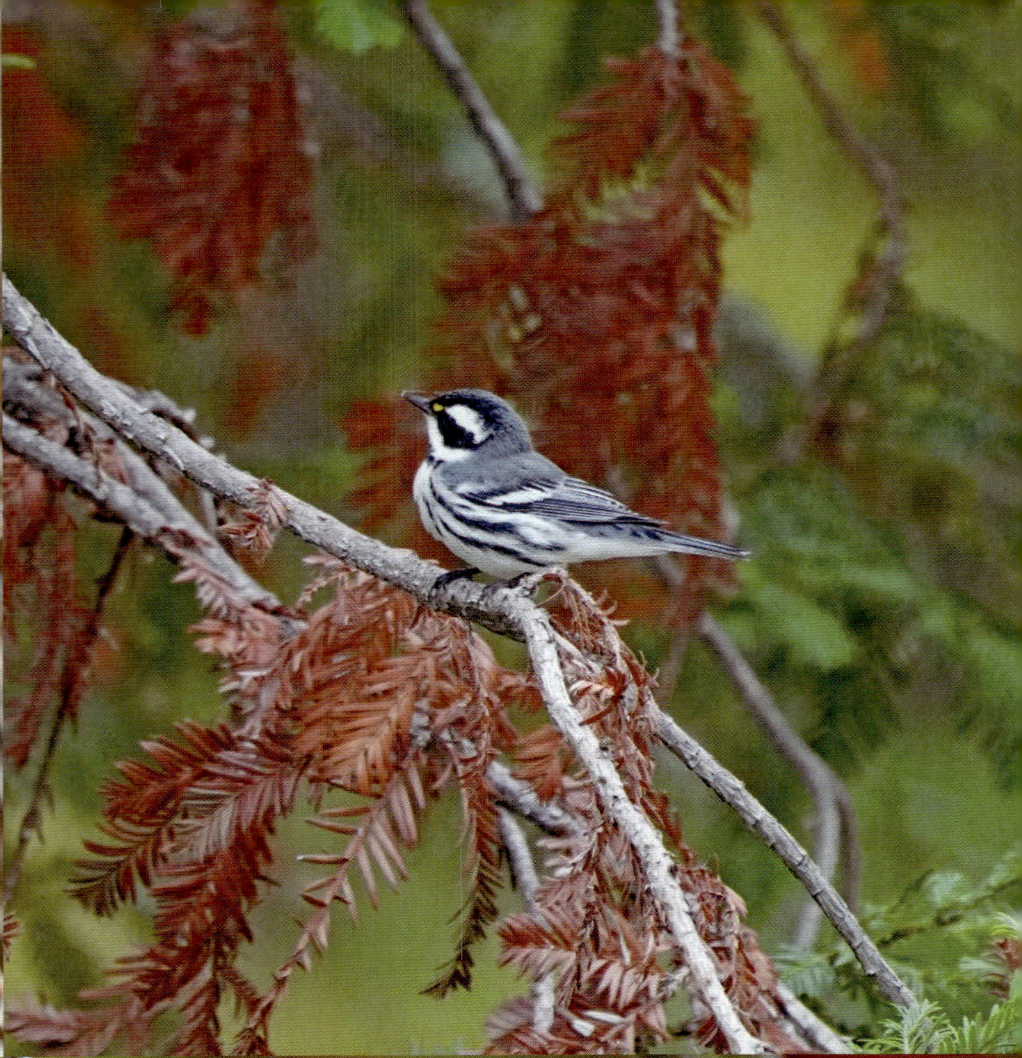
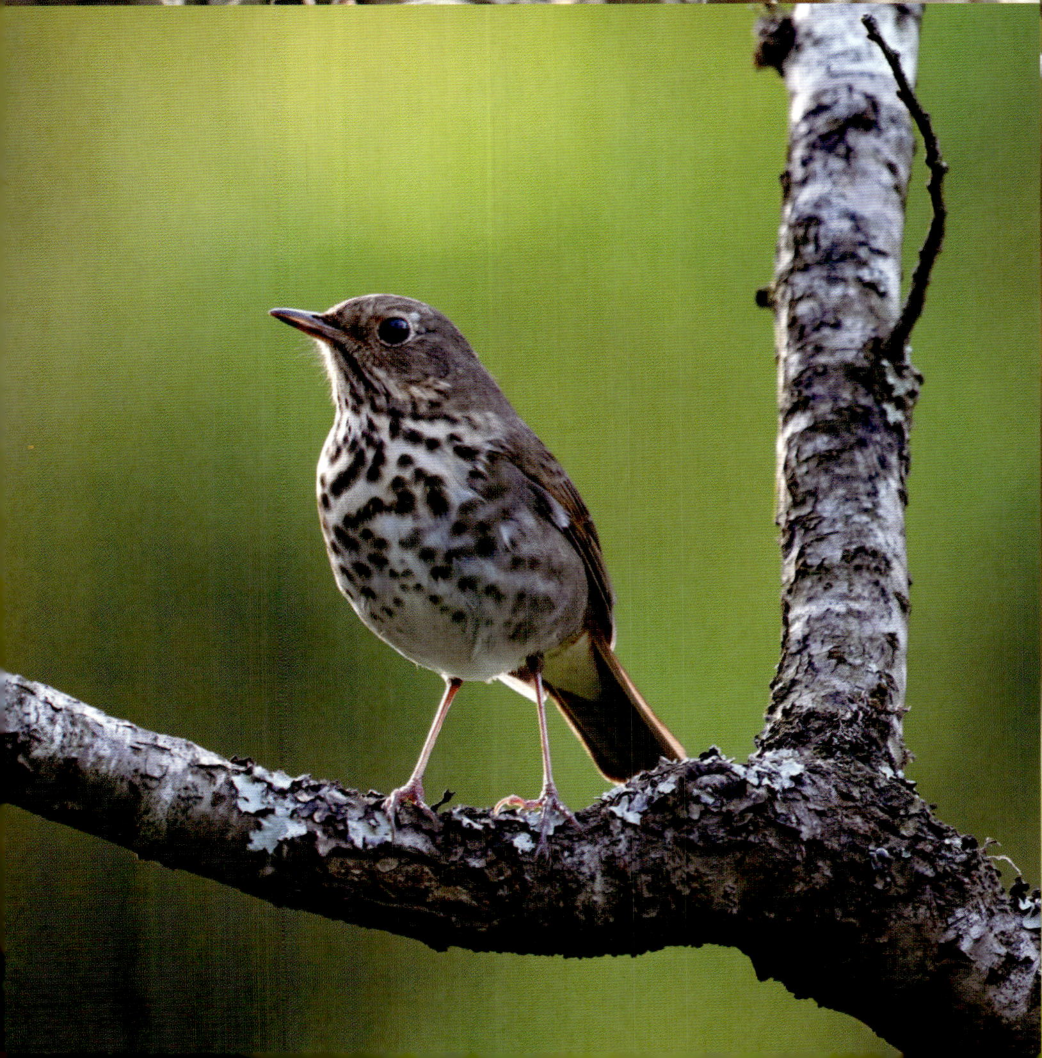
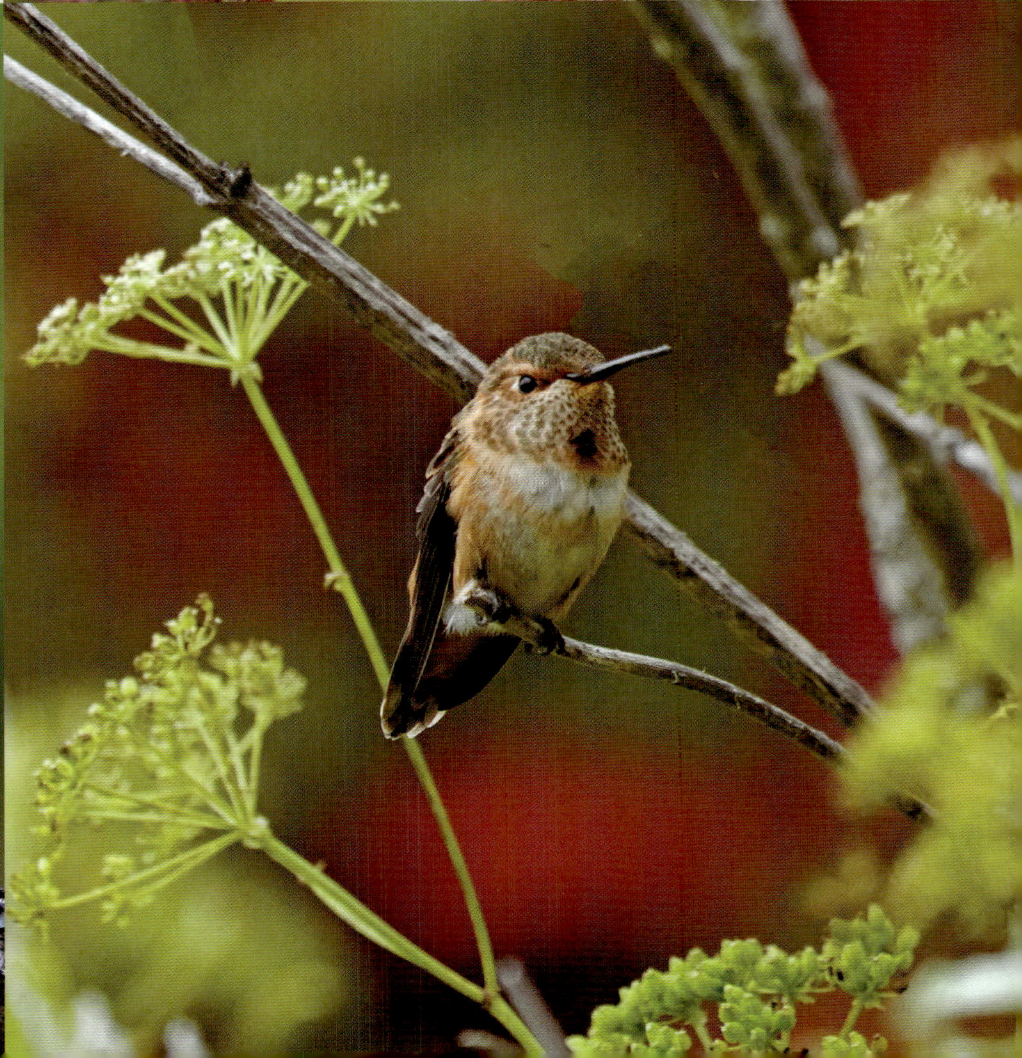

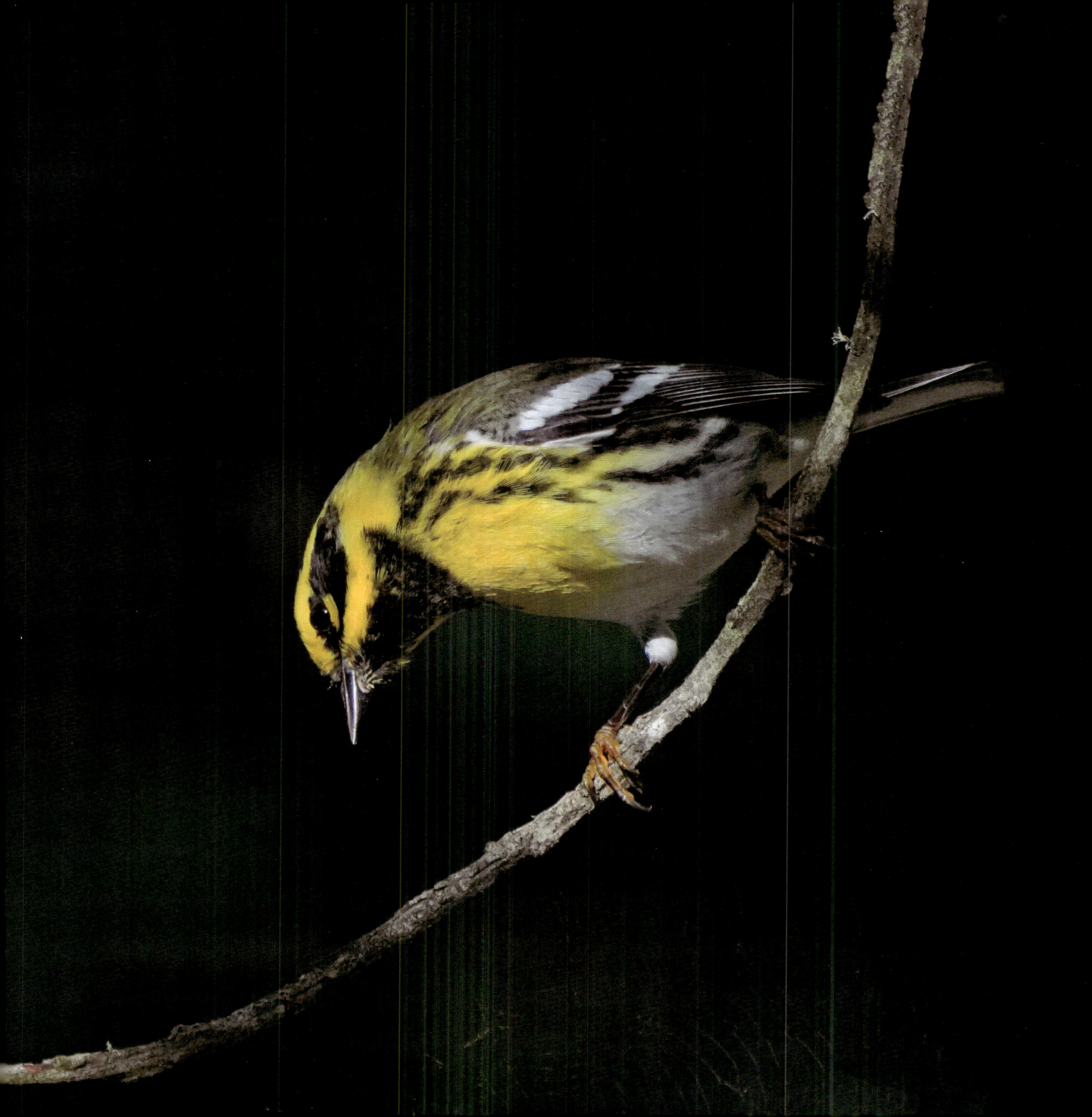

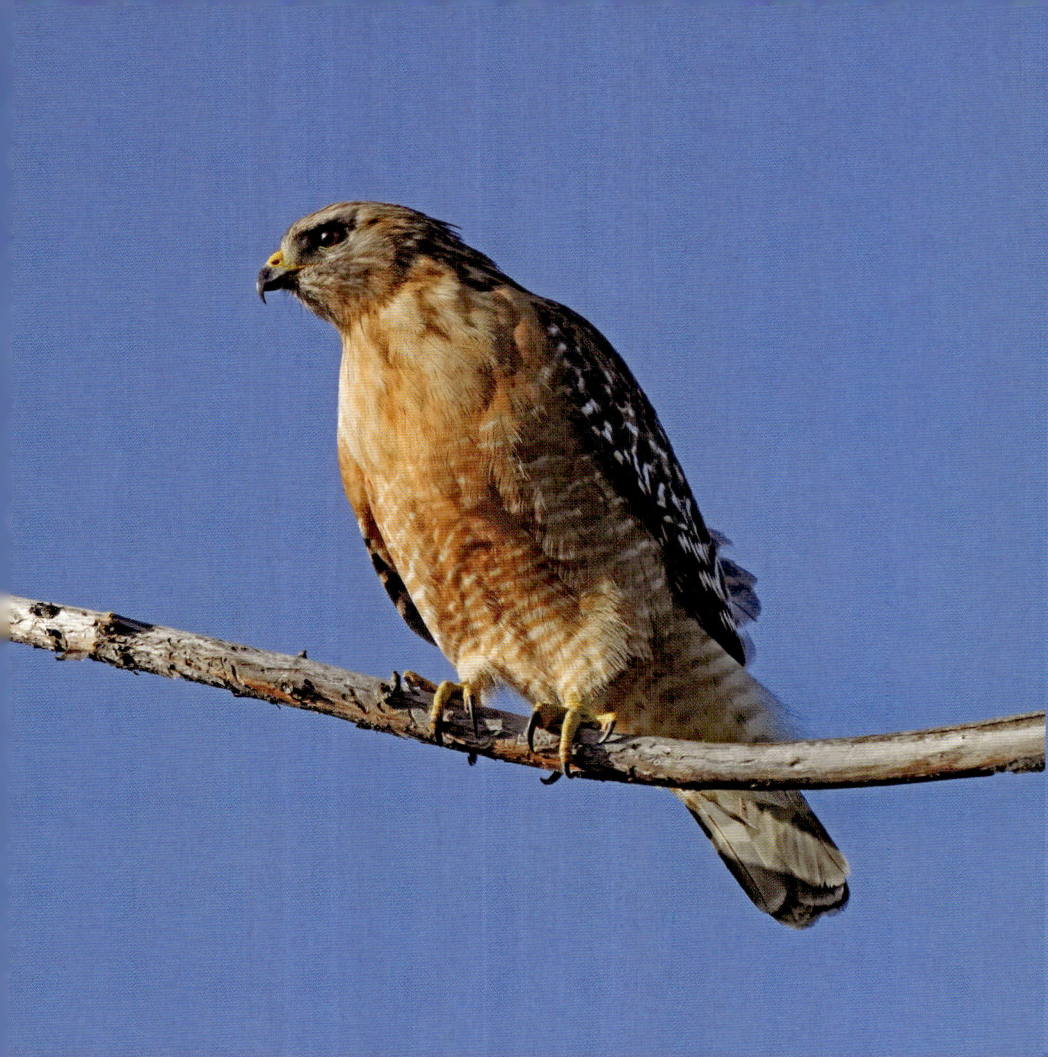
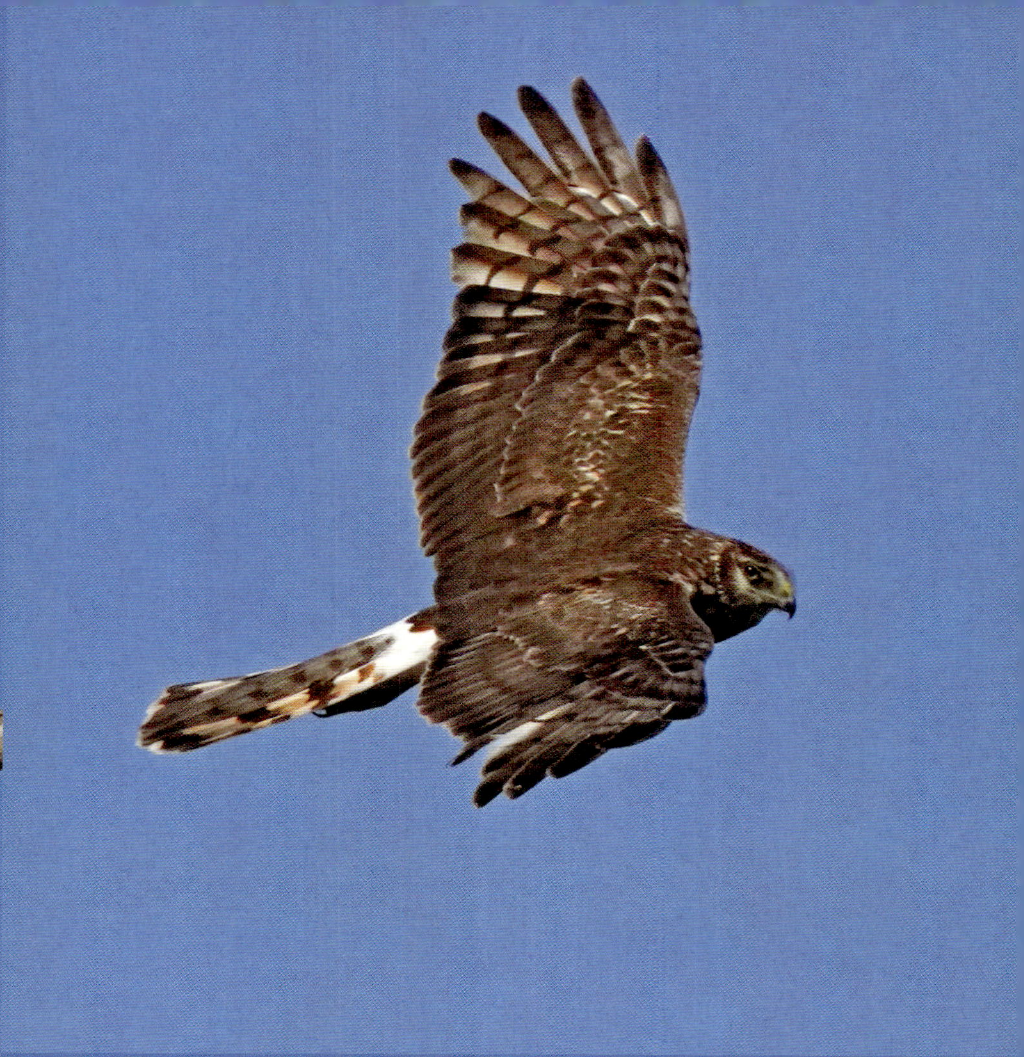
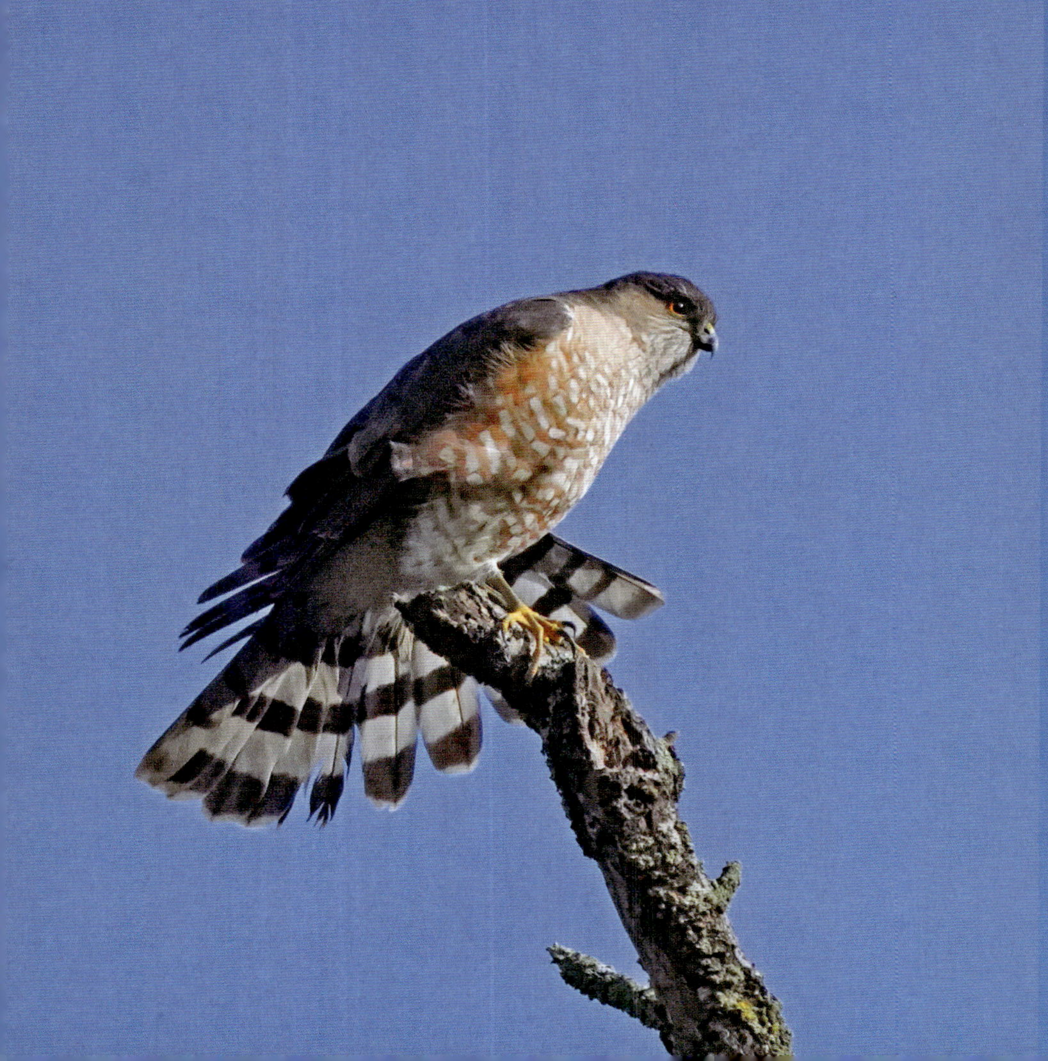
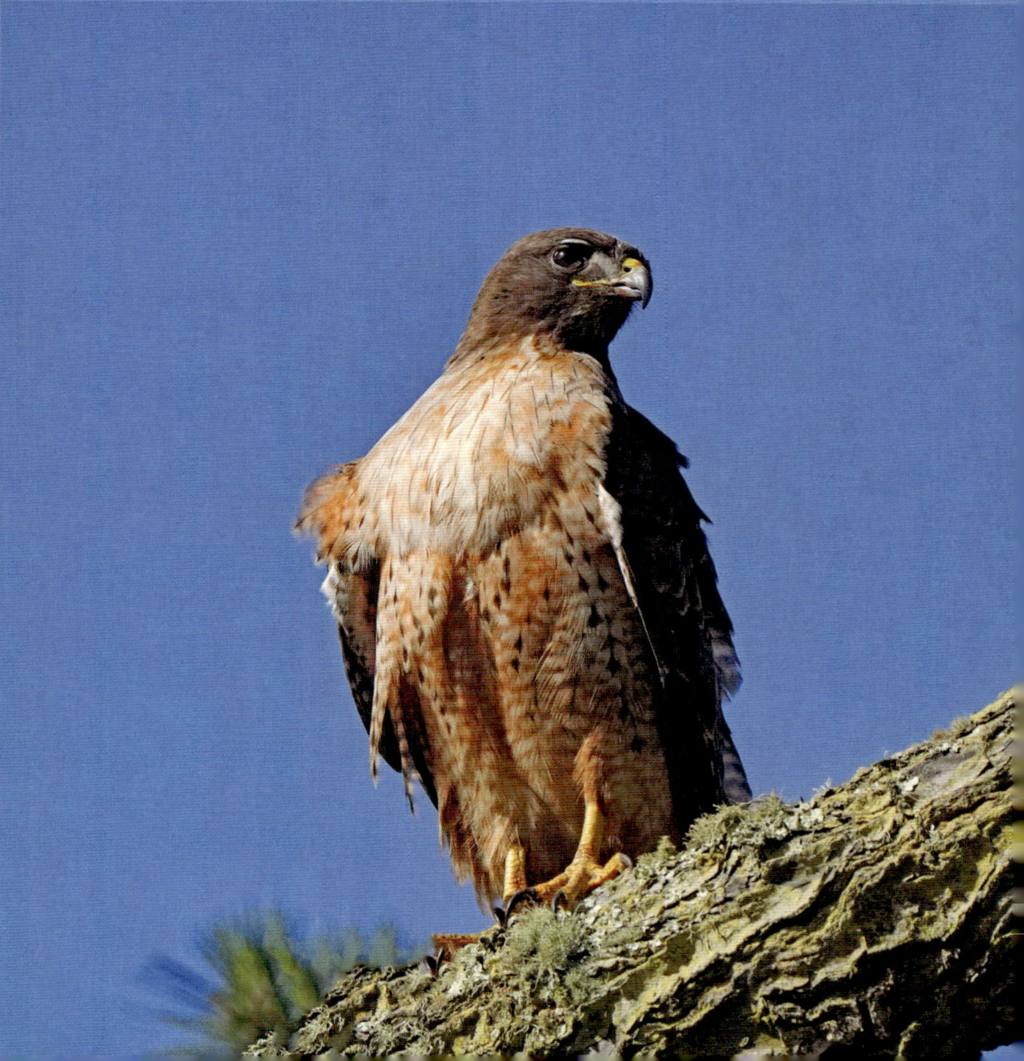

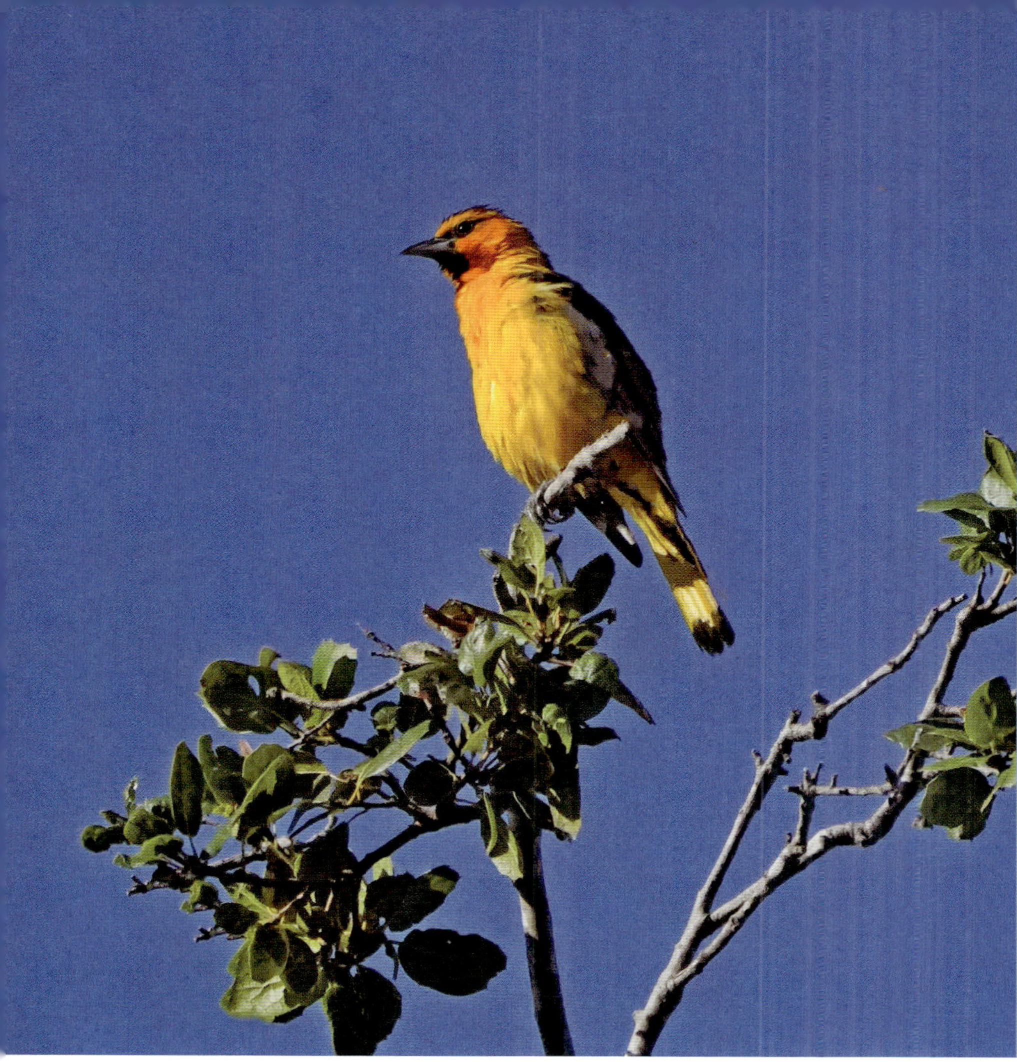

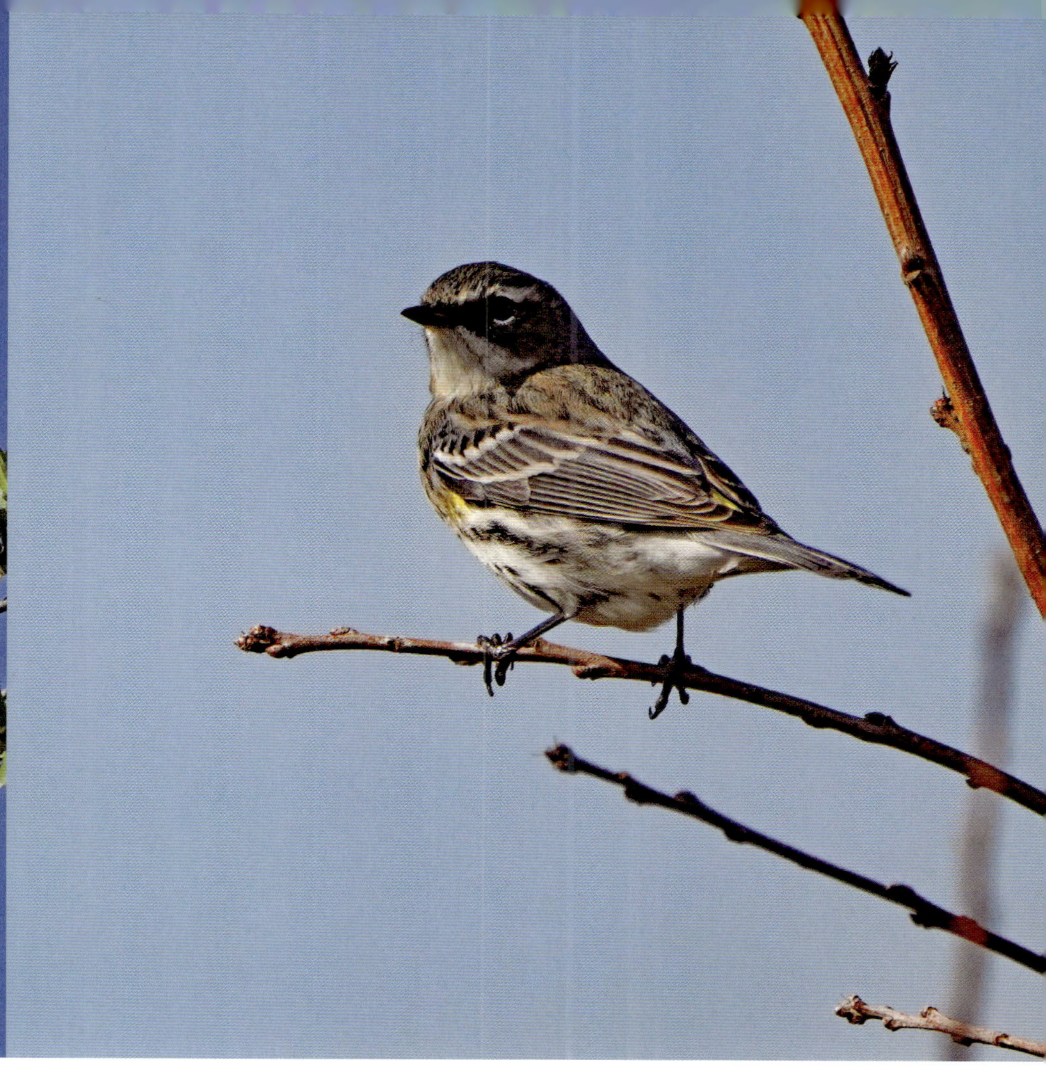

PRECEDING, CLOCKWISE FROM TOP LEFT: The Ruby-crowned Kinglet, Black-throated Gray Warbler, Townsend's Warbler, Allen's Hummingbird, and Hermit Thrush all travel migratory paths to reach our doorsteps in the Bay Area.

OPPOSITE: A quartet of aerial hunters showcases the diversity of raptors thriving across Bay Area landscapes. From the Red-shouldered Hawk calling from riparian woods and oak groves (top left), to the low-flying Northern Harrier quartering marshes and grasslands (top right), the agile Cooper's Hawk navigating forest edges (bottom left), and the ubiquitous Red-tailed Hawk patrolling open fields and woodlands (bottom right), each is a vital predator adapted to its niche.

ABOVE: The Bullock's Oriole (left) and Yellow-rumped Warbler (right) are among the 80 percent of migratory birds that navigate at night, in part to avoid diurnal predators like the Cooper's Hawk.

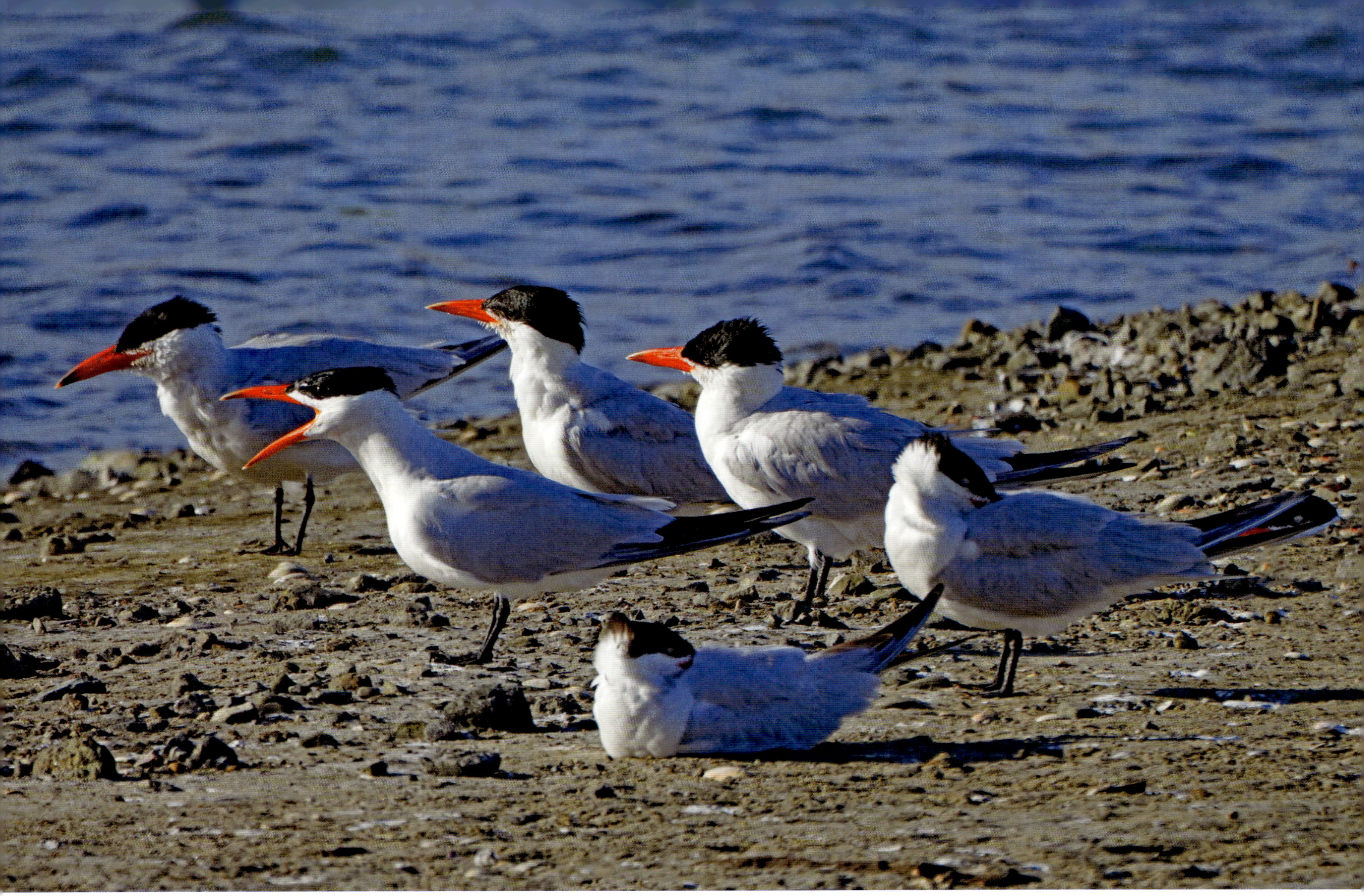

FOLLOWING: A Black-necked Stilt and an American Avocet (left) stand among their own reflections in the Corte Madera Marsh, a popular place to find wading shorebirds year-round. These birds belong to the same family and are both adapted to living in shallow water environments. Stilts feed by jabbing their prey; avocets use their curved bills to sweep through the water. The Whimbrel, Willet, Least Sandpiper, and Greater Yellowlegs (right, clockwise from top left) all flow into the Bay Area on different migration paths, the longest of which extend as far as the tip of Tierra del Fuego (for the Greater Yellowlegs) and the remote northern tundra (for the Whimbrel).

Caspian Terns (above) arrive in the Bay Area in the spring to breed, nesting in colonies on islands and isolated spits of land. In the fall, they migrate as far as South America.

Standing a foot tall, the Long-billed Curlew (opposite) is North America's largest shorebird, migrating to the Bay Area's estuaries and mudflats from the grasslands of the Great Plains and Great Basin. It is designated as Climate Endangered, and it is projected to lose roughly 60 percent of its climate suitability range in California by the year 2050.

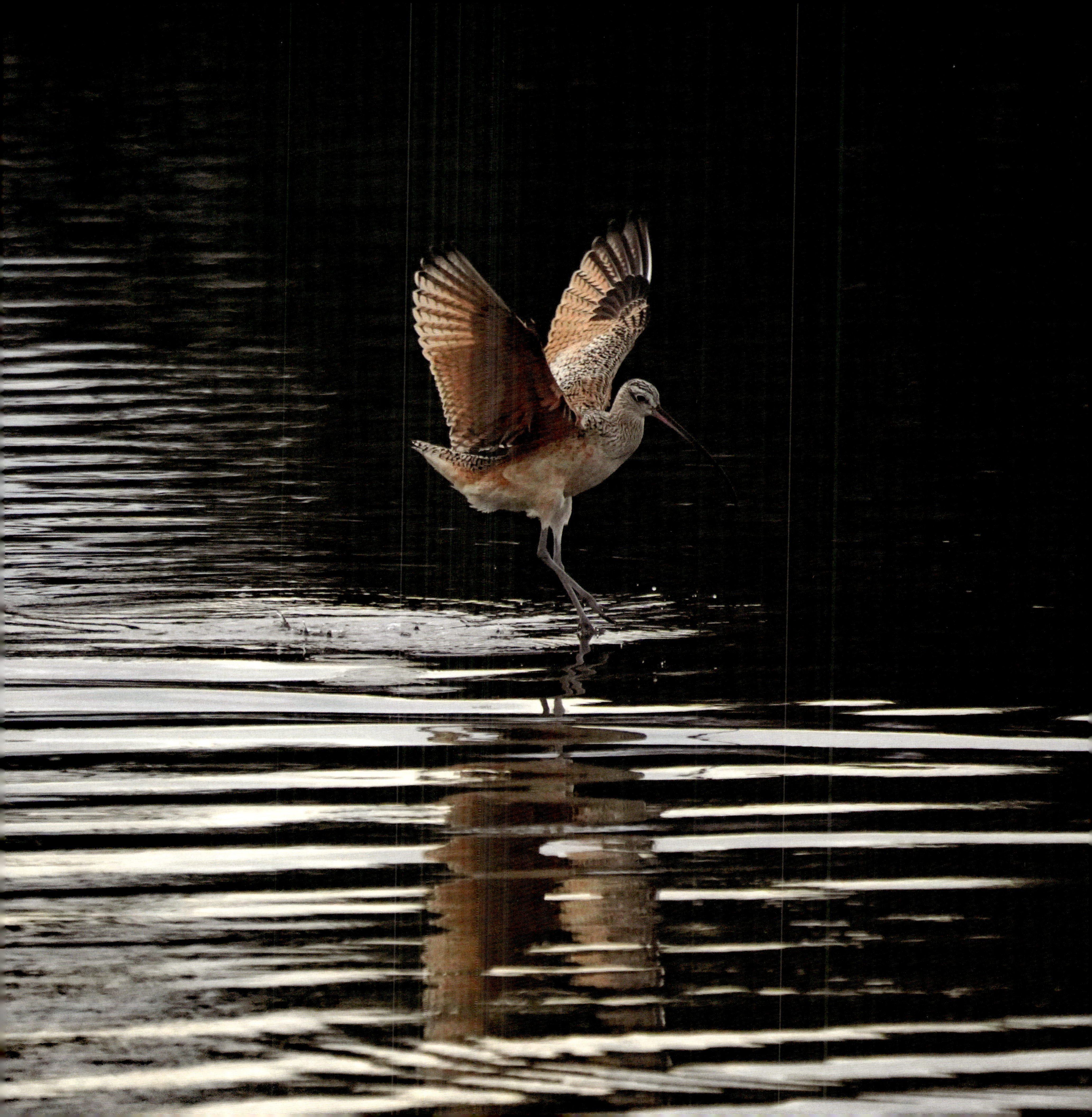

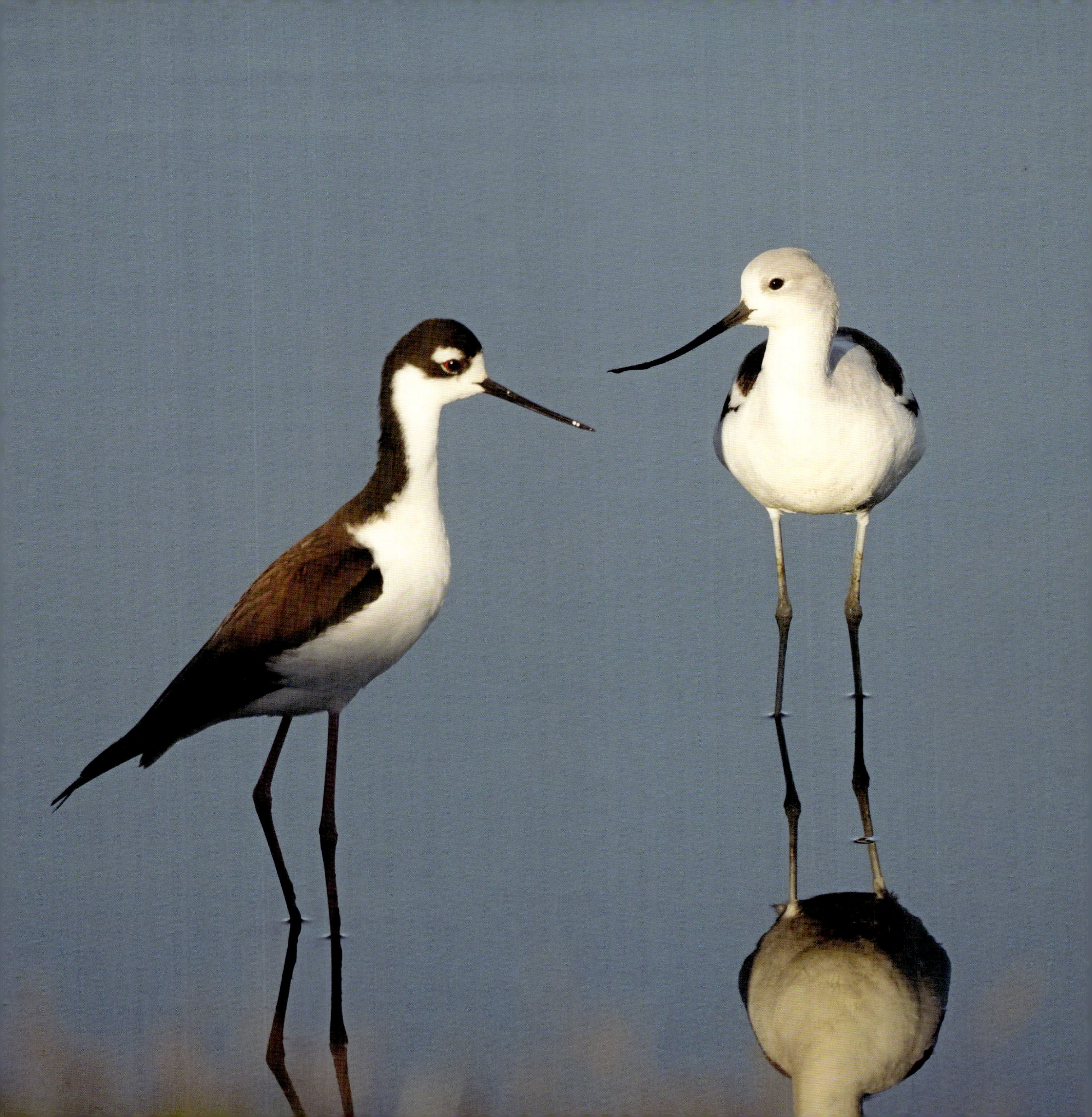

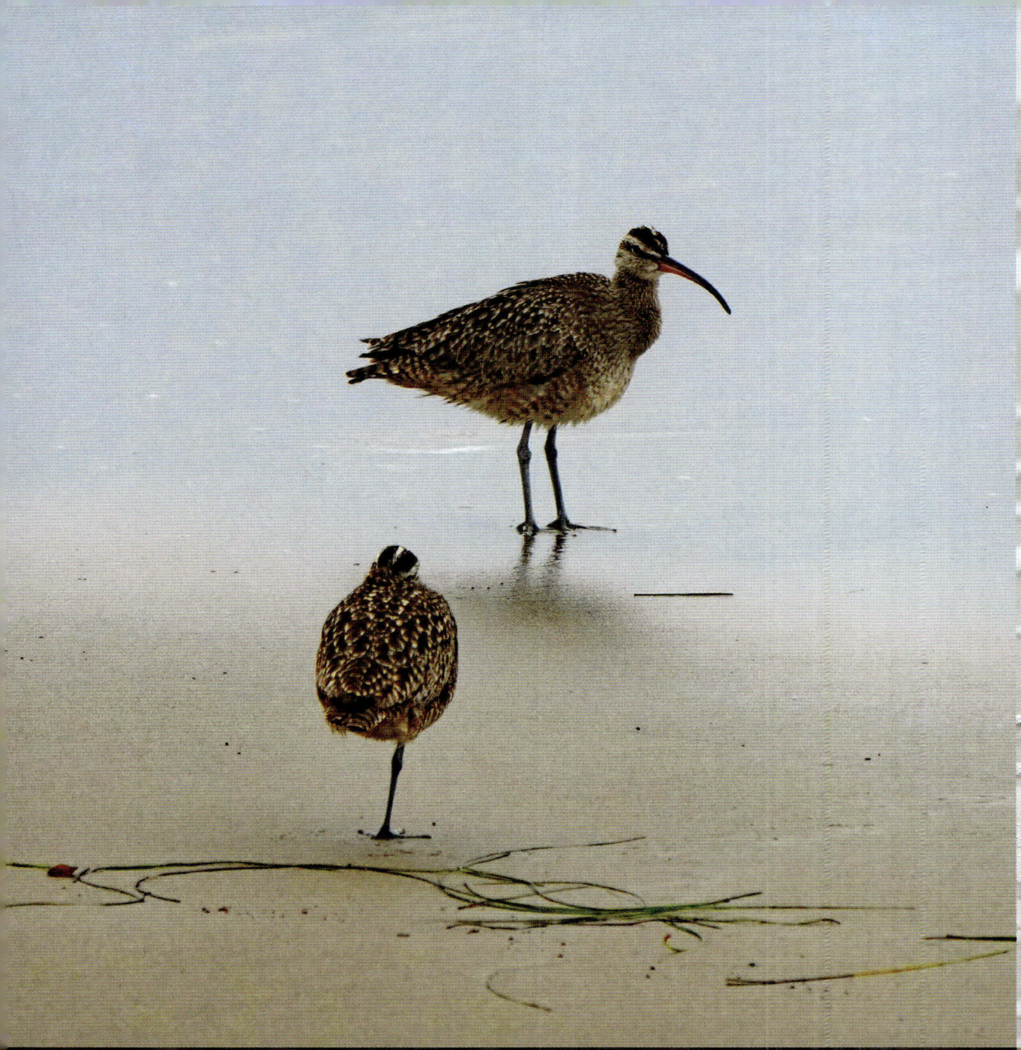
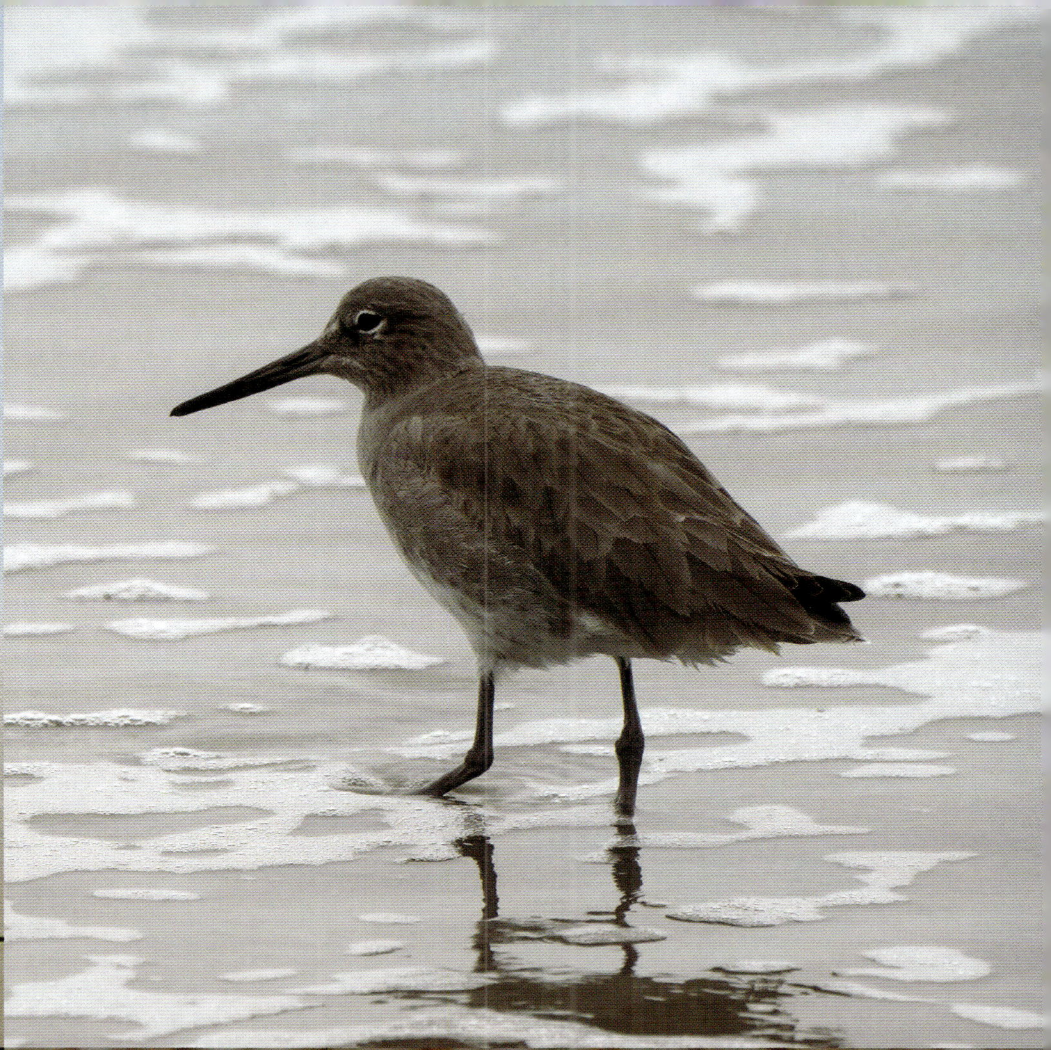
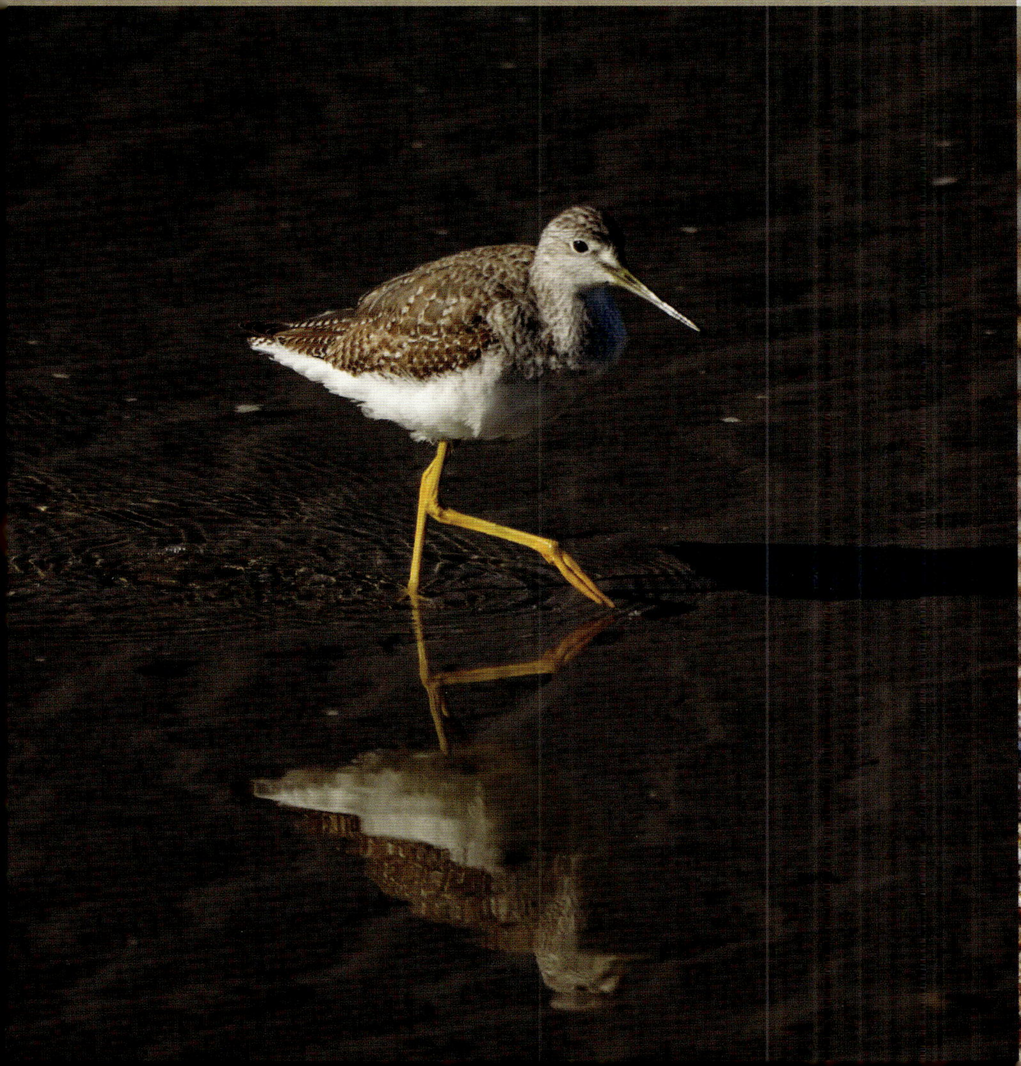
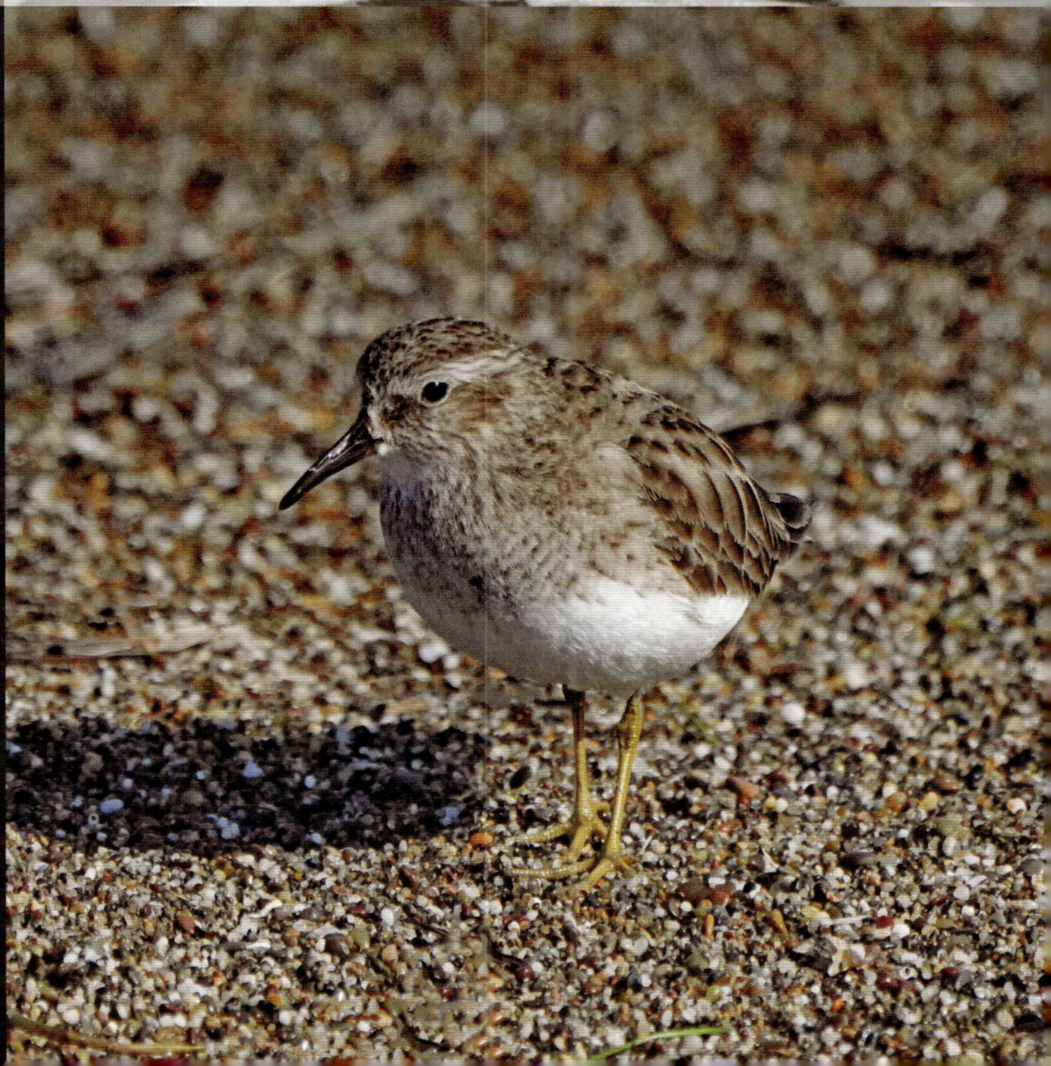

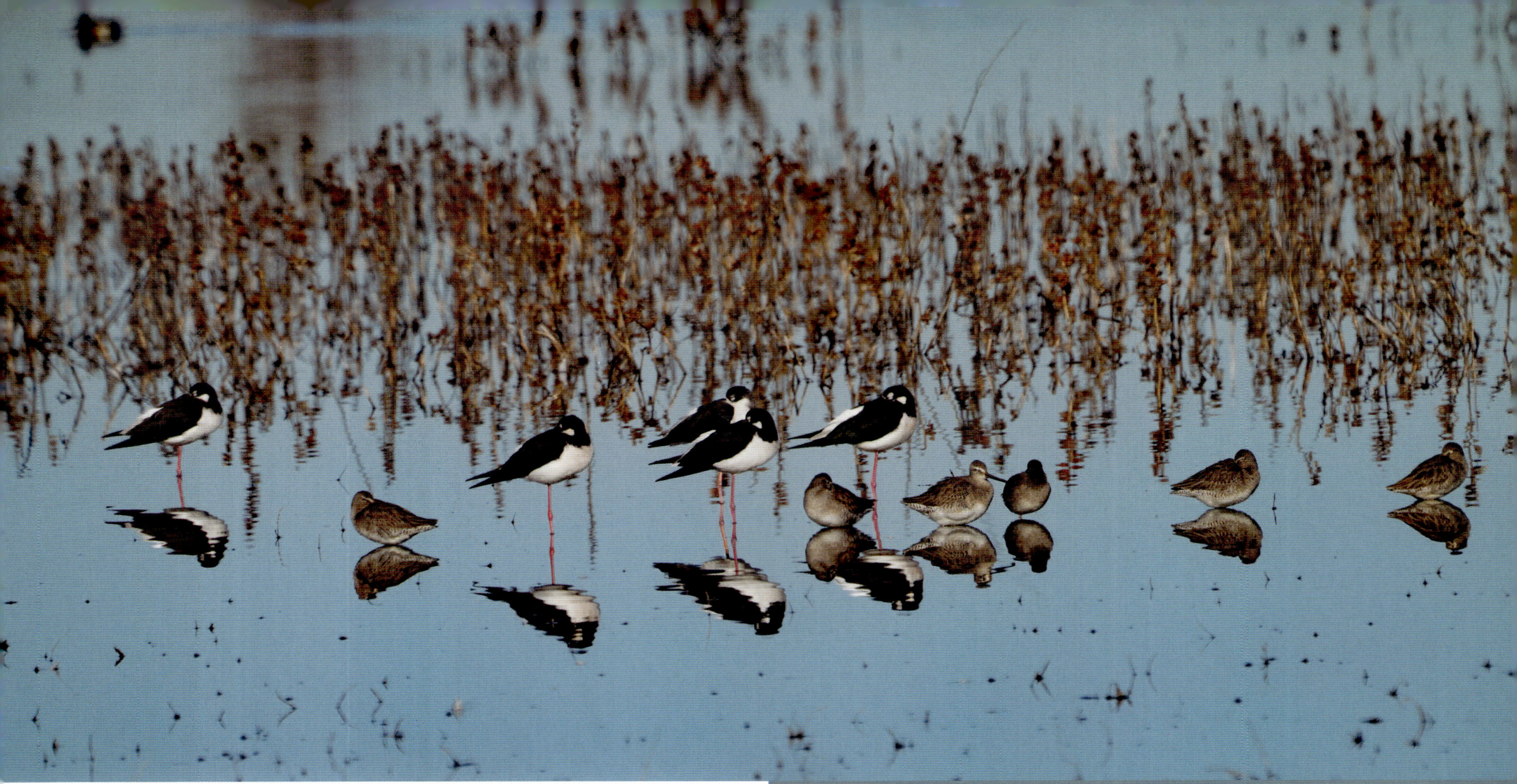

Where the Pacific Flyway funnels into the Bay Area, species intermix in shared waterways where food is plentiful and where the shallows protect them from predators in the night. Leggy Black-necked Stilts hunker among plump Long-billed Dowitchers (above).

A Greater Yellowlegs (right) passes through California wetlands on an annual migration that spans from the far tip of South America to the bird's summer breeding grounds in Alaska.

Young Brown Pelicans (opposite) skim overhead with leisurely wingbeats. These enormous, primitive birds spend summer and fall in San Francisco Bay, then fly south to their rocky nesting colonies in Mexico and the Channel Islands.

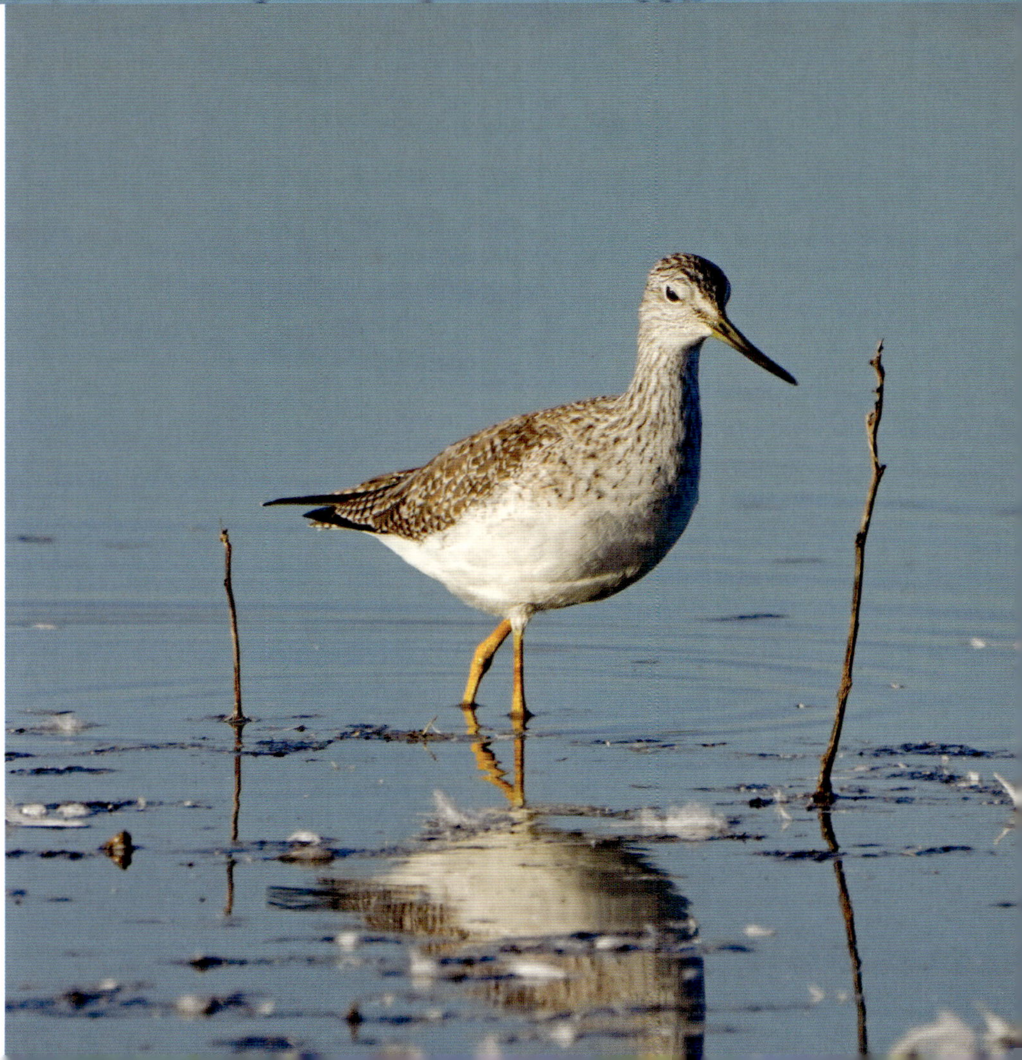

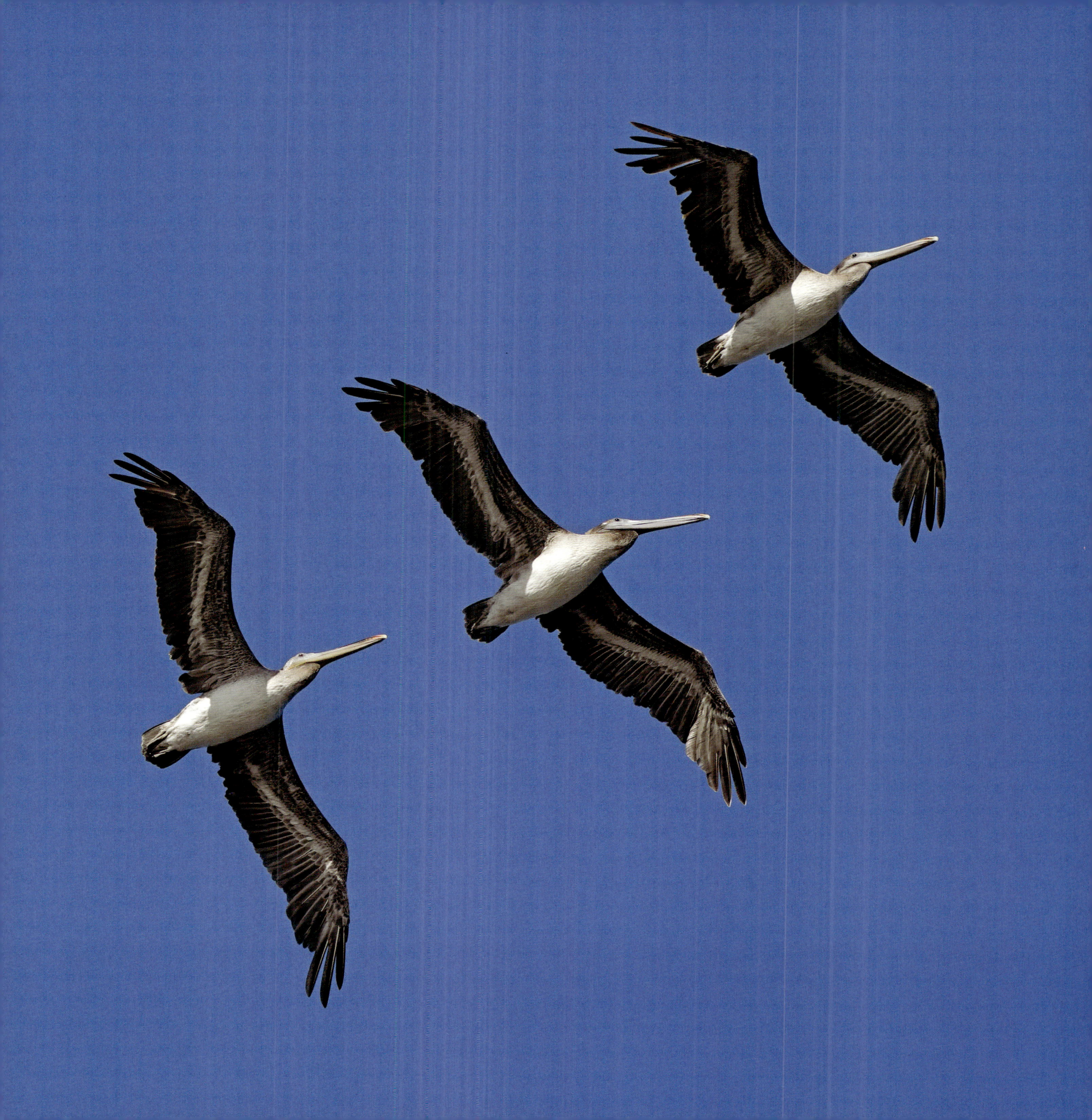

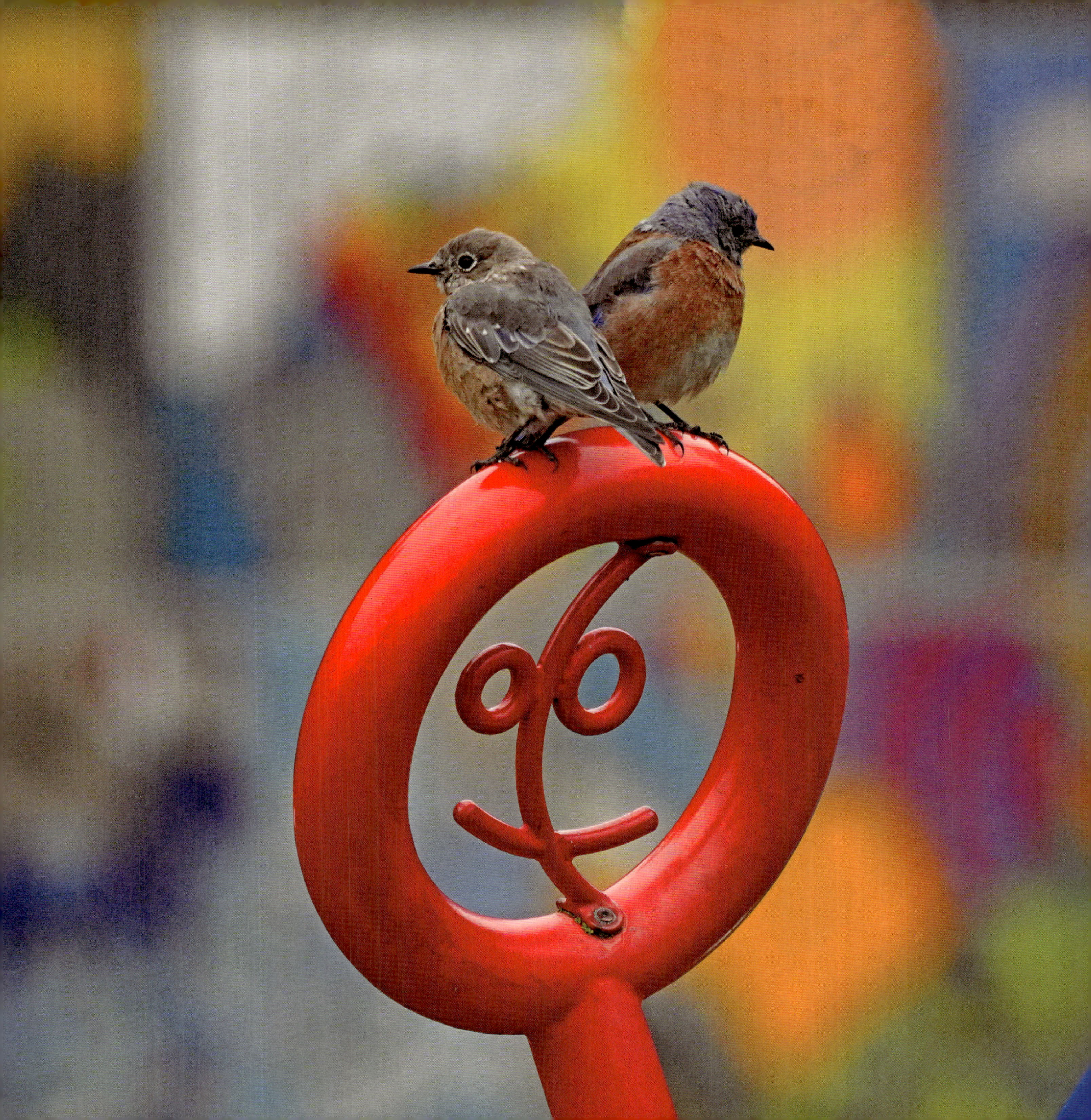

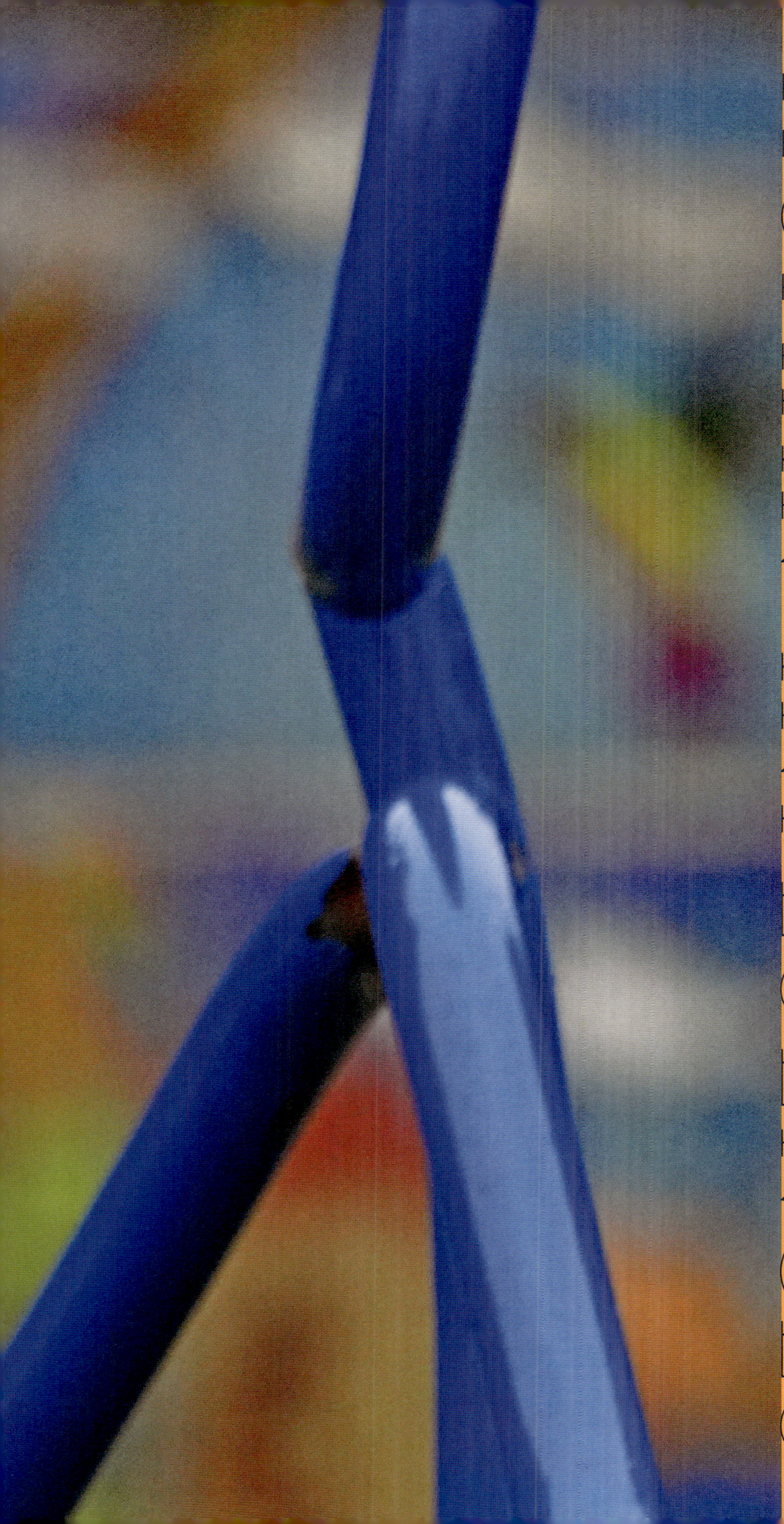

HUMAN INFLUENCES

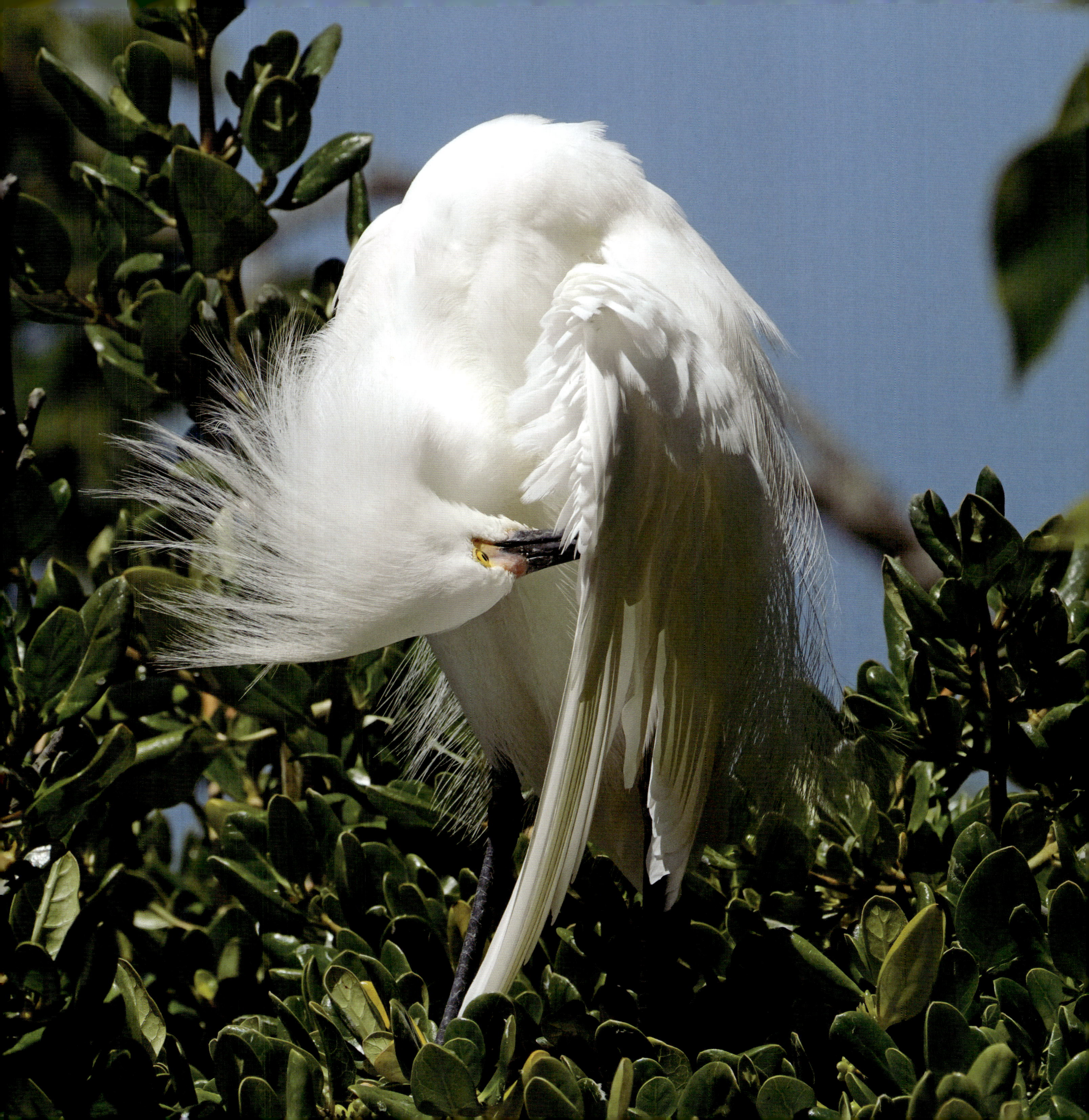

Birds and humans are more than neighbors; our lives and histories interweave. In the Bay Area, birds have filled people's coffers and collections and have fed our bodies and our spirits. They have been frontline harbingers of change, and illuminators of our hidden past. They reflect our actions back to us, and in the evolving story of conservation in the region, they may guide our actions too.

I can almost smell the wet riverbank where sedge roots grow, pale and smooth beneath their dark outer skin. I can nearly hear the crisp sound of willow shoots being scraped down to size. Here in a side room of the Grace Hudson Museum & Sun House, Hopland Pomo basket weaver Susan Billy shows me baskets that she has made out of these materials, perched alongside baskets that her family members have made, across generations. Some of the baskets are half my height, with open weave and tapered sides for funneling quail—an elegant trap for small ground-dwelling birds. Other baskets are woven tightly, with dark bulrush forming patterns against the pale weave of sedge root. "This is the quail pattern," Susan tells me, pointing out a repeating L shape jutting upward like a periscope. It evokes a clustered bevy of quail with their little topknots sticking up. Actual California Quail topknot feathers adorn several of the baskets in the collection, springing from the rims like silk fringe. Bird feathers of all kinds are an essential material in California basket weaving, and their patterns and colors kindle the imagination.

The collection, and Susan's own path to basket weaving, are a testament to the endurance of a globally celebrated basketry tradition, which outlived government efforts to distance Native children from their cultural heritage through forcible separation from their families and communities. Susan herself only learned basket weaving as an adult when, two years into college, she traveled to her father's homeplace and knocked on the door of her great-aunt, renowned Southern Pomo basket weaver Elsie Allen. It was the beginning of a sixteen-year apprenticeship.

Carrying her grandparents' tools and listening to Elsie's quiet instructions, Susan learned not just about baskets but about how a basket creates a relationship between the weaver and the world that surrounds her. The birds that Elsie used for basketmaking were always given to her, Susan tells me. Elsie disliked receiving loose feathers, and preferred to receive them still attached to the skin. This way, she could study how the feathers were arranged on the bird's body, so that she could lay them into the weave of a basket in the exact same shape that the living bird grew them.

PRECEDING: Two Western Bluebirds perch atop a whimsical sculpture on the shores of Spring Lake: a colorful affirmation that local recovery efforts to support these cavity-nesting birds are working!

OPPOSITE: The long feathers of the Great Egret and other heron species were once highly valued in millinery—until the Audubon Society pushed for powerful legislation to protect these birds in the Bay Area and beyond.

When the life of a basket is considered—the time it takes for bulrushes to grow, for students to learn from their mentors, for collectors to harvest and dry the materials, for birds to nest and fledge, for a basketmaker to receive their bodies as gifts—it becomes a story not just of a weaver sitting down to create an object but of a whole ecosystem. "It takes a thousand years to make a basket," Susan likes to say.

In San Francisco, at the California Academy of Sciences, a basket made by Susan's grandmother in the mid-1900s hangs in an exhibit. It is woven so tightly with bird feathers—lush green Mallard feathers, Crimson Tanager feathers, and, at the base, brilliant yellow meadowlark plumage—that none of the sedge weft shows through. Other baskets hang alongside it, tufted with quail topknots and radiant Acorn Woodpecker feathers.

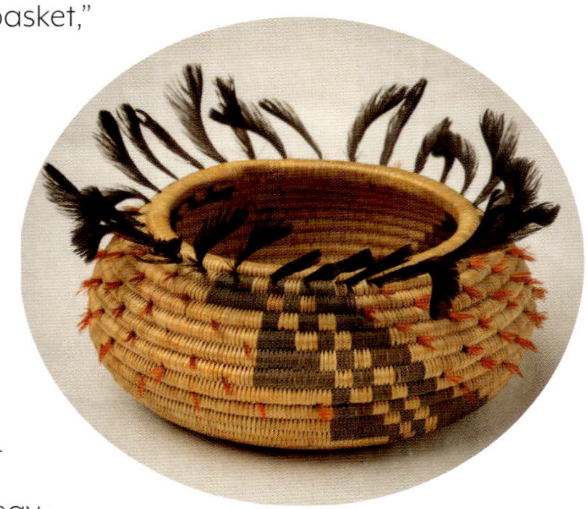

The academy's collections, dating back to 1853, are home to artifacts that reflect complex human relationships with birds, spanning centuries. The baskets here exemplify Pomo weavers' understanding of an interconnected world with birds at its beating heart. This depth of attention to birds isn't unique to Pomo culture (an umbrella which, in itself, includes more than seventy distinct North Bay tribes). In Ohlone ceremonial regalia, feathers from flickers would be stripped down and the quills arranged into bands in exactly the order that they grew on the flicker's wings. In a Coast Miwok origin story, Coyote plucks feathers from Turkey Vulture, Raven, and Crow, and plants them in different parts of the earth. These feathers turn into different communities of Miwok people. Ohlone craftsmen would carve the long, curved bones of herons and egrets into whistles used in ceremony. When winter migratory birds darkened the sky with their numberless wings, Bay Miwok people would bend and tie tule reeds into duck decoys for hunting. In a Rumsen Ohlone story from the Monterey area, Hummingbird brings fire back to the people—but only after scorching his throat, turning it the brilliant crimson that it is today. Creation stories and art, cuisine and ceremony all have evolved alongside birds for thousands of years across scores of Bay Area Indigenous cultures.

But for more than 250 years, colonizers, settlers, and cities have reshaped the Bay Area and recast birds in different light. In the early years, settlers told a new story about our birds, one of untappable abundance. In a cabinet in the academy's bird collections, a box holds a perfectly preserved pair of rope-soled "eggers boots" from the mid-1800s, from San Francisco's early boom years. Residents of the growing city picked the landscape clean of whatever protein they could find, and began to look farther afield for food. In the same year as the Gold Rush, an enterprising pharmacist set sail

A Pomo artisan from the early 1900s wove this basket (above) from willow shoots, bulrush roots, Acorn Woodpecker feathers, and dark California Quail topknot feathers. (Photo by Sharon Beals)

At the northern edge of San Francisco, near a Crissy Field parking lot, a Great Blue Heron (right) placidly defies the rules imposed by urban humans.

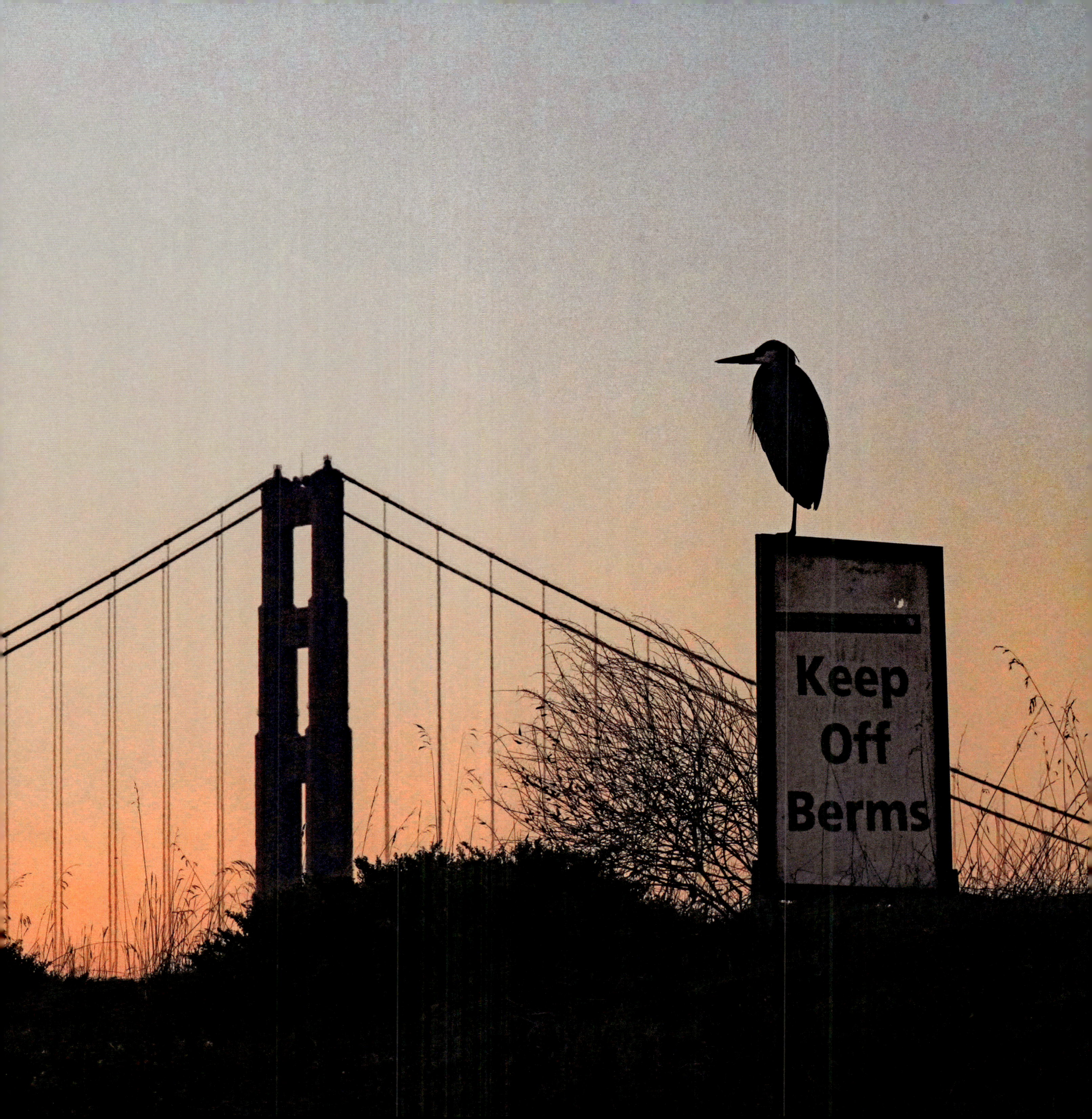
Keep
Off
Berms

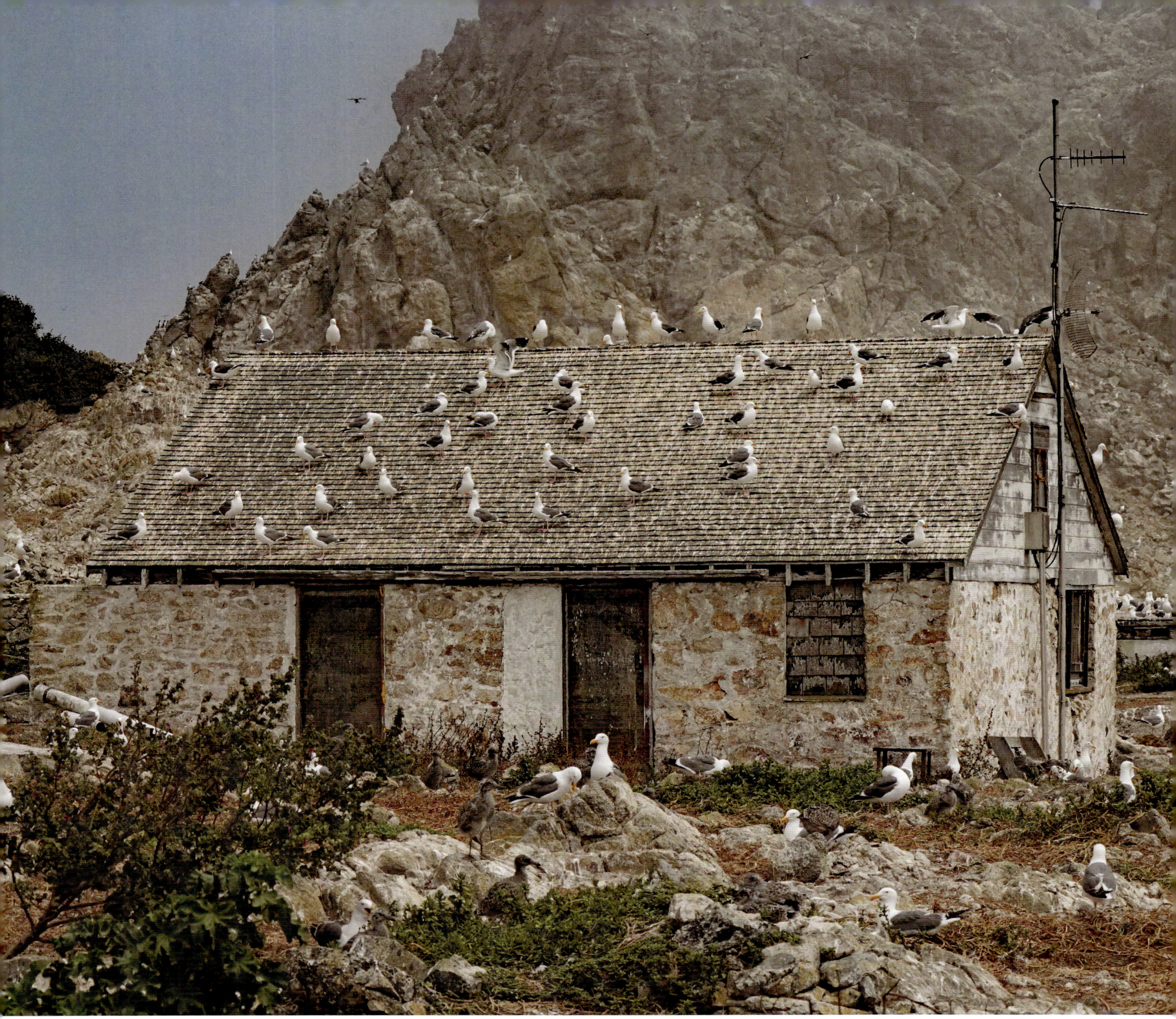

On the Farallon Islands, the old stone carpenter's shop used by lighthouse keepers in the early twentieth century has been annexed by nesting Western Gulls.

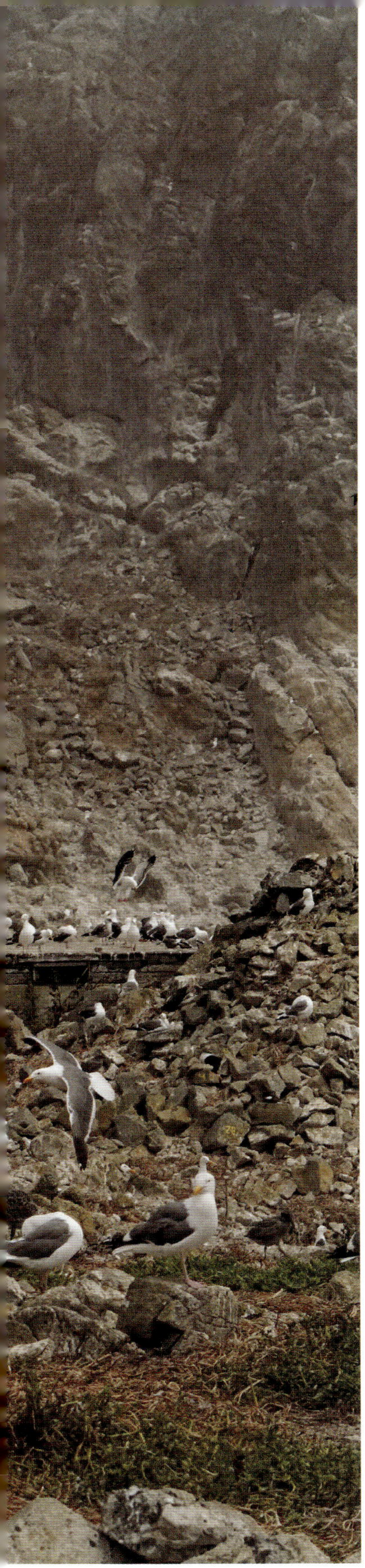

across the turbulent 25-mile crossing to the Farallon Islands, where a different kind of gold awaited: enormous eggs from Common Murres. Men returned to scale the steep cliffs wearing grippy boots and pocket-lined vests for egg collecting. Between the years 1849 and 1881, anyone enjoying a baked good in San Francisco likely dined on pastries containing the rich red yolks from Farallon seabird eggs. The birds nearly disappeared forever. At the academy, the rough-bottomed boots sit alongside a box of preserved murre eggs with shells as blue and mottled as a streambed.

Much has changed since the egging days on the Farallon Islands. Petaluma's chicken industry finally appeased San Francisco's hunger for protein. As the demand for feathered hats decimated egret and heron populations at the turn of the century, the Audubon Society formed West Coast chapters to defend birds and their habitats. The society fought to pass the Migratory Bird Treaty Act, which remains one of the strongest bird protection laws in the world. The Audubon Association—now the Golden Gate Bird Alliance—began guiding monthly birding excursions in 1917: the beginning of a century-long mission to enjoy, study, and protect birds in the Bay Area.

In the last century and a half, urban grids have crisscrossed the oak savannah of the East Bay and San Franciscos dunes. Bridges span the bay over gray currents once crossed only by birds and boats. Now, Peregrine Falcons dive like rockets down the sheer canyons formed by Financial District skyscrapers. City dwellers gather in Crissy Field to watch thousands of migratory shorebirds arrive in the fall and winter. A rowdy flock of red-headed parrots throngs over the Embarcadero and past the steep chapel of Mission Dolores—and they have developed their own distinctive San Francisco dialect! At dawn, birders stroll Golden Gate Park, binoculars tilted, just out of earshot of Fulton Street traffic.

Offshore, the Farallon Islands are now left to the birds. Instead of eggers, only a small, rotating team of biologists from Point Blue Conservation Science live year-round in the old lighthouse keeper's quarters. Curiosity, instead of avarice, guides the management plan. Since 1968, scientists have gathered data on the islands to better understand the web of life in the California Current, from microplastics to murres. It is one of the longest-running collaborations between a government agency and a nonprofit organization. Now, the Farallon Islands National Wildlife Refuge is the largest nesting seabird colony south of Alaska. Sometimes, the islands attract unlikely visitors too. Rare migratory birds—some blown off course by thousands of miles—take refuge on the Farallones, which are welcome crumbs of land in the expanse of the Pacific. On a single day in the spring of 2024, interns on the islands saw a Swallow-tailed Gull, arriving from as far away as the Galapagos, and a Blue Rock

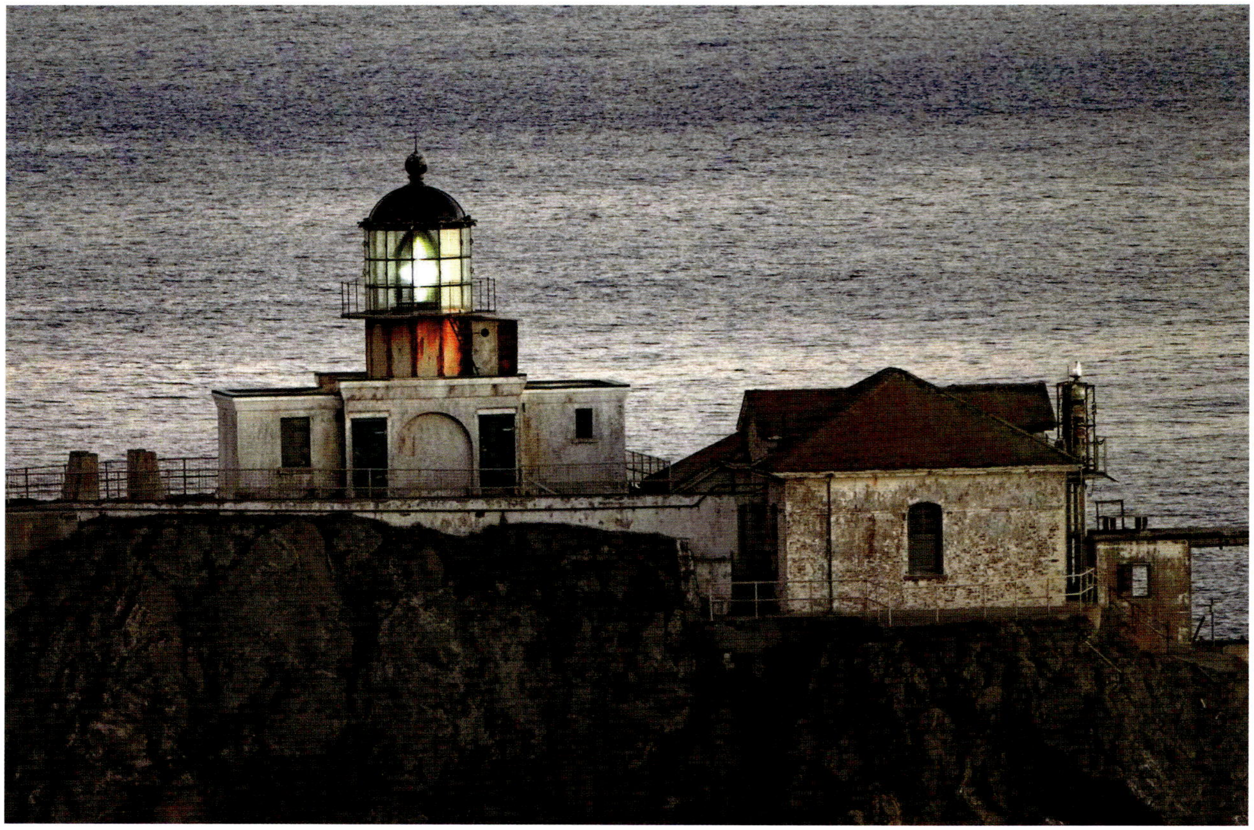

The third lighthouse ever built on the West Coast, the Point Bonita lighthouse has helped shepherd ships—from Farallon Islands egg-hunting boats to NOAA research vessels—through the perilous Golden Gate straits since 1855.

Thrush from Southeast Asia, which had only been sighted once before in the United States. "A once-in-a-century event," they wrote in their report.

"Vagrants" like the Swallow-tailed Gull and the Blue Rock Thrush are at the heart of the Bay Area's conservation story. In the early 1960s, members of the Western Bird Banding Association would trek to the outer coast of the Point Reyes Peninsula, looking for rare birds. Here, disoriented East Coast birds off route from their migrations turned back from the open Pacific Ocean and sought refuge. Much like the Farallon Islands, the long 10-mile hook of the peninsula jutting out into the sea was a rare, lifesaving perch surrounded by inhospitable waters. These vagrants—Yellow-bellied Flycatchers, Northern Parulas, Ovenbirds—captivated bird-lovers. It was a treat to find them so far from their range, especially alongside the rich populations of local and migratory birds more regularly found in Point Reyes.

A young seasonal ranger at the newly designated Point Reyes National Seashore spent the summer of 1963 roaming the peninsula, banding birds, and compiling a checklist of the species he found. That fall, he brought the list to a meeting with the Western Bird Banding Association, with the idea of founding an observatory on the coast. It felt urgent—in part because the dazzling diversity of birds felt rare and worthy of witness; in part because the birders knew how vulnerable that diversity was. In 1962, Rachel Carson had published *Silent Spring*, bringing to light the ecosystem-wide devastation caused by pollutants. "The birds," she wrote, "where had they gone? . . . It was a spring without voices."

That band of birders at the meeting scraped together a small budget and, in 1965, founded the Point Reyes Bird Observatory—now Point Blue Conservation Science—with the intent of understanding where the Bay Area's migratory birds were coming from and to track populations so that we all could better grasp fluctuations and figure out their origins. They established a field station at the southern end of the national seashore and made it their work to pay close attention. Among the founders of the organization was a young man named Rich Stallcup, who became the first resident bird bander at what is now the Palomarin Field Station. He spent the whole of his life working to understand the birds of the region. "He had the soul of a poet, the mind of a scientist, and the spirituality of a shaman," wrote biologist John Winter when Stallcup finally succumbed to leukemia in 2012. "He crawled around inside the mind of a bird and led you to a higher purpose through birding; an understanding of *your spirit*." Stallcup's birding philosophy influenced Point Blue's identity as the organization continued to grow, drawing in specialists and honing rigorous standards for how to collect consistent data across decades. What began as a ragtag community scouting for rare birds at the edge of the continent has now grown into a global information network, perhaps the largest ecological database in the world, with its hub right here in the Bay Area.

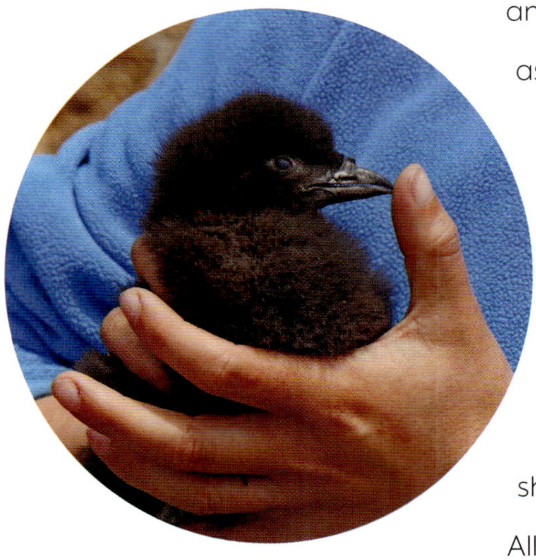

A puffy Rhinoceros Auklet chick. These birds disappeared from the Farallon Islands in the 1860s and reappeared in the 1970s, nearly 100 years later.

Beyond the science of conservation, opportunities continue to expand for locals of all kinds to find new ways of relating to birds that reweave them into our daily lives. And they bring us together too. Dozens of Bay Area birding groups gather at city greenbelts, at water treatment plants, and at beaches under the shadow of the Golden Gate Bridge in search of shared birding experiences. More than a hundred years after its inception, the Golden Gate Bird Alliance still leads walks and classes around the Bay. Every fall, the Central Valley Bird Club hosts a symposium for hundreds of bird-lovers, with guided field excursions in the California Delta, featured speakers, and illustration lessons. Each year, more and more young people show up to the event. "The symposium is as much about people as it is about our beloved birds," the club's website reads.

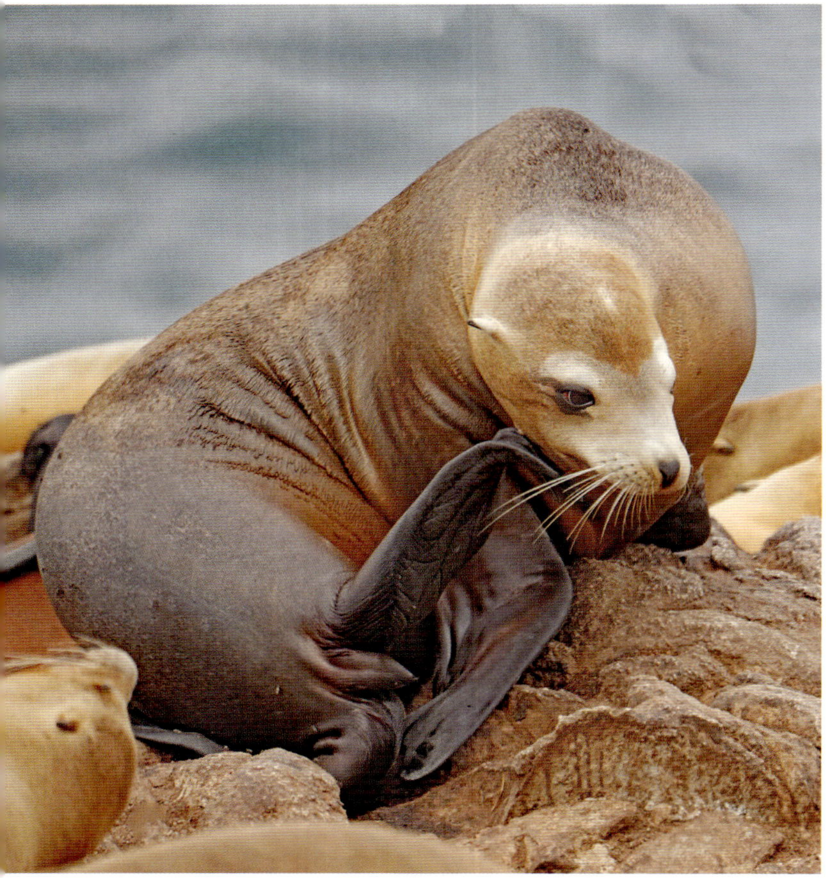

At festivals, in clubs, in Audubon chapters, and at annual bird counts across the broader Bay Area, learners of all ages come together to share what they know, to learn new ways of understanding what they see, and to look closely at the avian world together, in community with one another.

Each year, the Wild Wonder Foundation hosts an annual nature journaling conference, the largest international event of its kind. Participants bring notebooks, pastels, watercolors, scrapbooking supplies, and other tools of observation. Together, they share skills: how to capture complex textures, how to visualize data, how to illustrate a cross section of a marsh to accurately map what it looks like both above and below water. "Attention, observation, curiosity, and creative thinking are not gifts," writes Wild Wonder Foundation cofounder John Muir Laws, "but skills that grow with training and deliberate pratice." This movement harkens back to some of California's earliest interrelationships with birds: To understand ecosystems through slow and attentive working of pencils and brushes on paper is perhaps not so different from the art of Pomo basket weaving with willow shoots and feathers.

The Farallon Islands aren't just "for the birds"—as Northern Fur Seals recover from fur hunting, they compete for space with nesting seabirds, alongside California Sea Lions (above and below). The islands are a cacophony of avian squawks and the groans and rumbles of plump pinnipeds.

Even on the remote Farallon Islands, human influences interweave with wildness. Western Gulls (right) bring illicit treats home with them across long distances: the islands are littered with spareribs and chicken wings purloined from mainland dumps and restaurants.

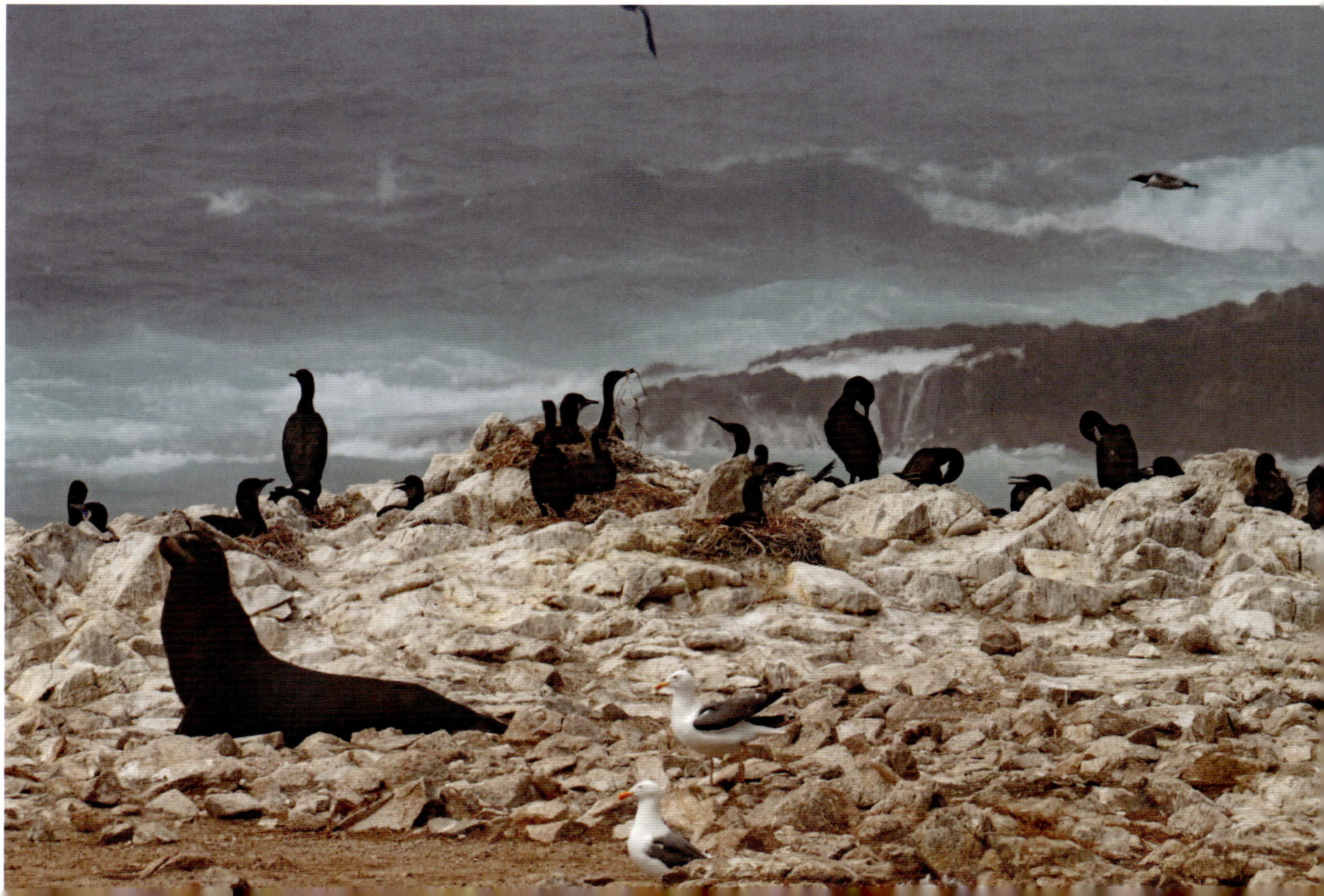

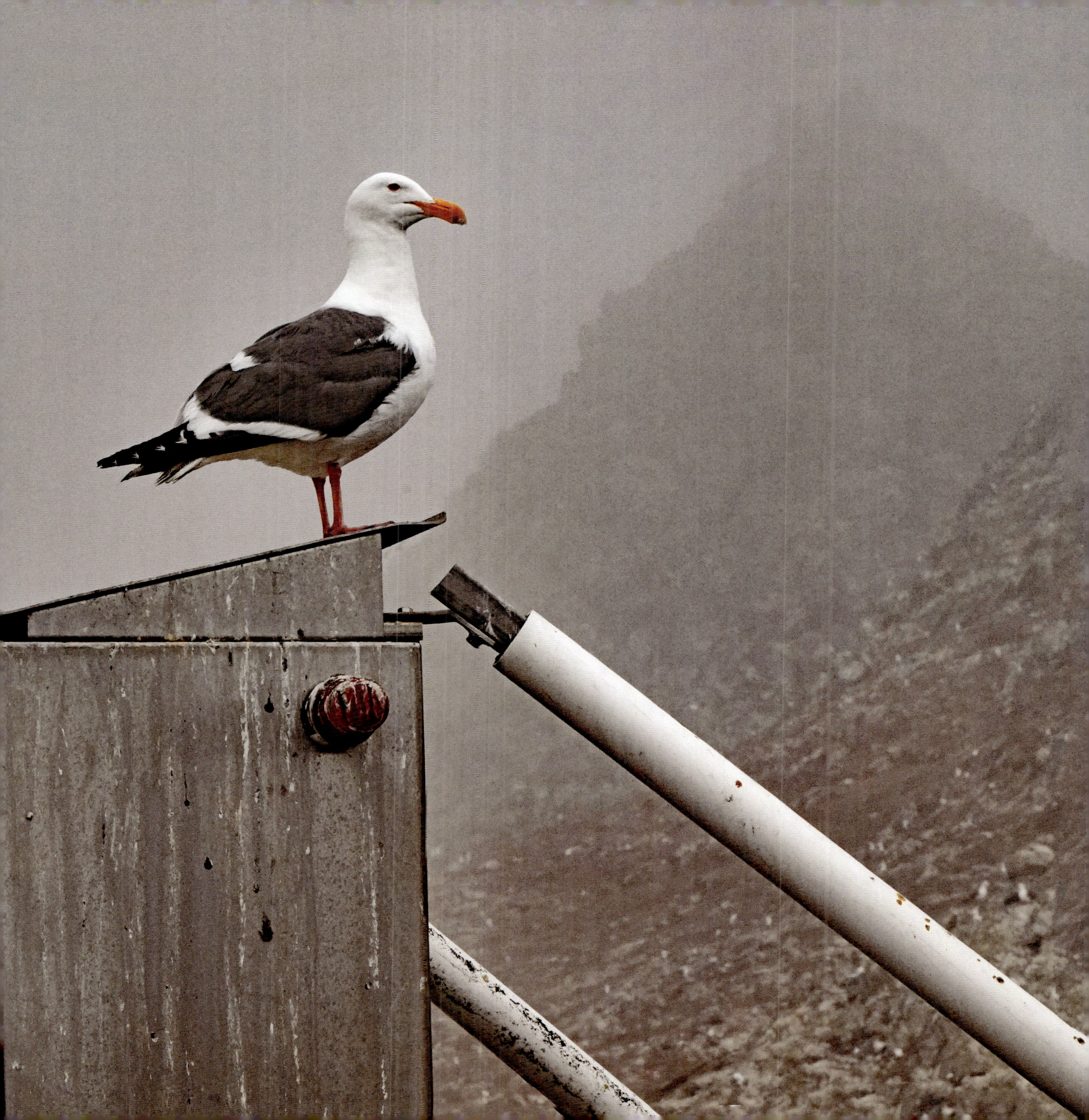

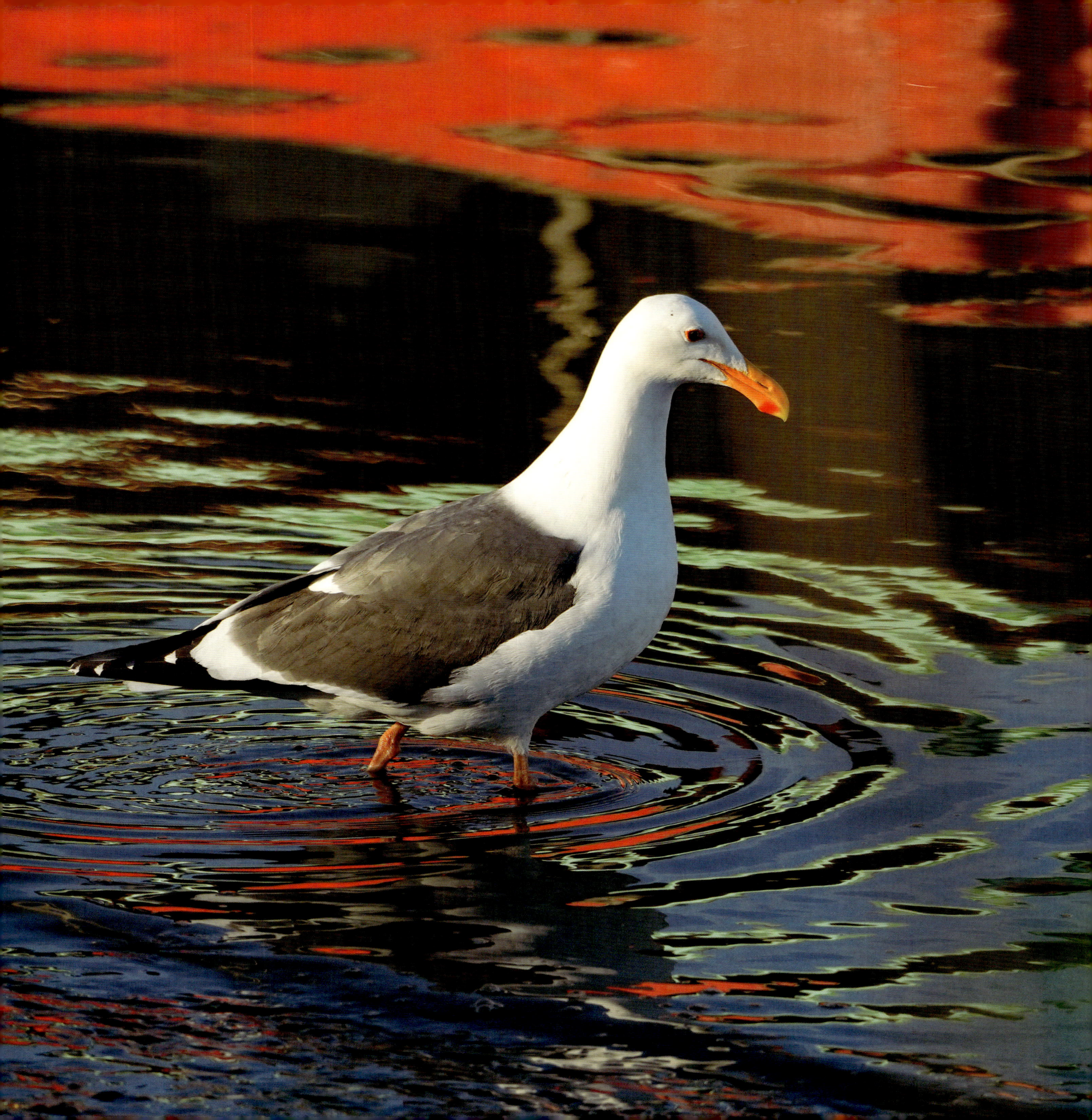

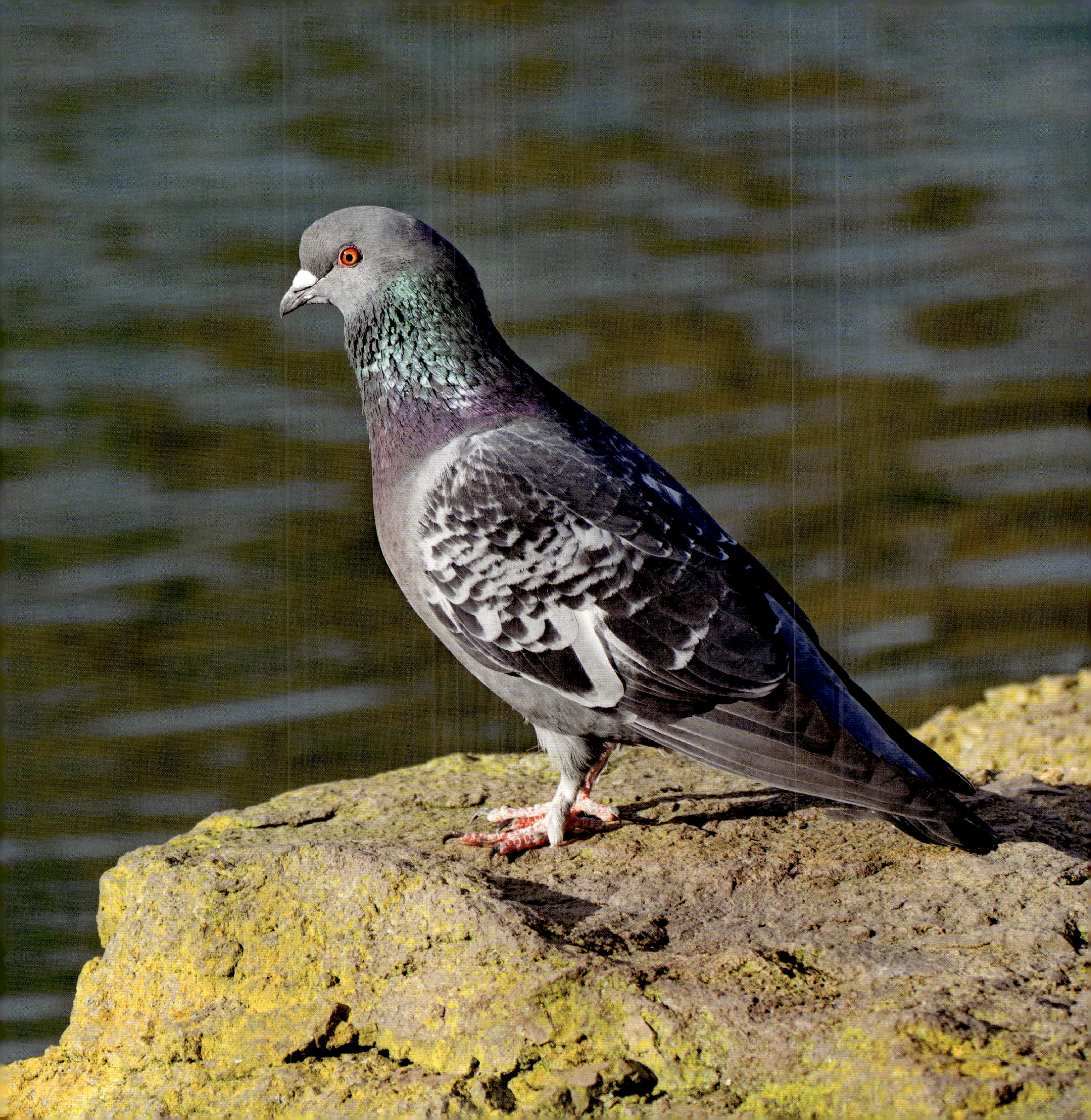

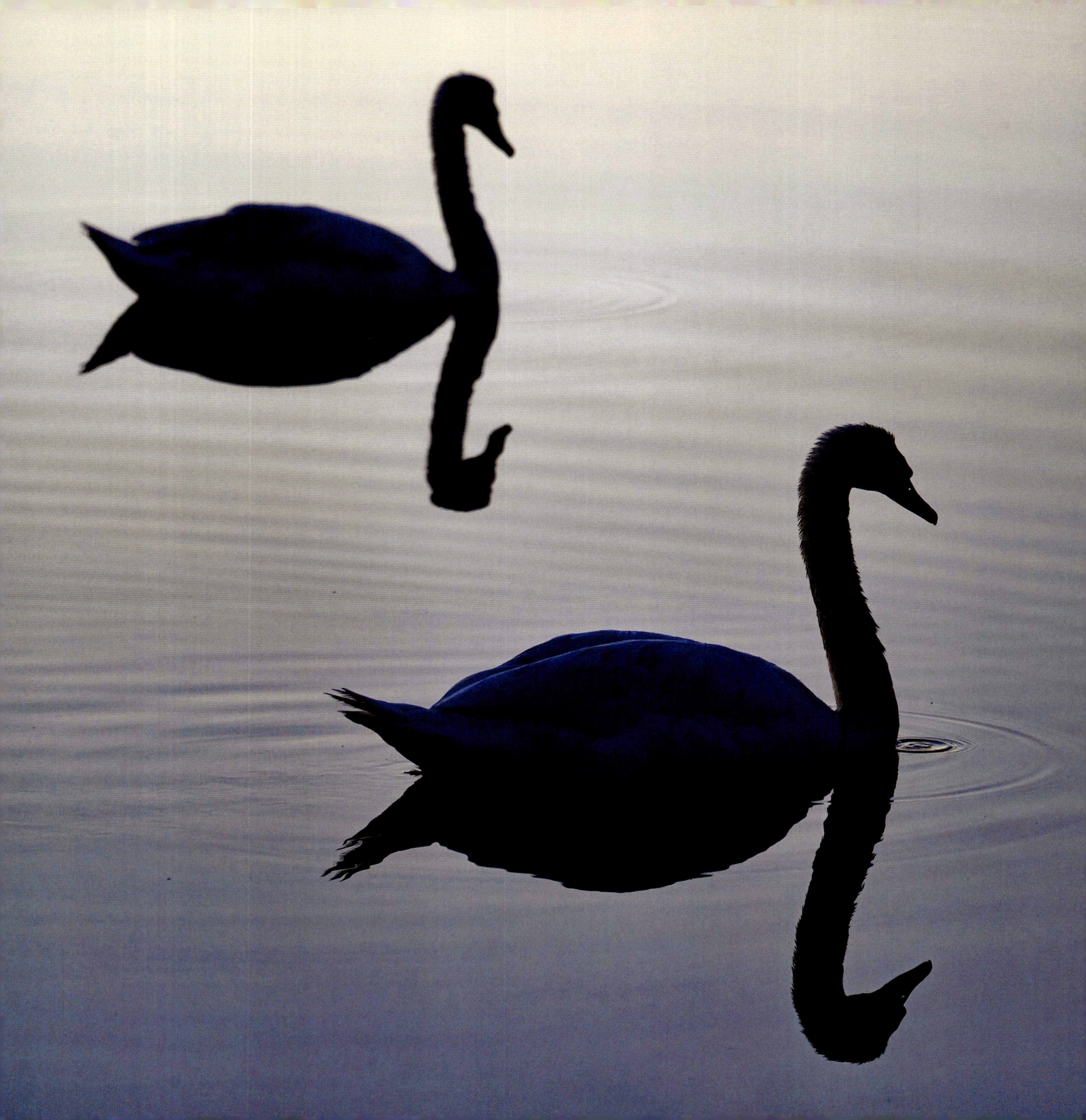

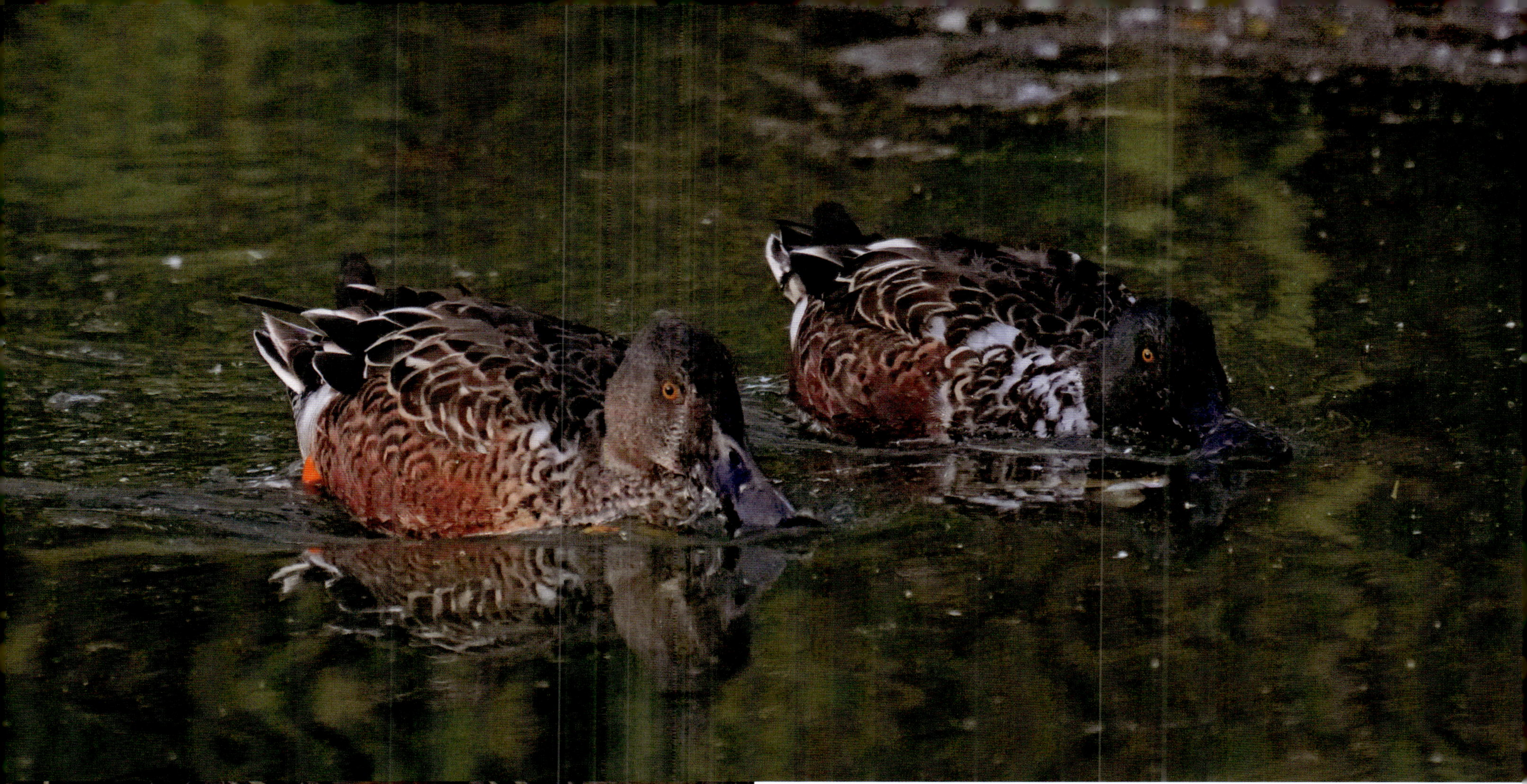

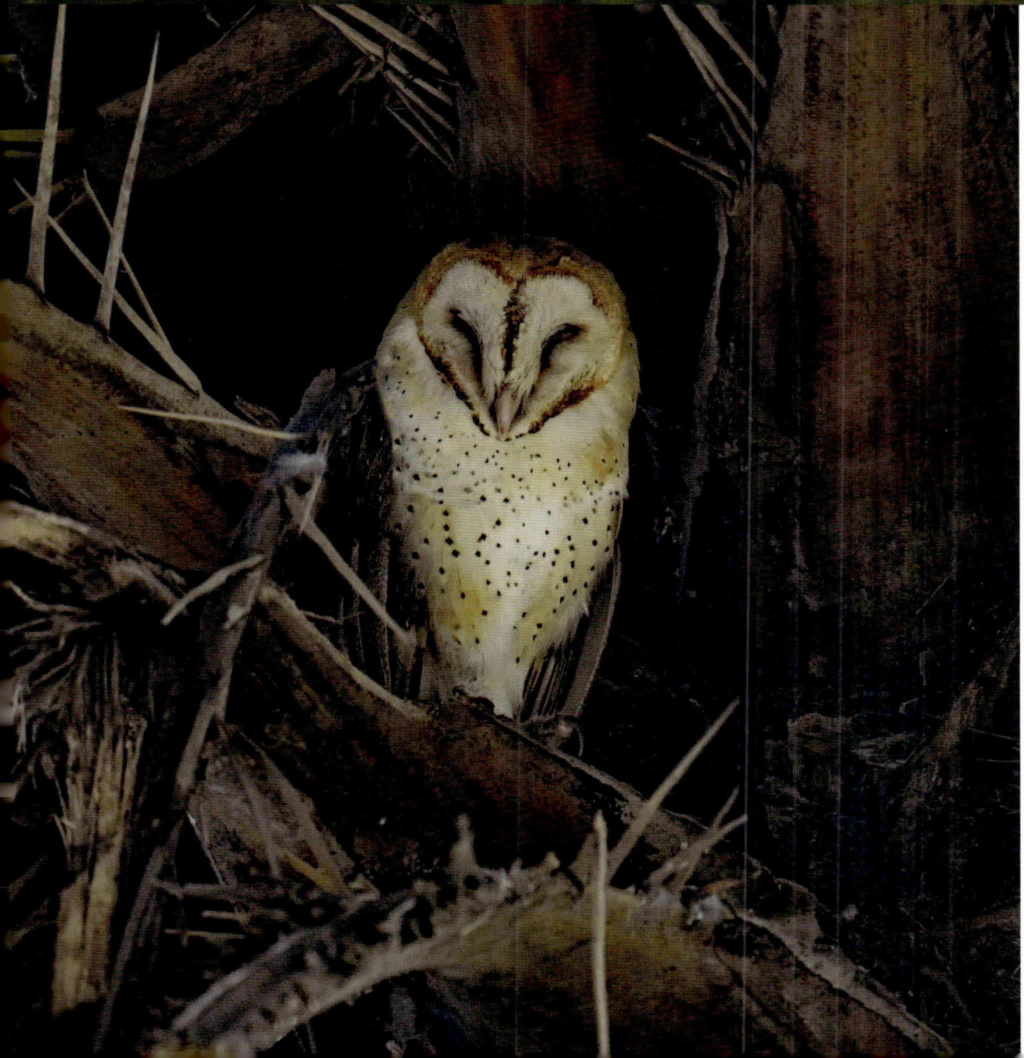

PRECEDING: Avian and human worlds meet where a Western Gull (left) wades in the reflection of a Sausalito fishing boat. The Rock Dove (right) is the wild ancestor of all domestic pigeons. Their strong sense of direction has endeared them to humans throughout history—with their remarkable homing capabilities, they have been used for communication in times of both peace and war. In 1933, during the construction of the Golden Gate Bridge, the San Francisco Racing Pigeon Club provided 250 carrier pigeons to broadcast messages of the groundbreaking ceremony all across California.

ABOVE: Over the course of more than 70 years, Las Gallinas Water Treatment Plant at the edge of San Rafael has evolved from a single-purpose water treatment facility into a human-created hub of environmental stewardship that thousands of visitors explore by foot and by bike every year. The plant maintains a freshwater marsh, irrigated pastures, storage ponds, and a saltwater marsh, which house hundreds of resident and migratory bird species, such as these amber-eyed Northern Shovelers (top) and Mute Swans (facing page).

LEFT: Forty miles north of San Francisco, Point Reyes National Seashore nearly became a patchwork of private subdivisions, highways, and golf courses in the late 1960s. Dedicated local activists steered conservation history and pressed President Nixon to sign a funding bill to ensure the seashore remained wild and accessible to the public and to over 450 bird species like this barn owl.

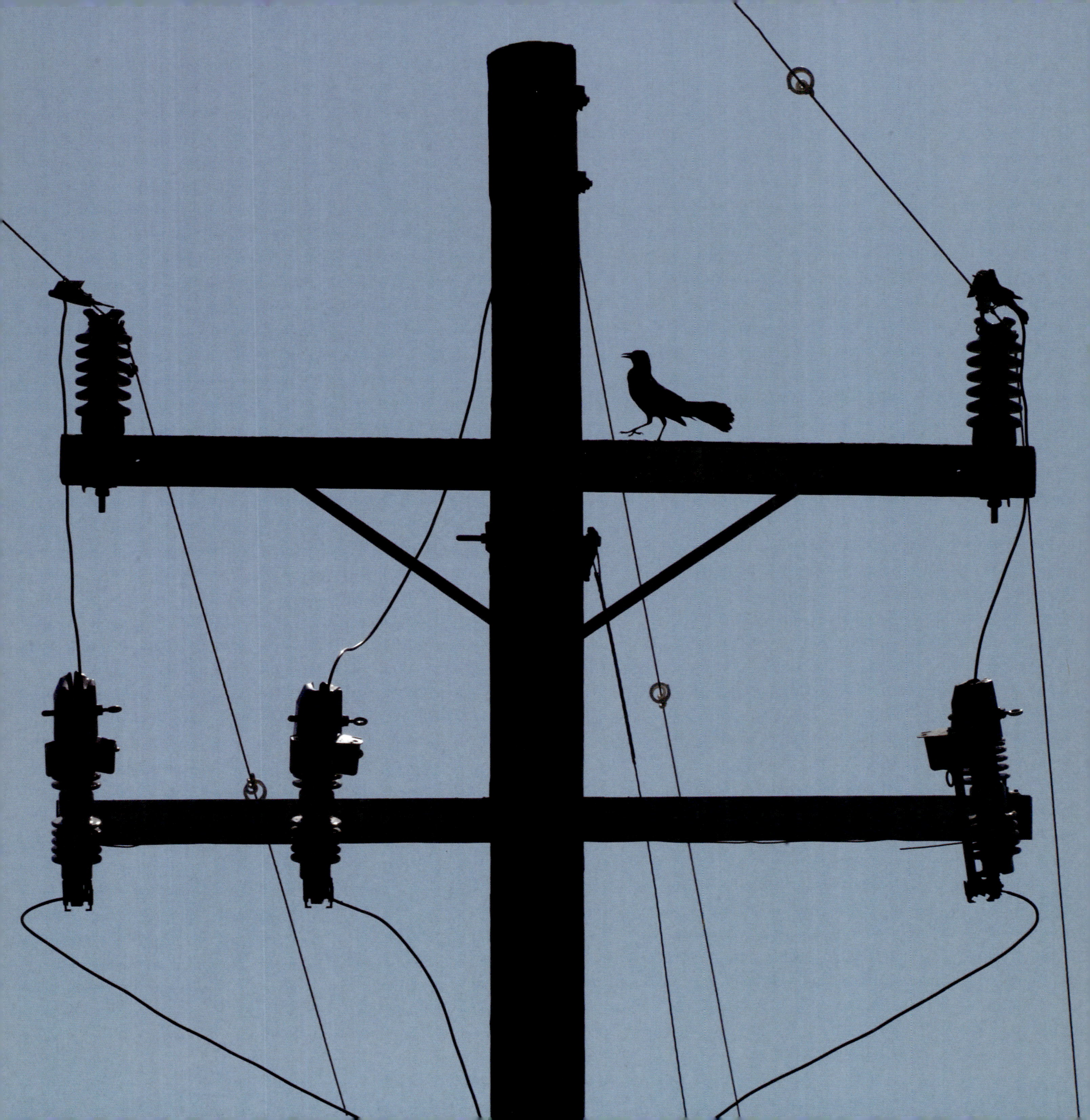

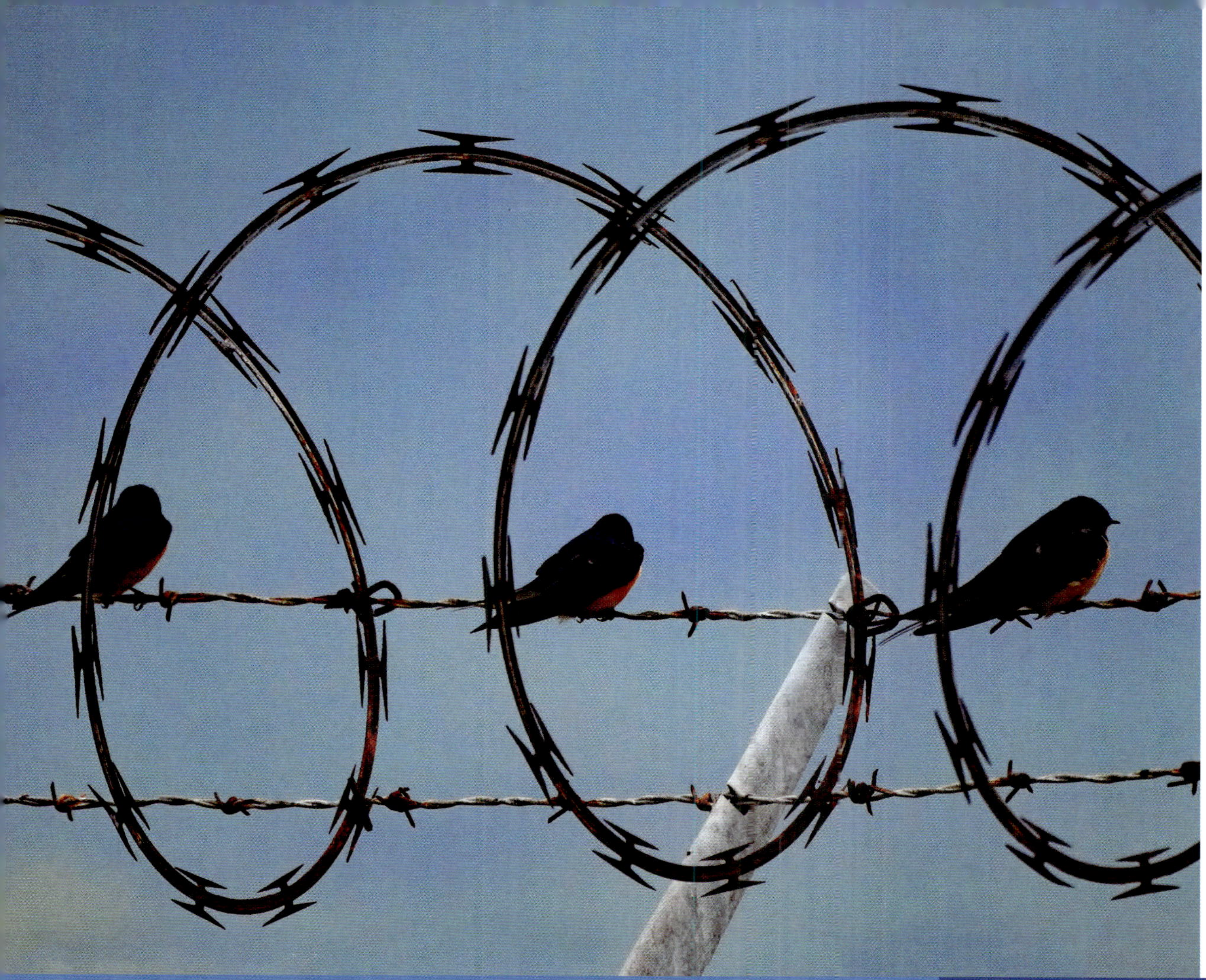

Birds on a wire! Feathered loiterers—a Great-tailed Grackle, Barn Swallows, a Common Raven, and a European Starling (clockwise from left)—contrast with the angular geometries of Bay Area power lines and fences.

FOLLOWING: Most commonly seen as a spring and fall migrant in the Bay Area, the Say's Phoebe (left) frequents open fields and ranchlands. The Lesser Goldfinch (right) is often attracted to farmlands and weedy areas with seeds. A Lesser Goldfinch captured on a ranch in California holds the record for the oldest known Lesser Goldfinch, at a record-breaking seven years!

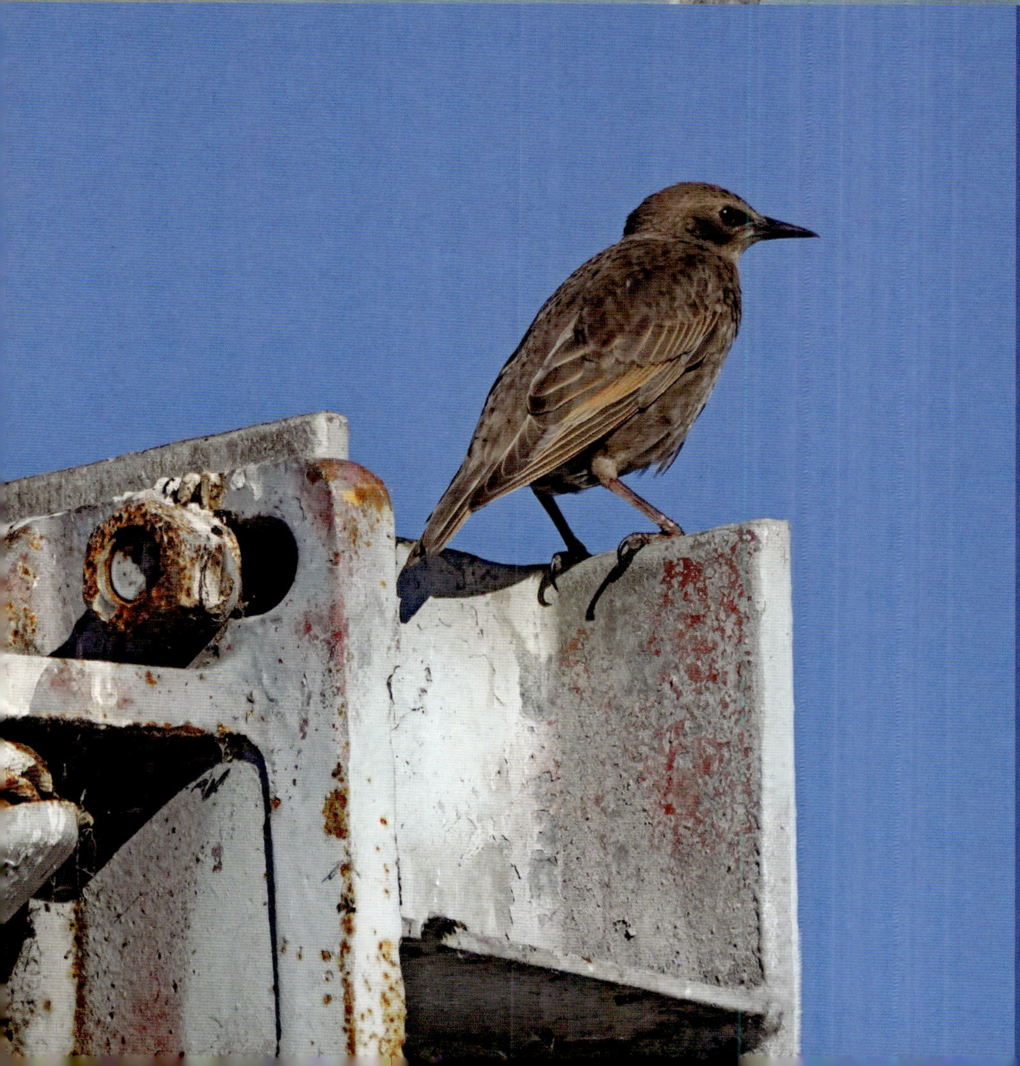

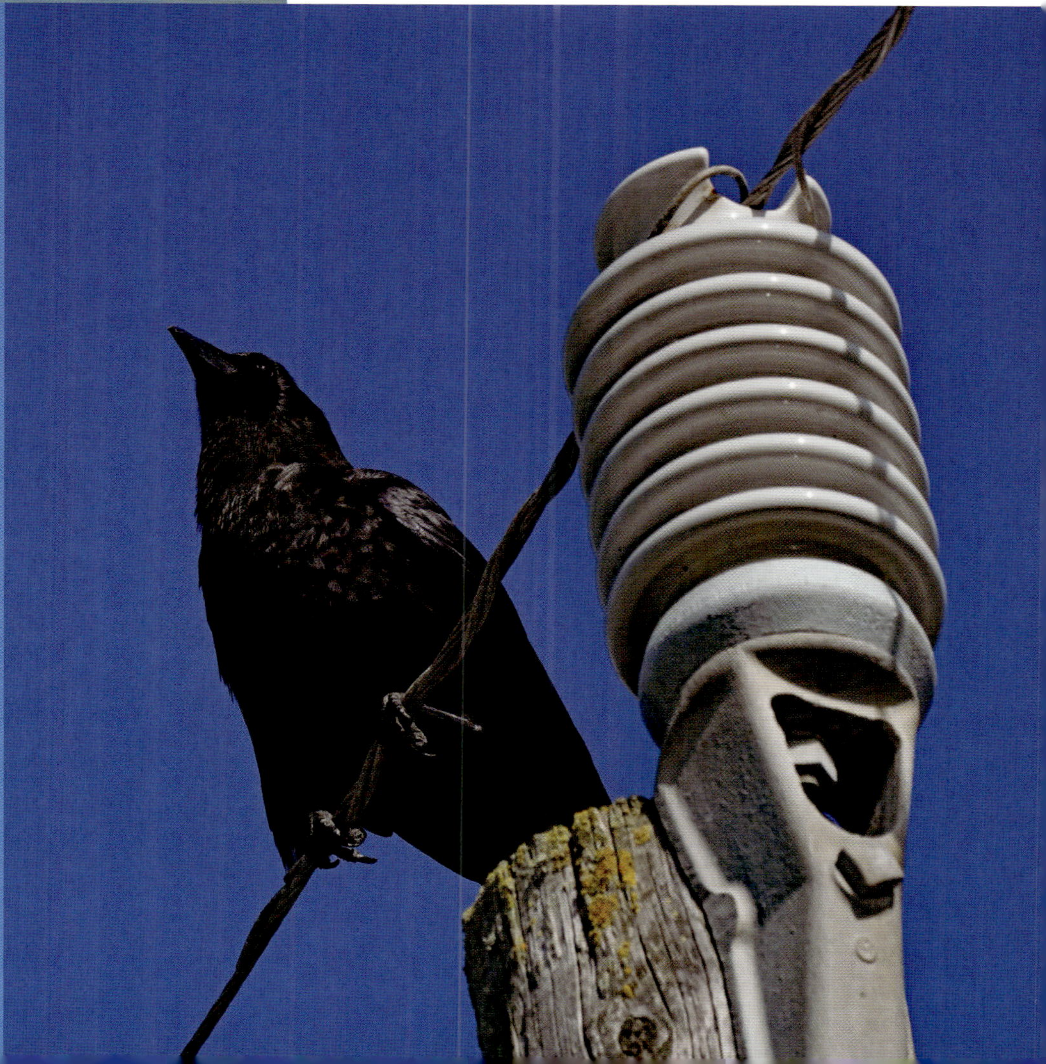

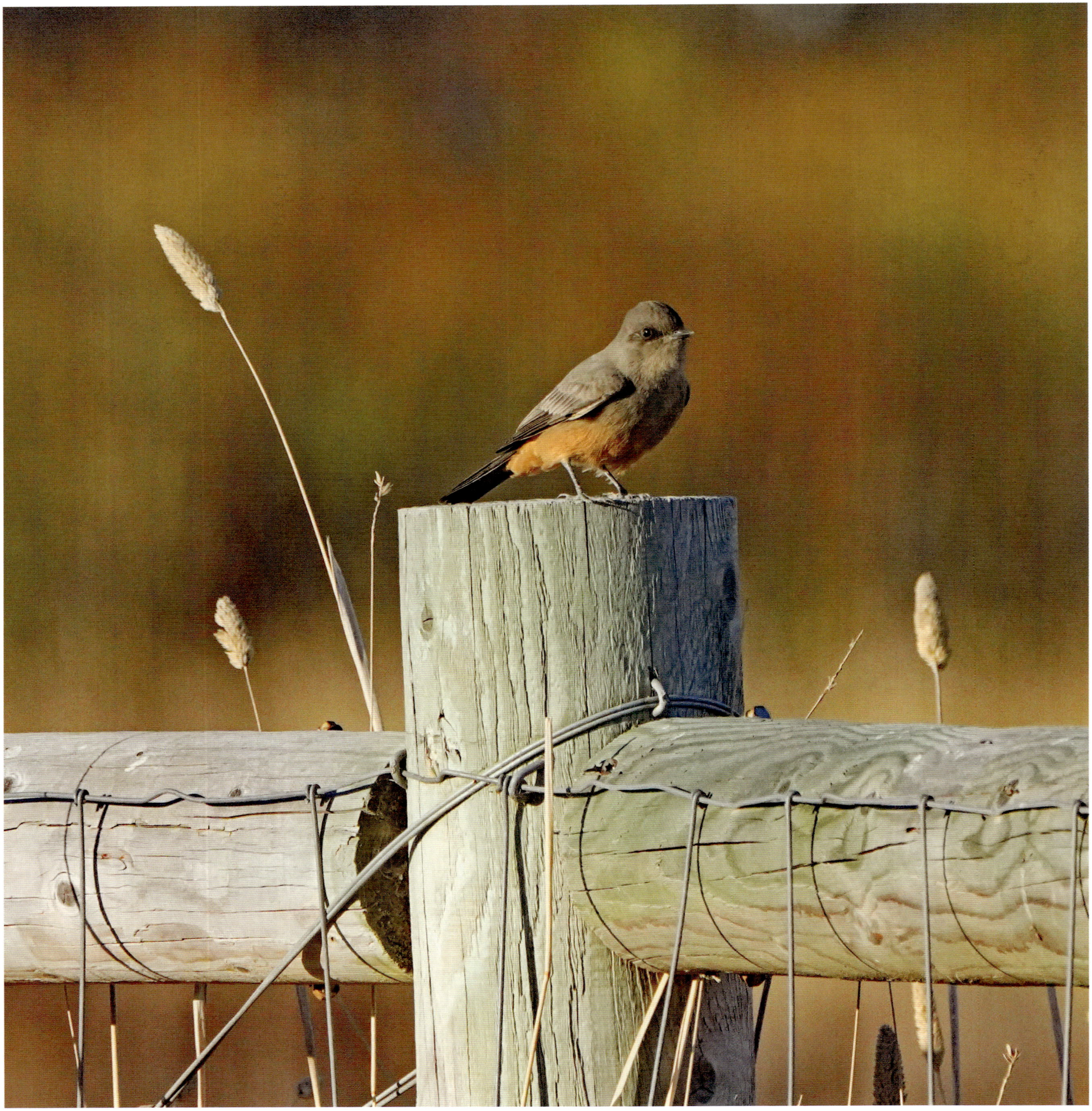

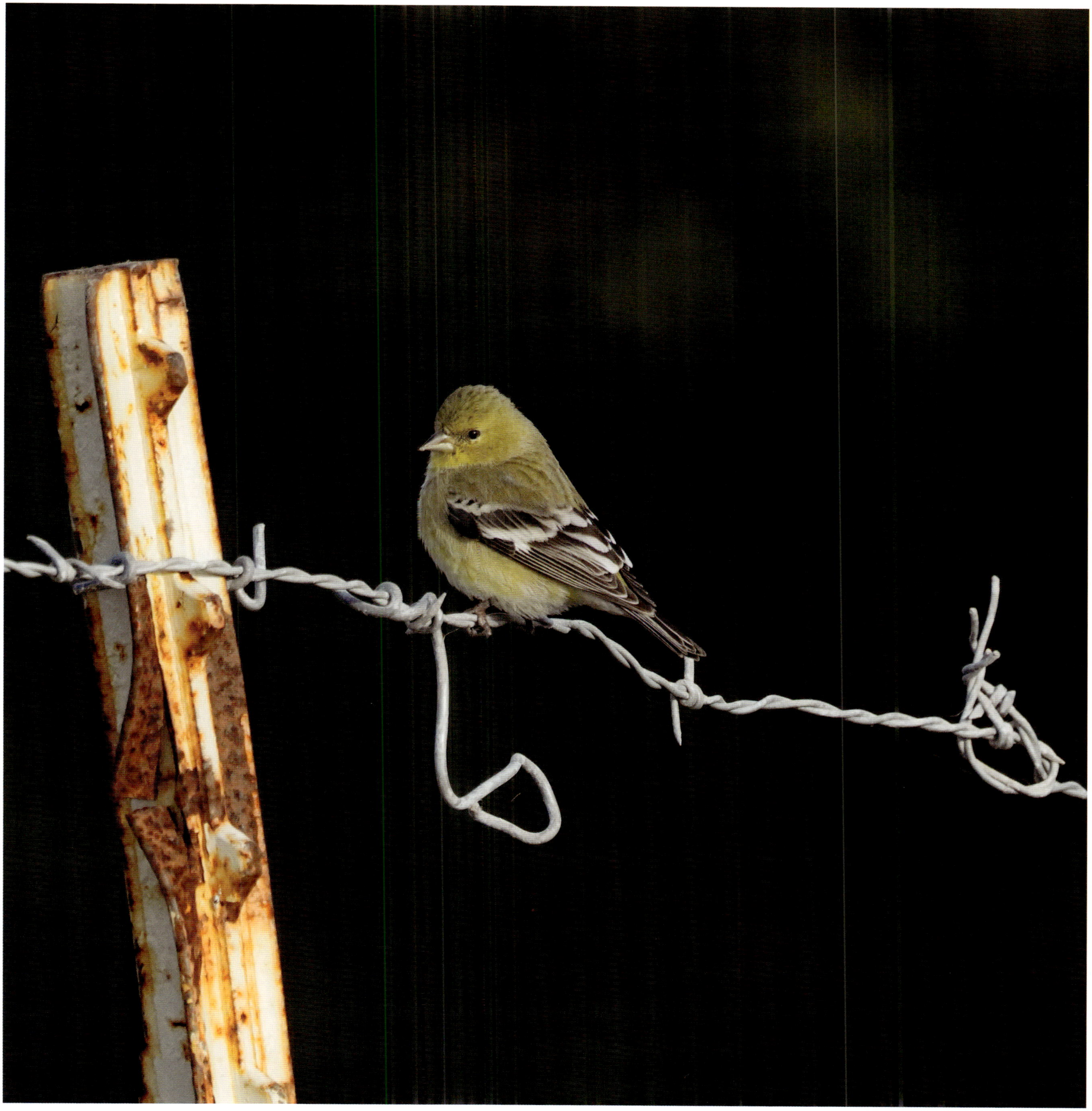

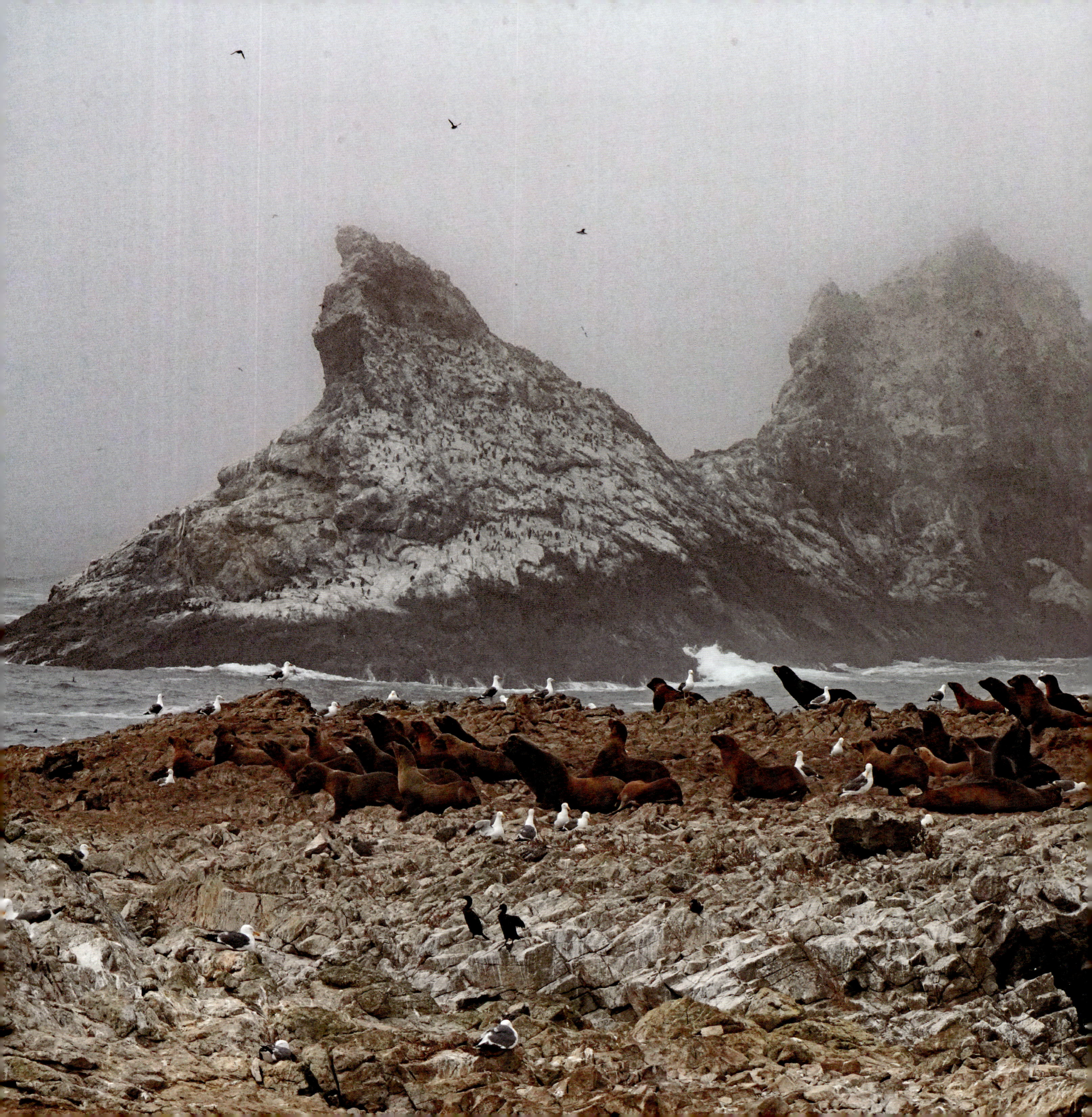

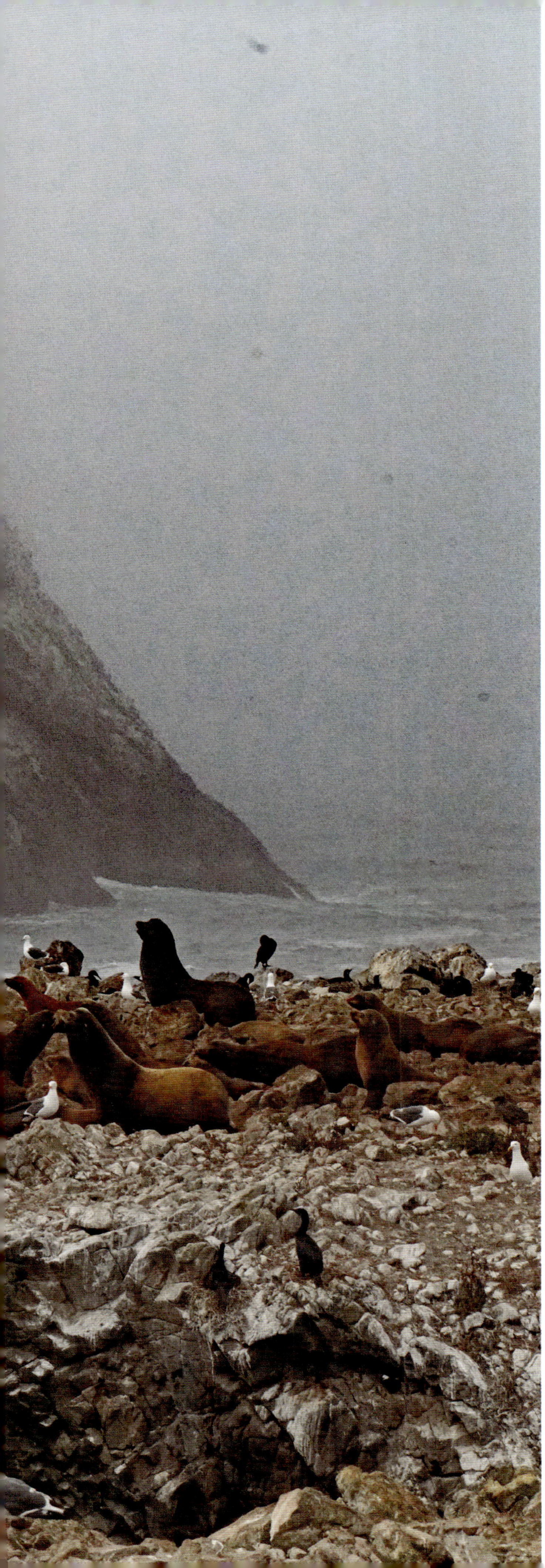

THE DEVIL'S TEETH

The *Salty Lady* tilts atop big oceanic swells, then pitches down each trough like a marble on a run. A Pink-footed Shearwater skims low across the rolling surface. A Tufted Puffin whirs ahead of the boat, feet splayed behind. "Five minutes out," our captain tells us over the loudspeaker. But ahead, there is only ocean, and a wall of fog. Gradually, a white line of surf materializes, where the sea breaks against unseen land. Then, out of the fog, near and sudden, rise the sharp cliffs of the Farallon Islands: the Devil's Teeth.

A similar scene likely greeted Doc Robinson when he ventured 25 miles offshore from San Francisco in 1849. He filled his boat with the Common Murre eggs that he found there, and the fortune he made selling the eggs from that single run was enough to start his own pharmacy in the city. Soon, the Pacific Egg Company claimed exclusive rights to collecting Farallon eggs. The company's eggers donned boots with rope soles for gripping the steep cliffs. They smashed every egg they could find—mottled green shells crushed among crimson yolks—so that when they returned they could ensure that the eggs they collected were fresh. In this way, they plundered half a million eggs a year, paring the murre population from 400,000 to just 6,000 birds. When competitors arrived to challenge the Pacific Egg Company's lucrative claim on the island, both sides took up arms. The Egg War of 1863 left four men injured and one dead, shot through the stomach.

Now, the island resounds not with gunfire but with the shouting of birds, and the occasional whine of a crane as it lifts the small boat that runs between the island and the larger resupply boat. This is how we arrive: raised in a skiff like a basket on a pulley. There are no docks on the sea-scoured Farallones. Ashore, we follow a narrow path toward the

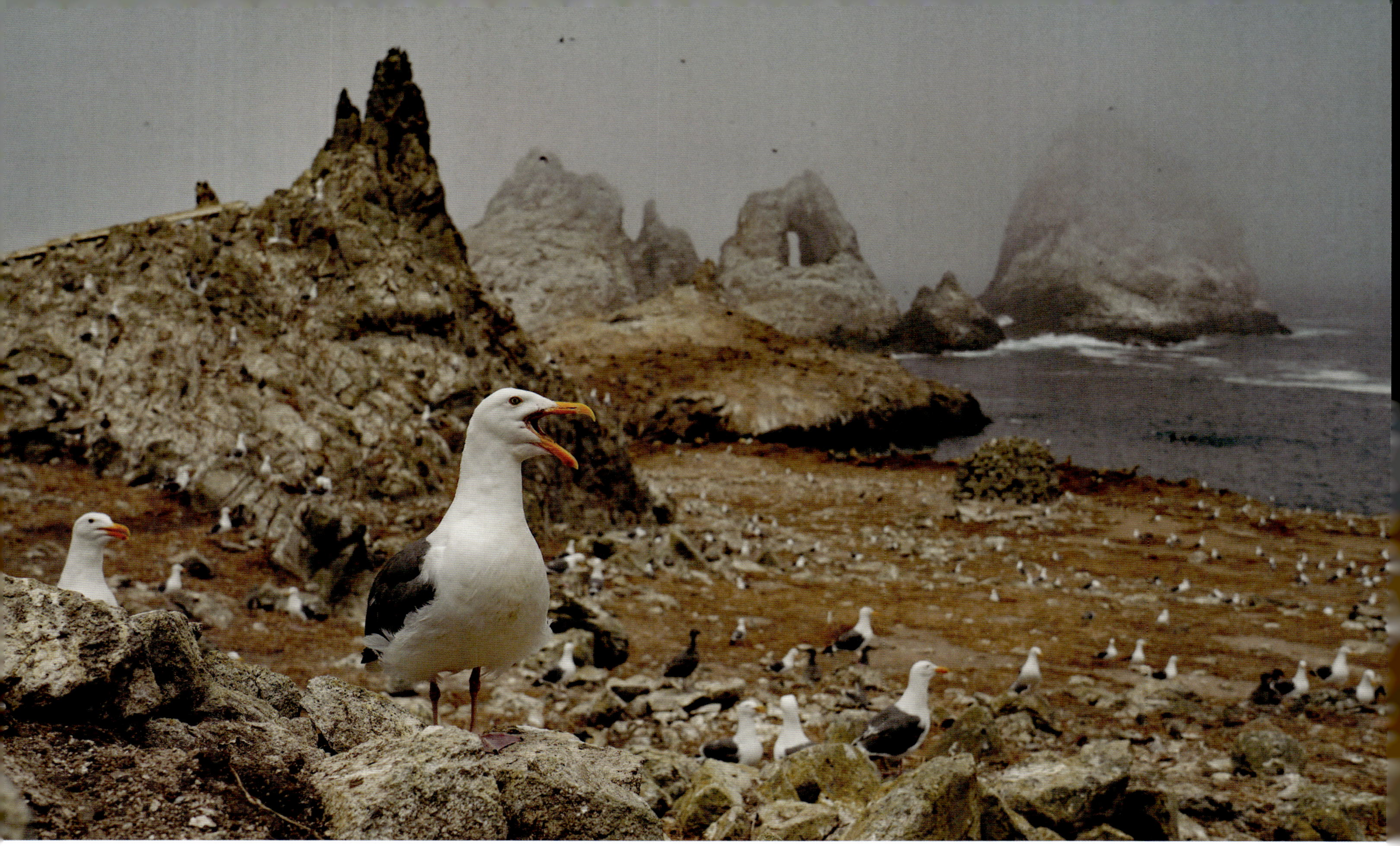

old lighthouse keeper's buildings—the only ribbon of land not thronging with birds. Twenty-five percent of California's breeding seabirds—a quarter million of them, recovered from the egging years—nest in this Martian landscape roughly one hundred football fields in size, and it sounds like it. Western Gulls shriek and crane their necks alongside gangly half-grown chicks. Common Murres cascade down the cliffs by the thousands.

In a kind of restitution for centuries-past wrongs, now when puffy Cassin's Auklet chicks hunker underground, they keep cool in ceramic nests that were commissioned by Point Blue in response to a half-century of climate data gathered on the islands. Temperatures have begun to reach record highs that can kill the birds in their nests. At night, Point Blue research assistants catch auklet parents in mist nets as the birds return from oceanic hunting expeditions in the California Current.

Alongside the rotating cast of apprentices, who have continuously monitored the Farallones for over 50 years, the birds themselves are playing an active part in data collection. When netted, the birds drop the fish they carry, and these—along with feces and other samples—are sent for analysis at a mainland lab. With nearly 60 years' worth of fish specimens, researchers have been able to go back and look at changing conditions in the California Current, and to find answers to questions that they wouldn't have known to ask when they first established the remote research station. For example, the term *microplastics* wasn't coined until 2004. Although the proliferation of plastics in the ocean and in our food might seem like a "now" story, the data shows a longer, more troubling truth: Microplastics have been present in the bellies of Farallon animals since the very first sample was taken in 1968.

These clues that the birds send back from the edge of the continental shelf have helped shape conservation policy, including a 1994 ban on gillnetting and the establishment of a Marine Protected Area around the islands in 2010.

Amanda Spears, Farallon biologist, tells us that after living for a season on the Farallones, scientists who leave the islands continue to dream of them for years. And, as we leave, it does all seem a bit like a waking dream: riding the skiff back to the boat, watching a skein of fog close around the cliffs until just the sound of the birds remains—as it may have sounded two centuries ago, before San Francisco got hungry.

PRECEDING: The Farallon Islands are sometimes referred to as the "Serengeti of the Sea"—the area is considered one of the most productive and biodiverse marine ecosystems on the planet.

OPPOSITE PAGE: Nesting Western Gulls fill the craggy landscape as far as the eye can see on the Farallon Islands on a foggy summer day.

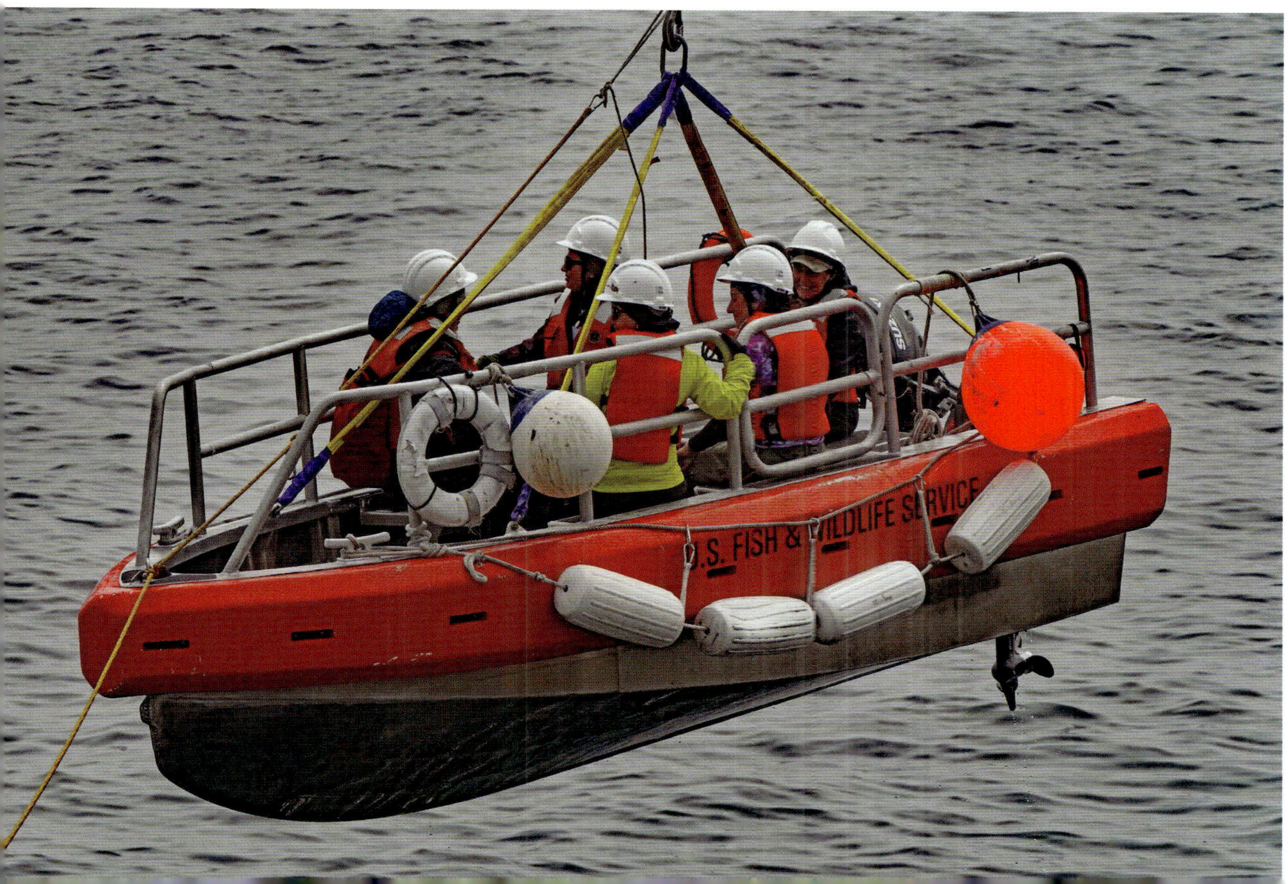

LEFT: In a place as sea-torn as the Farallon Islands, resident biologists rely on this crane as their only method for landing. Small boats are hauled up atop the cliff, and passengers disembark on high ground. For more than 50 years, the infrastructure to operate the crane has been supported by a remarkable public/private partnership between Point Blue and the Department of Fish and Wildlife, although the future of funding for this long-term monitoring site is tenuous.

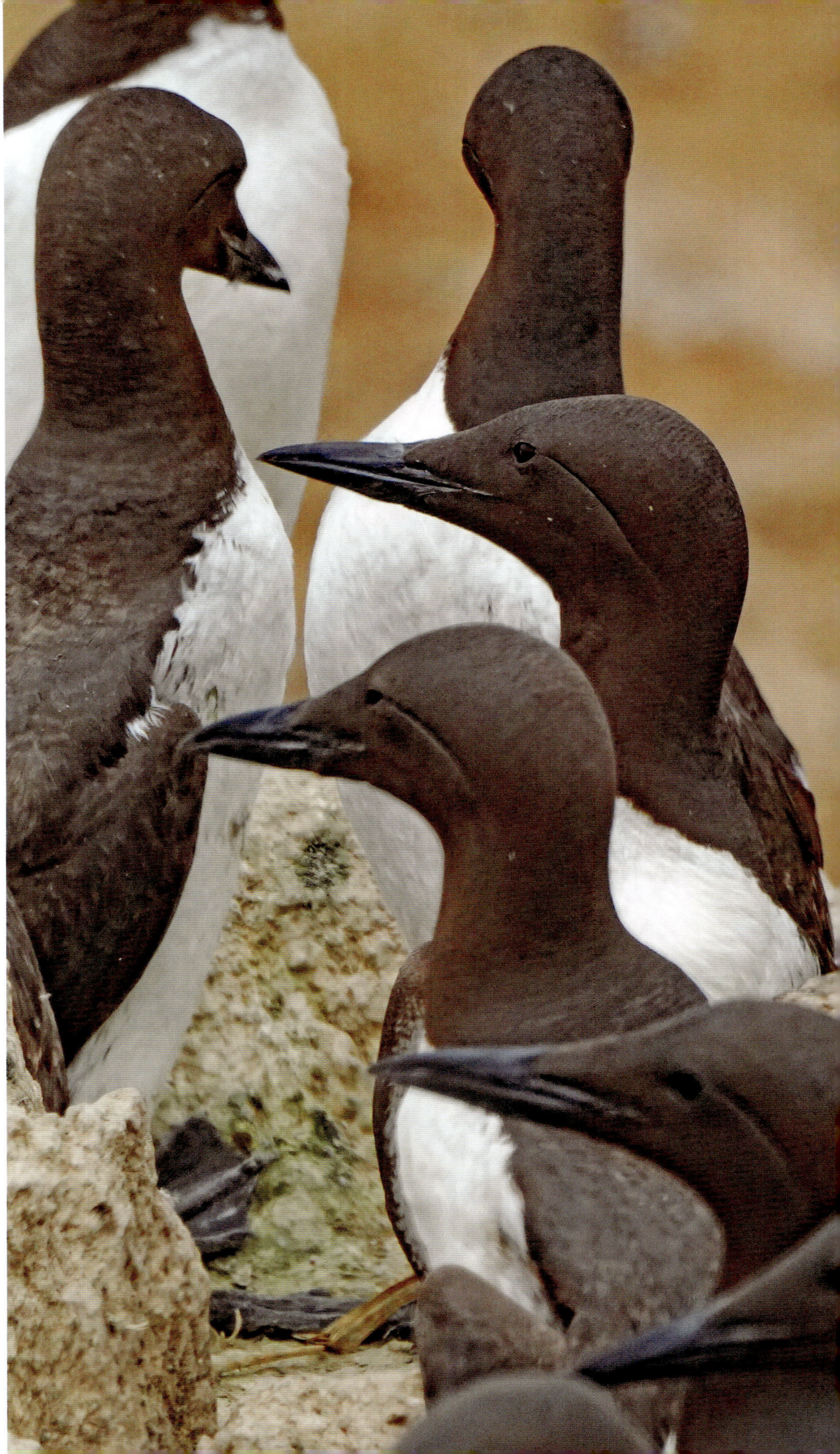

Open wide! Common Murres have long, sharp-edge tongues that they use to press fish against bumps called "denticles" on the roof of their mouths, helping them secure slippery prey until they can swallow it whole.

FOLLOWING: Both male and female Pigeon Guillemots (left) have striking red feet. During courtship, the male bird will prance in a circle around his potential mate, showing off his vivid feet. These little diving birds form lasting pair bonds, nesting in colonies on rocky shores like the Farallon Islands. Their breeding success can be highly variable depending on marine conditions. Black Oystercatchers (right) are true liminal animals, stalking the rocky shoreline between tides in search of mussels and mollusks. The sex of a Black Oystercatcher can be determined by looking at the patterns in their eyes—females have flecks scattered throughout their irises.

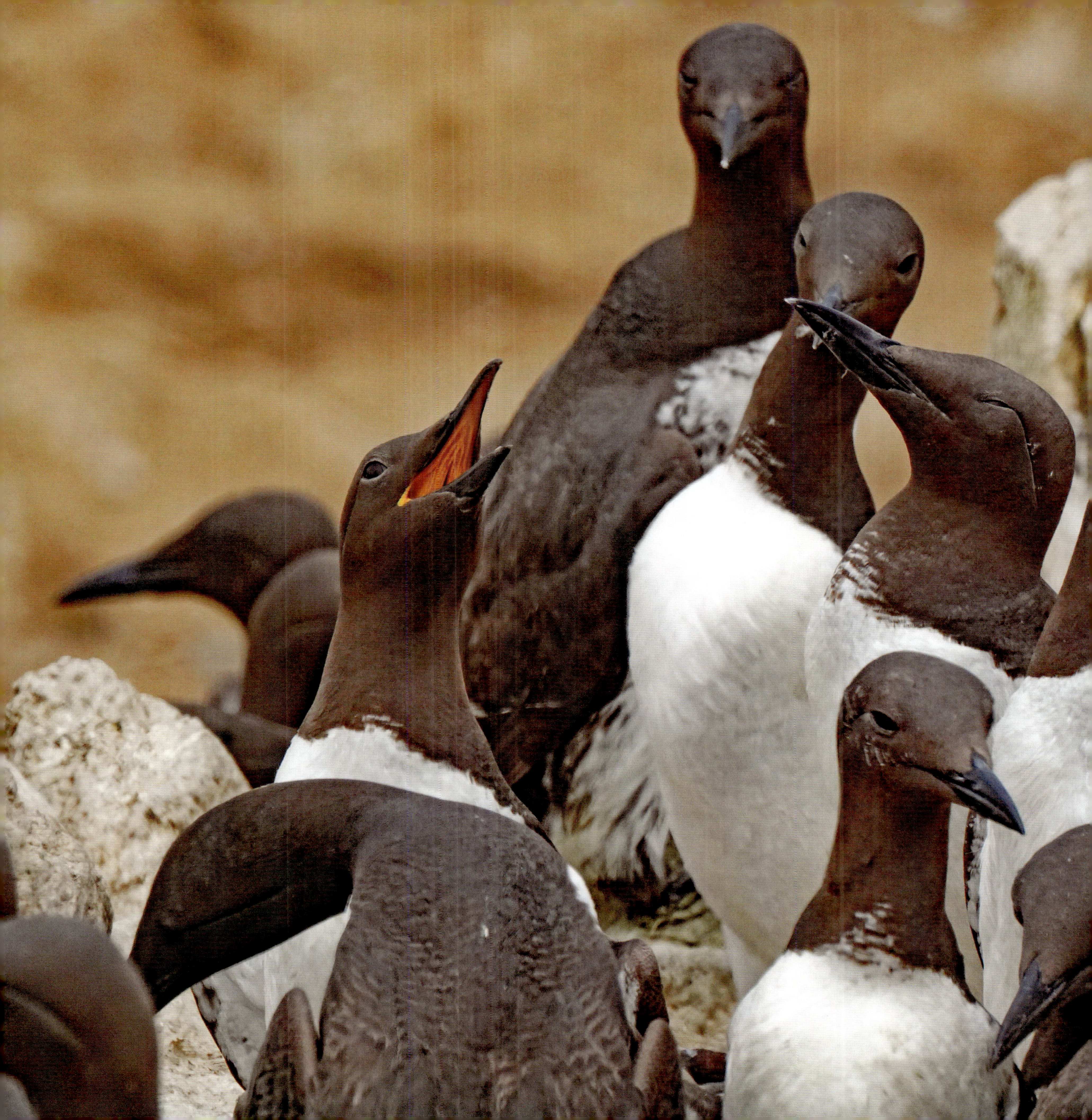

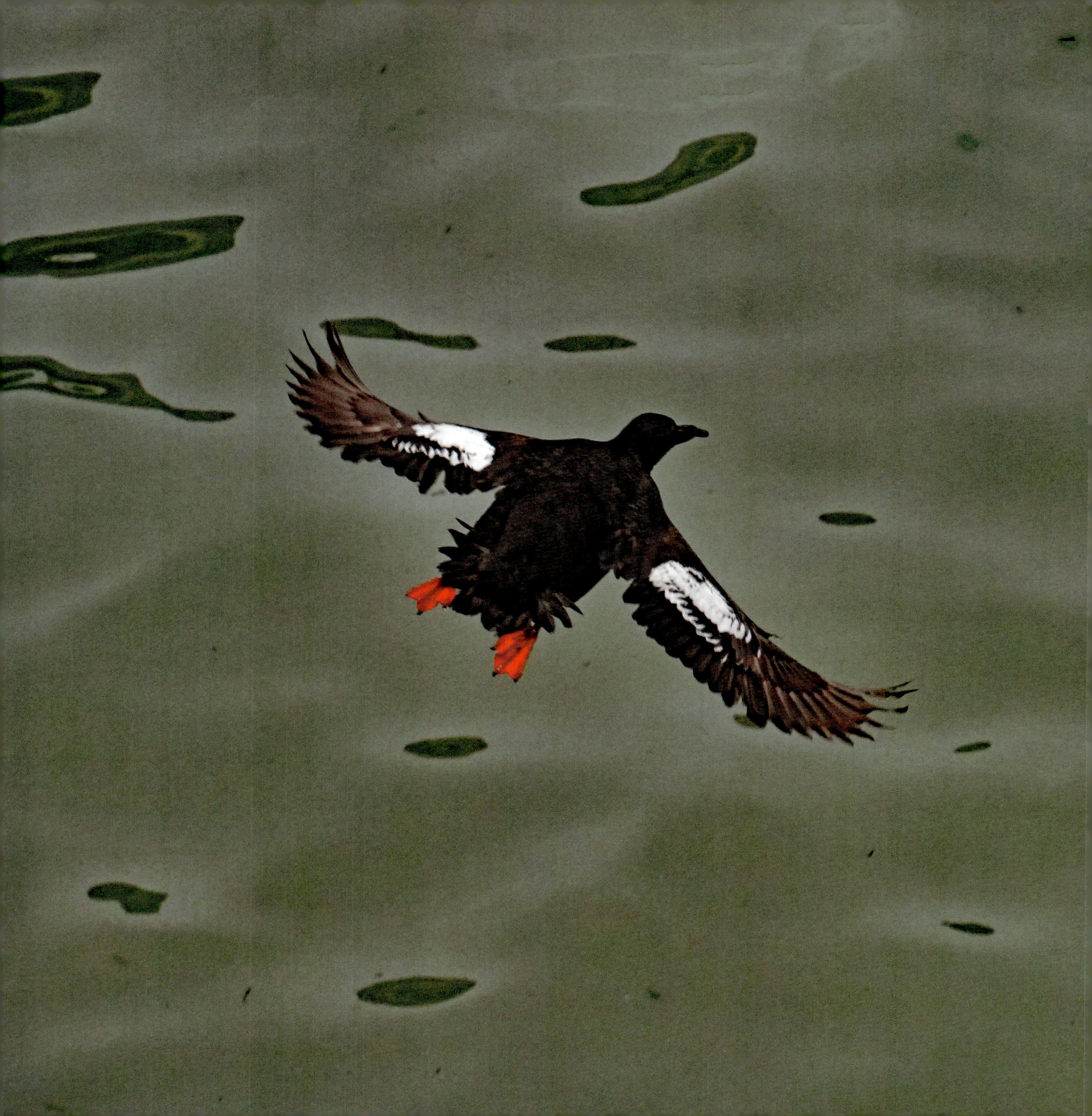

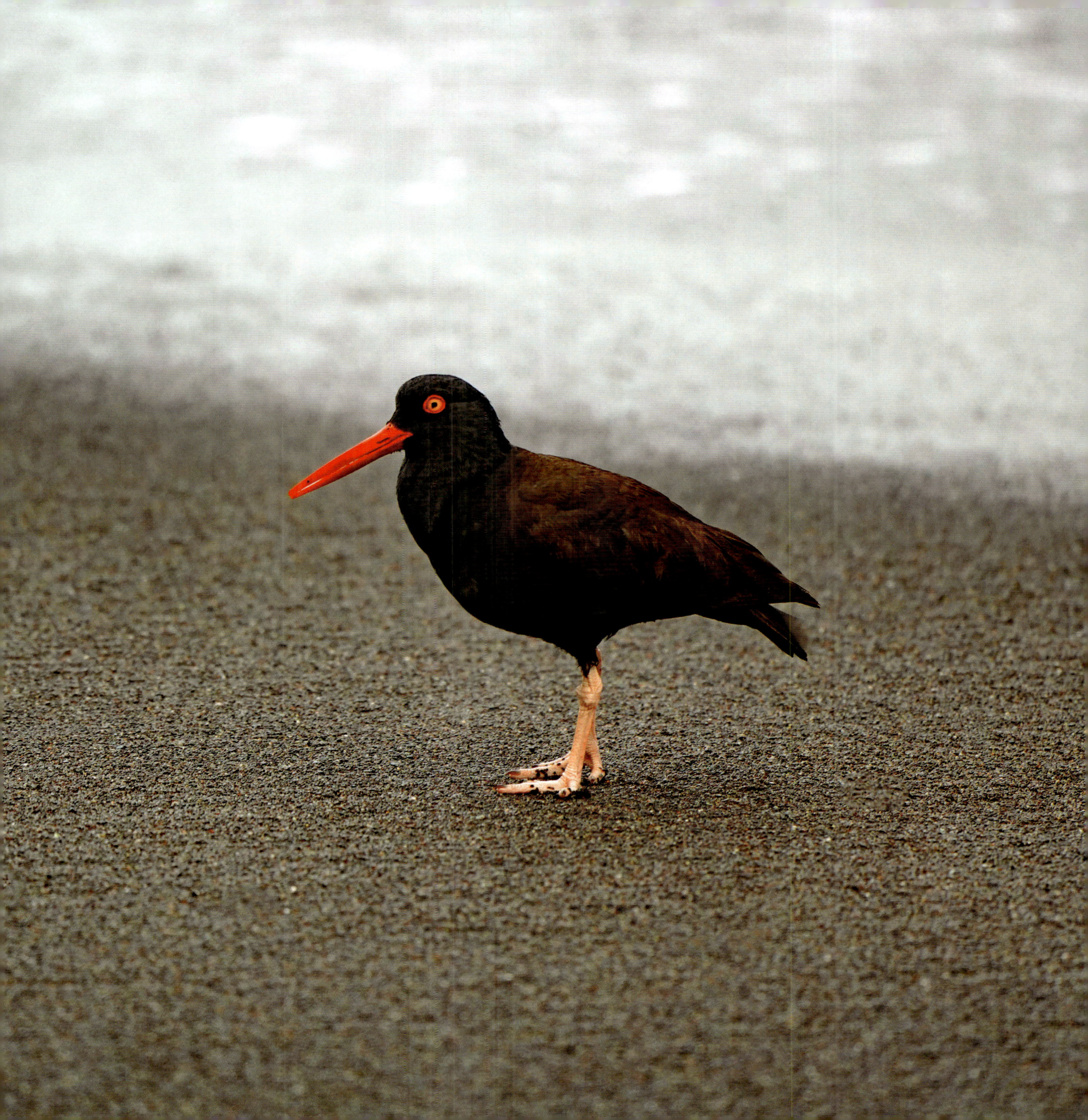

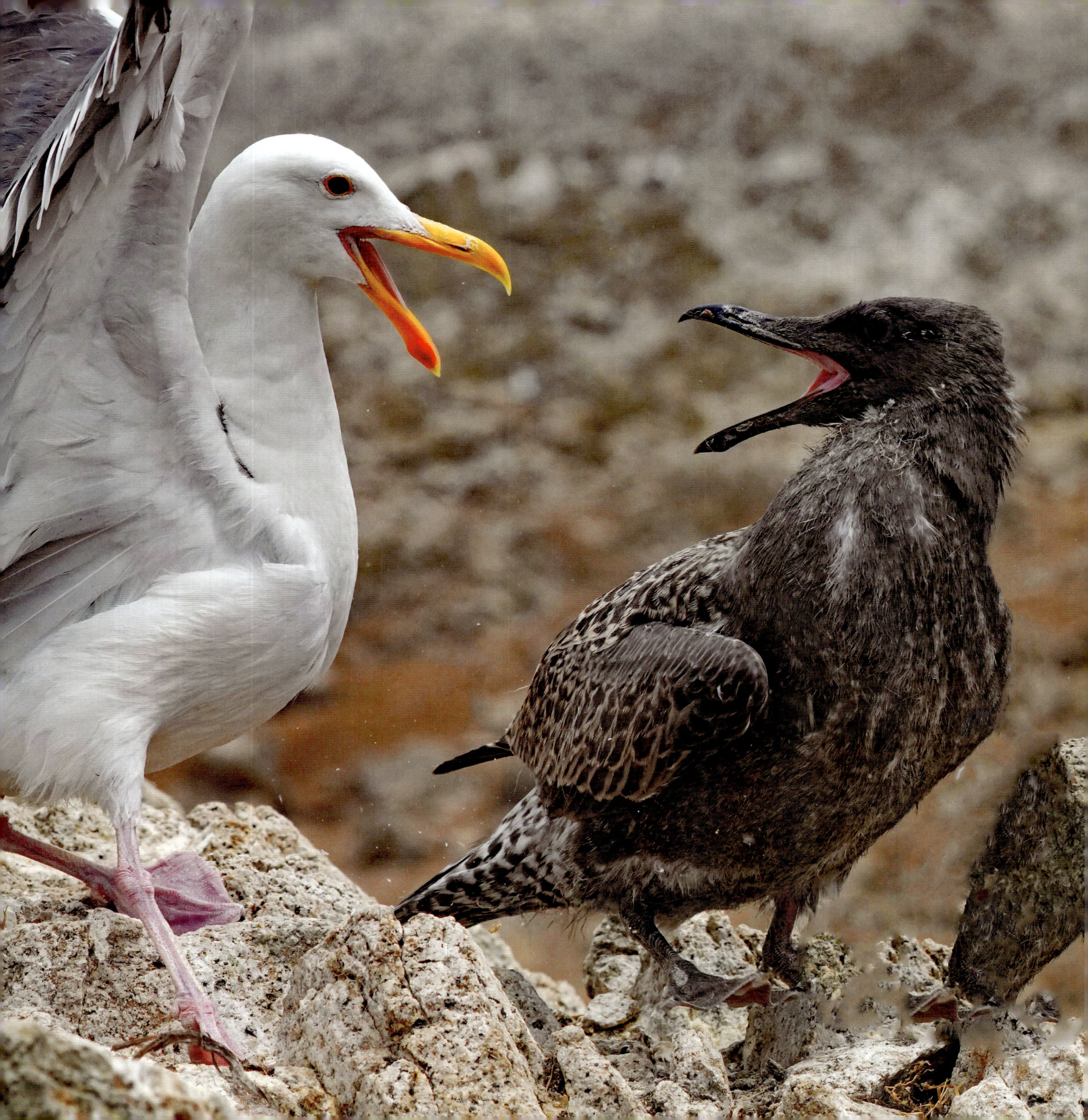

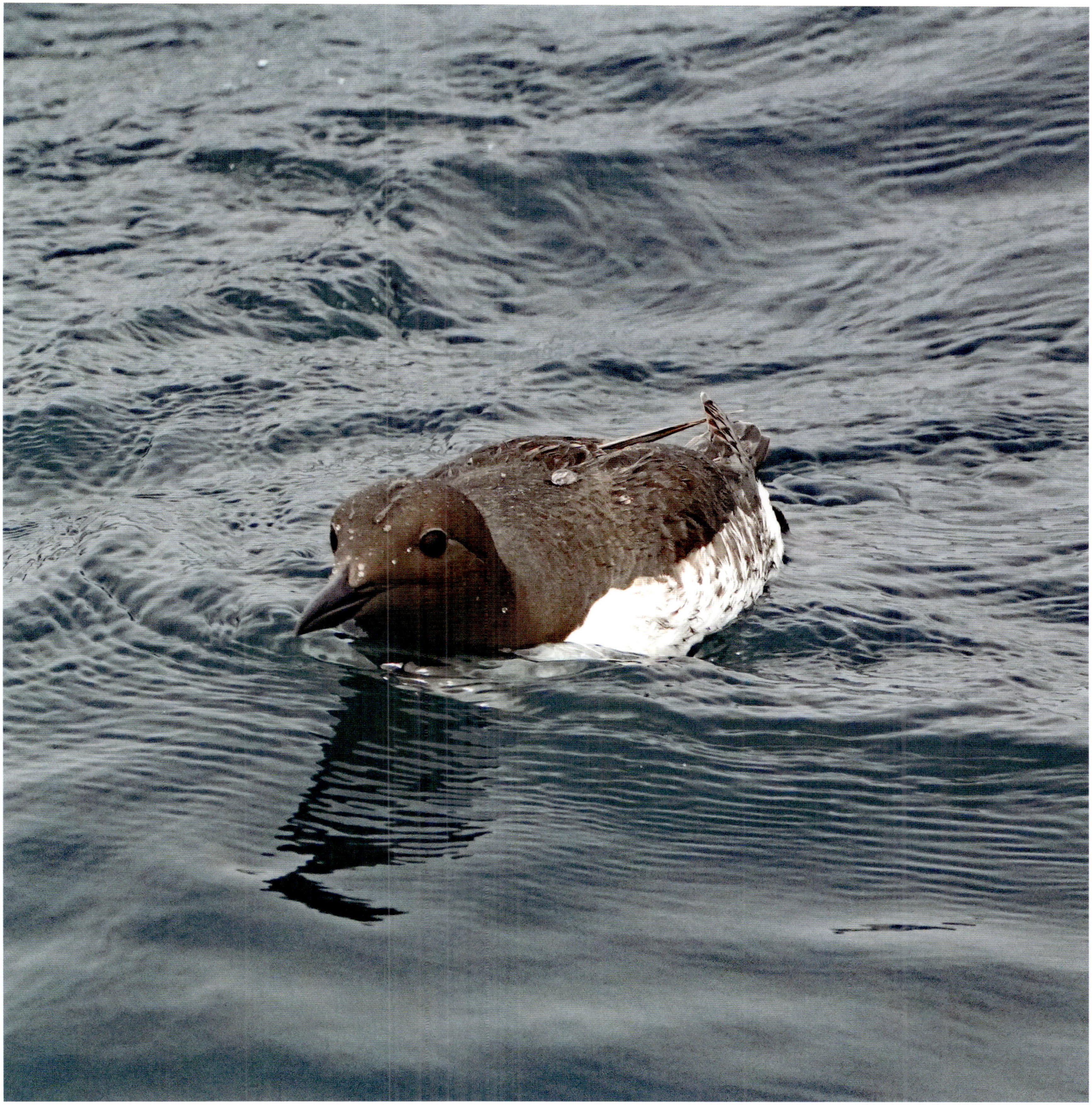

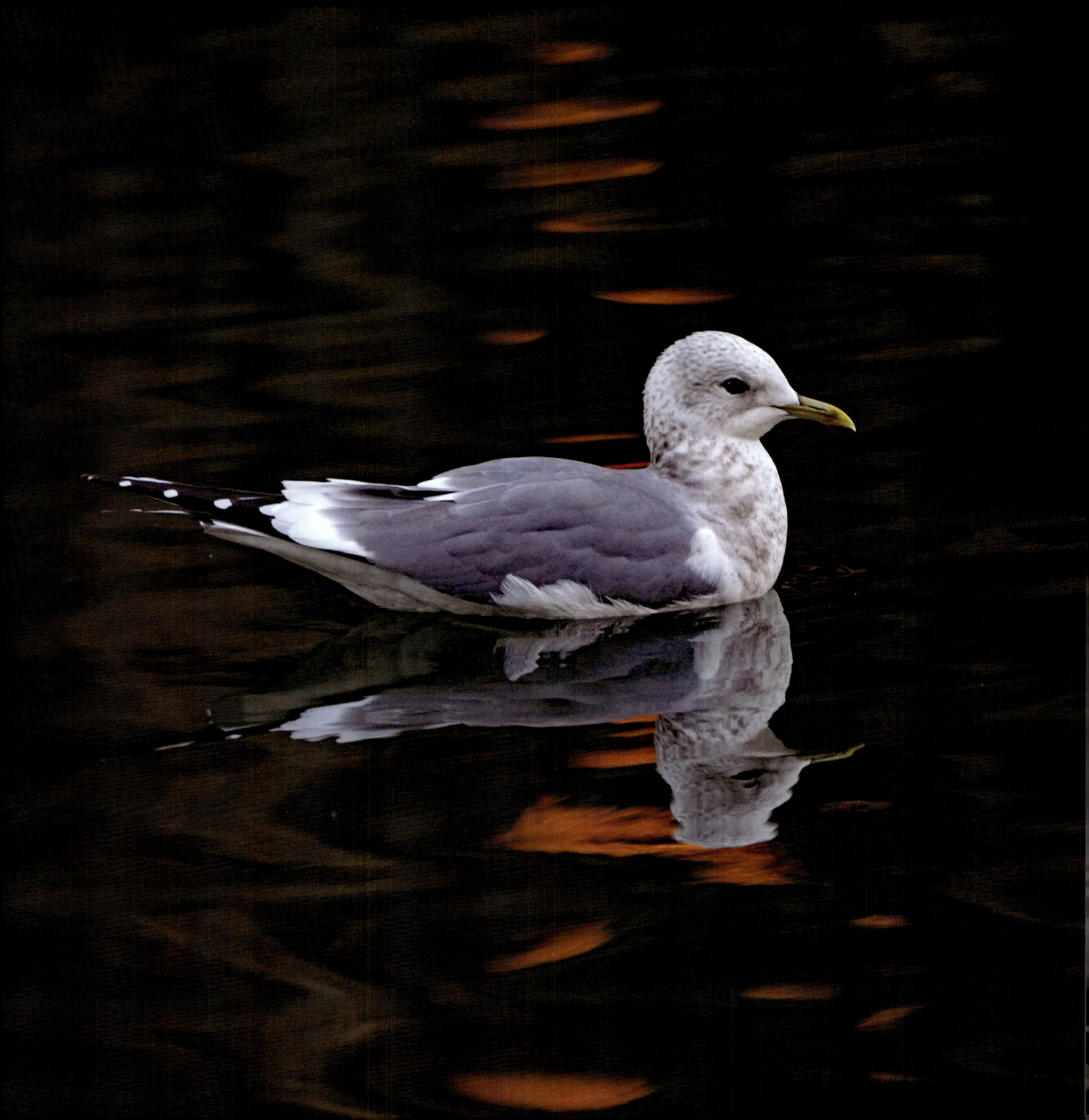

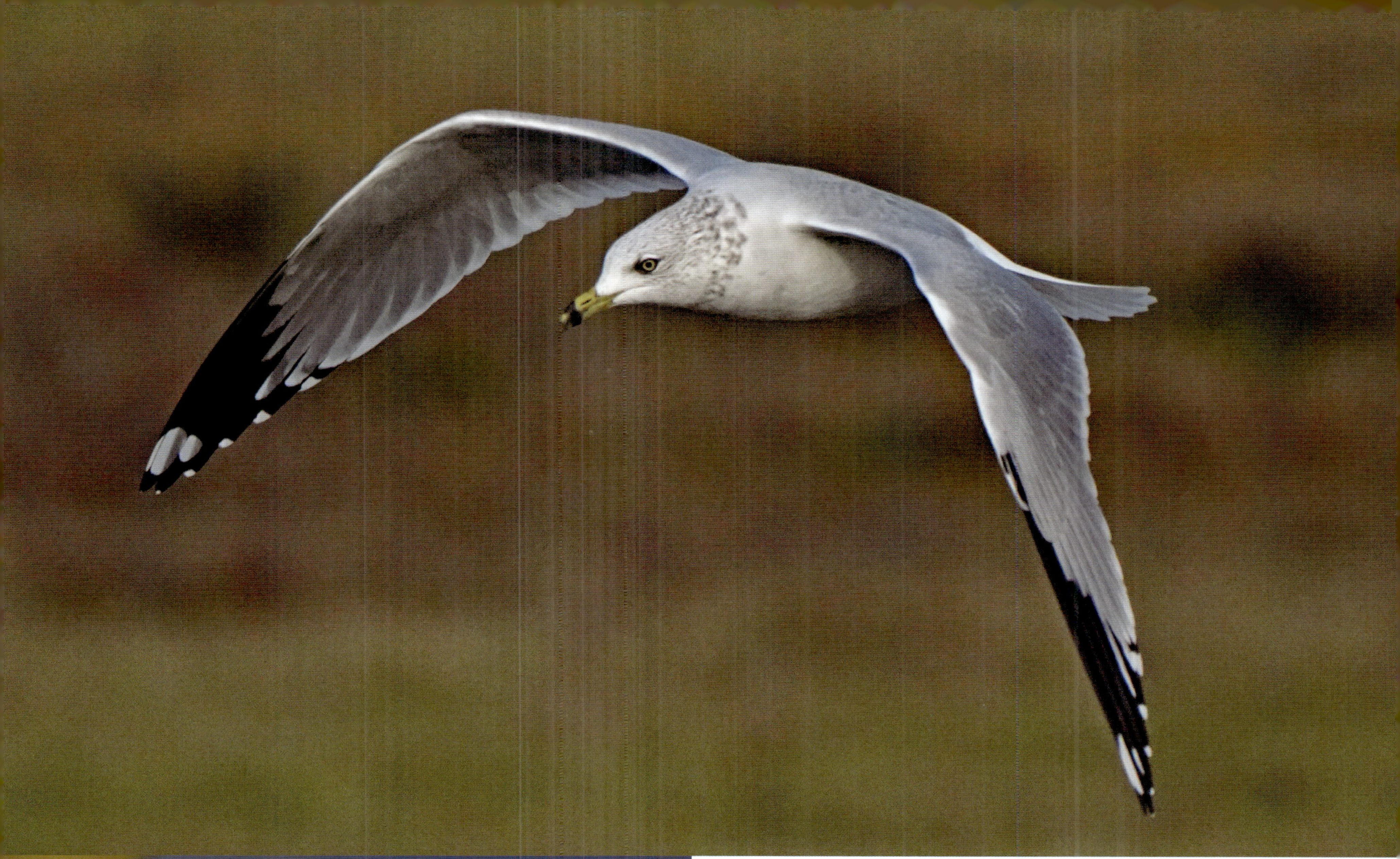

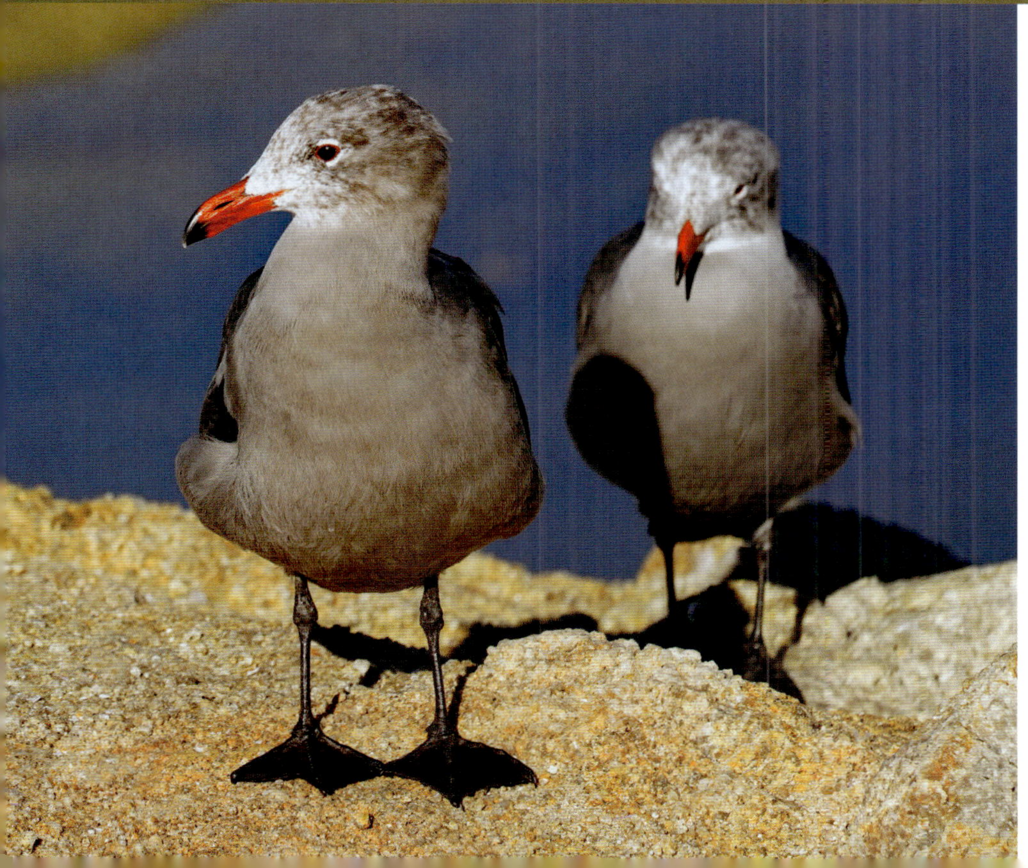

PRECEDING: A Western Gull (left) nesting on the Farallon Islands engages in a vocal dispute with its quickly growing chick. Common Murres (right) are adept swimmers, using their wings like flippers to navigate underwater, sometimes deeper than 200 feet.

OPPOSITE, ABOVE, AND LEFT: Gulls are a wildly diverse group of birds, with at least a dozen species inhabiting the Bay's coasts and wetlands at any given time of the year. Mew Gulls (opposite) migrate from their breeding grounds in Alaska and Canada, concentrating around river mouths, lagoons, and freshwater ponds in the winter. Ring-billed Gulls (above) breed inland and also find their way to our coasts in the winter, moving in large flocks, navigating using physical landmarks like major river valleys and coastal features. Unlike most birds that migrate south for the winter, Heermann's Gulls (left) breed in Mexico and then move north to the Bay Area for the non-breeding season.

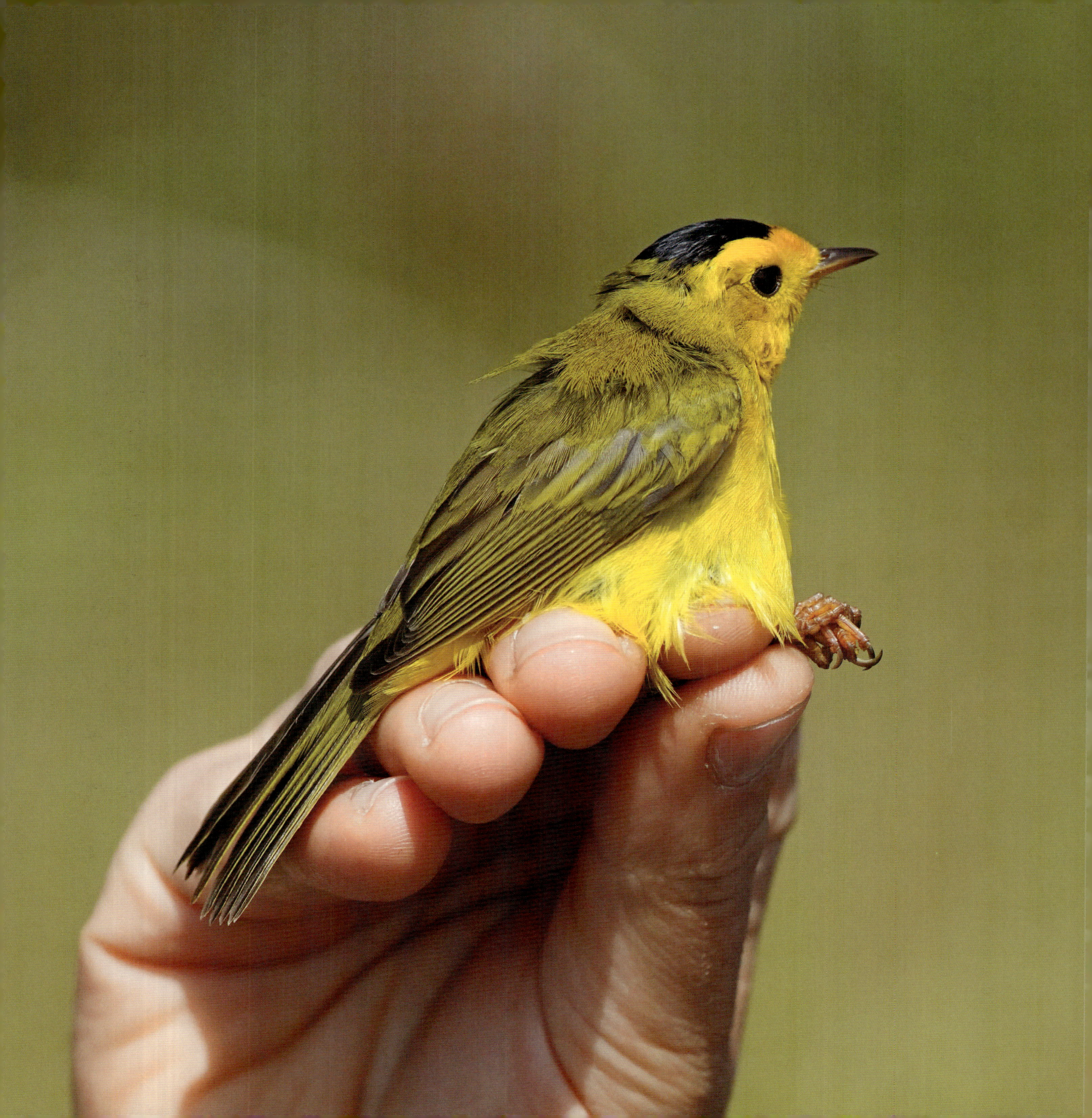

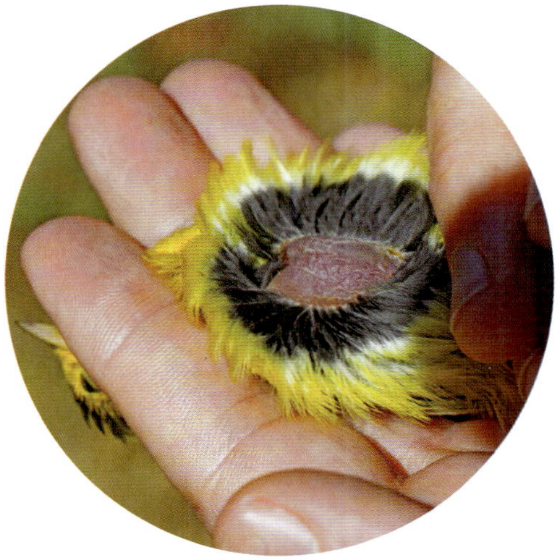

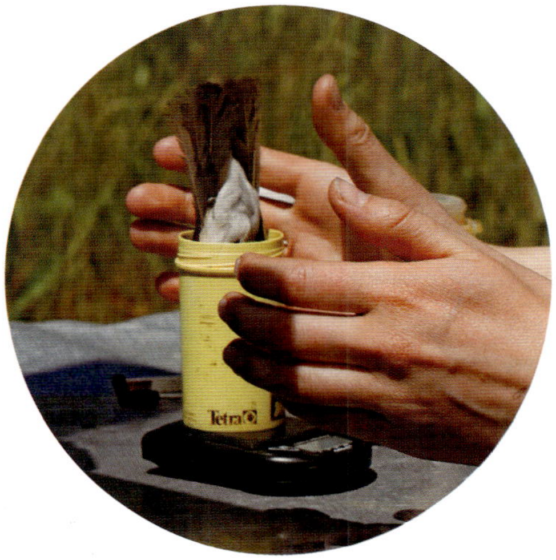

A Point Blue apprentice holds up a brilliantly colored Wilson's Warbler (opposite) for inspection at the Pine Gulch field site.

A view of the Wilson's Warbler's belly (above, upper photo) reveals the bird's featherless brood patch, indicating that it has an active nest nearby. Biologists work quickly to measure and record data so that birds can be released quickly back to their waiting nests.

Resourceful Point Blue scientists use film canisters, pill bottles, and other small vessels (above, lower photo) to gently contain the birds that they weigh.

CATCH, RELEASE, RECORD

A bird in the hand is warm and small, with a swift heartbeat that can be felt in the palm. Sarah Needles holds a Wilson's Warbler with its neck tucked gently between two knuckles. She blows briskly against its belly, revealing a tissue-thin pink layer of bare skin with warm blood pumping under it. This is the brood patch—where the bird has been warming her chicks. It identifies the bird as female: This, along with fifteen other data points, from wingspan to weight, is logged in a field notebook. Sarah crimps a lightweight aluminum band, engraved with a unique federal number, on the bird's leg and sets her loose to return to her nest.

The Pine Gulch study site is a bit like a fairy-tale castle: Every dawn, it appears amid the fog and bulrushes, and by midday it has vanished again. Point Blue Conservation Science apprentices like Sarah arrive at daybreak to rig diaphanous nets between tall poles and to lay out pliers, rulers, scales, and magnifying visors across a makeshift table. Radios crackling, they take turns walking circuits between their nets, brushing past minty-sweet arroyo willows on the narrow trails bordering Bolinas Lagoon. The nets—nearly invisible in the drenching green shadows of alder and box-elder—gently capture flying birds. Apprentices tease them from the threadlike weave with practiced fingers and tuck them, wiggling, into cotton bags to carry back for measurements. In this way, every bird netted at the field site is identified—each becoming a distinct character in an unfolding story about distant migrants, local birds, and the space they share here at the edge of the continent.

The main Palomarin Field Station, farther down the road, is less ephemeral. The shelves of the banding station are heavy with field journals filled with daily bird notes from the last half century of netting and banding. We crack one open and leaf backward through time, letting the data on the pages bring alive the taste of coastal fog, the ringing calls of Hairy Woodpeckers, and the bluster of wind that hastened scientists to go out and furl in their nets on one particular day or another. It's all here: the hardcover daily annals of the longest-running avian ecology field station west of the Mississippi.

Point Blue was founded to uncover the mysteries of migratory birds in Point Reyes: Where did they arrive from, and where were they headed? Initially, mist nets were used as an inventory tool to monitor which birds were present, like snapshots in time. Then, in May of 1986, the potential for mist nets as a technology changed dramatically. That spring, there was a striking, unexplained decline in the number of hatch-year birds caught in the nets—a 62 percent reduction compared to the average of the previous decade. It was a bigger loss than any seen before or since. Scientists were baffled—there had been no unseasonable freeze, no notable change in rainfall from the previous years. What there had been—an ocean away—was a catastrophic meltdown at Chernobyl Nuclear Power Plant. Palomarin scientists connected the dots. Nuclear fallout remains the leading hypothesis as to the cause of the unprecedented reduction of hatch-year birds in the summer of 1986. The event highlighted the power of long-term standardized ecological monitoring to detect unusual patterns and generate new questions.

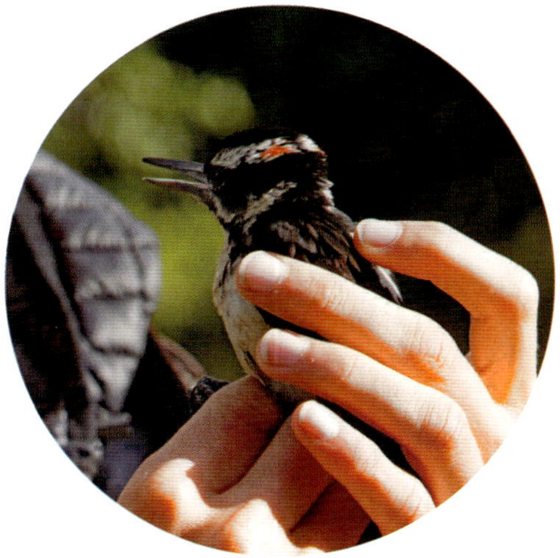

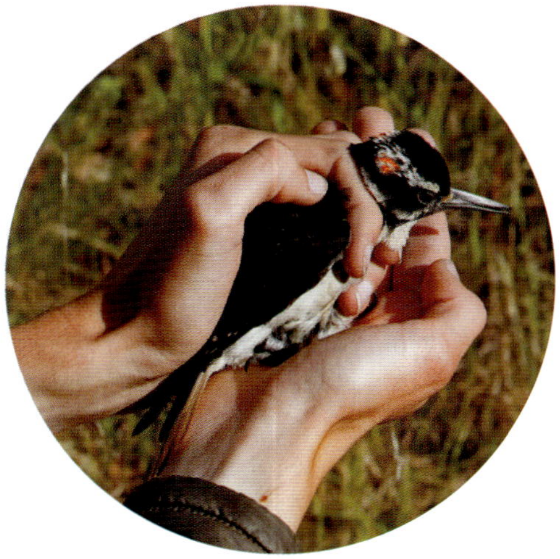

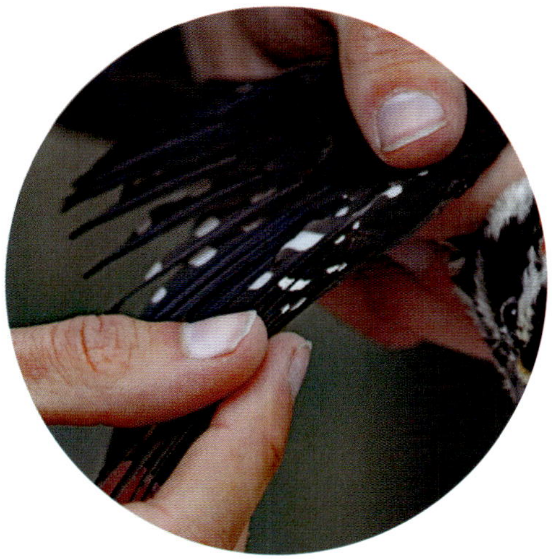

A series of photos reveals different views of a Hairy Woodpecker in the hands of a Point Blue apprentice. The feathers on woodpeckers' wings can help biologists age them quite precisely. This gives rich information about the age structure of populations, including how many young there are, and what the survival rate is from year to year. As these numbers fluctuate, scientists can begin to pinpoint causes in the surrounding ecosystem.

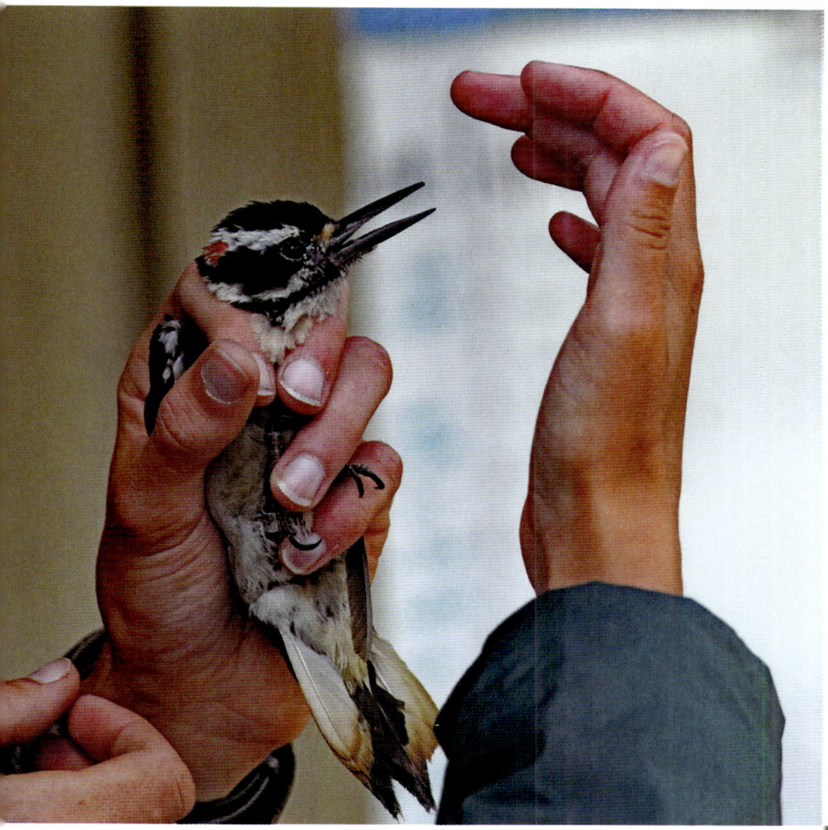

Since its inception, the Palomarin program has continued to reveal the interconnectedness of birds and their environments. Just as a gauzy spiderweb catches vibrations, Palomarin's delicate mist nets can reveal unseen global tremors. As new challenges arise—climate disruption, conservation of international migration corridors, shifting fire regimes—these nets will serve as an enduring link between the natural world and the data that can help us protect it.

Anecdotally, Point Blue bird-banding biologists report that woodpeckers (left) with their powerful beaks, are the likeliest family of birds to draw blood when held with bare hands.

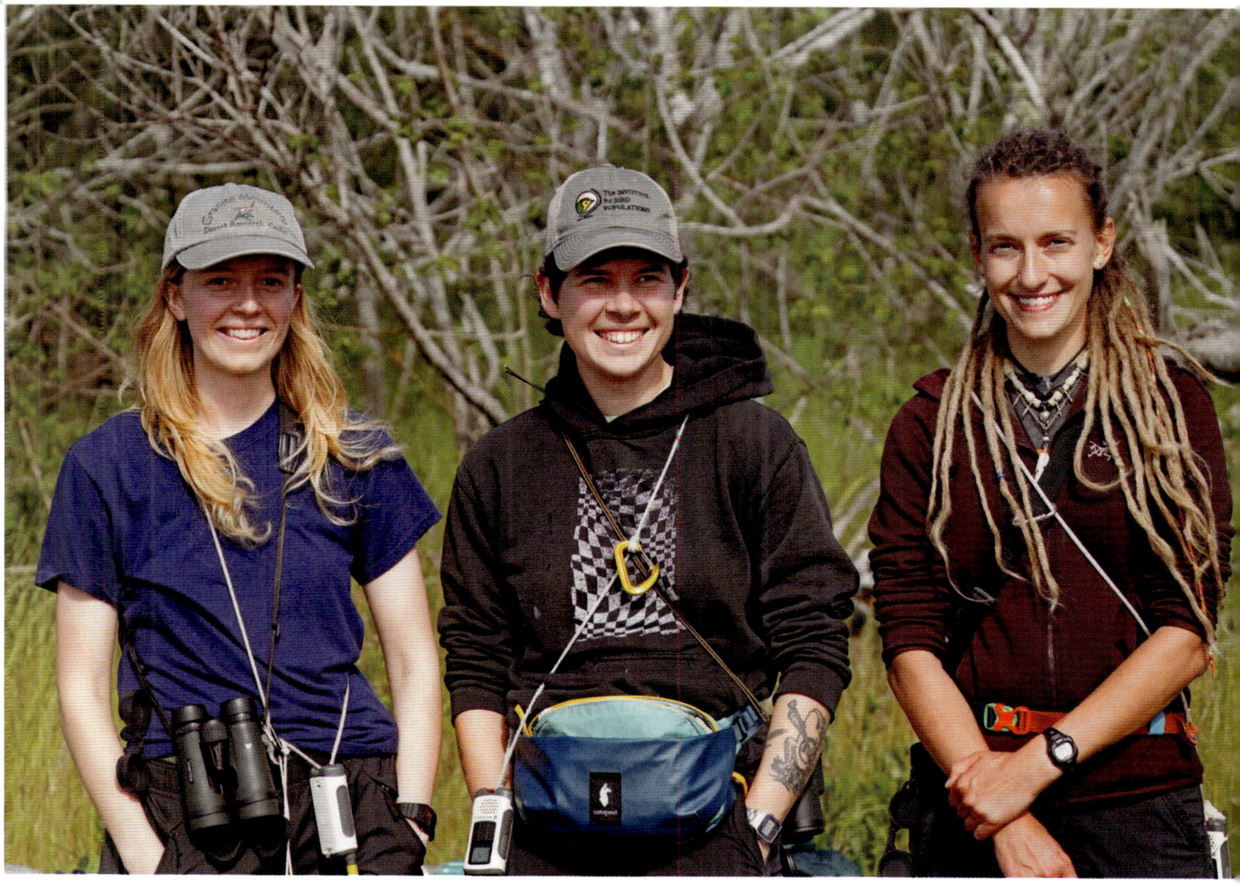

Point Blue apprentices Sarah Needles, Fen Conway, and Larissa Babicz (right) are part of a lineage of more than 750 apprentices who have worked at the Palomarin Field Station since its inception in 1965—many of whom have gone on to become leaders, scientists, and educators for local and global conservation movements.

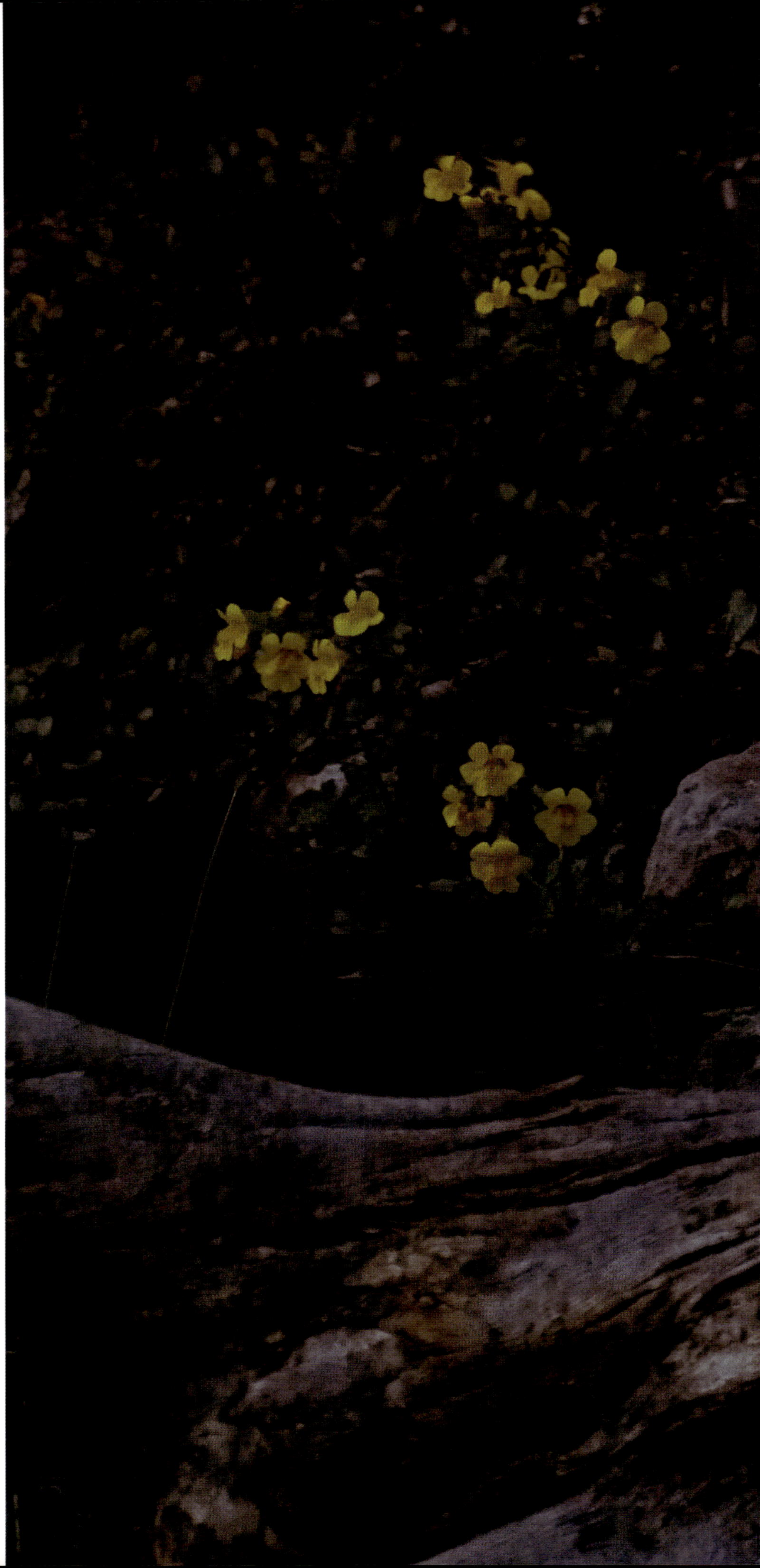

A juvenile Black-crowned Night Heron claims a solitary moment among some yellow lakeside monkey flowers in Golden Gate Park—a quieter perch than the crowded rookery where he likely hatched among a tangle of other heron nests. The Black-crowned Night Heron was unanimously voted to be Oakland's official city bird in 2019—the official resolution cites the herons as embodiments of Oakland's "tenacious and dynamic community no matter how many times they are displaced."

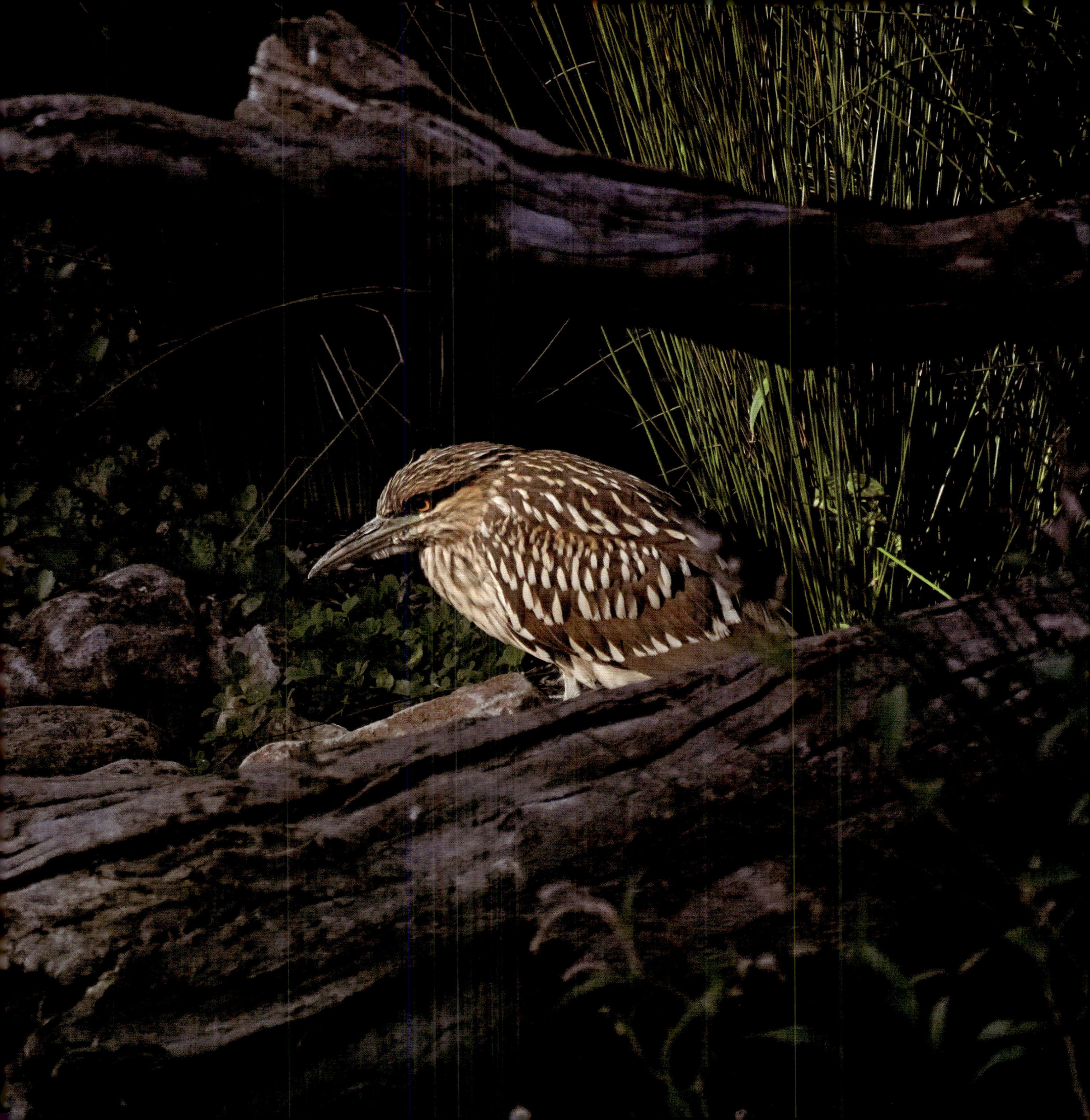

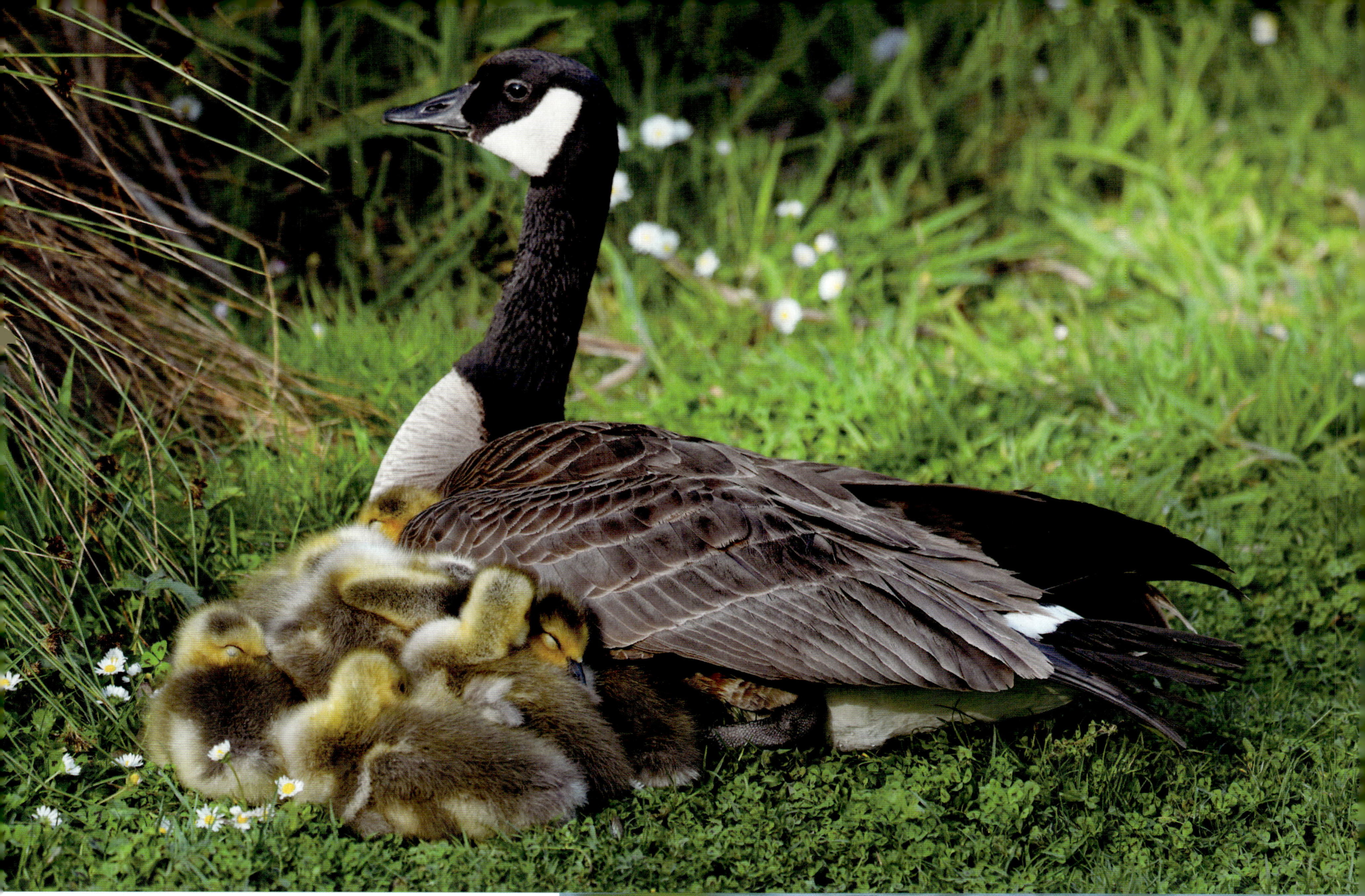

In the 1970s, the Marin County Breeding Bird Atlas was a colossal local survey which, over the course of four field seasons, meticulously mapped out locations and numbers of every breeding bird species nesting in Marin County. At the time, only two breeding Canada Geese were counted in the entire county. Now, a new edition of the Breeding Bird Atlas is underway, with 2025 as the final survey season. Even without official final numbers, it's clear that Canada Geese (above, right, and opposite) have succeeded wildly in taking new territory and raising families in the mild climate of the Bay Area in the past half-century!

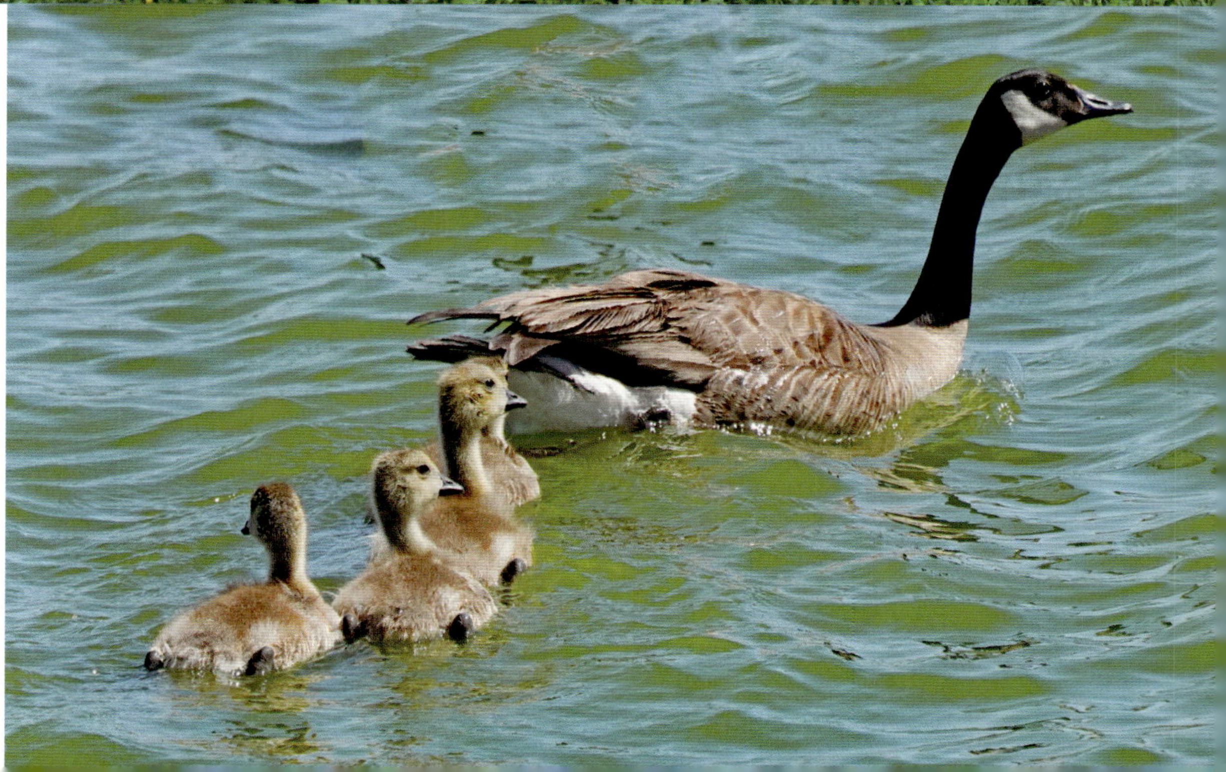

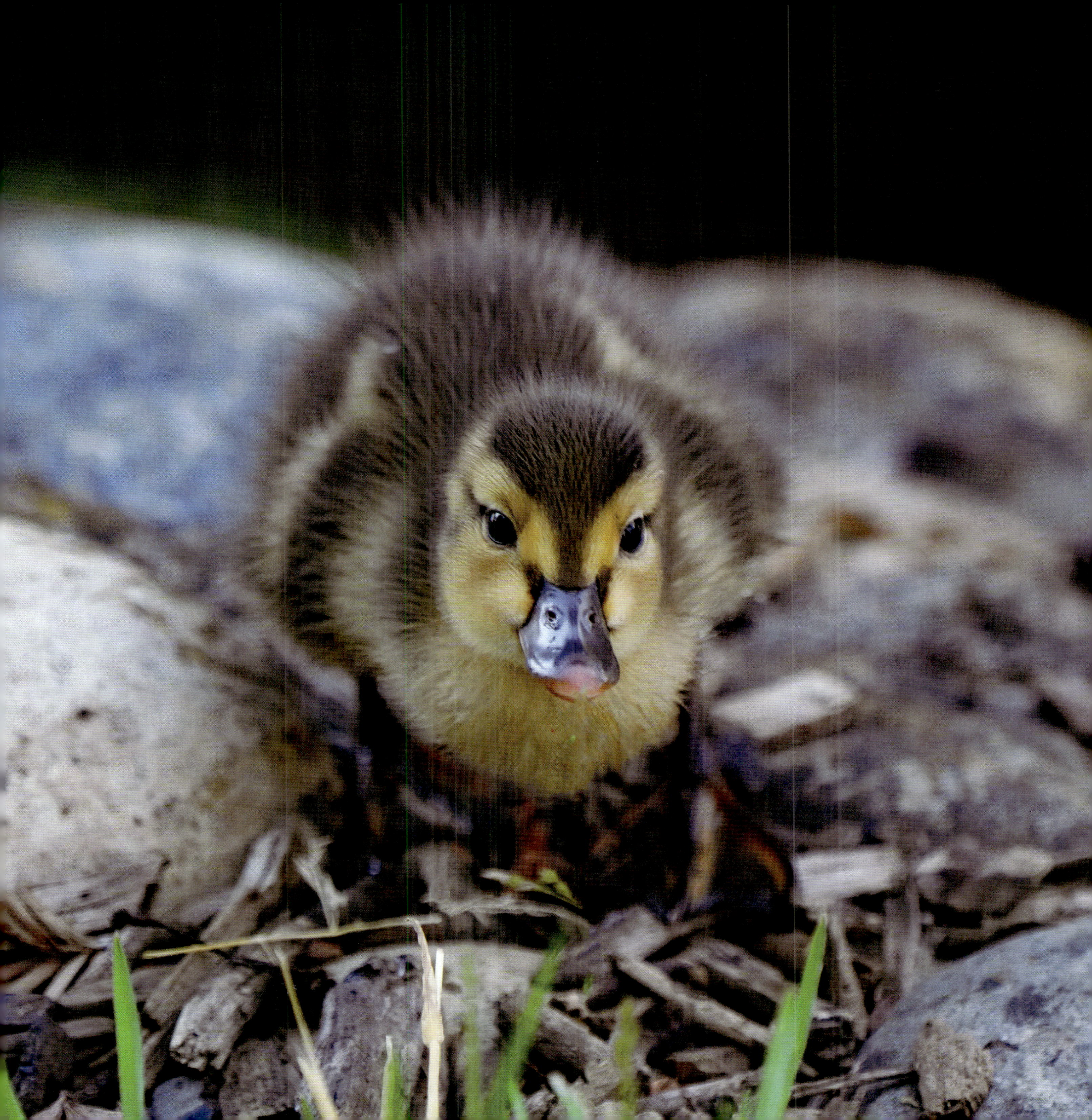

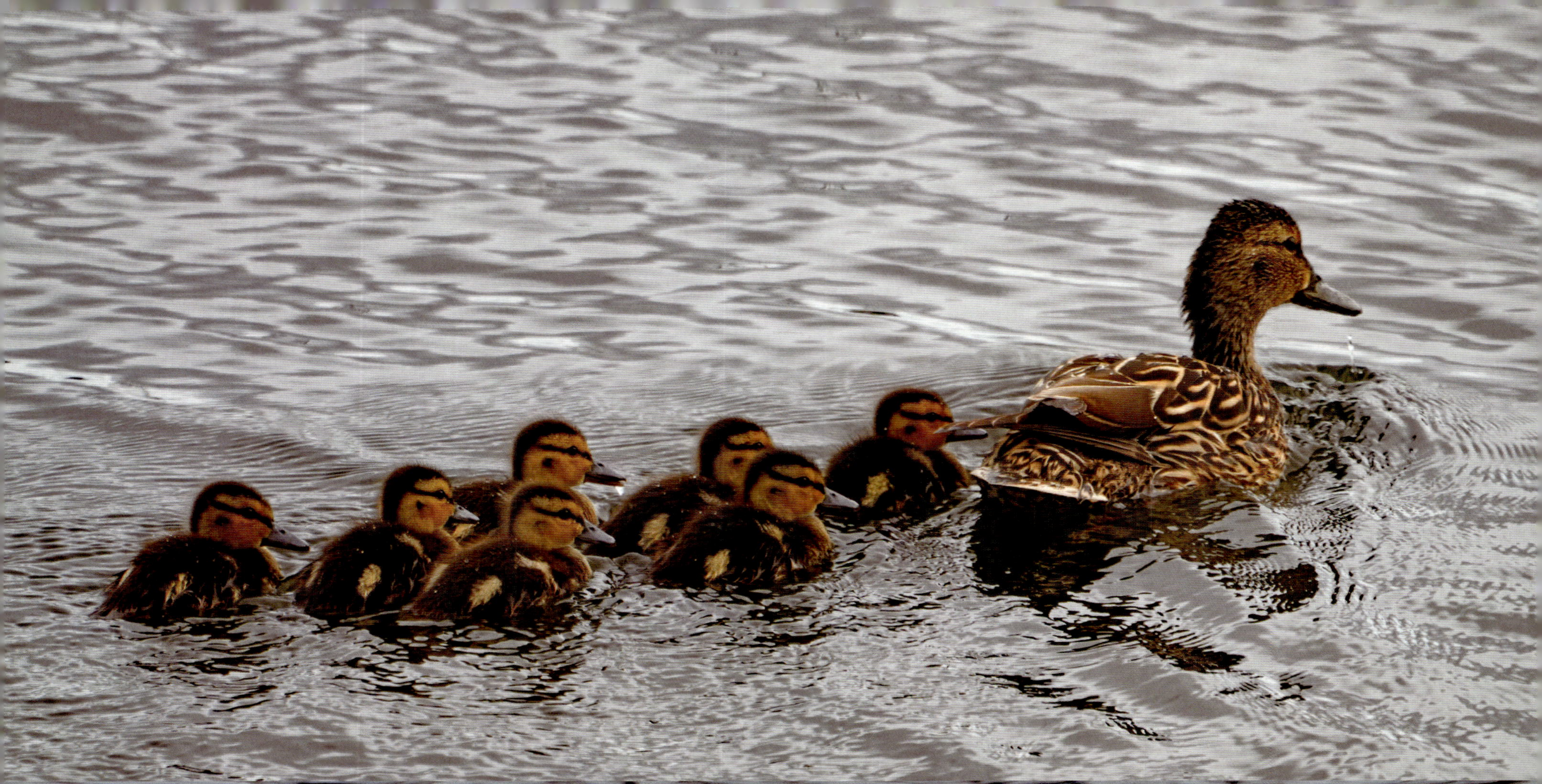
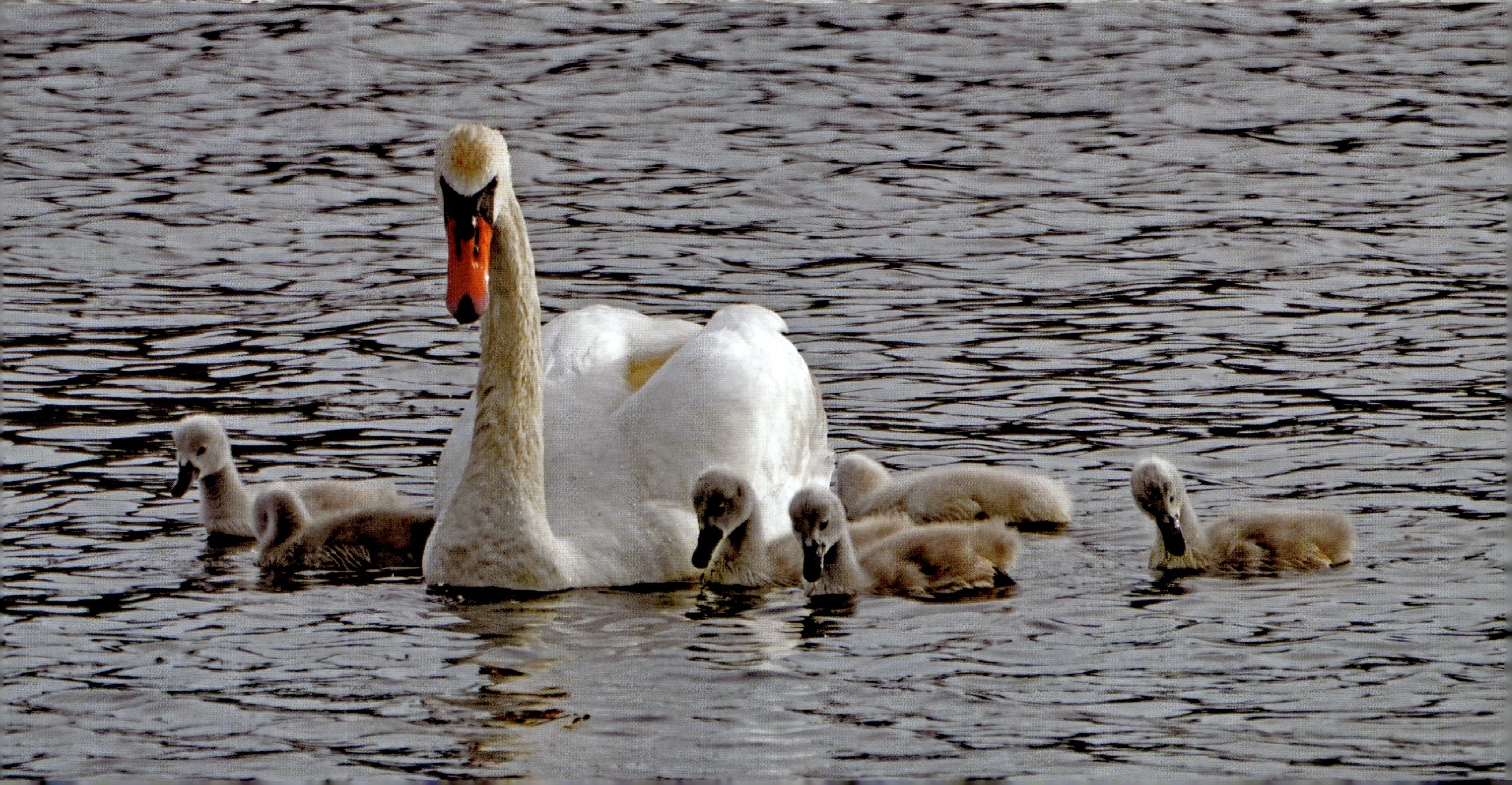

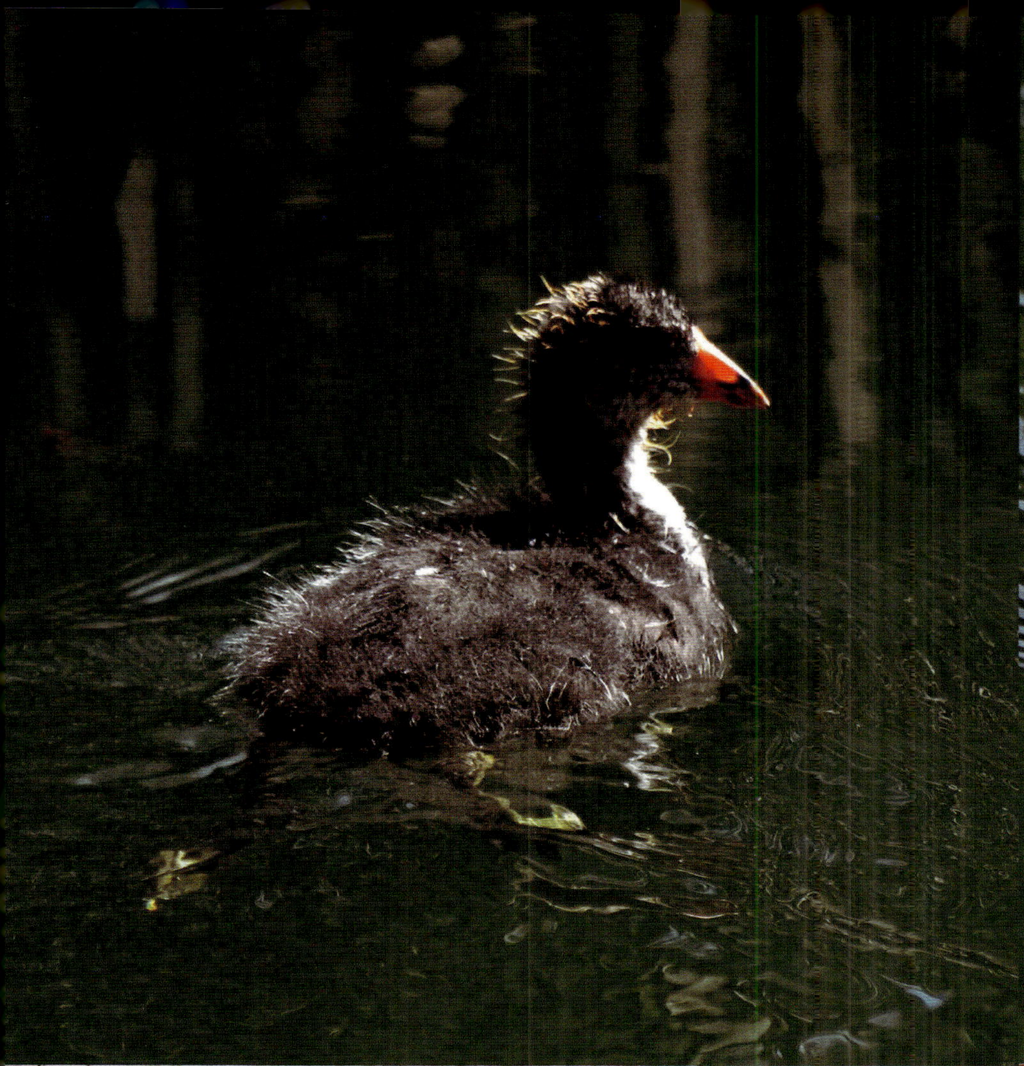
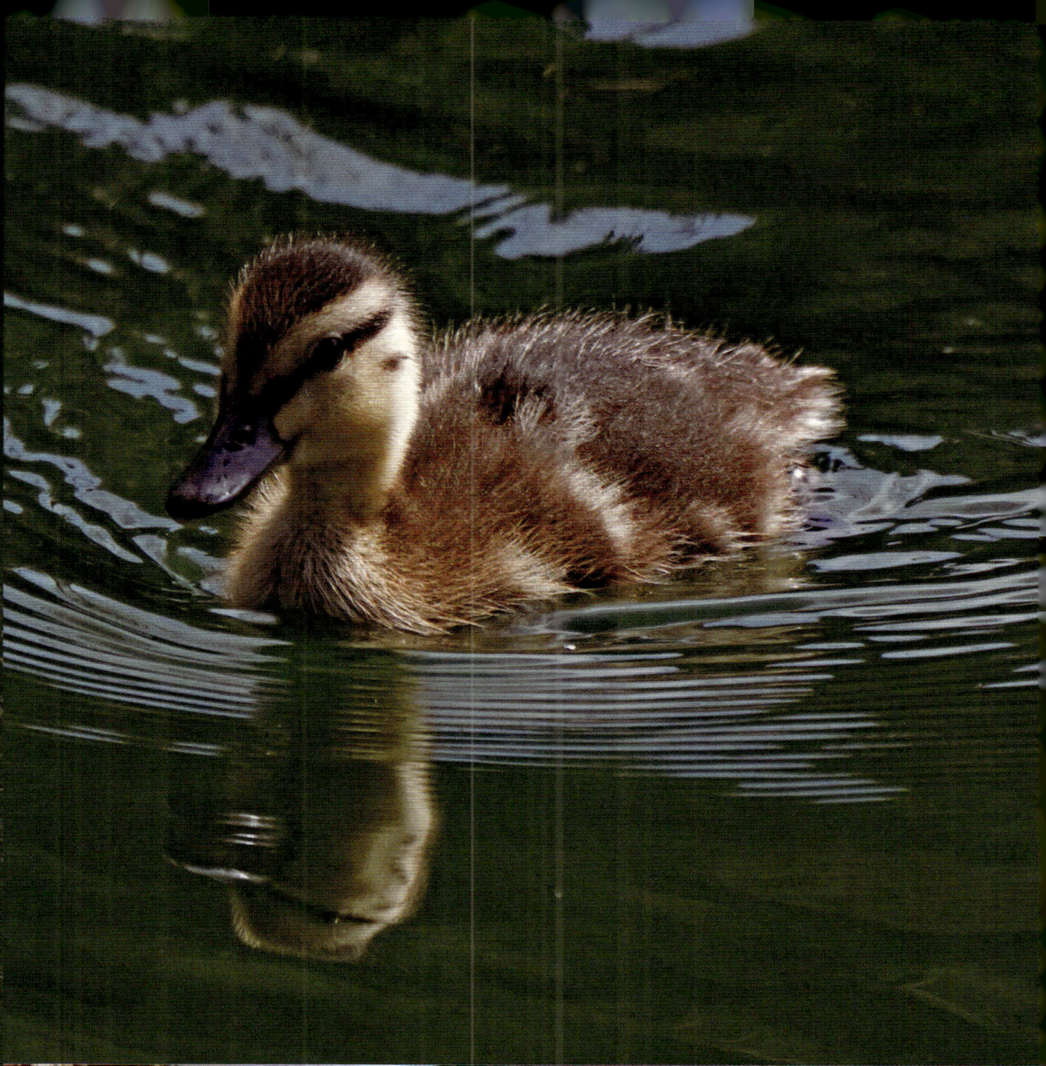
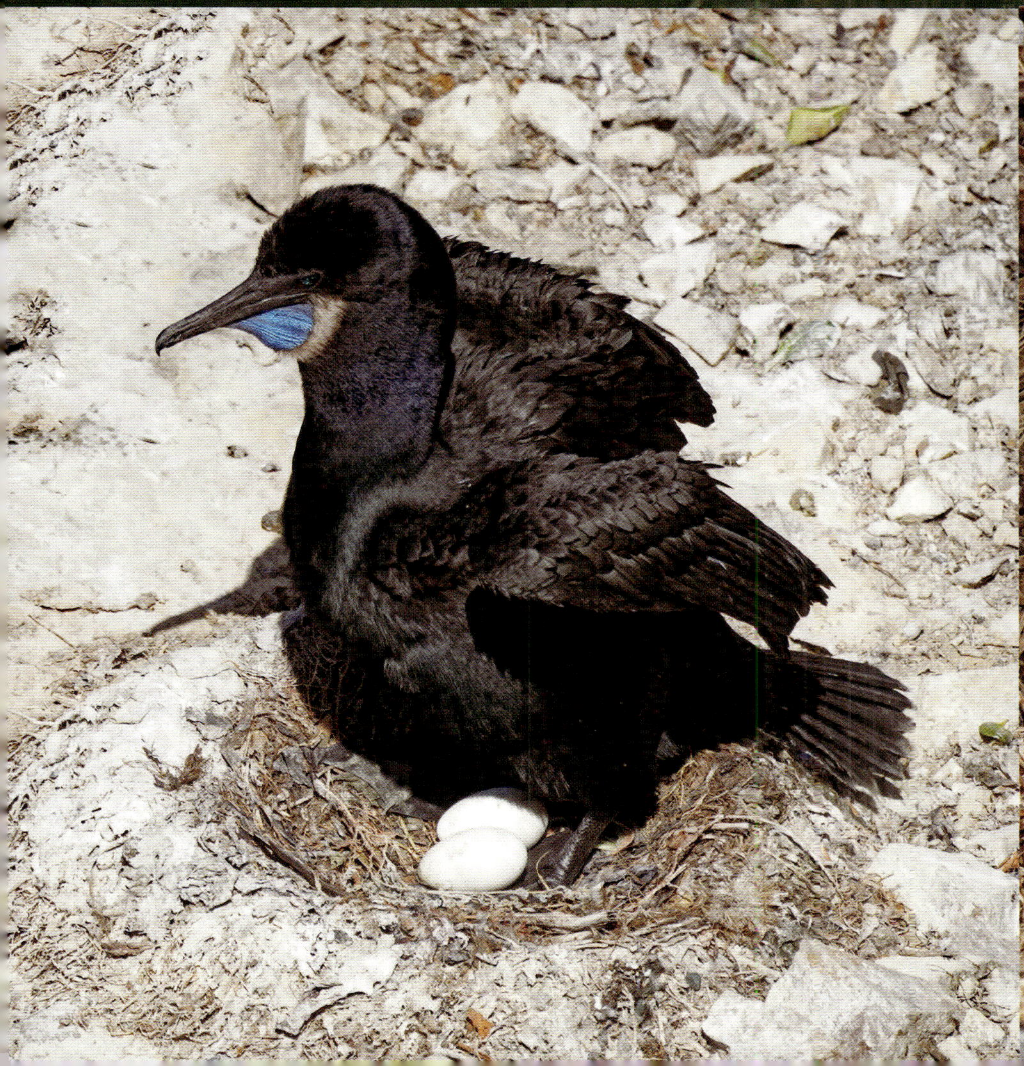
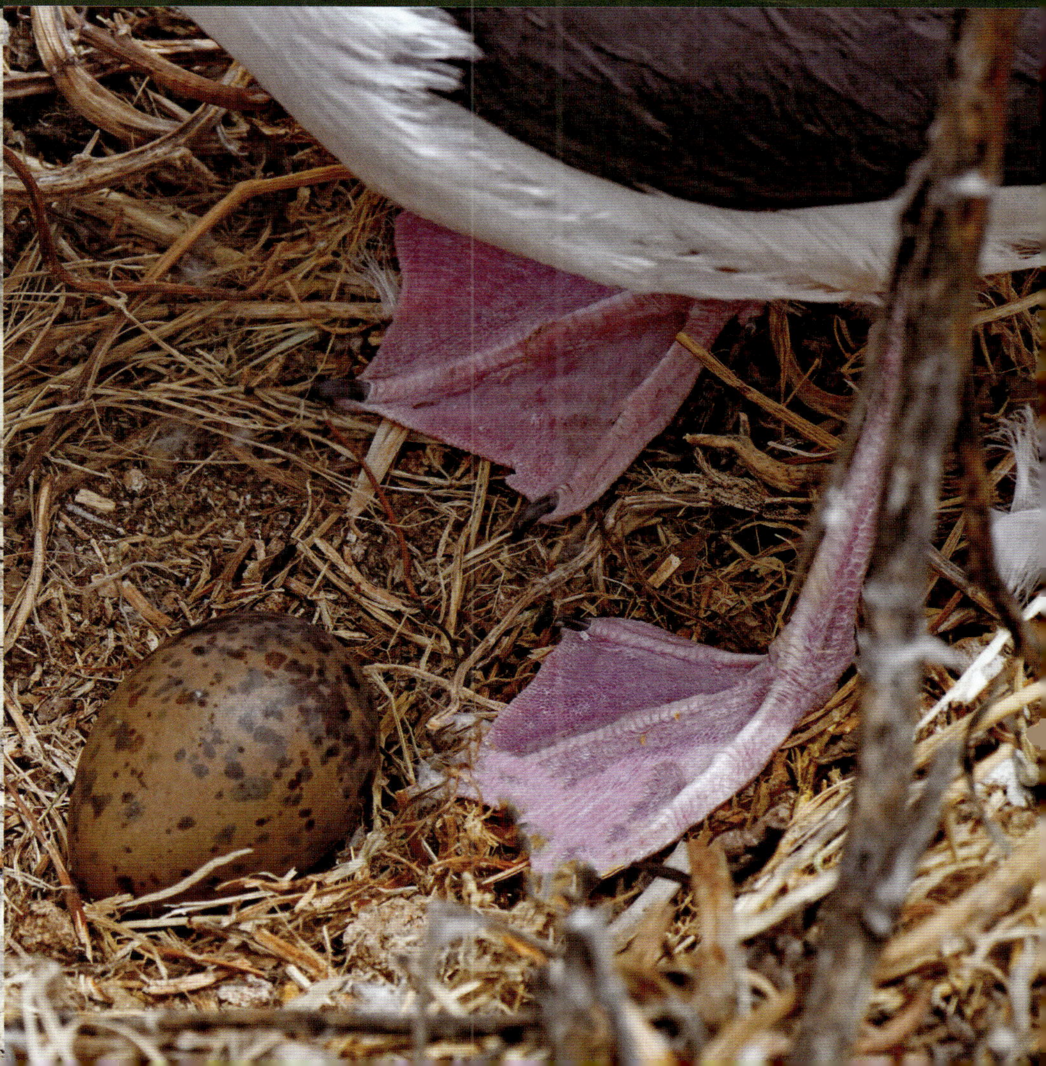

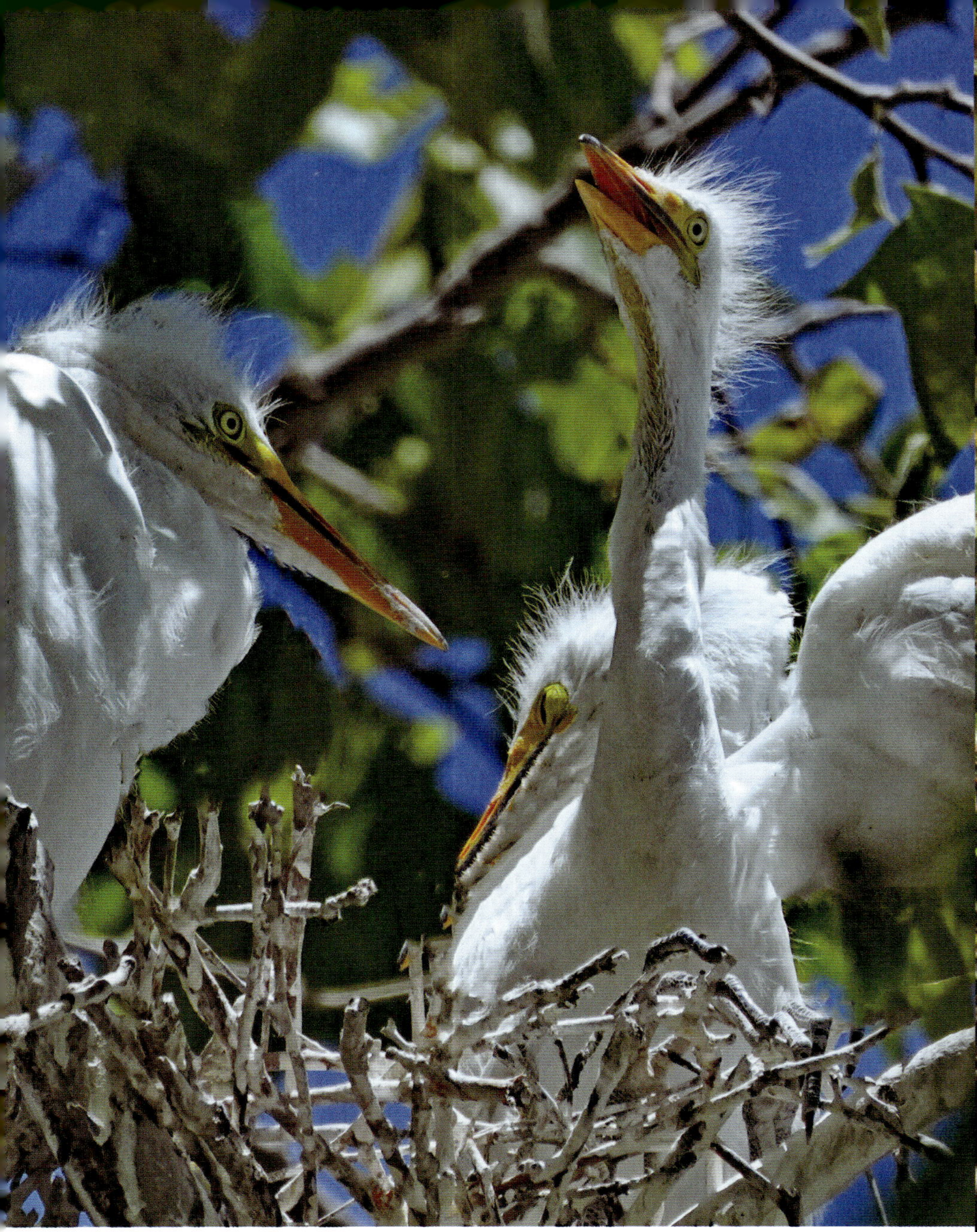

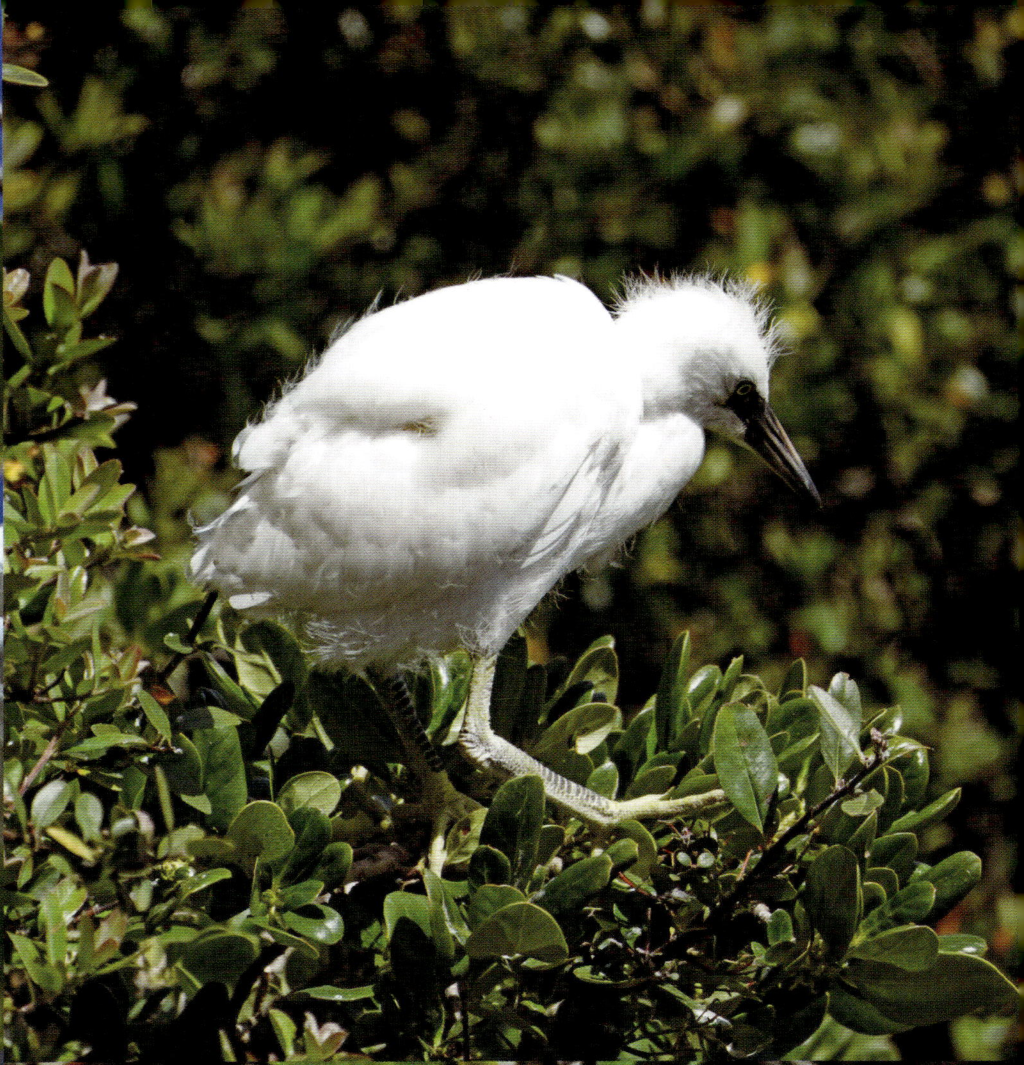

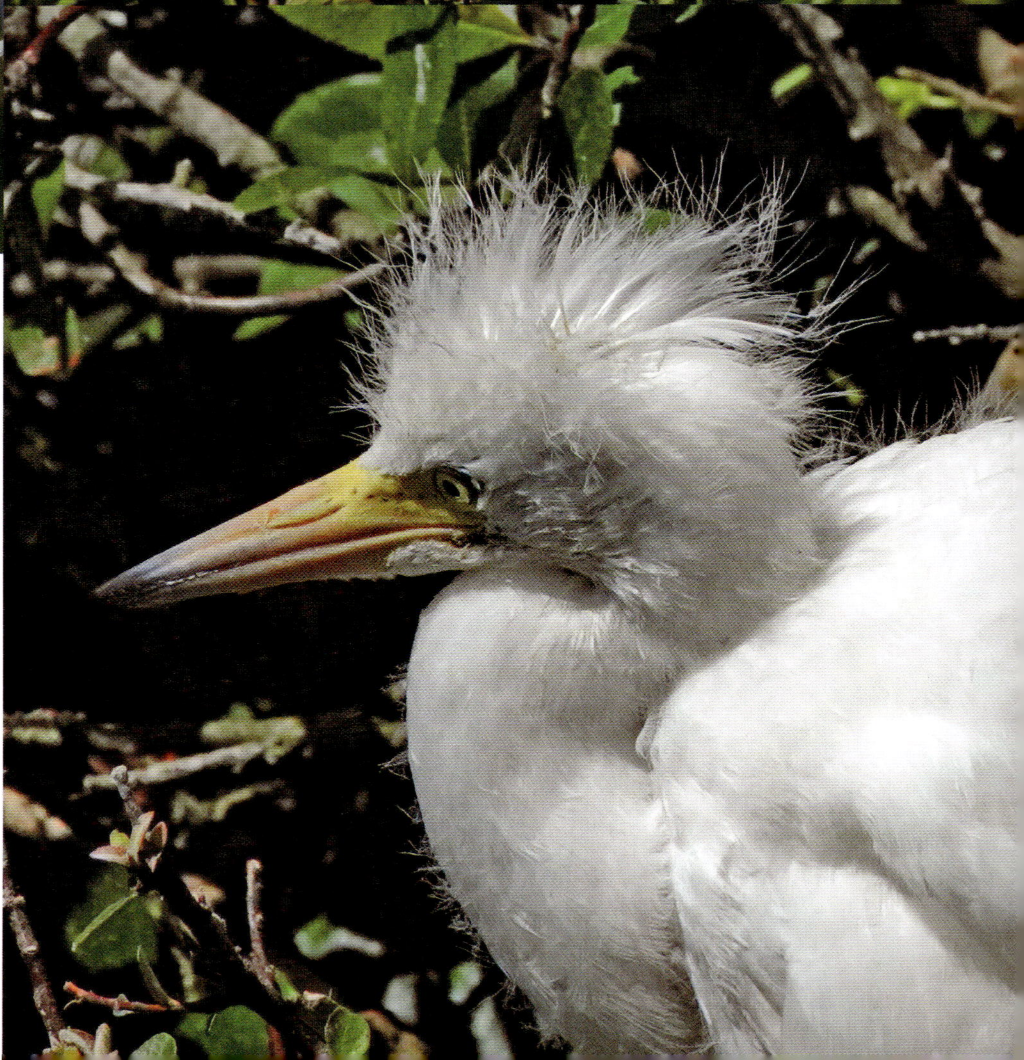

PRECEDING: Baby boom! Clockwise from top left, a mother Mallard with babies in tow, an American Coot chick, a Mallard duckling, a nesting Western Gull, a Brandt's Cormorant incubating her egg, and a family of Mute Swans all exemplify the Bay Area's richness as a breeding mecca for birds large and small.

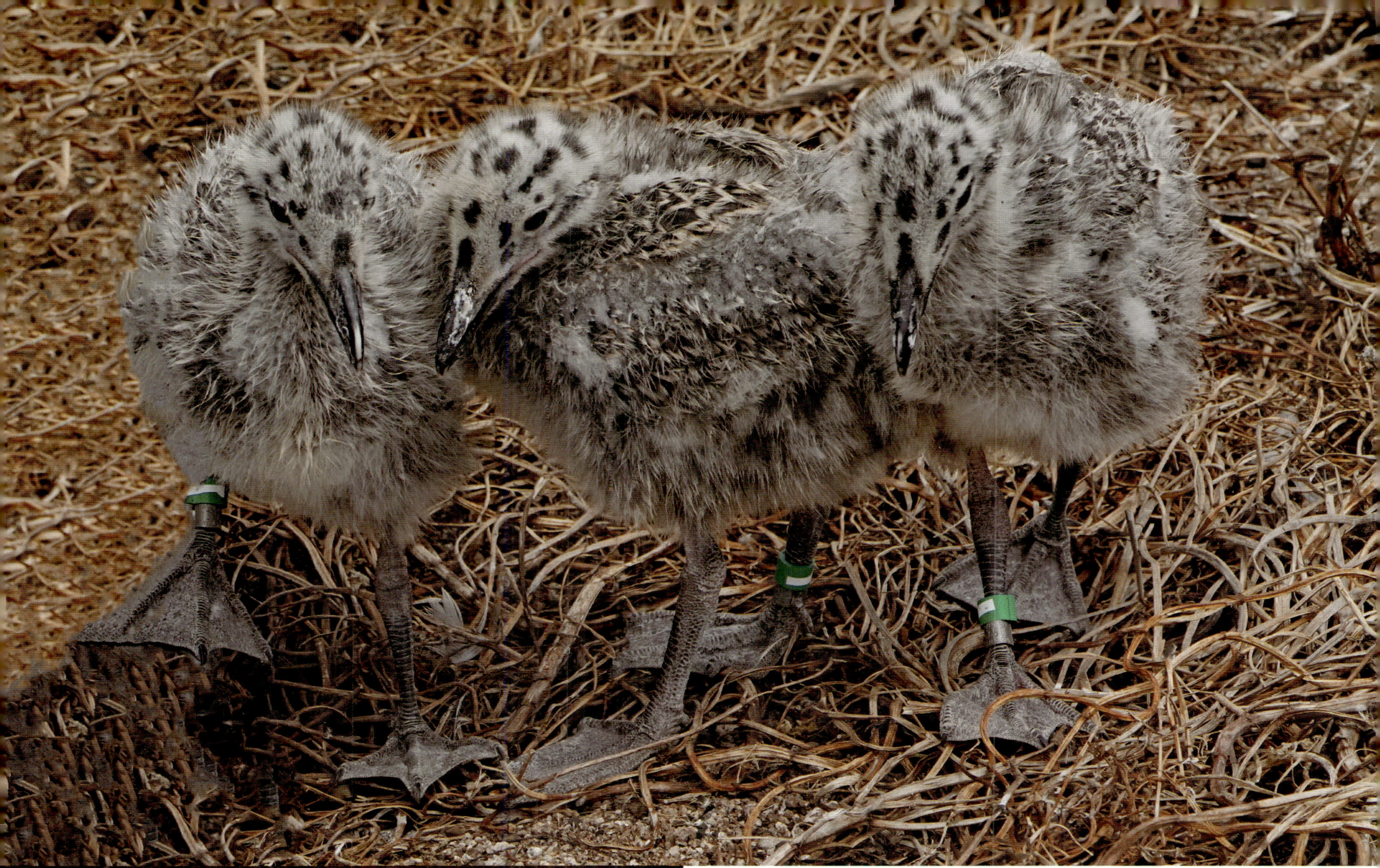

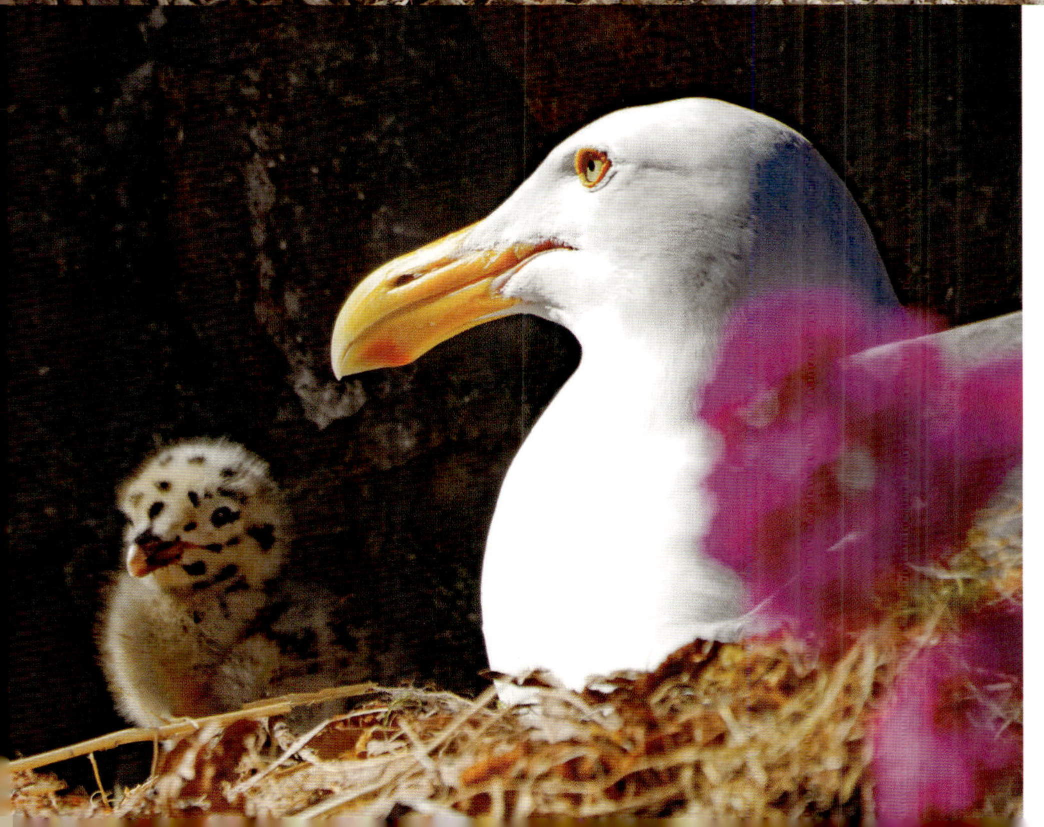

OPPOSITE: On the western shore of Alcatraz Island, Great Egrets (top left, bottom right) share nesting habitat in the shrubby trees with Snowy Egrets (top right). After leaving the nest, fluffy chicks of both species don't fly right away; instead, they hop and scramble in the nearby branches—a developmental behavior called "branching."

ABOVE AND LEFT: Western Gulls nest directly on the ground in isolated places where predators like foxes and coyotes can't reach them: islands, headlands, and abandoned seaside structures. In colonies where females outnumber males, two females will sometimes form pair bonds.

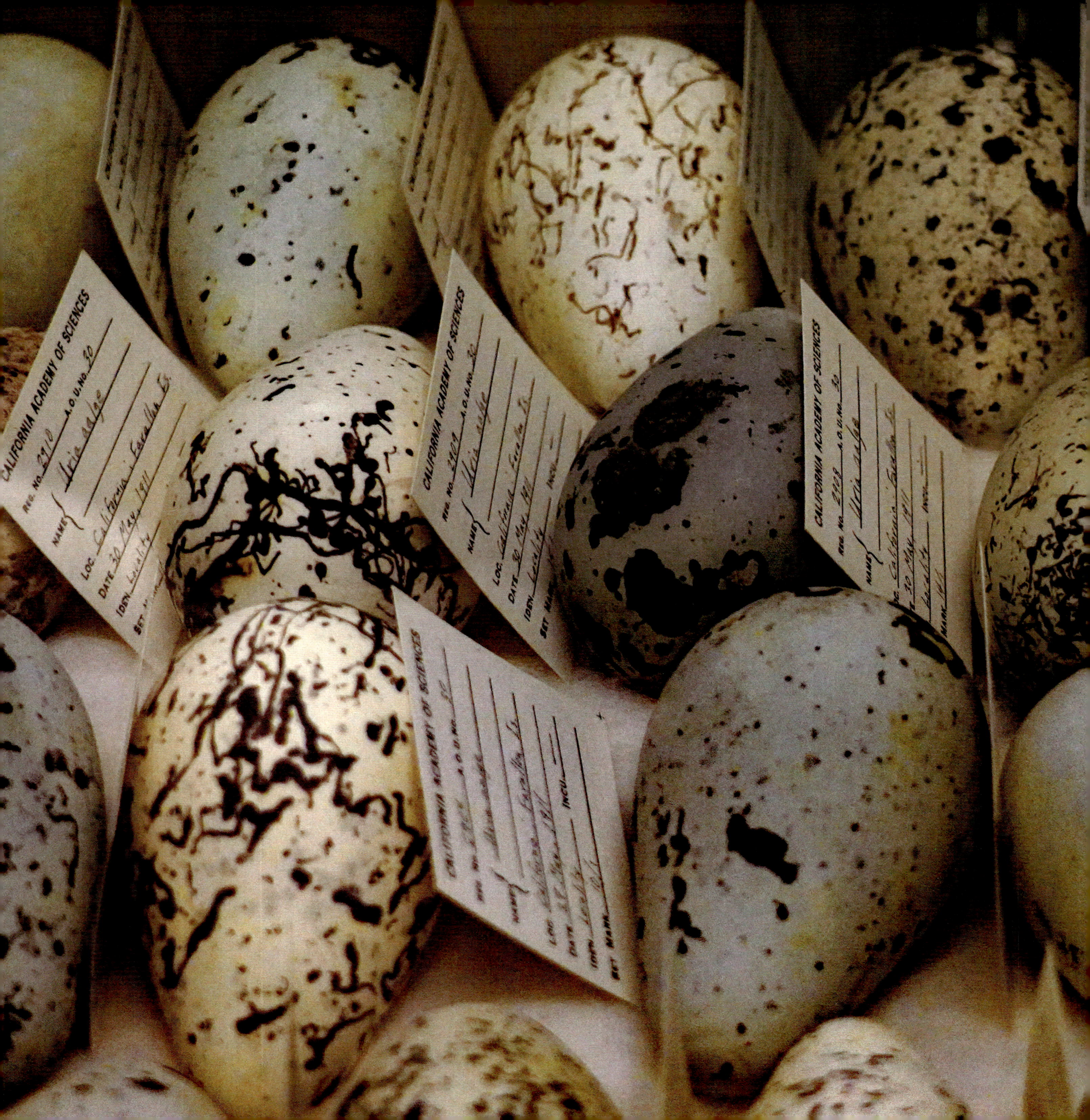
CALIFORNIA ACADEMY OF SCIENCES
CALIFORNIA ACADEMY OF SCIENCES
CALIFORNIA ACADEMY OF SCIENCES
CALIFORNIA ACADEMY OF SCIENCES

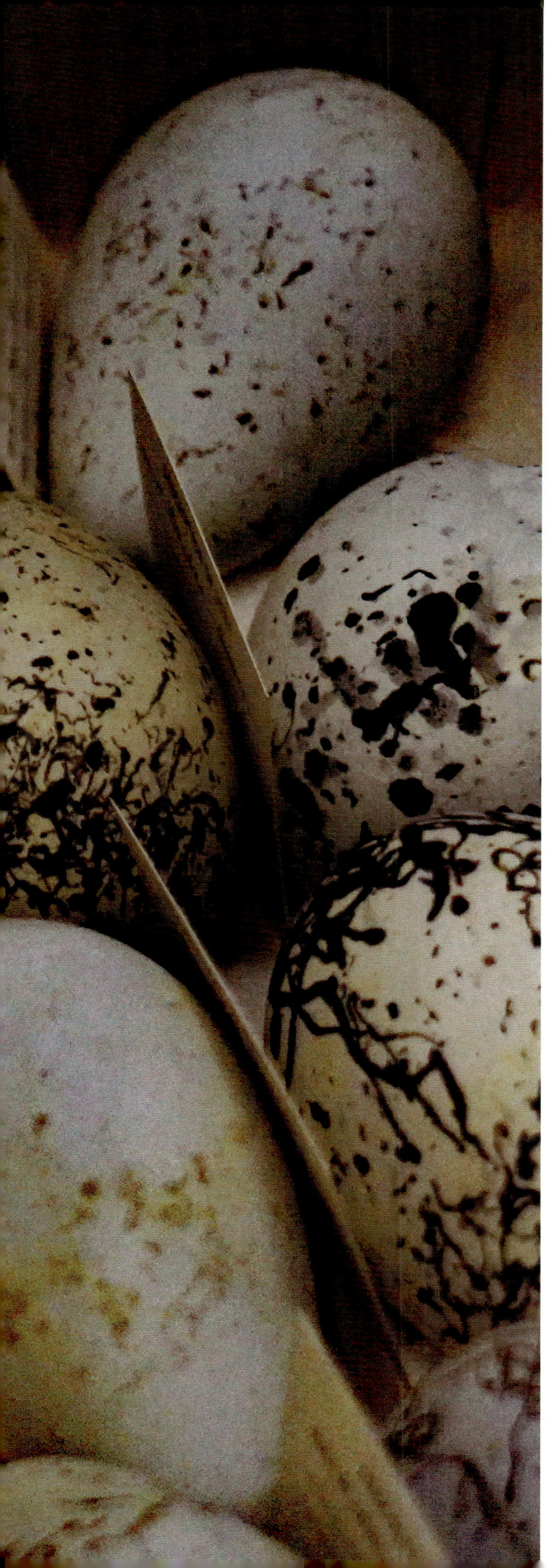

A LIBRARY OF BIRDS

It takes enormous restraint not to reach out and pet the birds: perfect owl specimens with plump, downy chests, so life-like that they could be sleeping; silky Anna's Hummingbirds preserved on sticks like a drawerful of cherry-red lollipops. The birds lie neatly on their backs, legs criss-crossed under their bellies and tied with twine. Tags—the oldest ones, from the nineteenth century, darkened with oil from the birds' bodies—mark when the animals were collected, where, and by whom. Some of these birds are "type specimens"—used by taxonomists as the very first models upon which descriptions for the whole species were based. Nearly a hundred thousand birds are nestled in the stacks. As we open and close drawers, their feathers lift and settle.

The California Academy of Sciences (CAS) was founded in 1853 to showcase the "wealth of the West," says Jack Dumbacher, curator of the Academy's Ornithology and Mammalogy Department. California had only joined the United States three years prior. The rapid acquisition of gold and land undergirded by the systematized genocide of California's Indigenous communities defined this era. It was a time of gunfight duels in the streets of San Francisco (this was how one of the academy's first board members died) and a time of gentlemen naturalists voraciously collecting flora and fauna new to Western science. The CAS collection is a result of this effort spawned equally by curiosity and colonization; it has become one of the West Coast's largest collections of birds—a collection that now serves as a lens on the past and a cipher for the future.

The drawers of the CAS bird collection contain more than just birds—the library is also home to nests, historical human artifacts, and these intricately patterned Common Murre eggs collected from the Farallon Islands in the early twentieth century. (Photo by Sharon Beals)

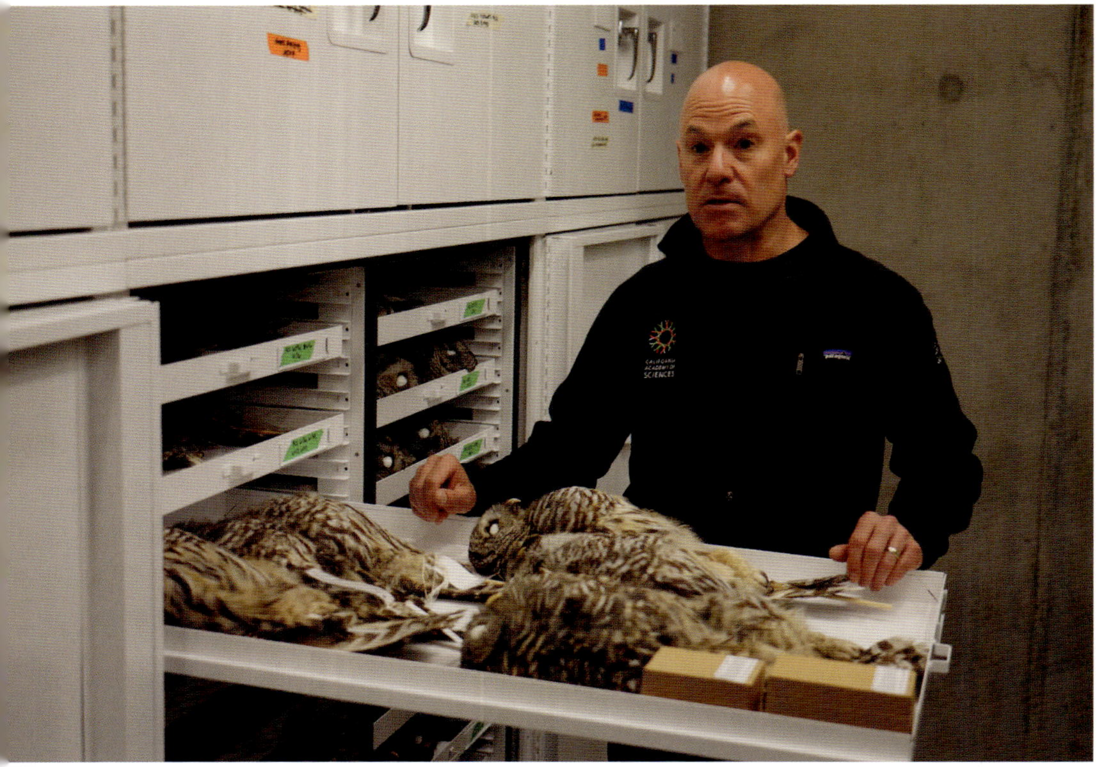

Jack Dumbacher (above) shows visitors several of the CAS Barred Owl specimens. Jack's research on the relationship between the endangered Northern Spotted Owl and the invasive Barred Owl has helped inform a complex debate on the West Coast regarding the interplay of land management, deforestation, and interventions to protect endangered species.

The specimens here help answer questions that field science cannot. Isotopes from birds collected in different years give insight into changing migratory patterns and reveal how birds' diets change over time. Feathers accrete heavy metals and can expose the hidden paths that mercury and other "forever chemicals" take into San Francisco Bay. Rodenticide found in the preserved livers of owls unlocked an odd mystery—the highest levels of poison were turning up in owls not at the edges of shopping centers and urban development but deep in pristine forests. A satellite follow-up of the study revealed that the specimens with the highest rodenticide levels were hunting at the edges of illegal marijuana grows. "Access to these owl specimens allows us to explore the health of the entire regional forest system," says Jack, and "to build a concrete scientific case for increased forest monitoring and species protection before it's too late."

Jack pulls open a drawer of football-sized Barred Owls, with white cotton tucked into wide eye sockets, and narrow beaks curving out of whiskery feathers. The museum has an especially big collection of Barred Owls—interlopers from the East that have been displacing endangered Spotted Owls in forests all across the West. CAS is parsing their genomic story to better understand how they hybridize with native Spotted Owls—something the founders of the academy could never have imagined, nearly a hundred years before the discovery of DNA.

In its onsite labs, CAS has freezers for DNA samples, PCR machines, CT scanners, x-rays, and structured light scanners. Each technology unlocks secrets from specimens whose labels may first have been written by candlelight. Researchers from other institutes borrow specimens like books from a library, with inventive new methods of inquiry. Some need claws from hummingbirds; others study eyes, stomachs, musculature, bones. "You have to balance using the specimens for whatever questions society needs answers to now, and preserving them so that they make it to the future and can answer tomorrow's questions," says Jack. "We can't even imagine yet all the ways that the specimens might be useful."

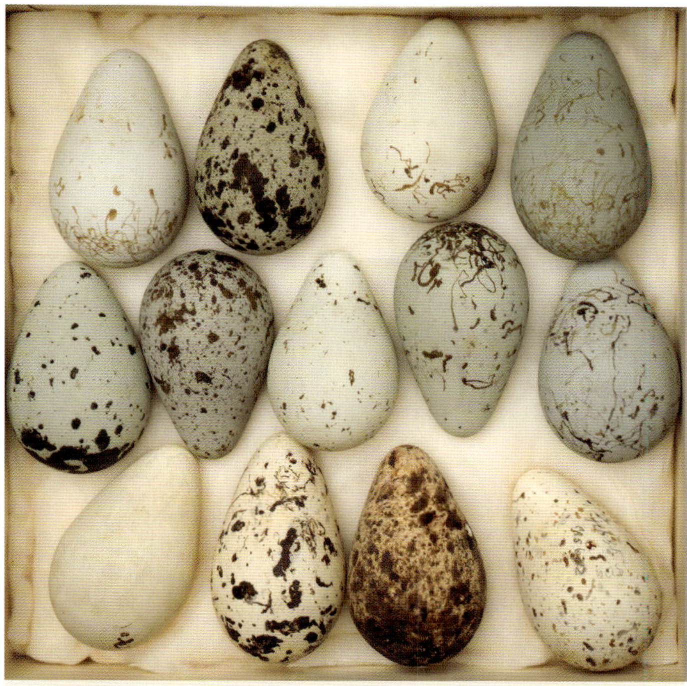

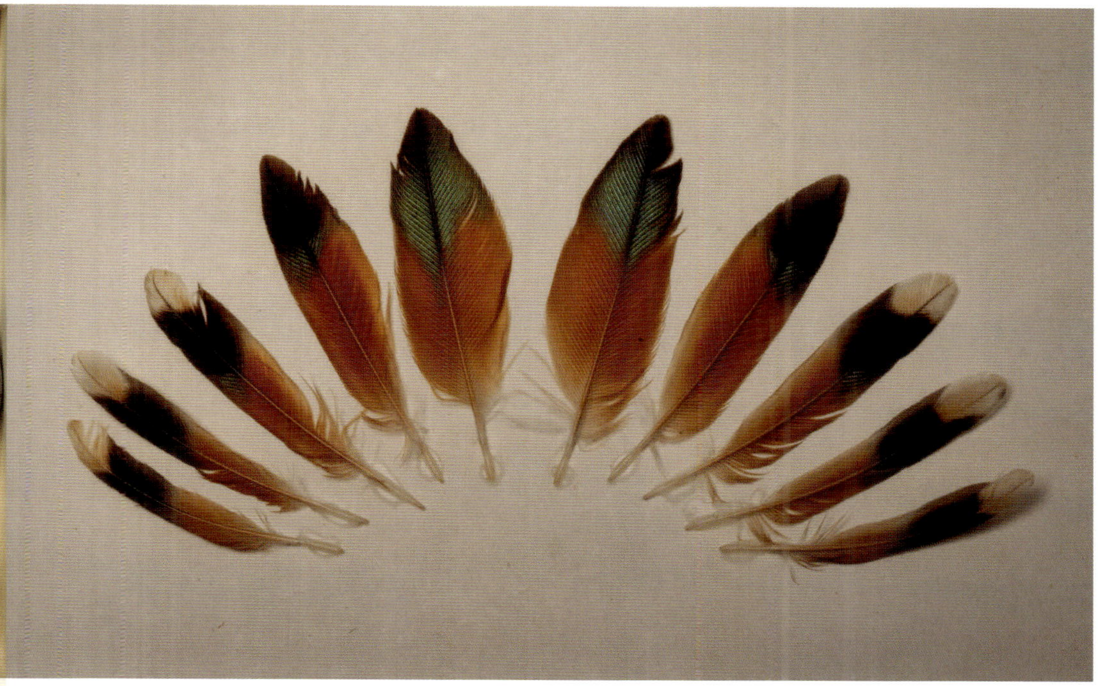

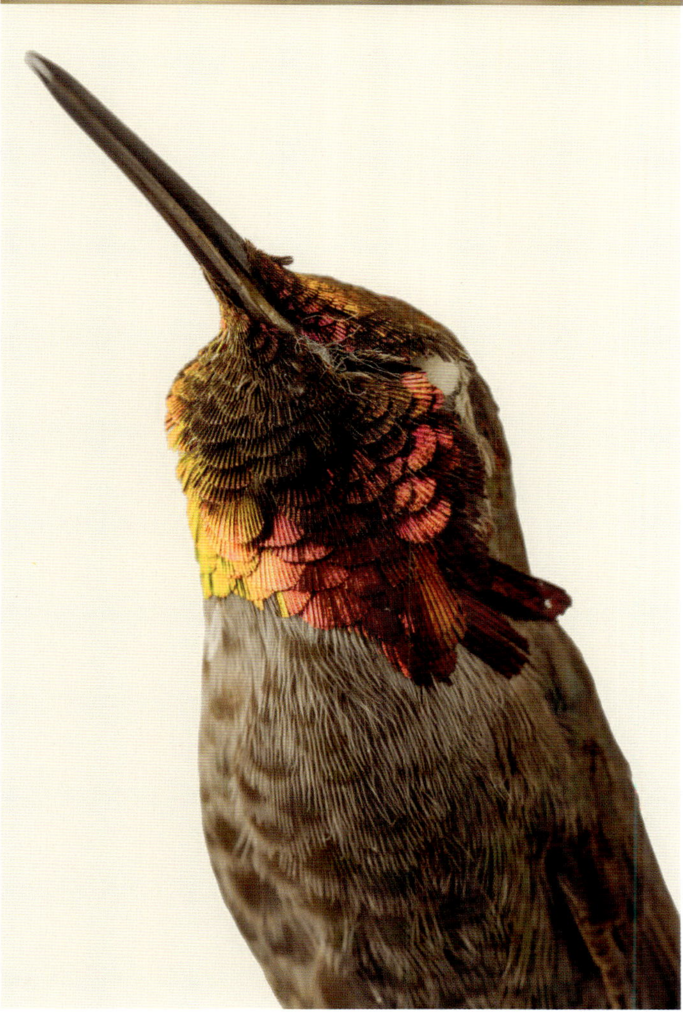

Each Common Murre egg (above left) has a streaky pattern as well as a color—turquoise, stone-brown, cream, pale blue—that are unique to that individual, like a fingerprint. The CAS collection of Common Murre eggs captures some of this mesmerizing diversity. (Photo by Sharon Beals)

The beautifully preserved tailfeathers (above) of a female Allen's Hummingbird are laid out side by side. The shape of the tailfeathers is one of the only ways to accurately differentiate between Allen's Hummingbirds and Rufous Hummingbirds, whose ranges overlap in the Bay Area. (Photo by Sharon Beals)

A close-up look at an Anna's Hummingbird (left) specimen reveals the subtle textures and gradations of iridescence on the bird's neck. (Photo by Sharon Beals)

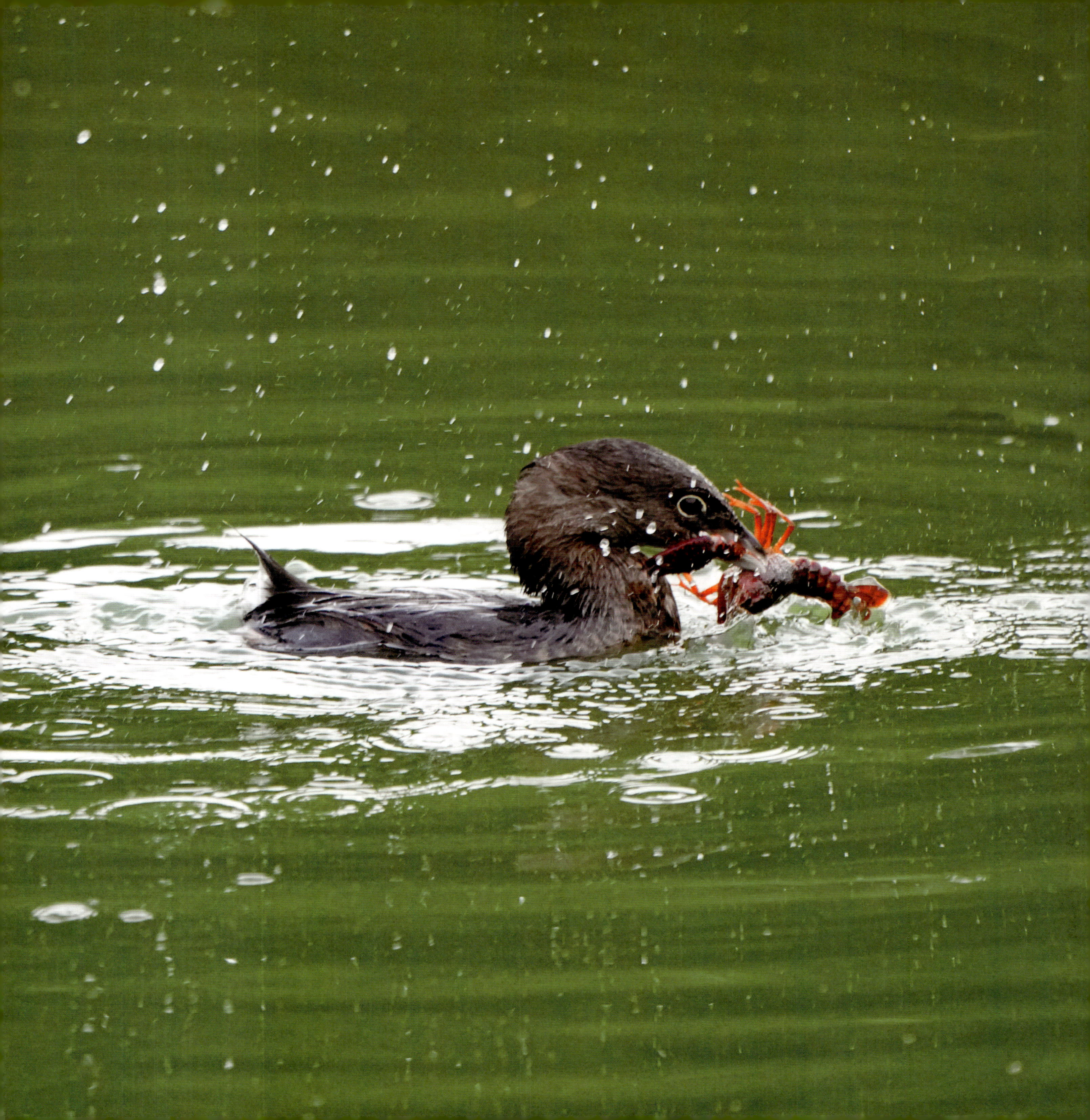

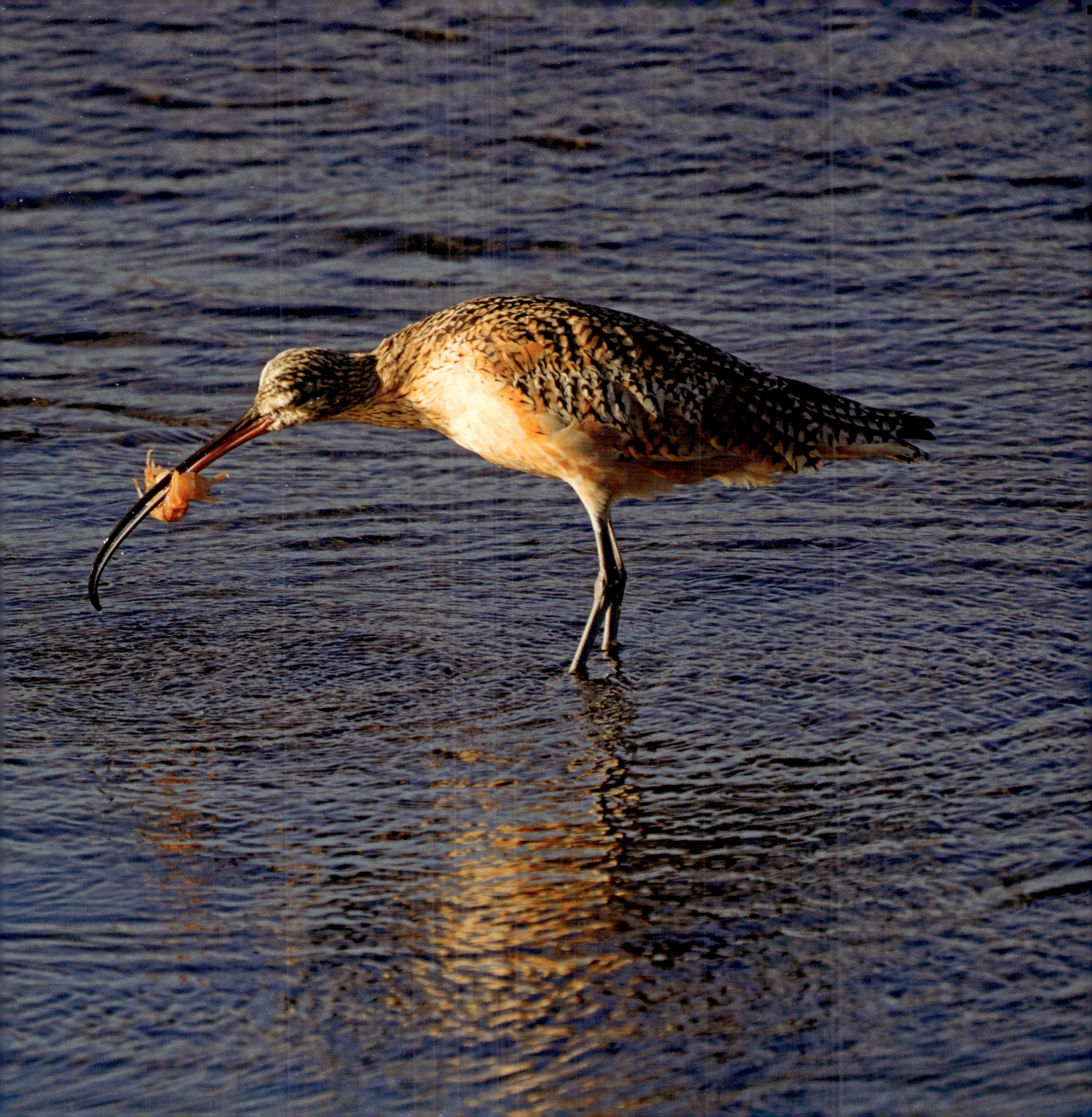

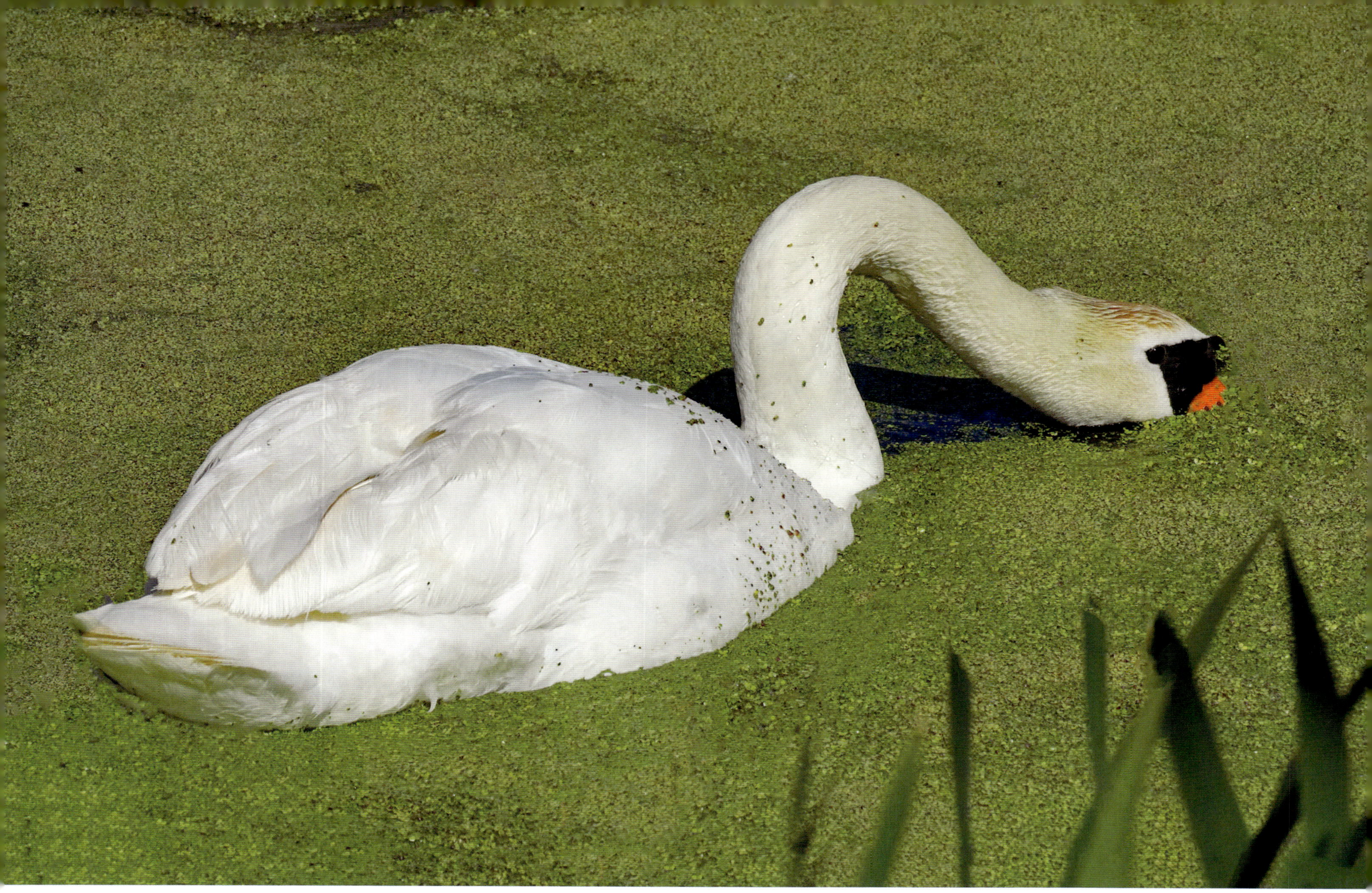

PRECEDING: a Pied-billed Grebe (left) and a Long-billed Curlew (right) both claim crunchy crustacean snacks. The food that birds eat helps scientists understand how isotopes move through an ecosystem. By analyzing isotopes in bird feathers, for example, researchers can trace the bird's movement and diet history, revealing the sources of their food and the paths of elements through the food web. This information is crucial for understanding processes like migration, and the transfer of energy—or unseen toxins—between different species in an ecosystem.

THIS PAGE AND OPPOSITE: Bay Area birds' diets are diverse and often specialized according to their habitat and beak shape. Clockwise from left, a Mute Swan grazes on algae, an American Kestrel eviscerates a small bird, a White-tailed Kite clutches a rodent, a Common Raven carries off a rabbit drumstick, and a Pigeon Guil emot slurps down a gunnel.

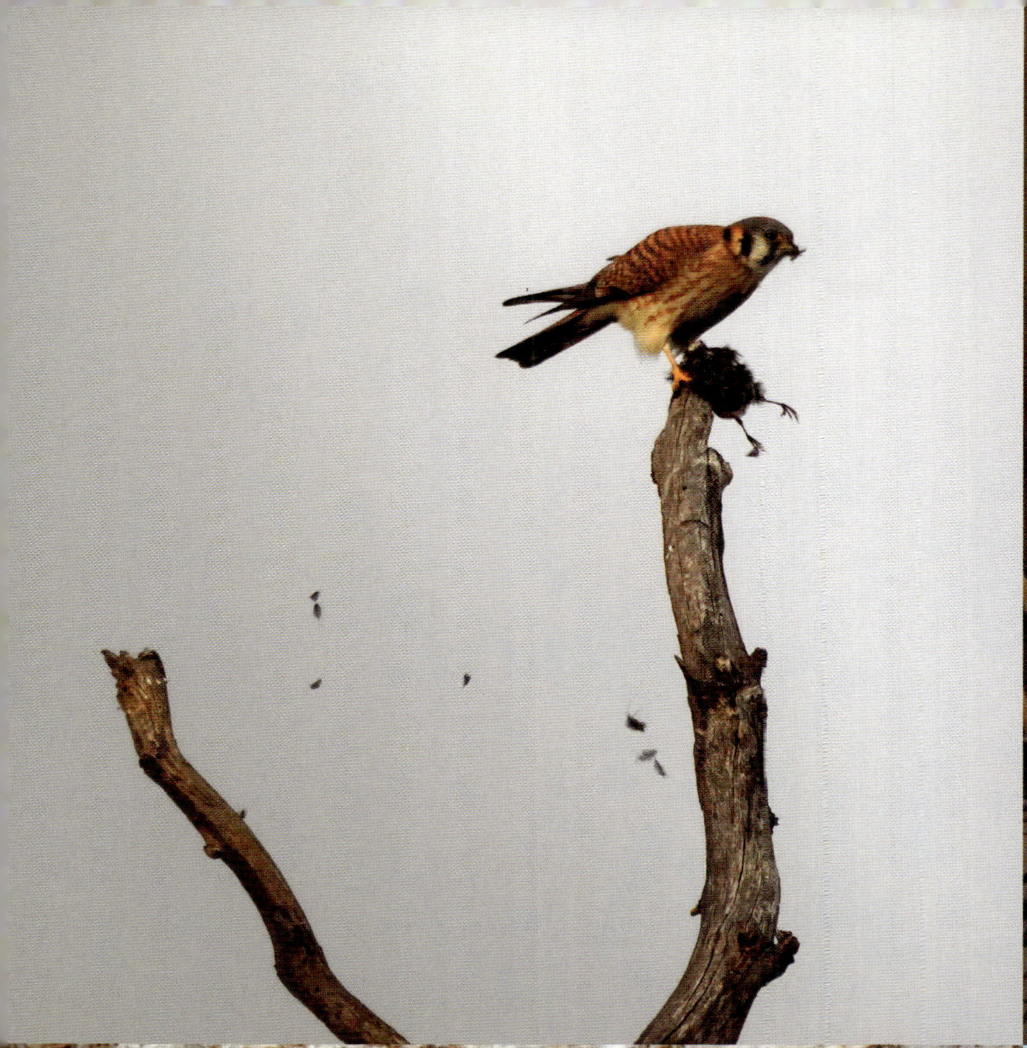
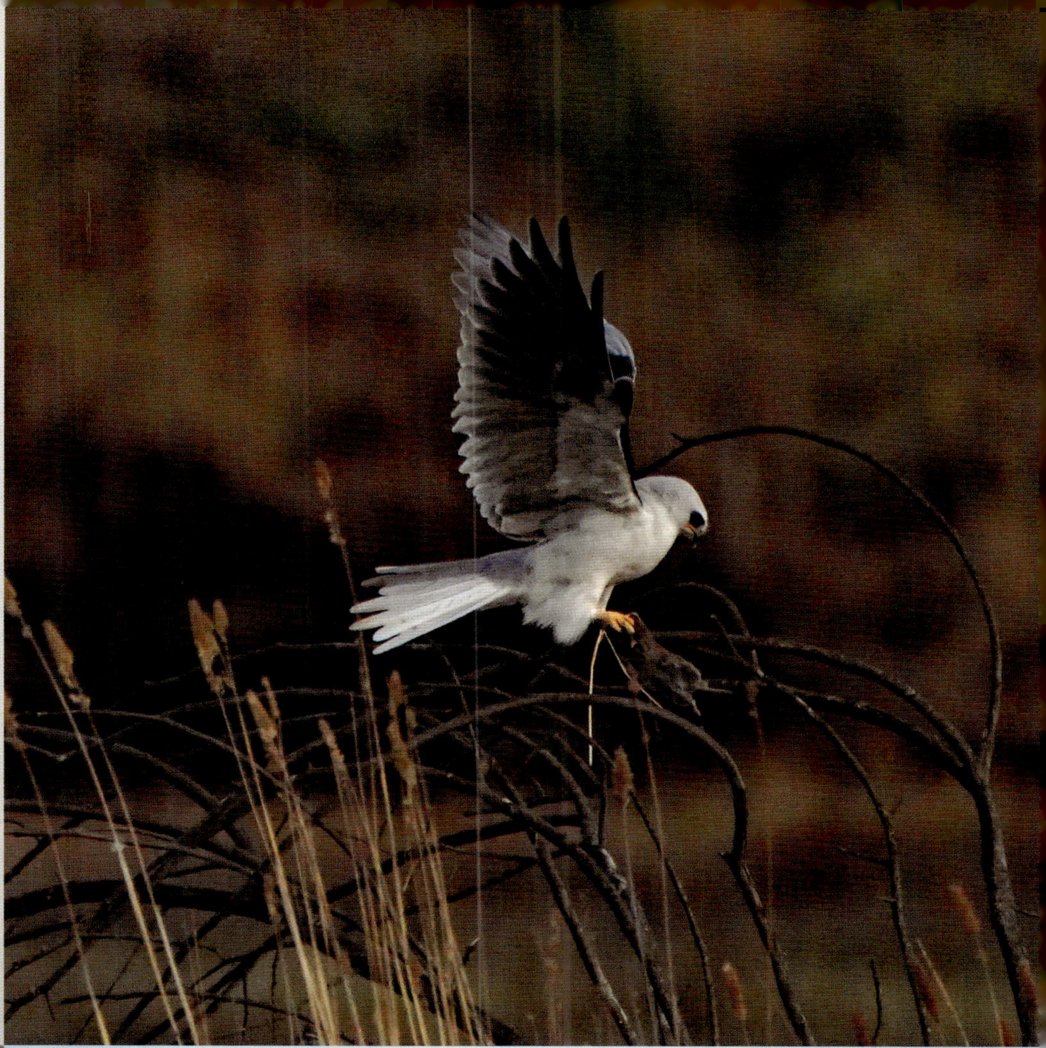
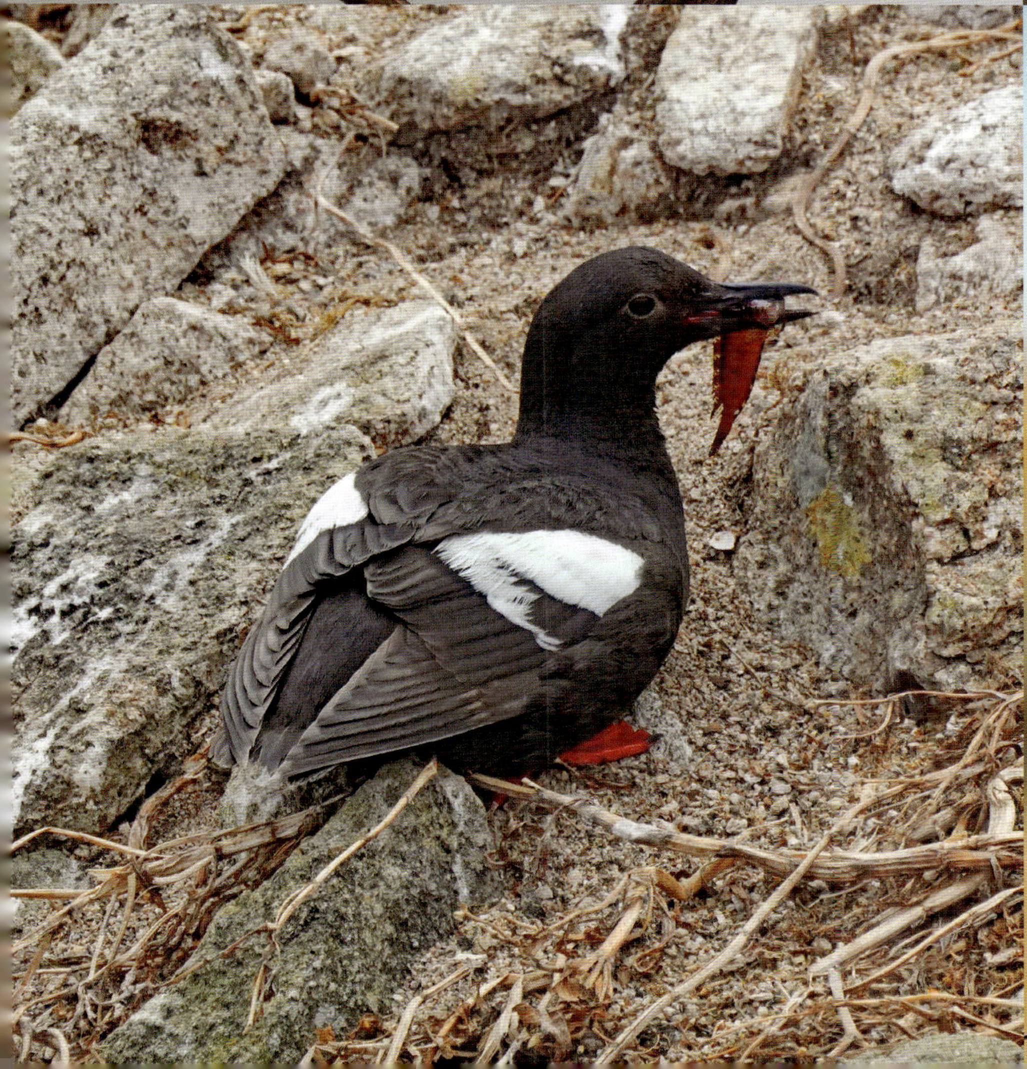
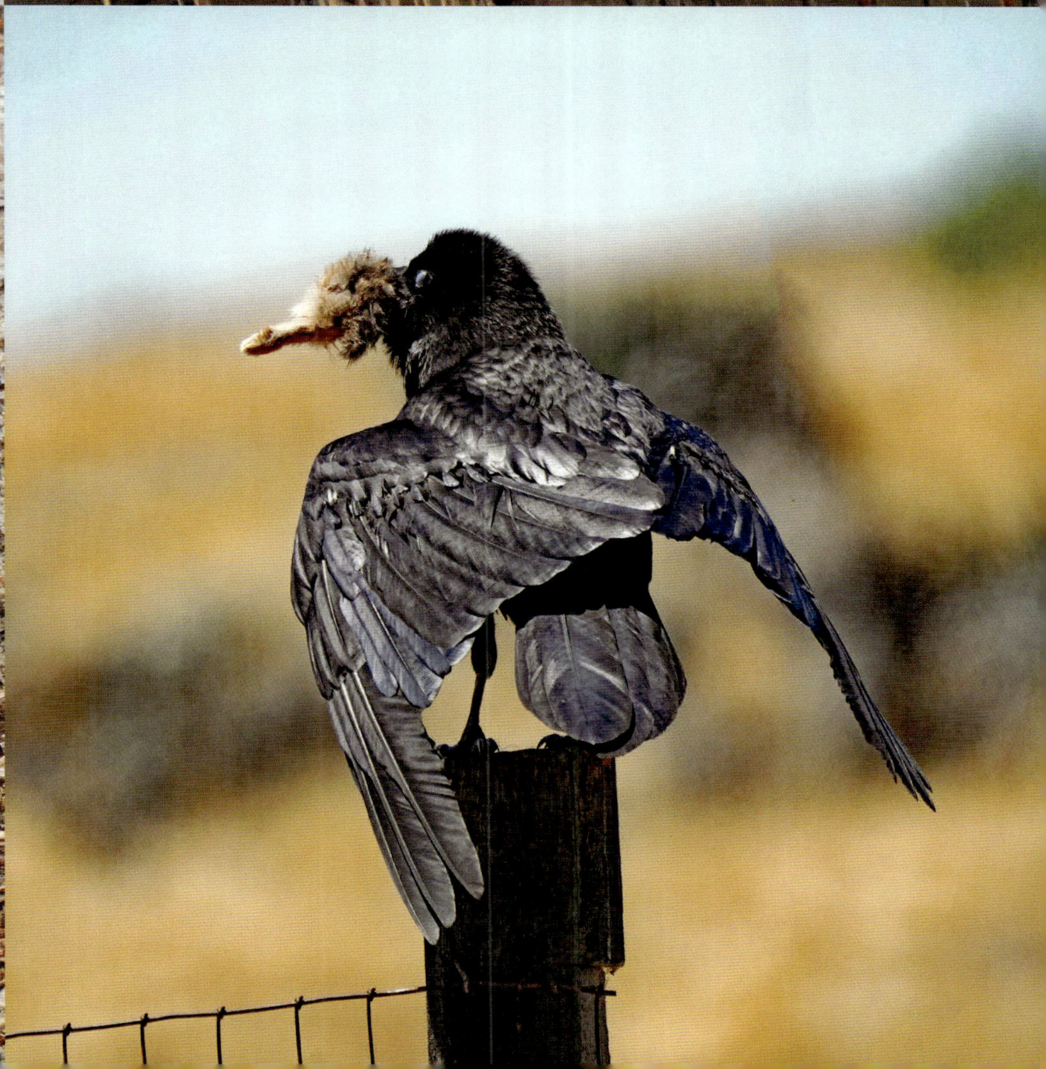

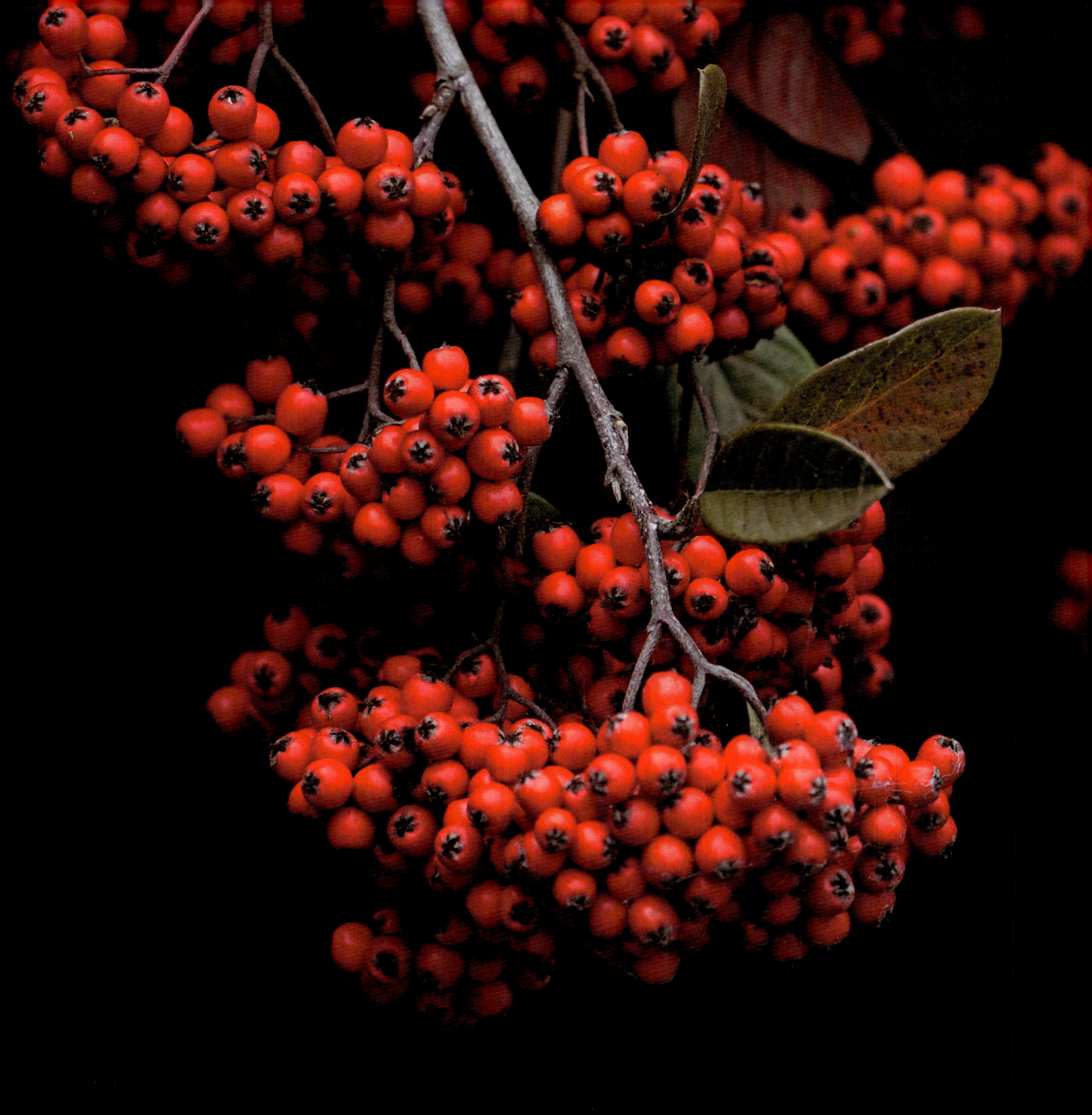

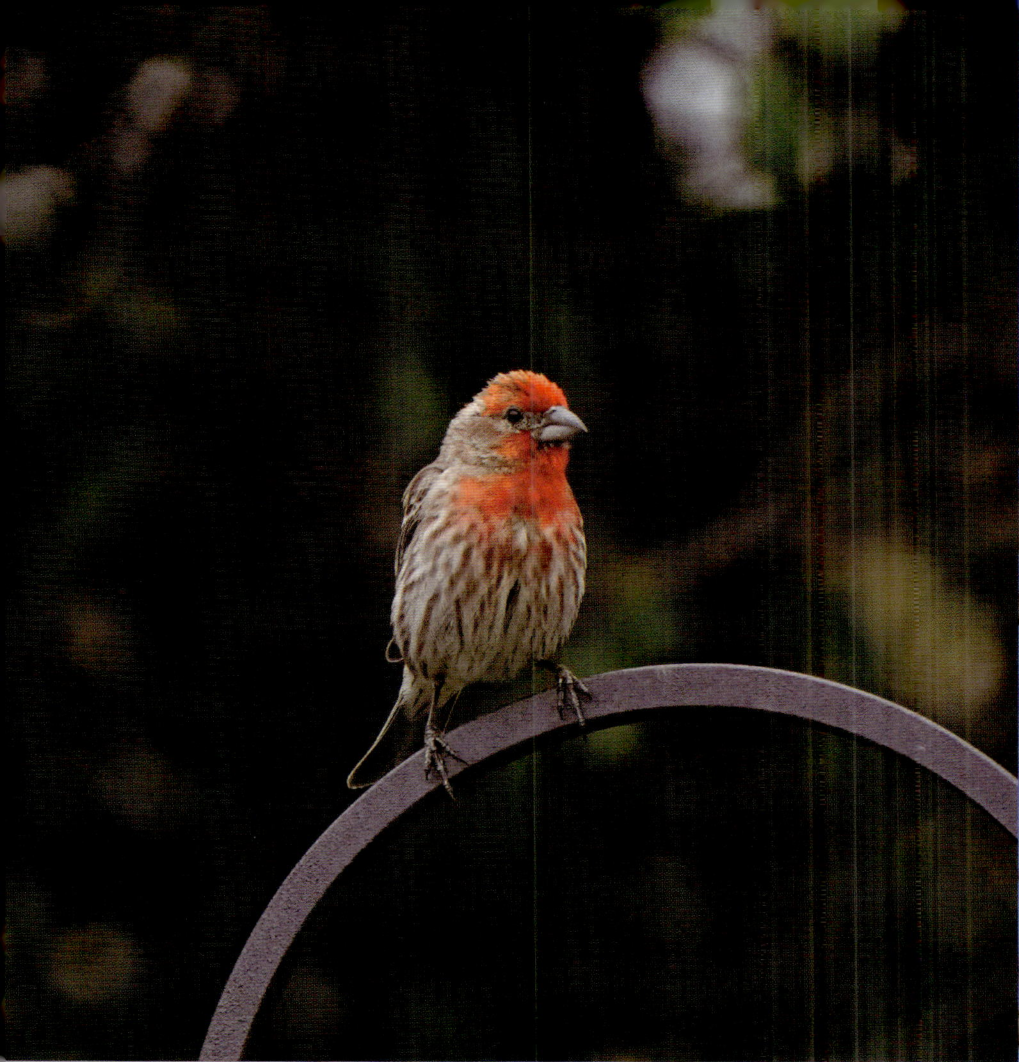

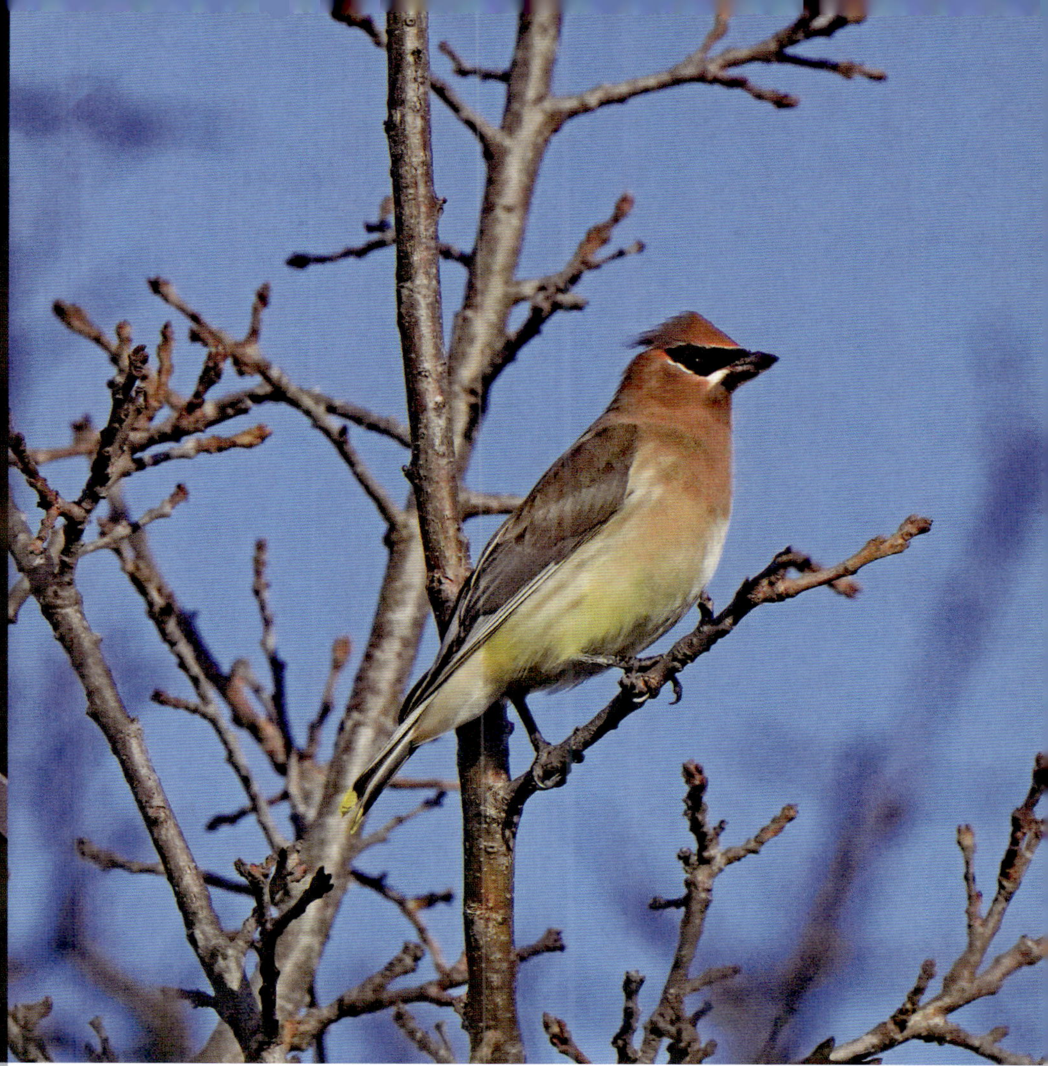

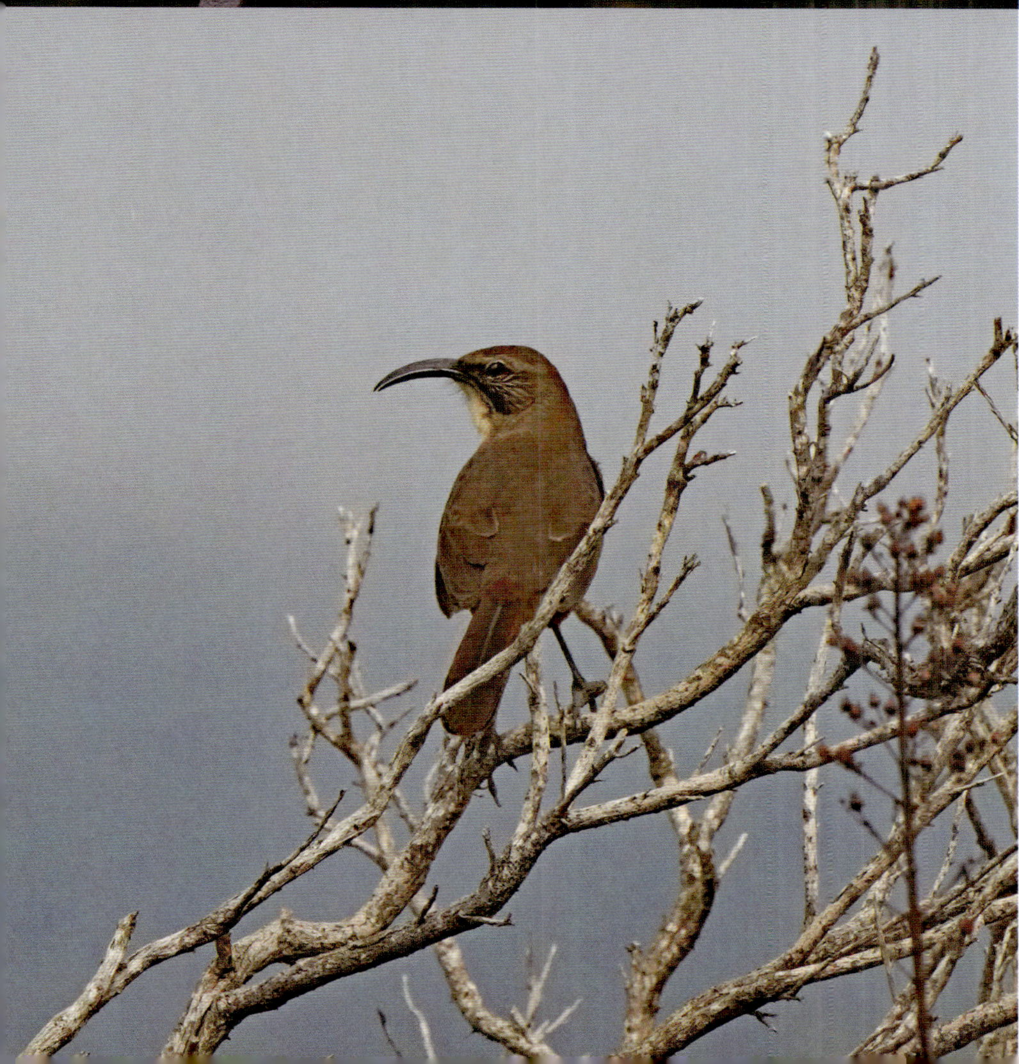

Many of our decorative backyard shrubs like this cotoneaster bush (opposite)—along with native Bay Area berries like toyon and manzanita—are magnets for fruit-loving birds including the House Finch (top left), the Cedar Waxwing (top right), and the California Thrasher (bottom).

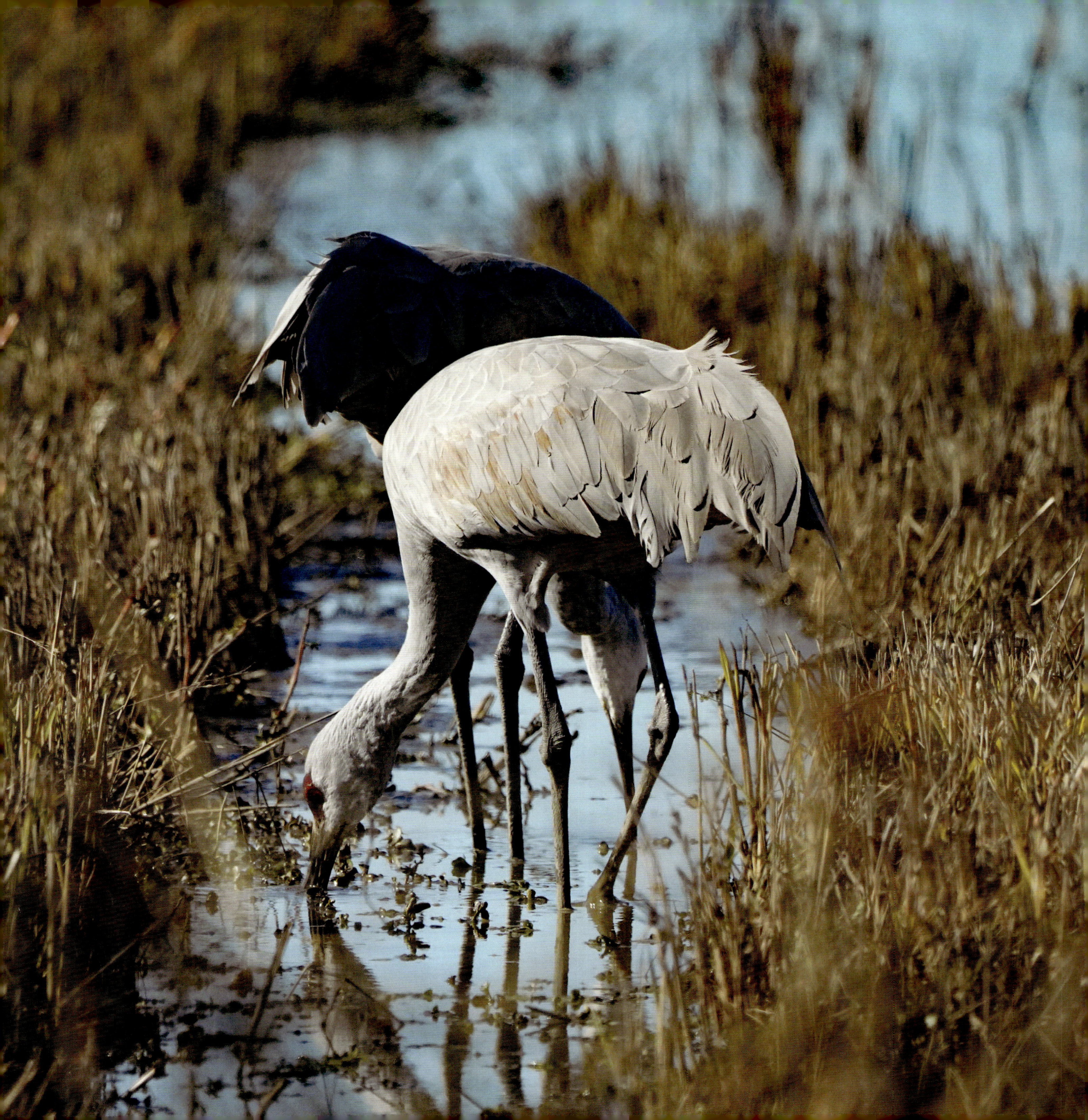

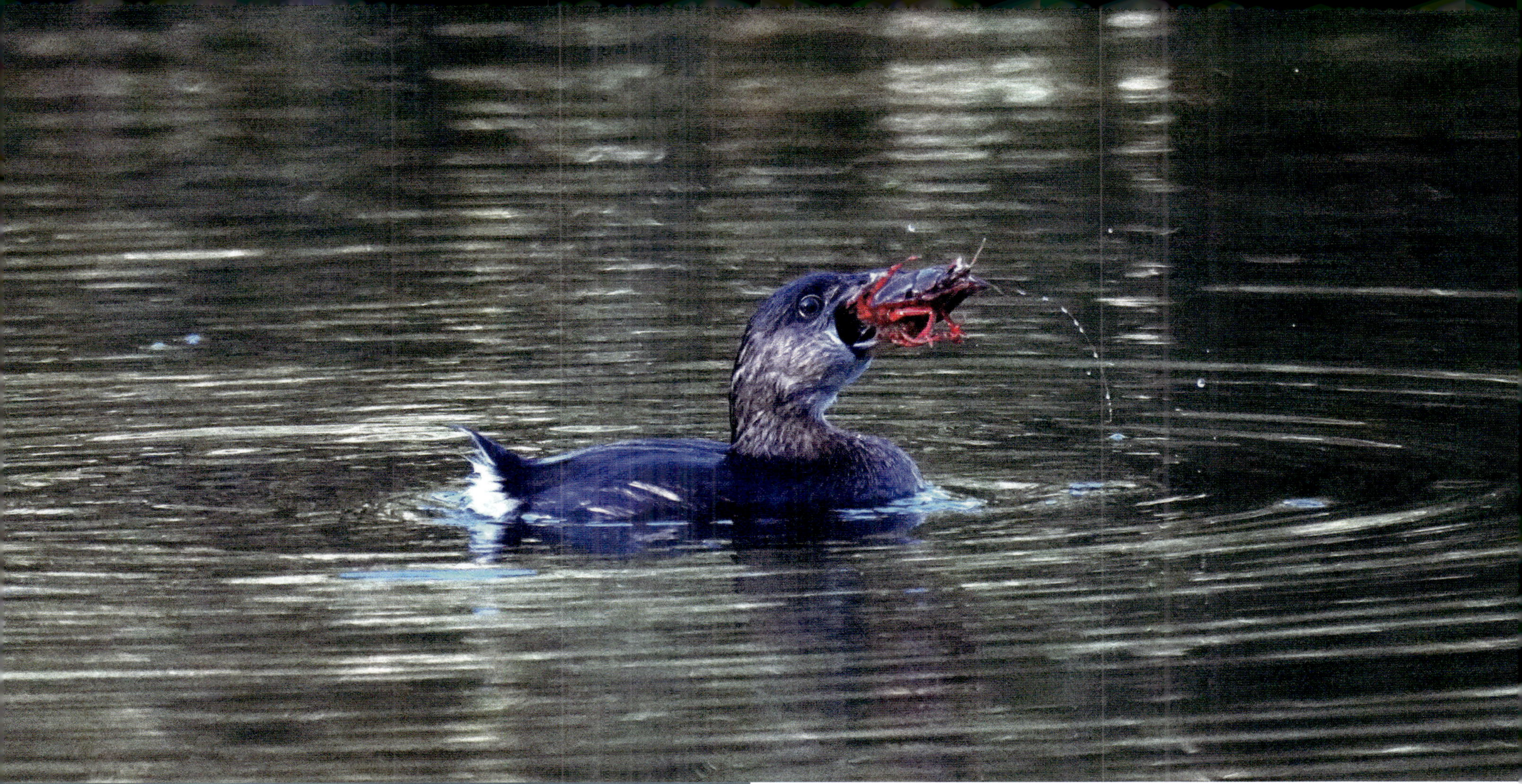

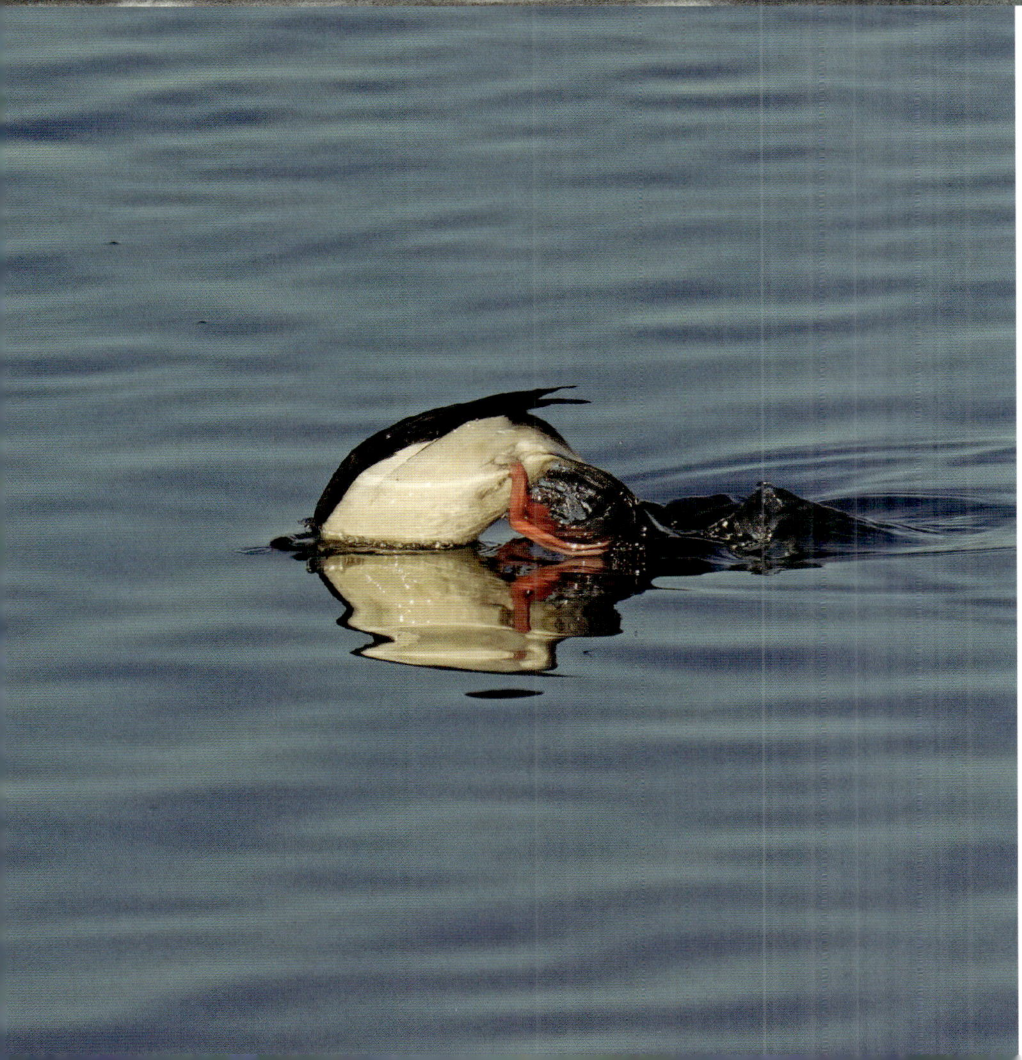

Where they overwinter in the Central Valley, Greater Sandhill Cranes (opposite) forage primarily on waste grain in corn, rice, and wheat fields. The crane species that come to the California Delta are entirely dependent on state and federal refuge lands and private agricultural lands for winter roosting and foraging habitat.

A tenacious Pied-billed Grebe (top) tries a new angle for swallowing down a large crayfish.

A male Bufflehead (bottom) takes a dive in search of invertebrates on the pond floor. His legs are positioned farther back than those of dabbling ducks, for better propulsion.

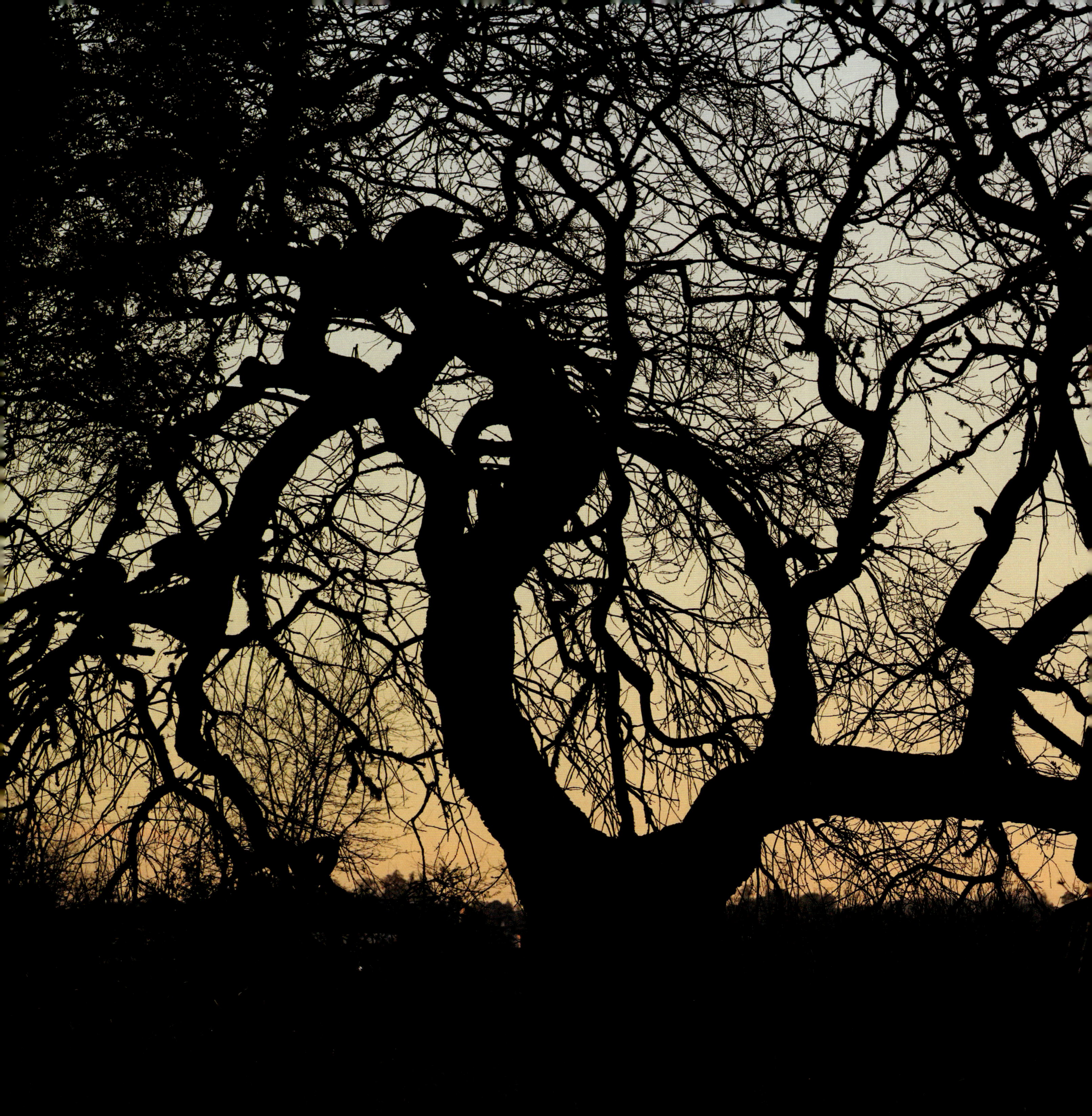

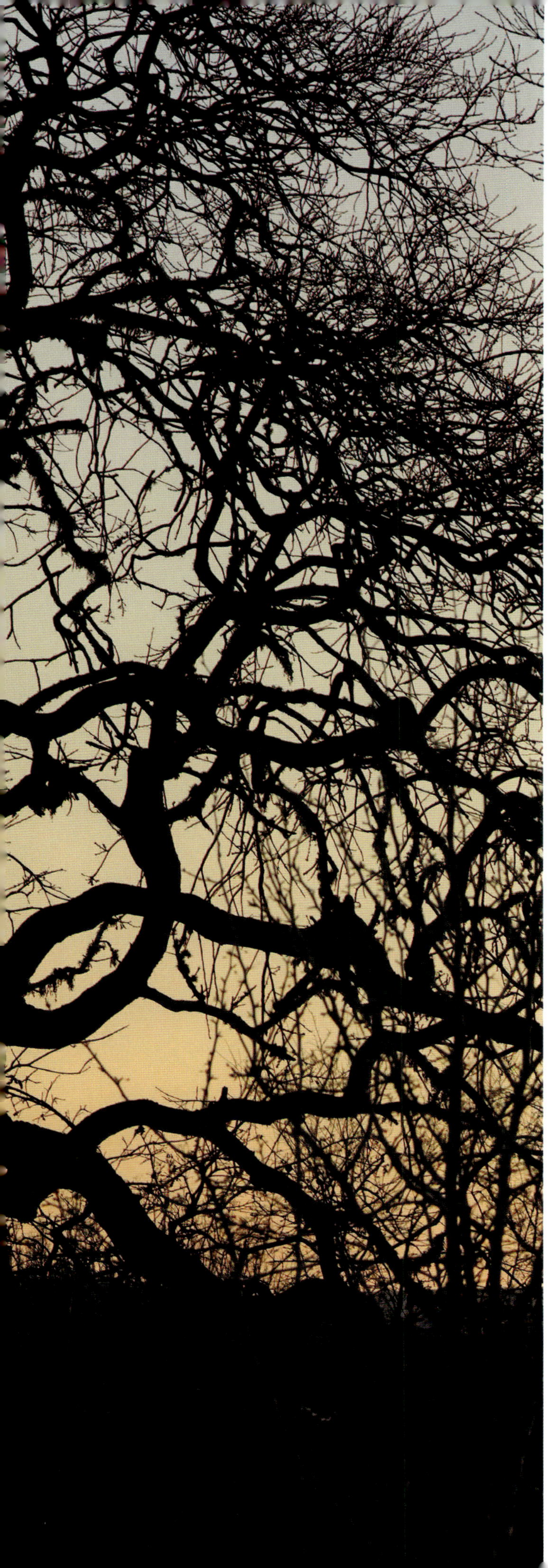

THE ART OF ATTENTION

Science and the creative arts share a few common roots: observation, attention, and patience. The field notebooks of Point Blue bird banders contain data on the birds trapped and released: their wingspan, how worn their flight feathers are, the ossification of their skulls as seen through the thin tissue on a bird's head. But the cover of this season's data logbook, filled with numbers and measurements, is also embellished with a vibrant sketch of a Pileated Woodpecker. "I'm especially interested in woodpeckers," says Larissa Babicz. Among her gear kit—a magnifying visor for looking more closely at birds, rulers for choosing which size of band to use on a bird's leg—pastels and pencils are some of her sharpest tools for understanding the birds she studies. We listen to the call of a Swainson's Thrush, as fluid and clear as a drop of water. Shortly, we are joined at the Pine Gulch Field Station by the Point Blue Conservation Science board chair Anne Chadwick. While the apprentices measure birds, she, too, sketches in her notebook.

Later, Larissa shows me another piece of her art: a poster-sized pastel landscape that captures every bird she saw over the course of a monitoring season. Wrentits nestle among the leaves of poison oak; a quartet of hummingbird species hovers near blooming sticky monkey flowers. The piece is not representative of what a camera might catch in a single snapshot of the landscape. Instead, the art evokes the diversity of an ecosystem from spring to fall, concentrated in a single frame. I linger in front of it, imagining the passing of time, feeling my chest fill with the wonder of so many birds in one place.

After our visit to Pine Gulch, we continue into the tiny West Marin community of Bolinas, where we follow a narrow alley to the back door of Keith Hansen's studio. The local artist has a video tripod set up by his window, overlooking a courtyard where a Scrub Jay preens in the late morning and Eurasian Collared Doves sidle along the eaves of a neighboring building. "Anytime I see something I want to paint, I can freeze the frame to study it more closely," Keith says. He is a steadfast observer, and looking out this single window, he has documented 231 different species of birds. His studio has become an eBird hotspot, with strangers walking in off the street, following the glow of the bird-monitoring app on their phones. What makes this little corner of Marin County so special isn't simply that there's a higher diversity of birds here than in other nearby places; it's a hotspot because someone here is paying close attention from the same still point, across time and seasons.

In Keith's studio, the Beach Boys sing on the local radio station, and a dazzling vase of feathers stands by his drawing desk. He lays out his latest project on a table: bird after bird, rendered in soft graphite and rich brushstrokes, drawn for a new app called Flock. It's the sister to the popular Merlin app, which enables users to identify birds by sight and by sound, he tells us. "But Merlin is all about ID; this will be about community. It's about sharing your sightings wherever you go." As users add to their lists, their digital journal will fill not with words but with images, just like a naturalist's field journal. Today, Keith is working on flycatchers. Eventually, he'll paint every bird in North America.

When we leave Keith's studio, Anne Chadwick rejoins us for lunch across the road at Smiley's Tavern. Over lemonade and tacos, she opens her field notebook and shows us her sketches from her morning at the Pine Gulch Field Station. I recognize the Swainson's Thrush that the apprentices netted earlier—the one with the song like dripping water. In her notebook, Anne has captured the bird in swift, accurate pencil strokes. We can see its wing feathers on the page, just as clearly as when they were fanned open by the apprentices to assess whether it was a juvenile or an adult bird. (Juveniles, we learned that morning, grow their feathers swiftly and a little shoddily; the feathers of adult birds are more robust.) "Creating art in the field helps me slow down and truly observe," Anne tells us. She has been taking lessons with South Bay artist John Muir Laws, whose growing community of field-journaling students has begun collaborating with Point Blue to create infographics that bridge ecology and art.

"It enhances curiosity, and it really helps to understand a bird's form and function," Anne says of her own field-journaling practice. She talks about the new strategic plan for Point Blue, which centers on what she calls "connected conservation"—the new insights that can emerge at the crossroads where art meets science. I think of Larissa, distilling sprawling seasons into a single page, and of Keith, connecting observers to a whole continent of birds and to one another. Anne is onto something. This kind of careful attention isn't just pretty; it's powerful.

PRECEDING: Look closely next time you go walking in the oak savannah: The gnarled and fractal branches of the Valley Oak support hundreds of different kinds of insects and animals, including more than 60 nesting bird species.

TOP LEFT: Point Blue apprentice Larissa Babicz holds up a Brown Creeper at the Palomarin Field Station.

RIGHT: Larissa Babicz is the artist behind *Birds of Palomarin*, an intricate pastel rendering of the species she encountered during her field season as a bird banding apprentice for Point Blue.

BOTTOM LEFT: A Nuttall's Woodpecker moves among the mossy microcosm of an oak branch. These woodpeckers are true California natives, with a range extending just barely into northern Baja California.

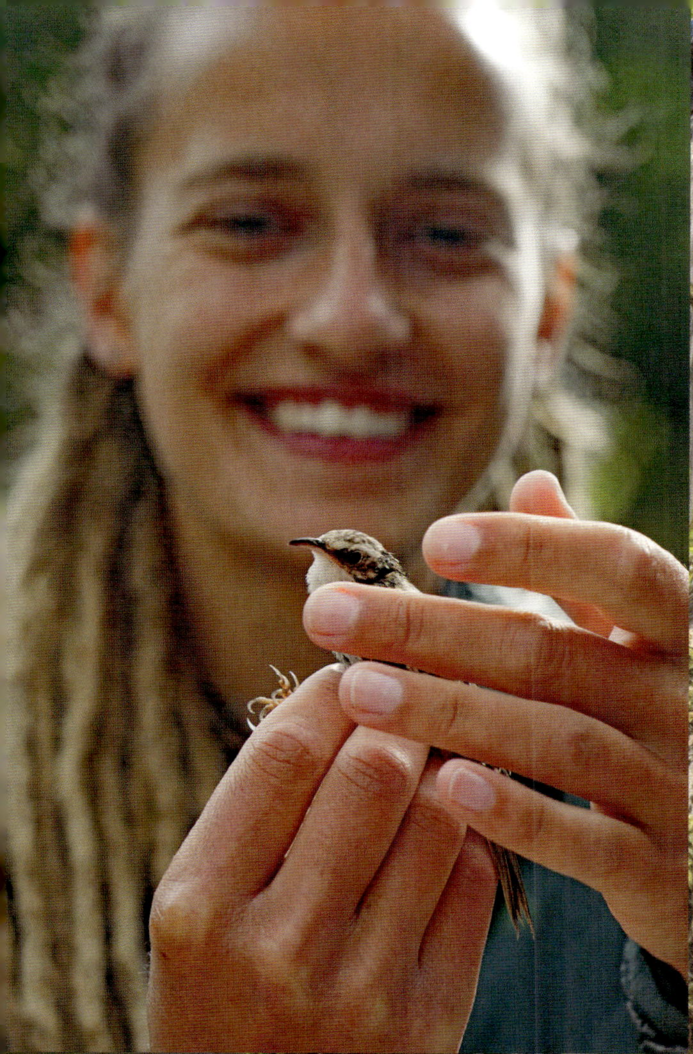

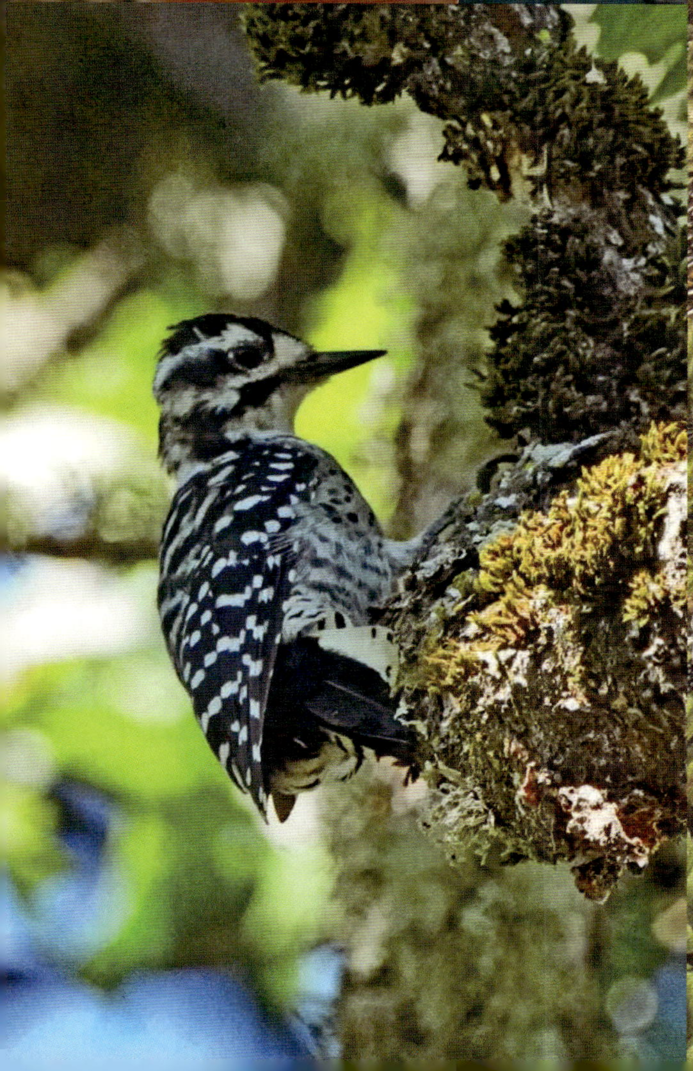

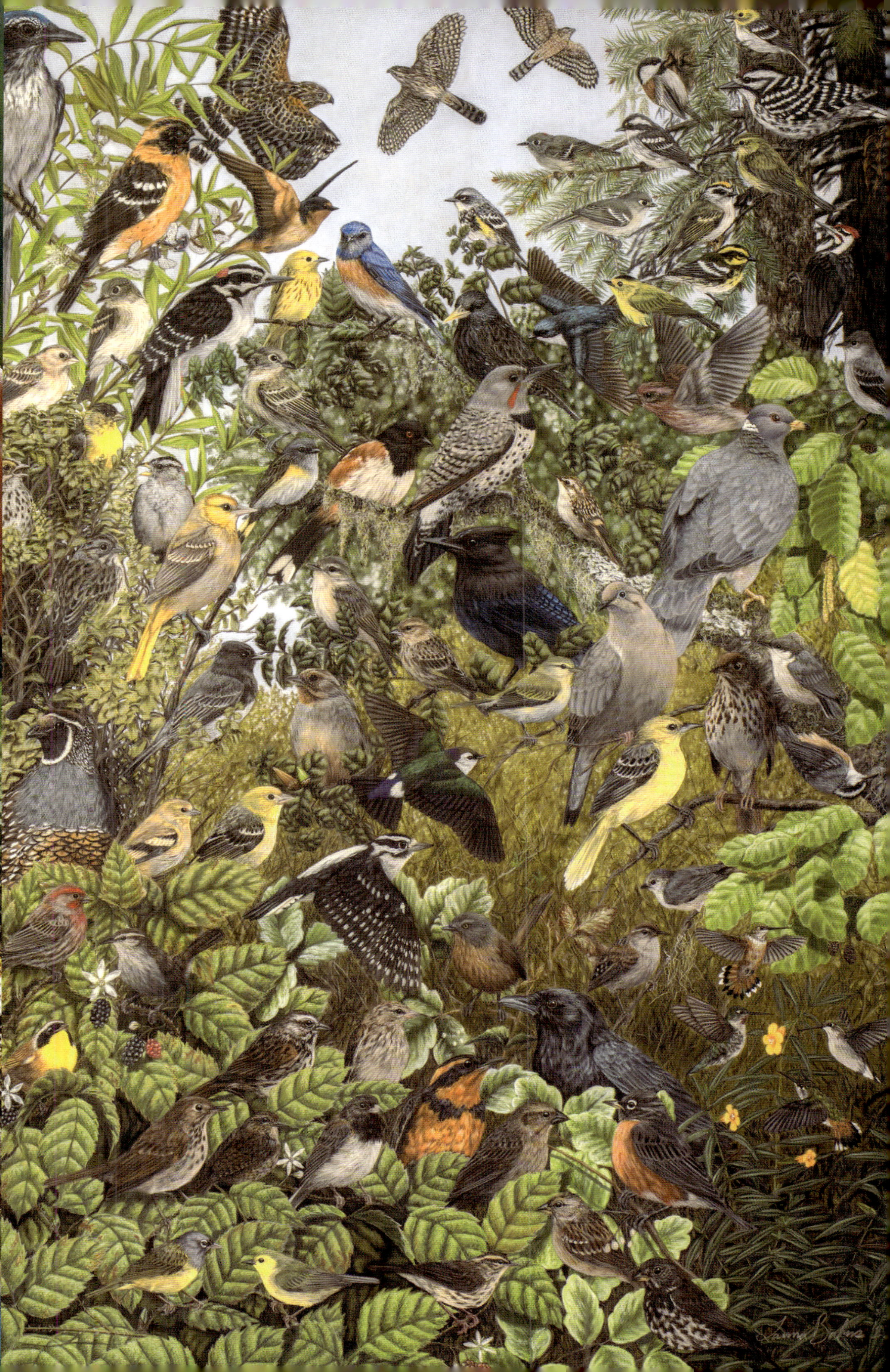

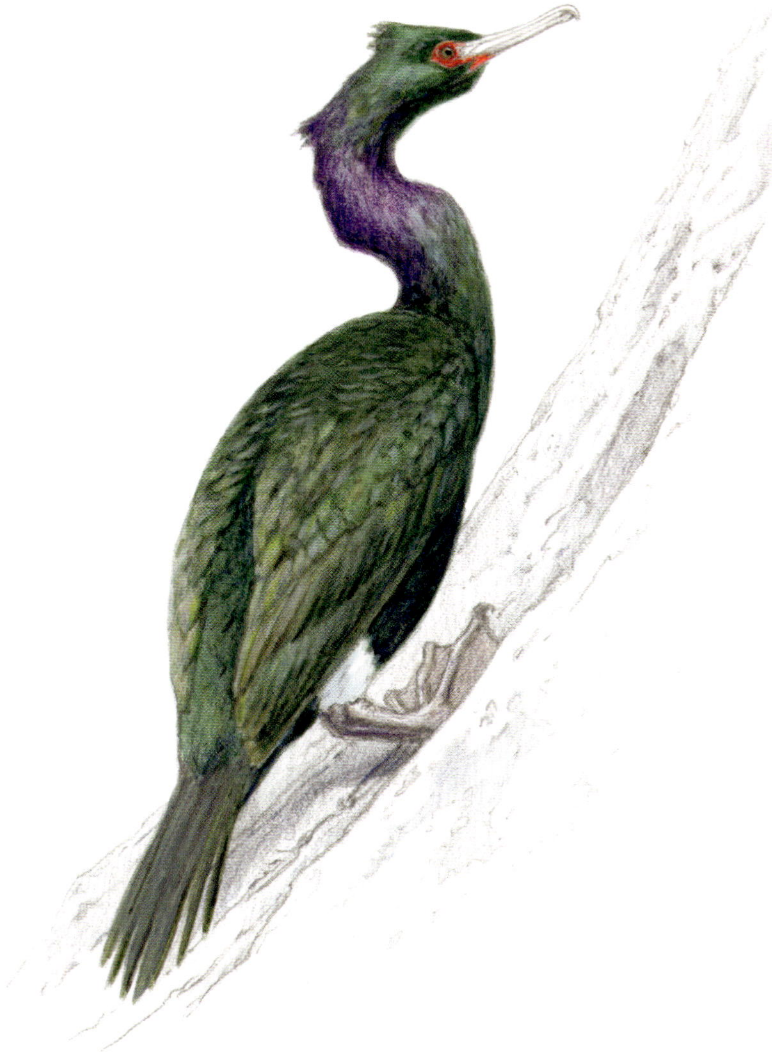

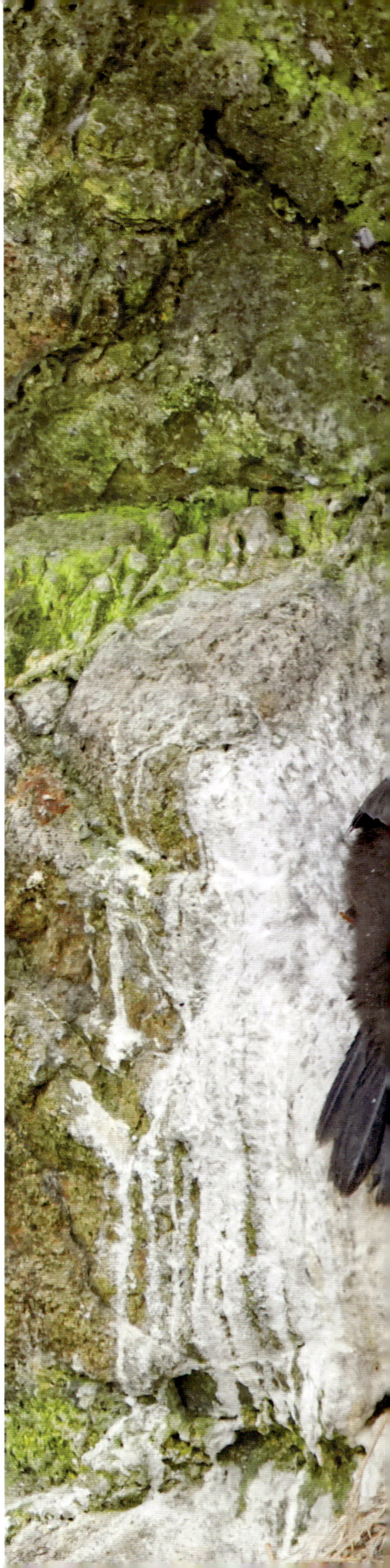

Pelagic Cormorants (right) perch along a steep cliff on the Farallon Islands, all striking different poses. The goal of a science illustrator is to distill the behaviors and physiology of an entire species into a single image that represents that bird. Keith Hansen's colorful illustration of a Pelagic Cormorant (left) for the cover of his book *Birds of Point Reyes* (Heyday, 2023) does just that: carefully mirroring and averaging out the details of the living birds.

FOLLOWING: As shimmering as this European Starling (left) appears, it likely is even more colorful in the eyes of other birds, which have a fourth type of cone cell in their eyes, allowing them to see ultraviolet light. Combined with colored retinal filters made of oil, which act a bit like a filter on a camera lens, bird eyes can perceive a mesmerizing color palette entirely unseen to us. The throat, or gorget, of an Allen's Hummingbird (bottom center) shines brilliant red—but only when the light catches it at exactly the right angle. The feather's structure refracts and scatters light back to our eyes, creating a blazing iridescent effect. Anna's Hummingbirds' (top, bottom right, bottom left) feathers shimmer a thousand different subtle colors as the tiny birds hover and zip. This iridescence is rooted in crystal-like nanostructures so complex that scientists are continuing to unravel their secrets.

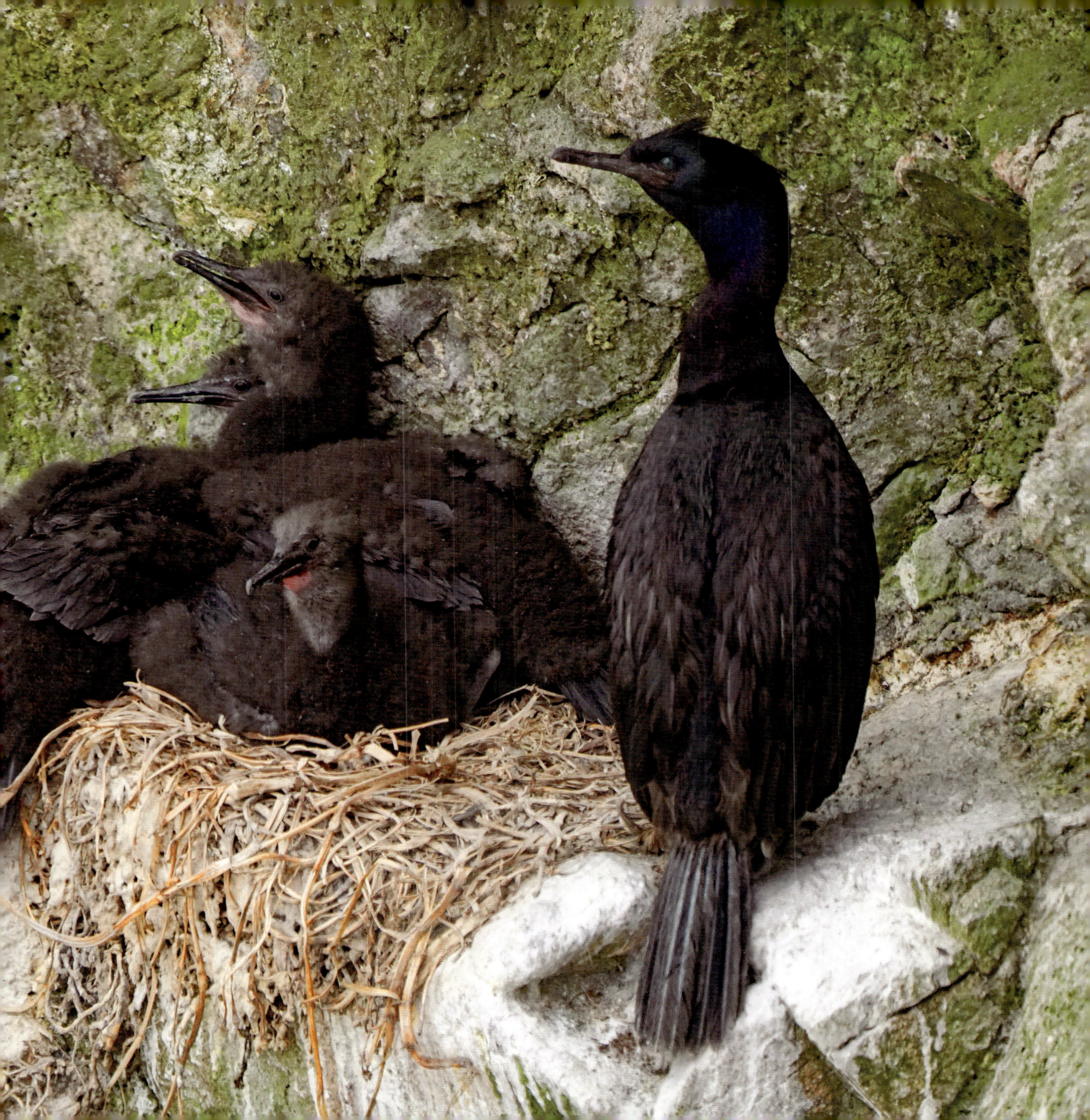

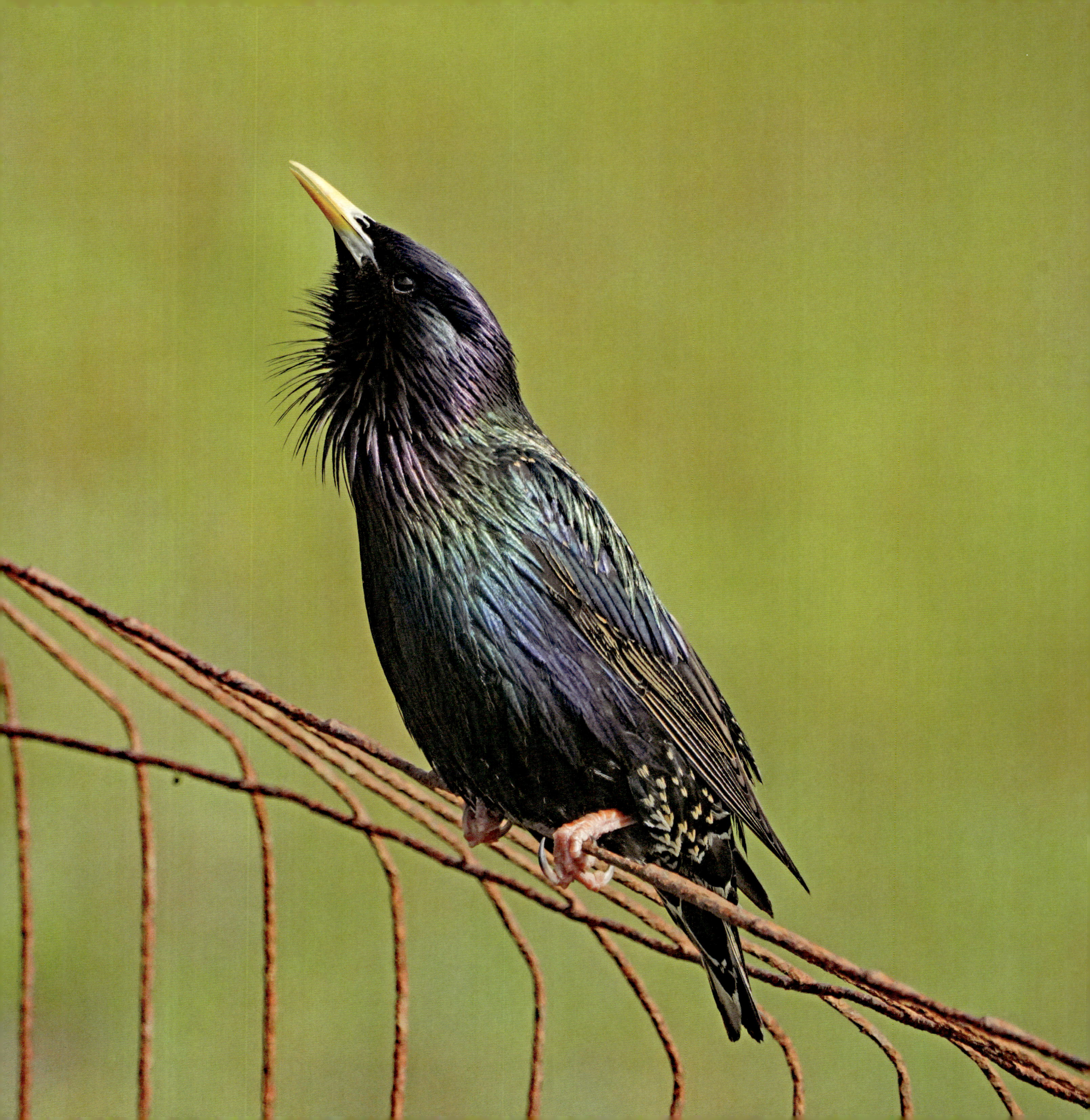

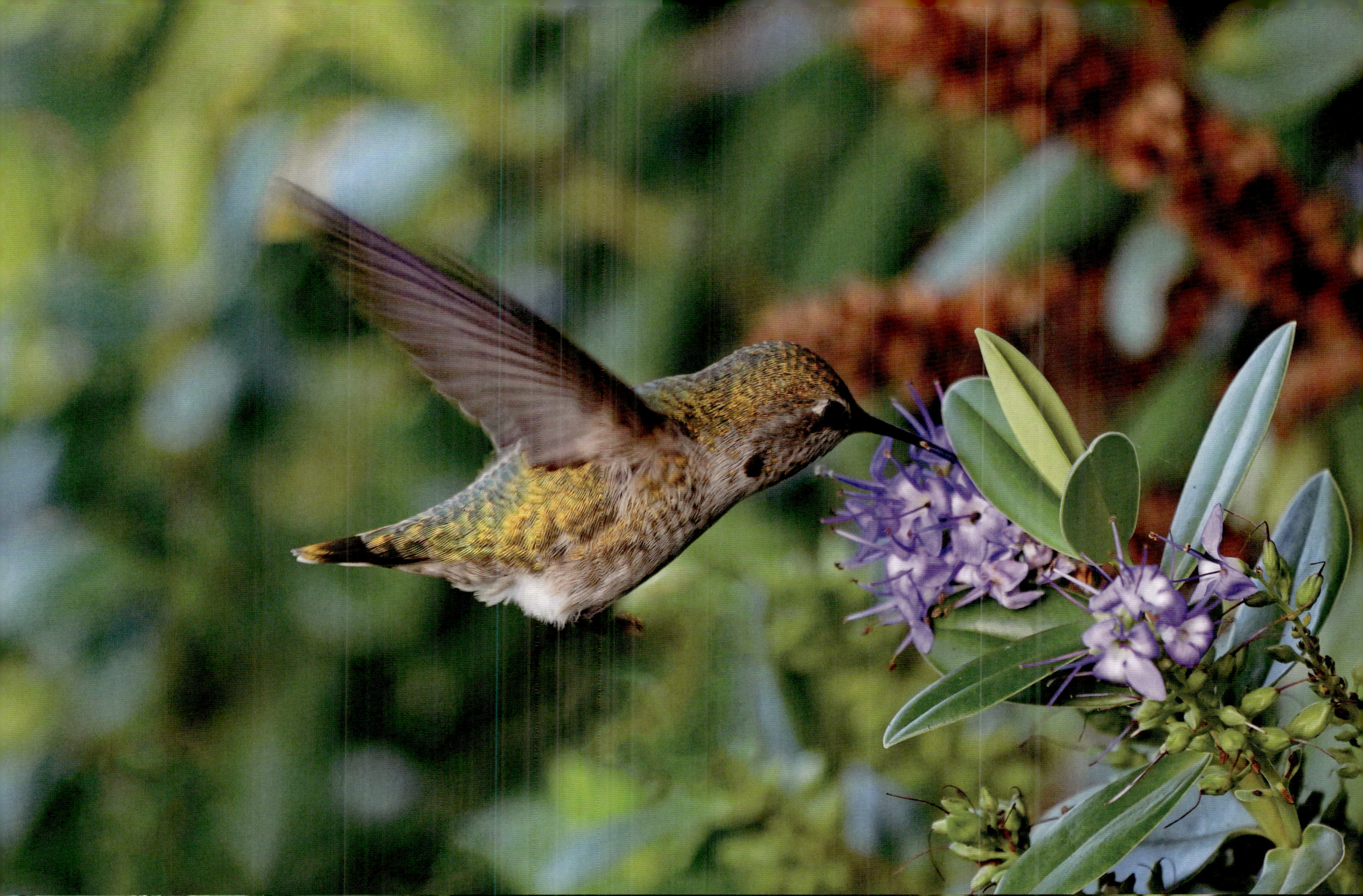
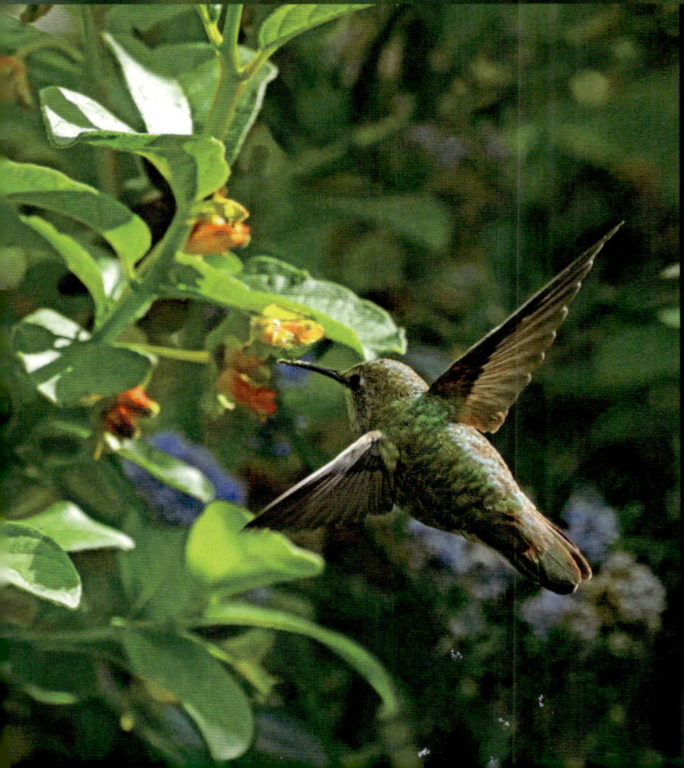
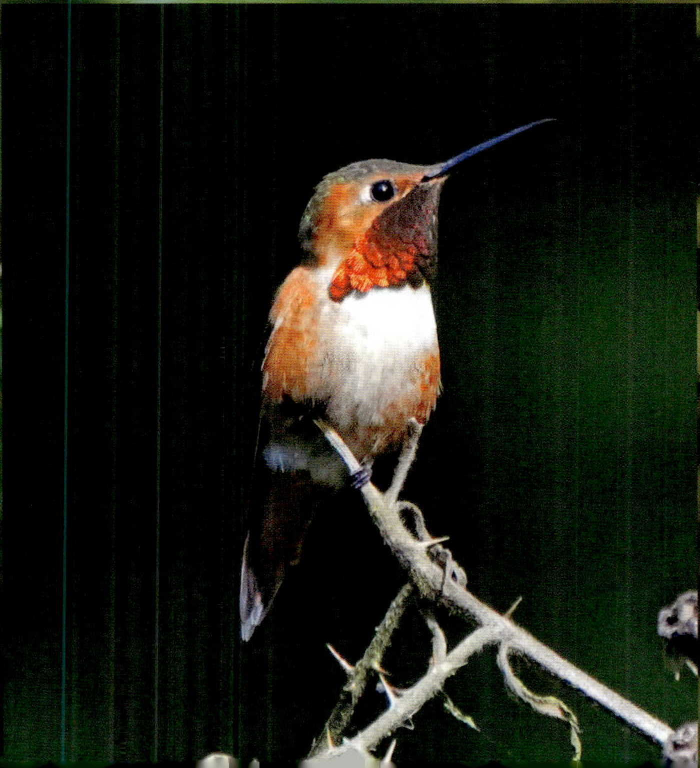
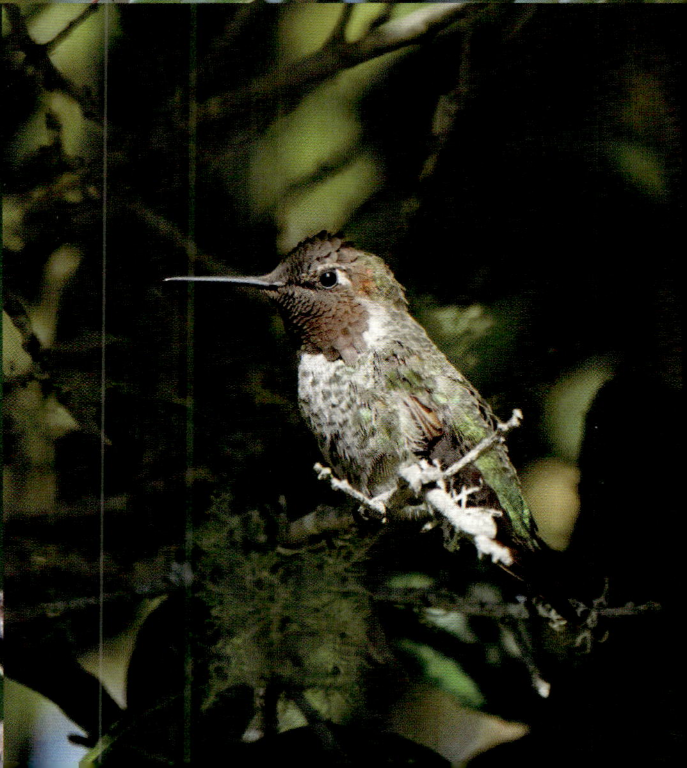

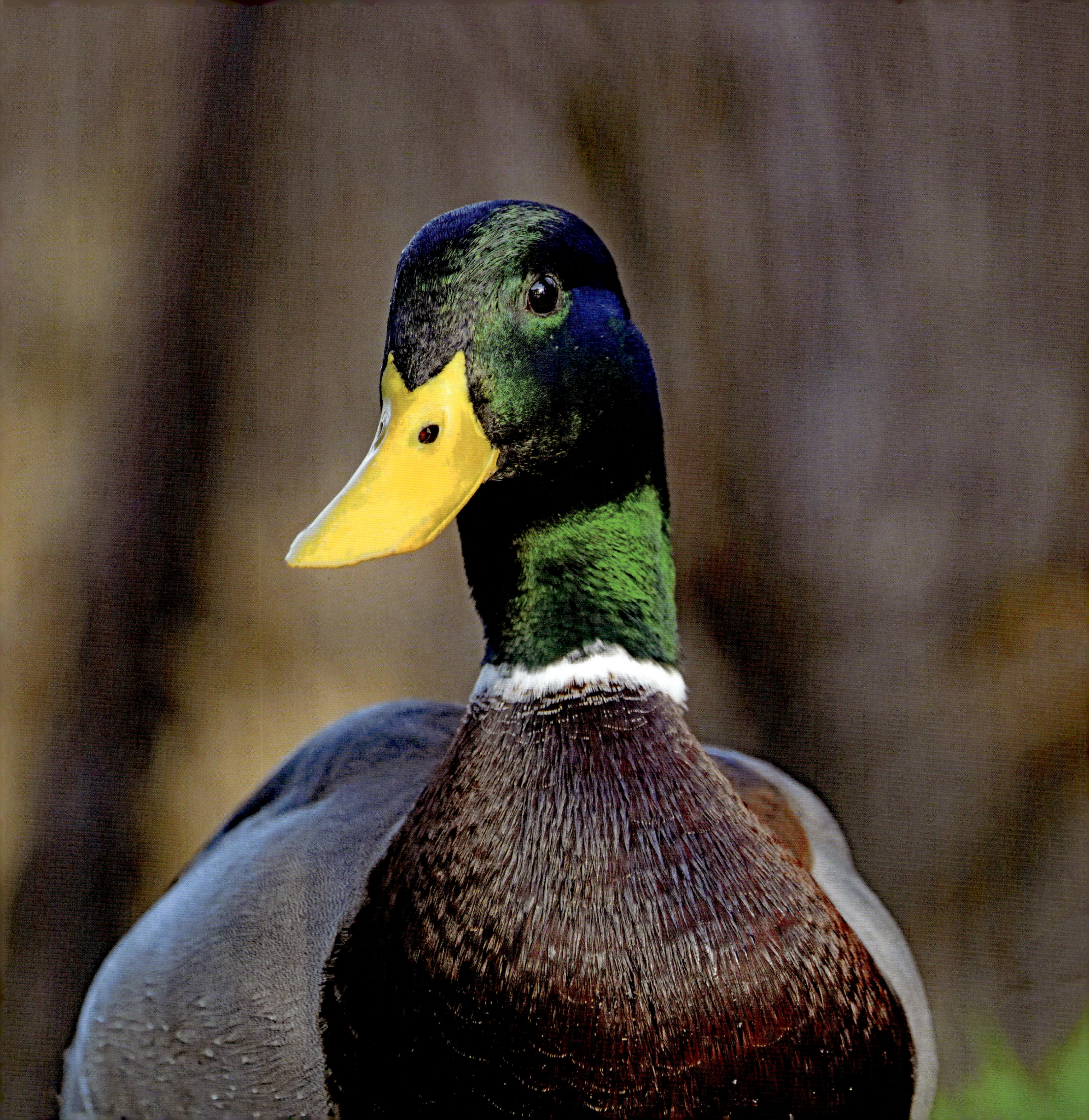

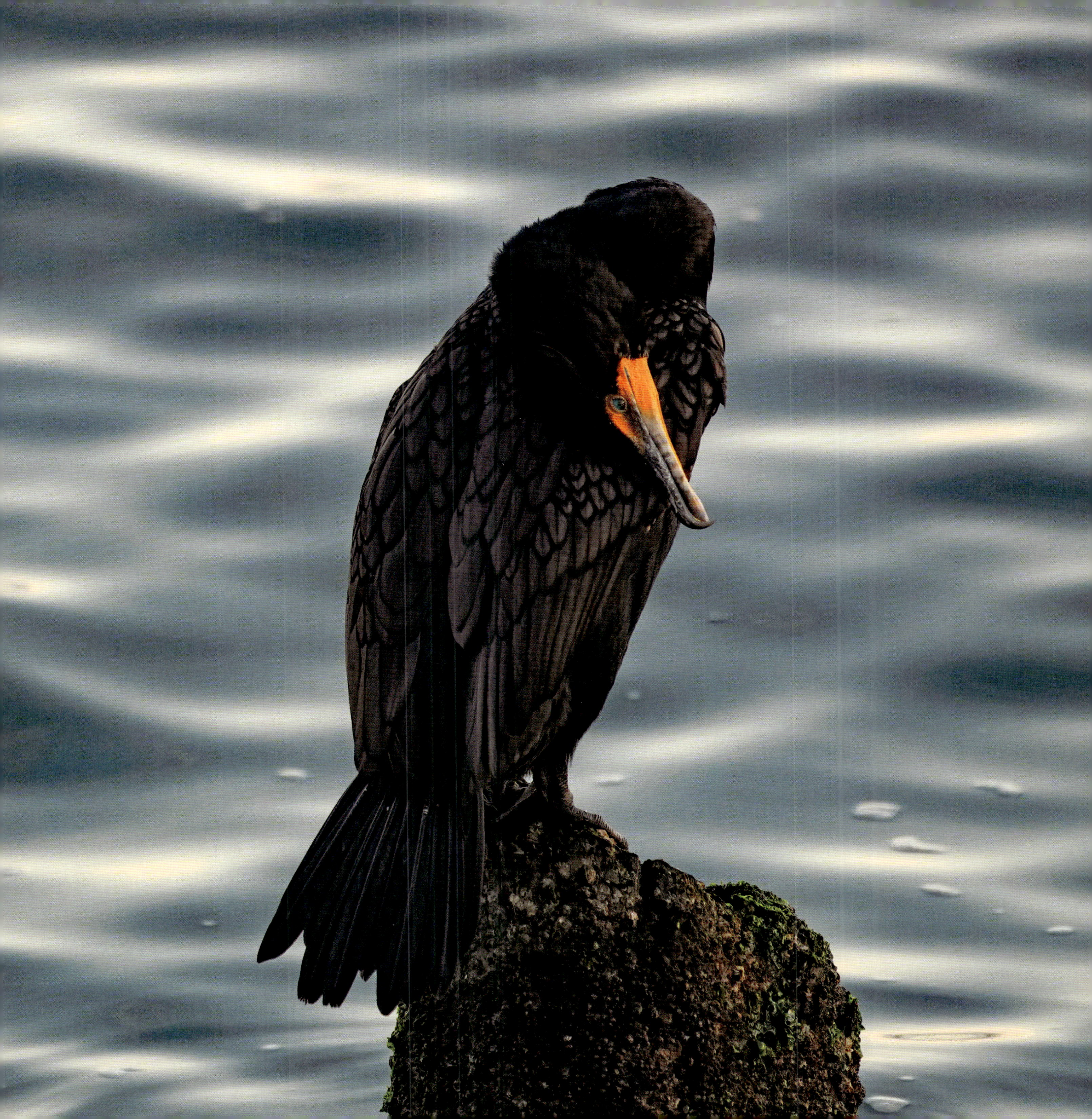

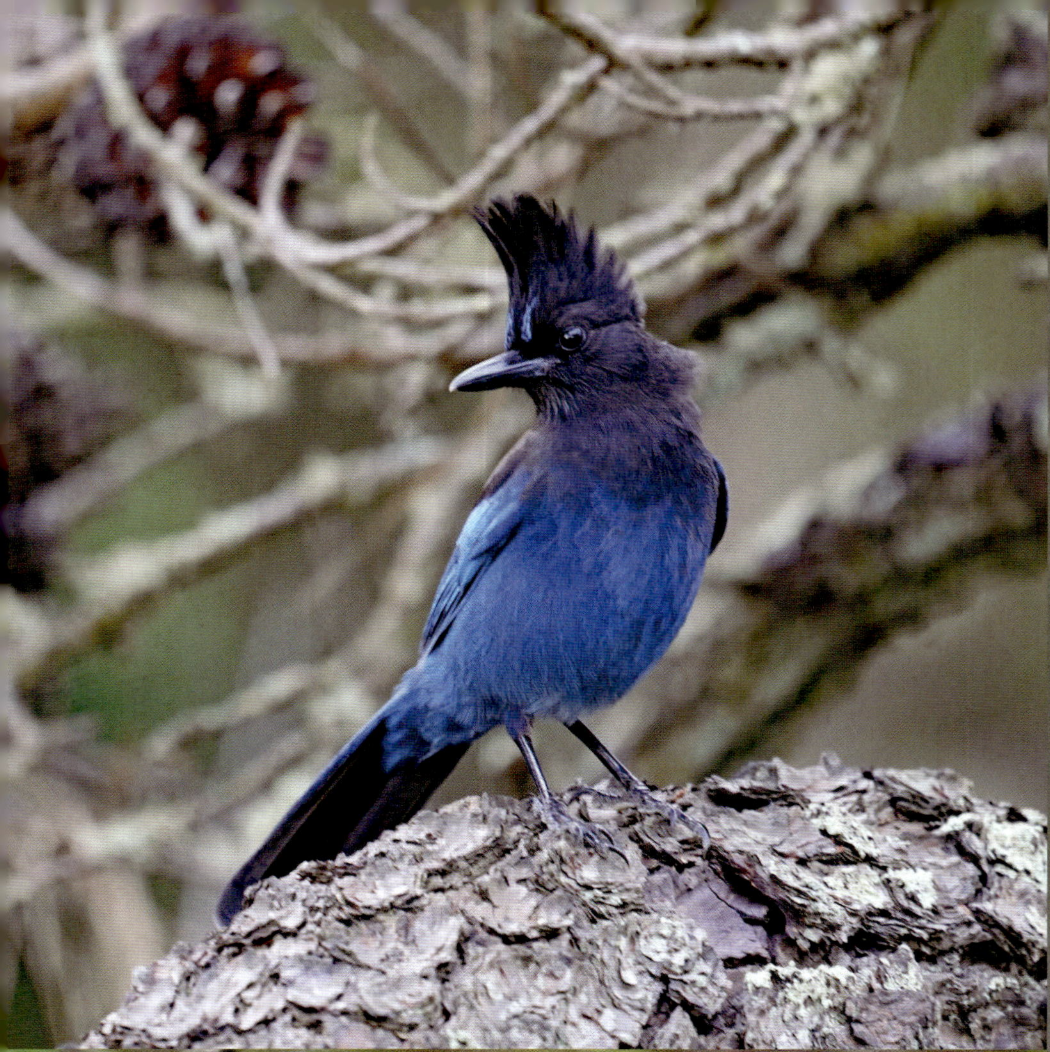
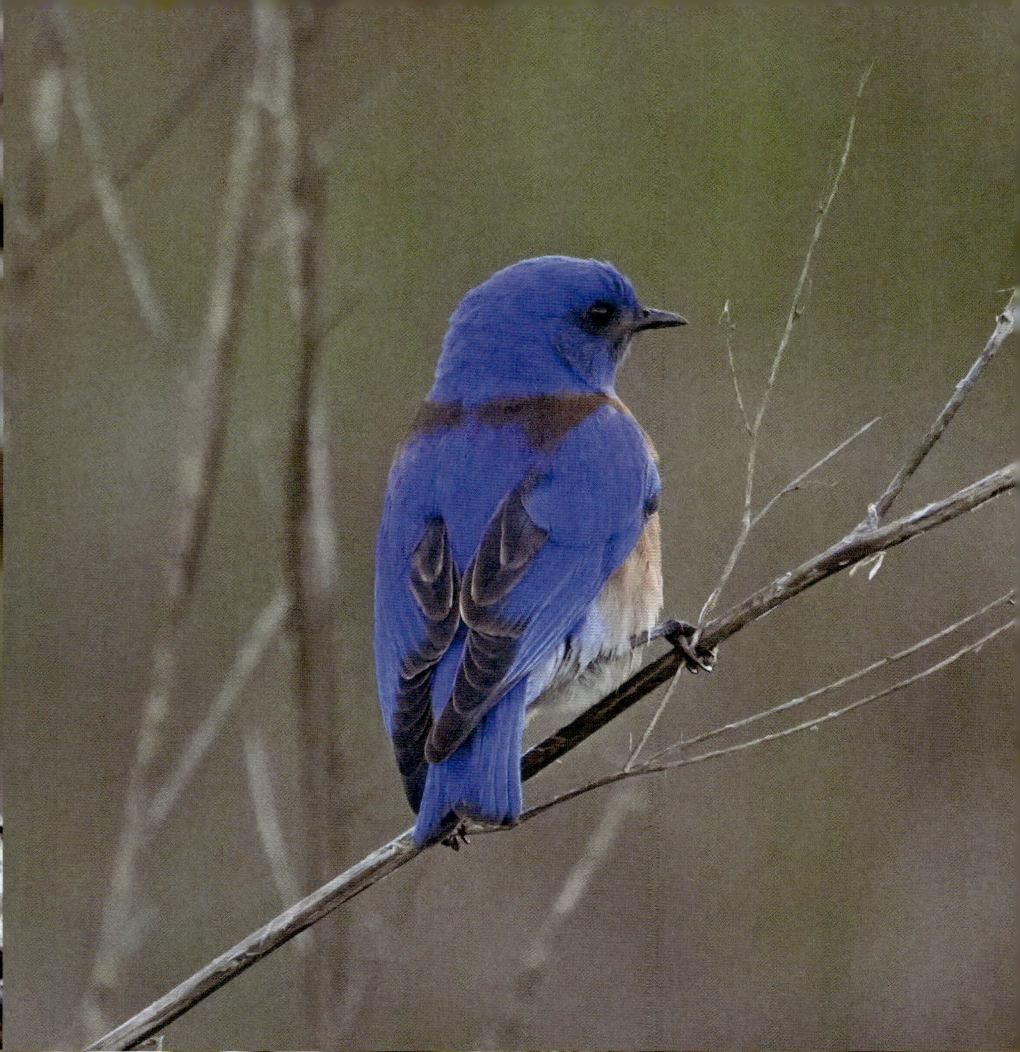
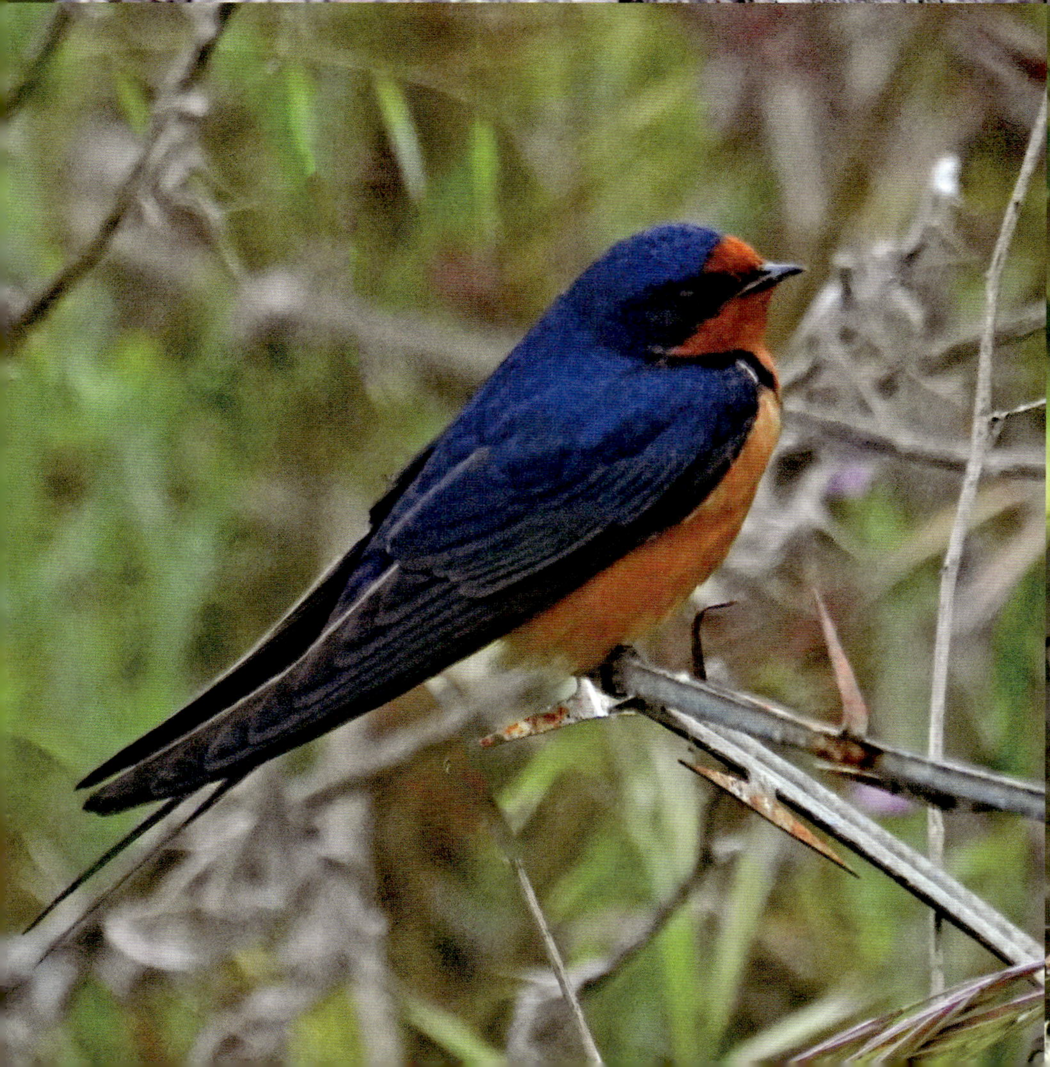
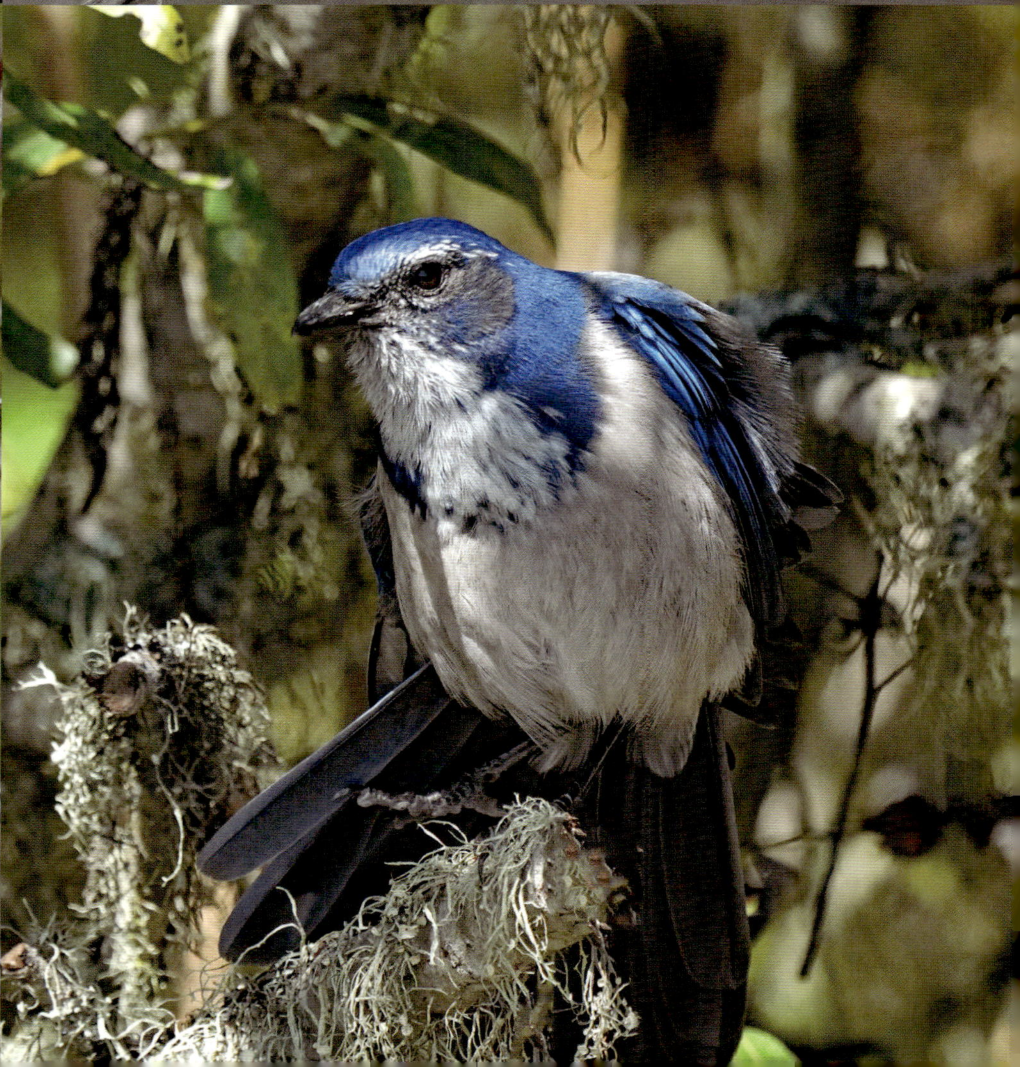

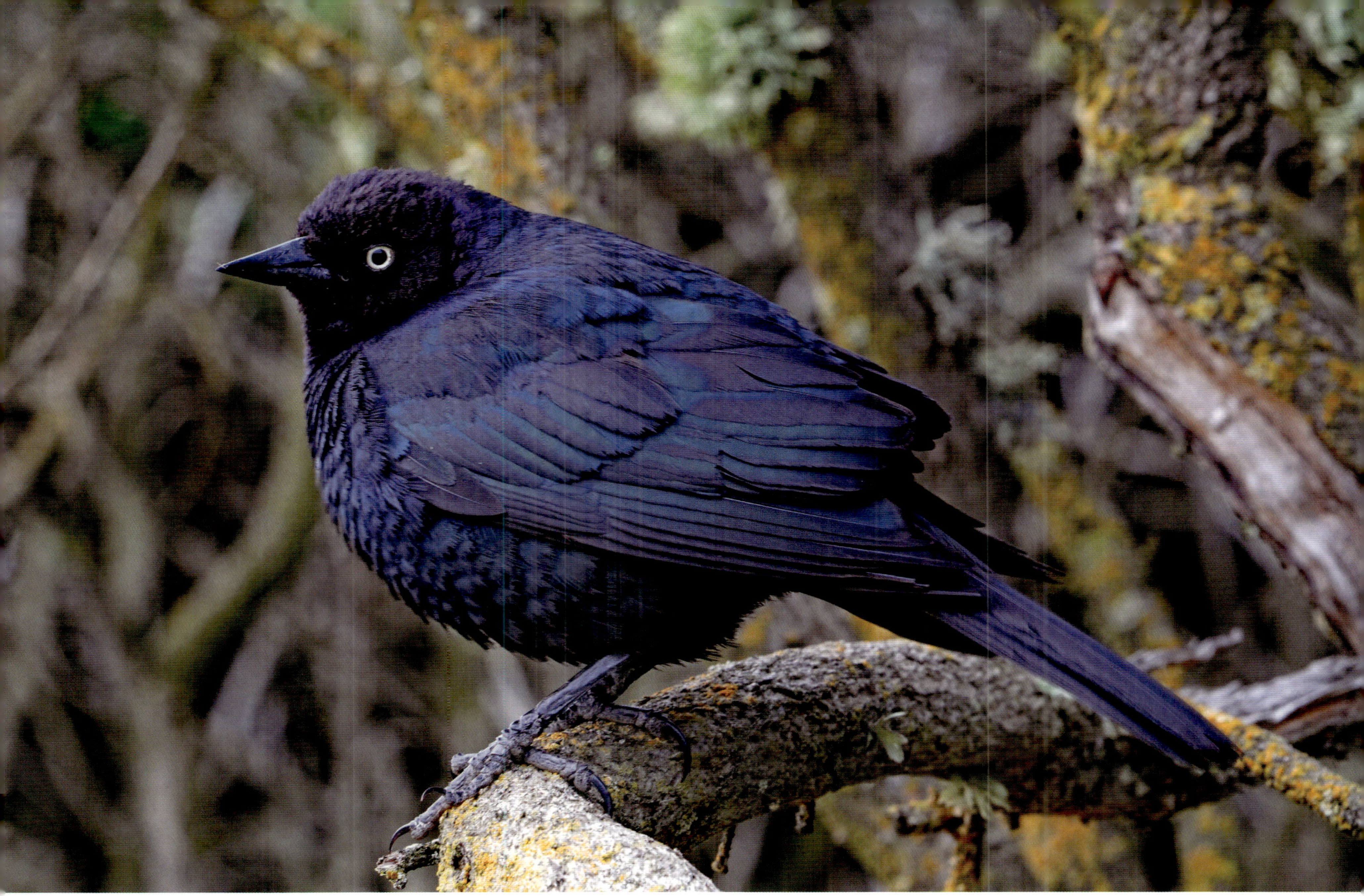

PRECEDING: As iridescent birds like this Mallard (left) move, nanoscale structures within their feathers' tiny branch-like filaments—known as barbules—interact with light to amplify certain wavelengths depending on the viewing angle. The intensity of the Mallard's iridescent green head feathers is related to the level of testosterone—higher levels of this hormone result in brighter feathers, especially during mating season. Many of the rich, dark colors we see in birds are pigment-based, especially hues like blacks and browns. A Double-crested Cormorant (right) displays beautifully textured deep black plumage, which is the result of melanin—the same pigment that determines a person's hair and skin color.

OPPOSITE, CLOCKWISE FROM TOP LEFT: A Steller's Jay, Western Bluebird, Western Scrub Jay, and Barn Swallow all appear brilliantly blue in the light, but look at a jay feather in the shade and you'll see only the underlying muted brown pigment. When light strikes these birds' feathers, the blue we see is caused by air pockets in the feather reflecting the blue wavelengths back to our eyes, while the other wavelengths are absorbed.

ABOVE: The luminous look of this Brewer's Blackbird results from a combination of dark pigment and refractive microscopic structures in the bird's feathers that create a glossy, shimmering appearance—another form of iridescence.

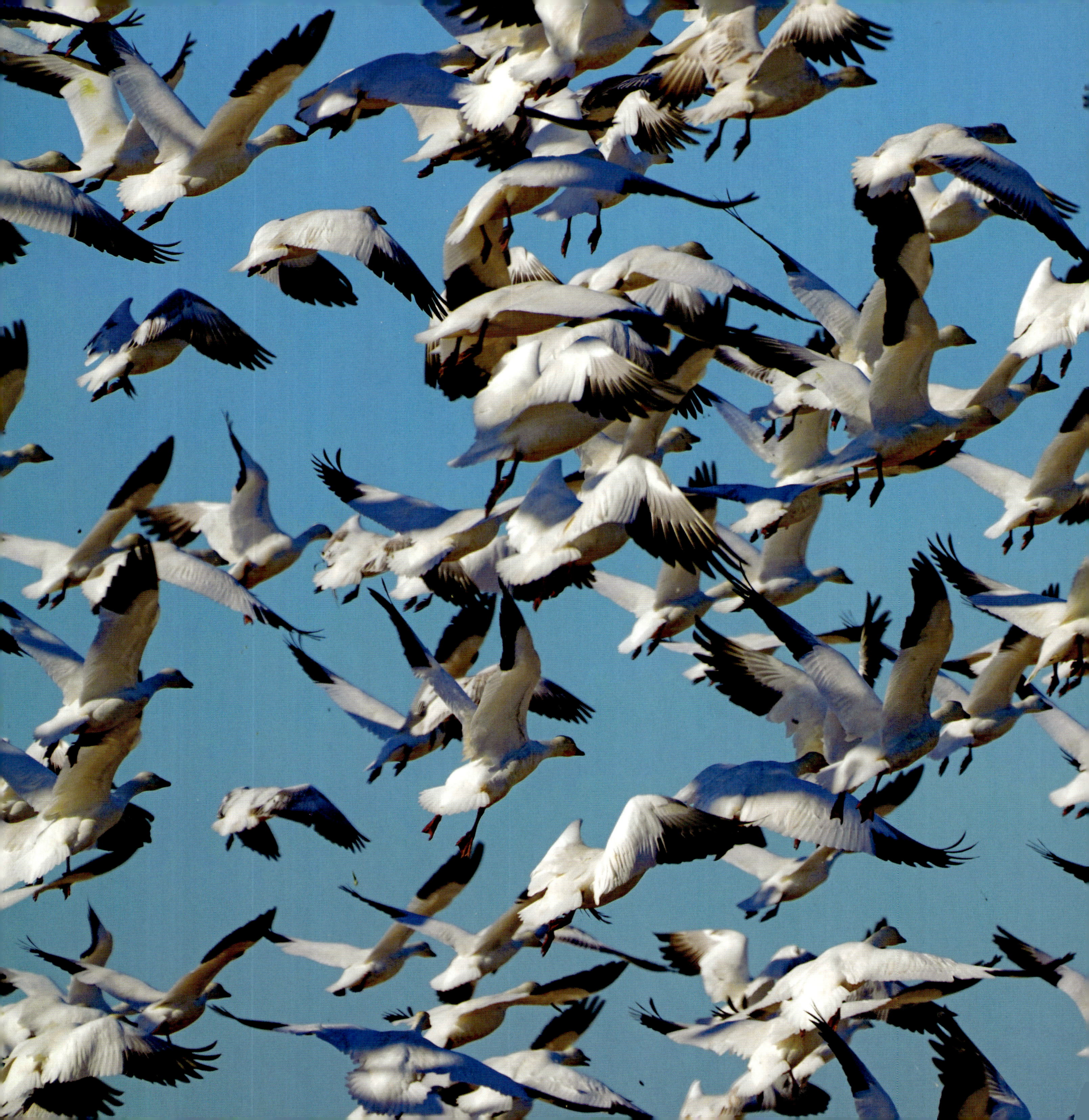

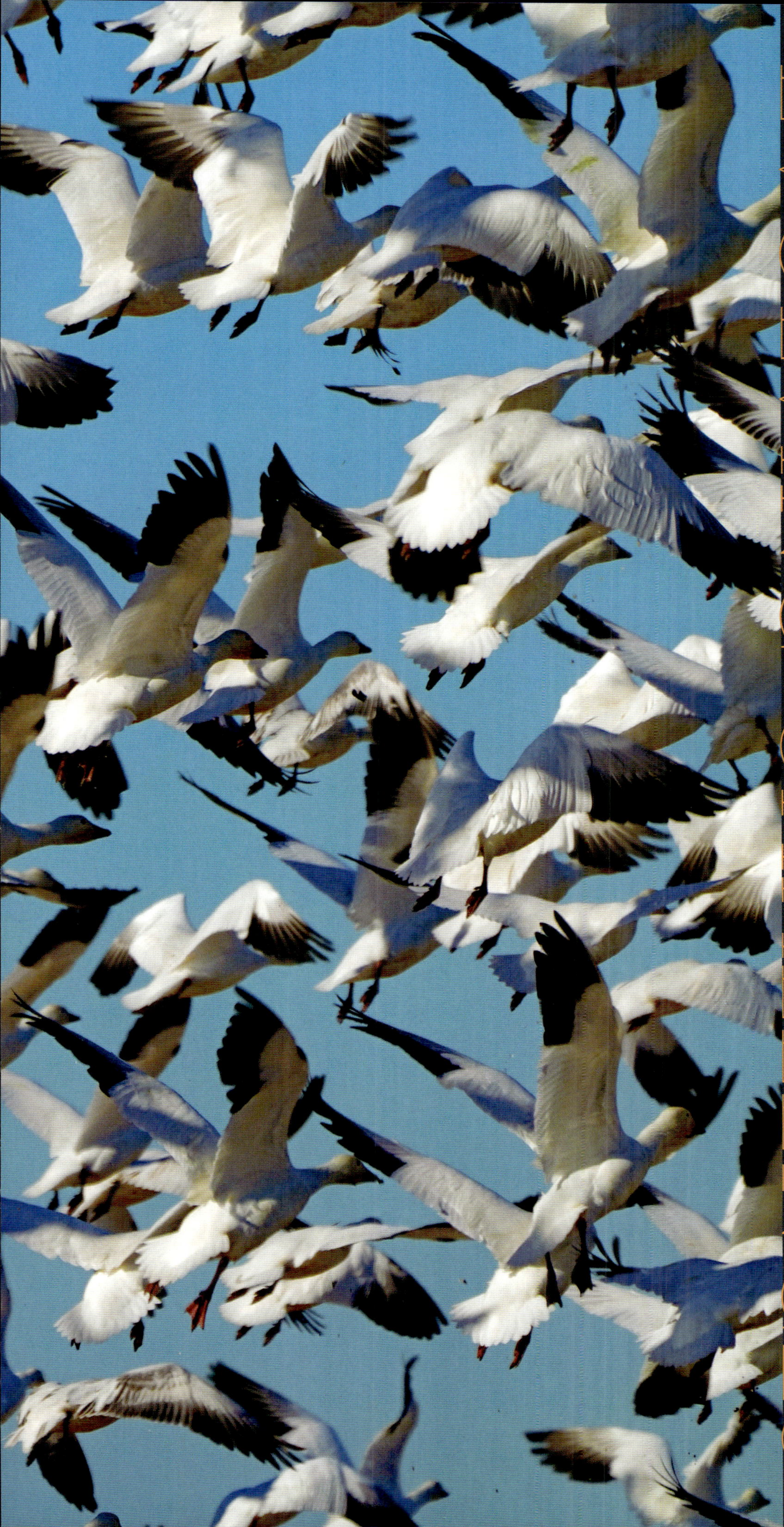

EVOLVING CHANGES

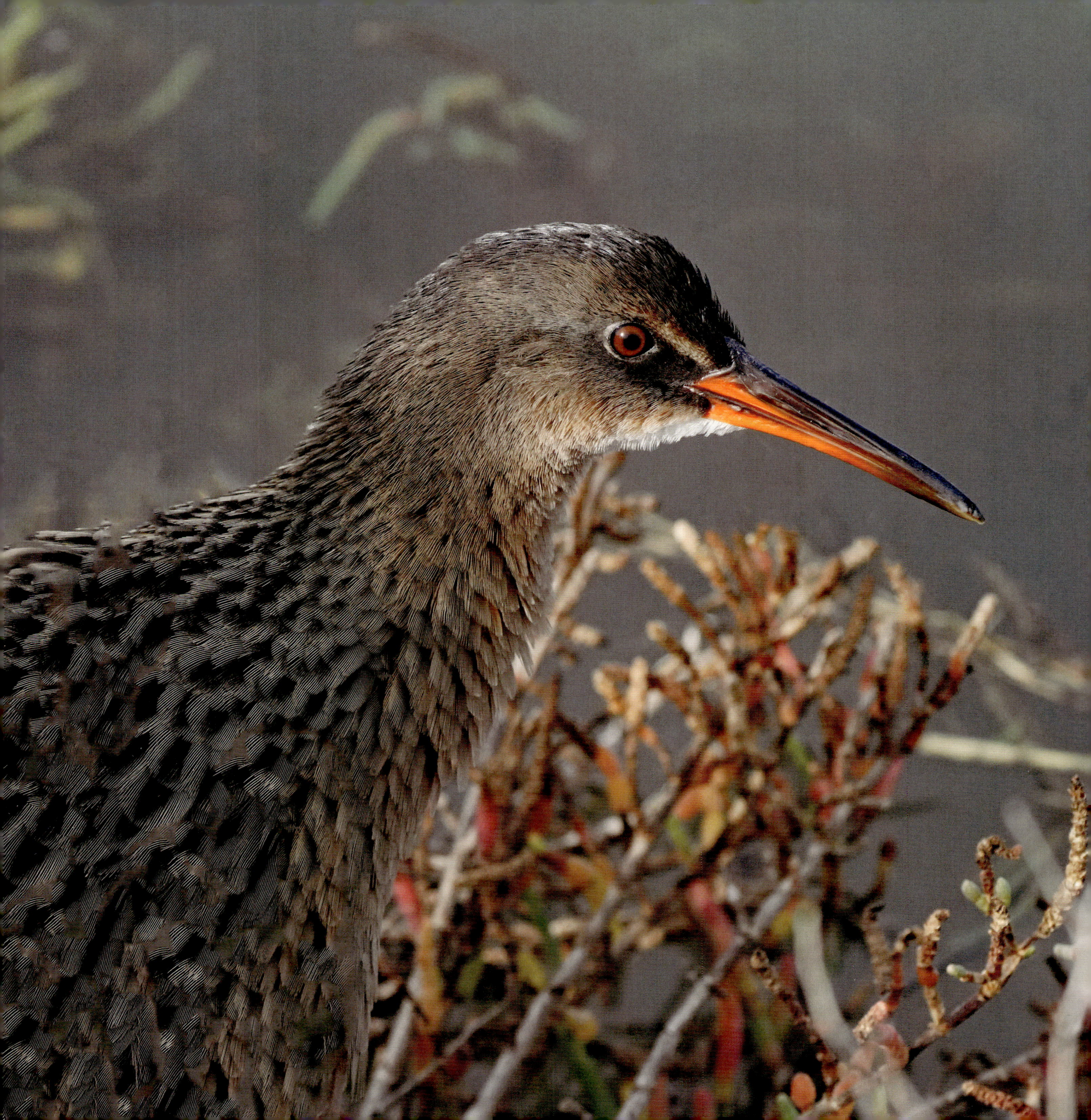

In twenty-first-century conservation, answers that are true are seldom simple. As birds and humans together look toward a complex shared future here in the shadow of the bridge, solutions that serve all species will require diverse perspectives, creative partnerships, and a commitment to paying careful attention, every day.

In Arrowhead Marsh, planes landing at the Oakland International Airport cast shadows over wading godwits and dowitchers. The tide slowly rises among stands of cordgrass. A small gaggle of urban birders follows a long boardwalk that ends at a stage-like platform in the heart of the marsh. They wait for the star of the show to take the spotlight: the endangered Ridgway's Rail, *Rallus obsoletus obsoletus*.

These small wading birds with striking orange beaks and somewhat pear-shaped bodies were once a common sight in San Francisco Bay. In 1897, five thousand of them were legally killed in one week alone. That number is greater than their entire remaining population. But habitat loss, even more than hunting, has driven the rails toward extinction. As the Bay Area was colonized—by Spain, by Mexico, and by the United States—the Bay's marshes were filled and drained. Ranches spread out across Ohlone and Miwok tribal territories. Salt ponds covered vast stretches of the South Bay—the last of which can still be seen from an airplane window, pink and orange, as you descend into San Francisco International Airport. Ninety-seven percent of the California Delta's tidal marshes were diked and planted with crops to feed a hungry state and a hungrier global market. Hydraulic mining in the Sierra Nevada sent literal mountains of sediment downstream, choking the remaining tidelands under more than a meter of mercury-laden mud. Across this altered landscape, urban development then encroached further and further into the shoreline.

Today, the greater San Francisco Bay is among the most heavily modified estuaries in the nation. We've drained and built on 95 percent of the Bay Area's freshwater wetlands and tidal marshes. Turning what were considered unhealthy, useless wastelands into productive space was considered a great accomplishment. But wetlands filter contamination out of urban runoff. They protect against erosion from storm surges, floods, and sea-level rise. They replenish groundwater—no small service at a time when drought conditions continue to preside over an ever-hotter state. And our Bay wetlands harbor species that have almost disappeared from the world for good: not only the Ridgway's Rail but also the threatened Black Rail, three endemic species of song sparrows, and the endemic Salt

PRECEDING: A whirling flock of Snow Geese takes to the air above winter fields in the California Delta.

OPPOSITE: An elusive Ridgway's Rail prowls among pickleweed during a King Tide at Muzzi Marsh. These birds have helped unite whole communities across the Bay Area—in 2008, nine Bay Area counties collectively voted to tax themselves for the San Francisco Bay Clean Water Pollution Prevention and Habitat Restoration Measure, which annually helps fund projects to protect and restore San Francisco Bay.

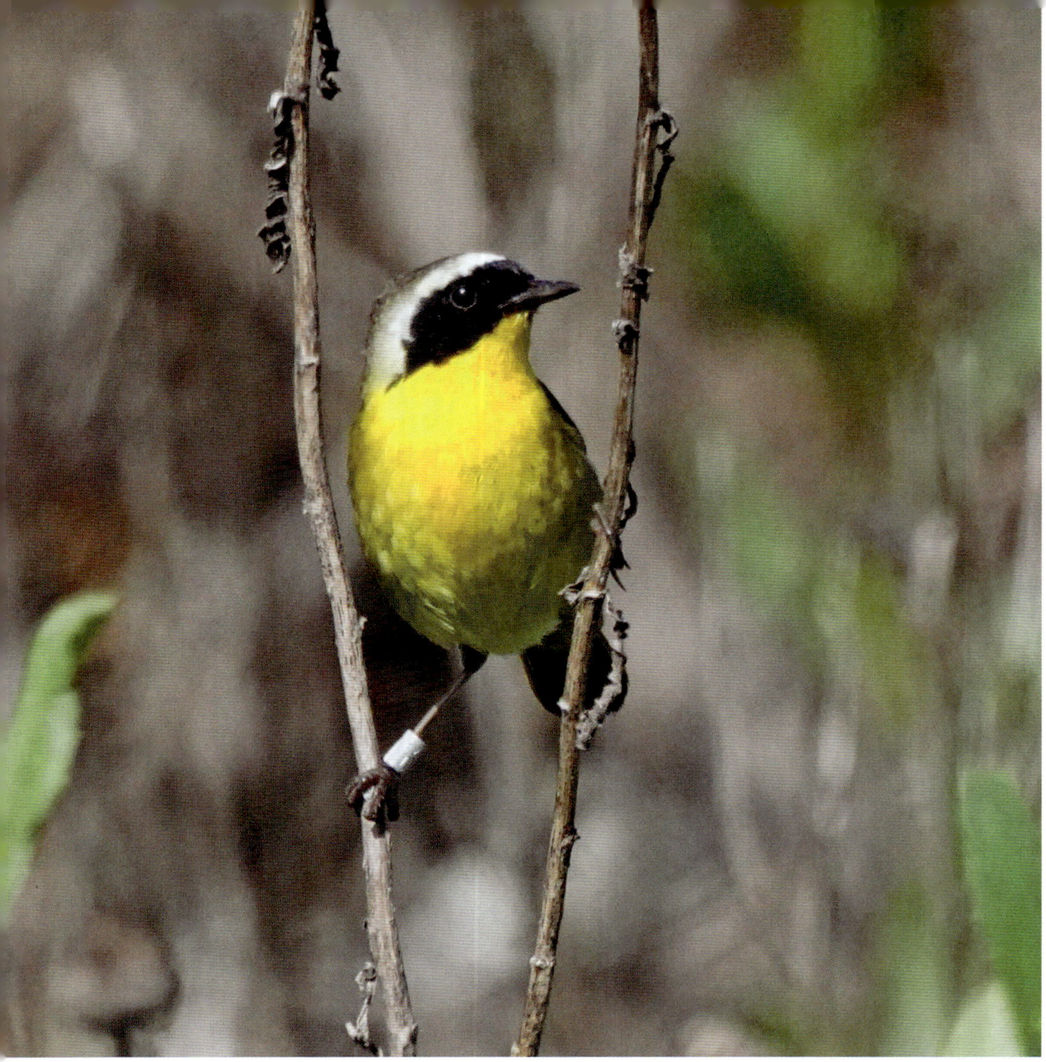

Marsh Common Yellowthroat, with a striking black face mask and a clear, high song as bright as a tapped wine glass.

As conservationists watched the last of the Bay Area's wetlands disappear, they rallied to come up with a plan to save them. In the 1990s, a vast interagency coalition, including representatives from local, state, and federal agencies, academia, and the private sector, collaborated to come up with the San Francisco Bay Area Wetlands Ecosystem Goals Project: a long-term vision for a healthy and sustainable baylands ecosystem, with hundreds of pages of actionable recommendations. In 2016, the group released a new publication, this time with substantial climate change data and a plan to address rising sea levels.

Today, Julian Wood, San Francisco Bay program manager for Point Blue, guides us deep into Arrowhead Marsh. We've been looking for Ridgway's Rails all summer, and this park, wedged between the airport car rental strip and the Oakland Arena, is one of our likeliest places to find them. Far off to the northwest, we can see the spires of the Golden Gate Bridge and the purple silhouette of Mount Tamalpais. At low tide, various rail species descend on long legs into the drained muddy channels to forage. The rails shelter and nest in the high places that remain when the tide rises: hummocks of cordgrass, pickleweed, native gumplant with buttery yellow blooms. The root networks of these plant communities, and their spongy peat underlayer, help absorb floodwaters and hold sediment in place when the ocean pushes higher.

The Salt Marsh Common Yellowthroat (above) is unique to the San Francisco Bay Area and is the only warbler endemic to the tidal marshes. More than 80 percent of their population has disappeared in the last century.

Marbled Godwits (opposite) come through the region on a migration that connects the Great Plains, California, Mexico, and Central America; threats along any of these stopovers represent a threat to the entire web of connection. Point Blue's science work extends across 14 countries, in recognition that conservation challenges in the Bay Area are not only local, but global.

Julian helped come up with the revised Bay Goals document, and much of his daily work is in helping land managers and city planners imagine solutions for the rising sea levels that are already encroaching on the densely populated bay. "Everyone knows how to build a seawall. It works, for a time," Julian says. "But the nature-based solutions are new for a lot of these planners." With California's 30x30 initiative (an executive order to conserve 30 percent of California's land and ocean by 2030), there's a lot of money available for restoring marshlands and for implementing other solutions rooted

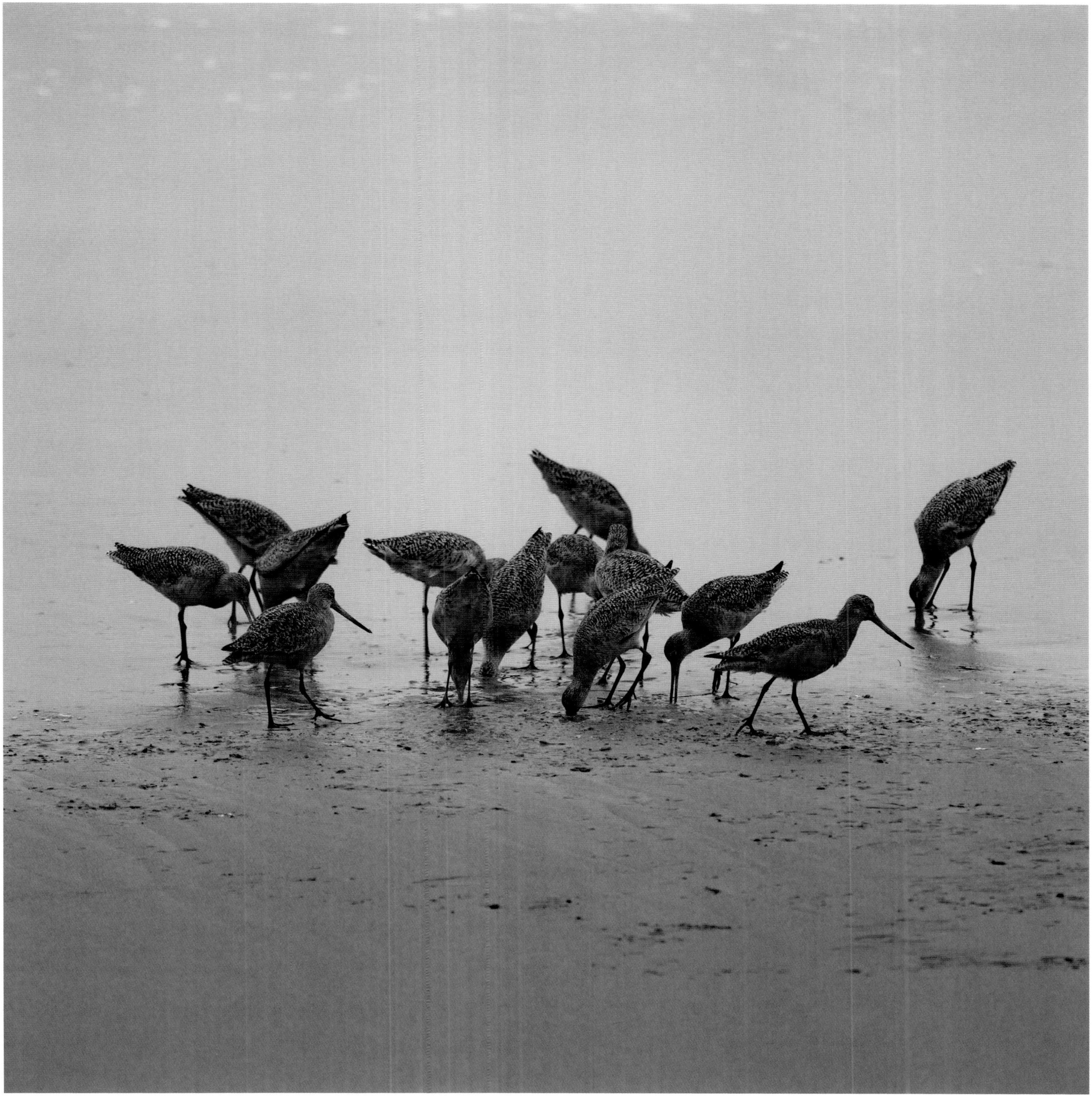

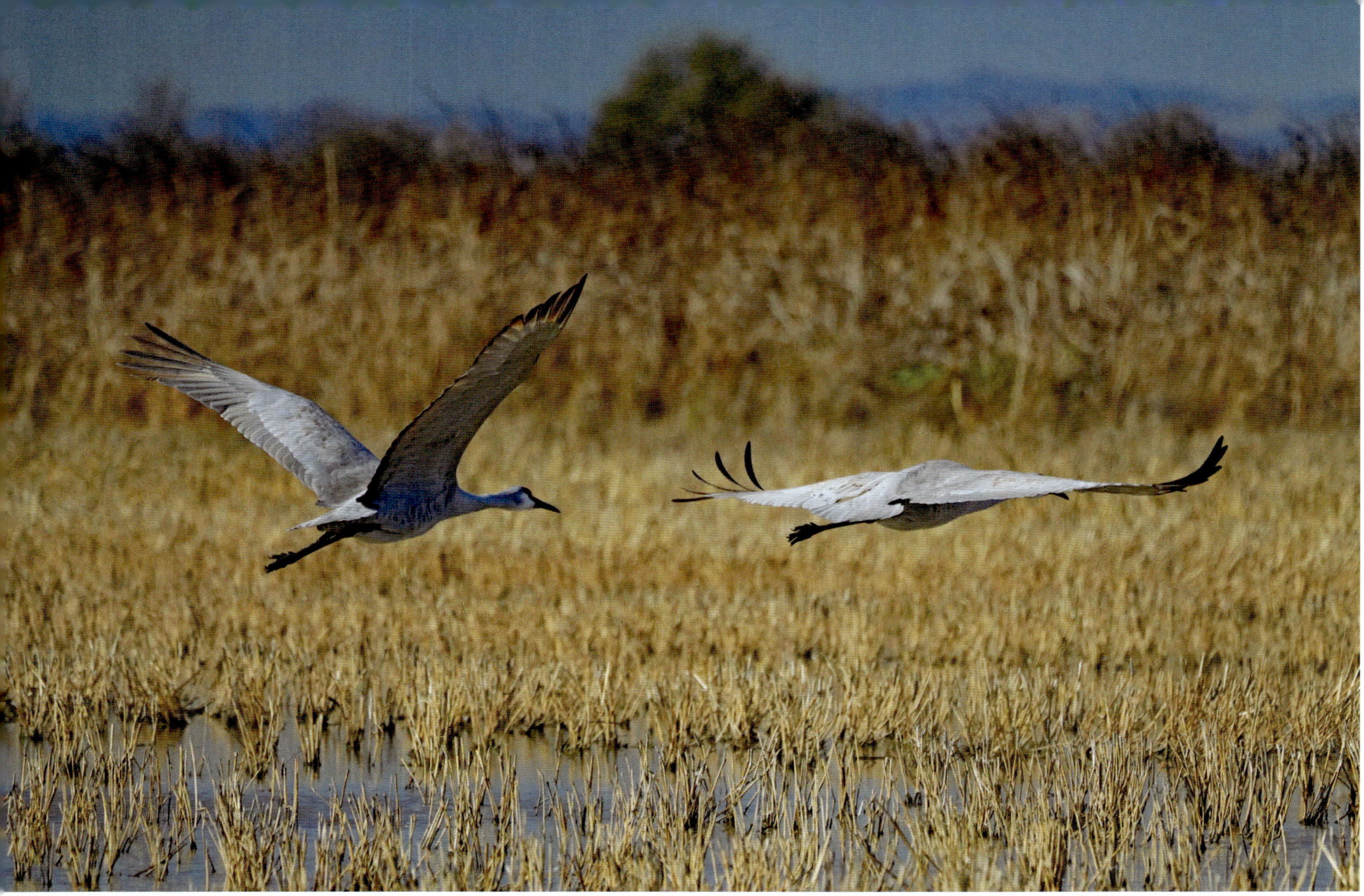

The California Delta was once a vast, brackish marsh and tidal plain, home to Tule Elk, Pronghorn Antelope, and Grizzly Bears. Greater Sandhill Cranes (above and right) continue to return to this region on their winter migration, and local farmers are working alongside biologists to find ways to make the now-agricultural landscape nourish both birds and people alike.

in ecosystem services. Projects like this turn out to be compelling puzzles for bright engineers interested in designing organic systems that will outlast concrete and steel.

Engineers and city planners are among a diverse—and perhaps surprising—cast of characters defending California bird habitat, which also includes military land managers, ranchers, hunters, and farmers. Conservation issues in California don't tend to be black and white—decisions to support birds' continued existence where they've lived for millennia also have to take into account urban growth and housing, economies, natural disasters, food production, other threatened species, water rights, defense, and recreation. In addition, climate change is shifting where birds can make their homes, as they adapt to yet another human alteration of Bay Area ecosystems. In tackling these complexities, Californians are forging unlikely partnerships.

One such story is playing out across California's rice farms. In place of the primeval maze of sloughs and river floodplains, over 7 million acres of irrigated farmland now fill the Central Valley. Follow Highway 80 north toward Sacramento and you'll find farm stands selling sacks of fresh walnuts, fragrant peaches and golden jars of honey. Central Valley farmers provide a quarter of the nation's table food and 80 percent of all the almonds in the world. Over two billion pounds of rice from the Central Valley are exported every year, feeding markets as far away as Tokyo and Istanbul.

Those same rice fields are also unlikely refuges for birds. The Pacific Flyway funnels 20 percent of the entire nation's waterfowl population through the Central Valley—over five million waterfowl and a quarter million shorebirds. In the winter, when rice farmers flood their fields to support decomposition before the next growing season, birds throng to these shallow waters. They feed on the

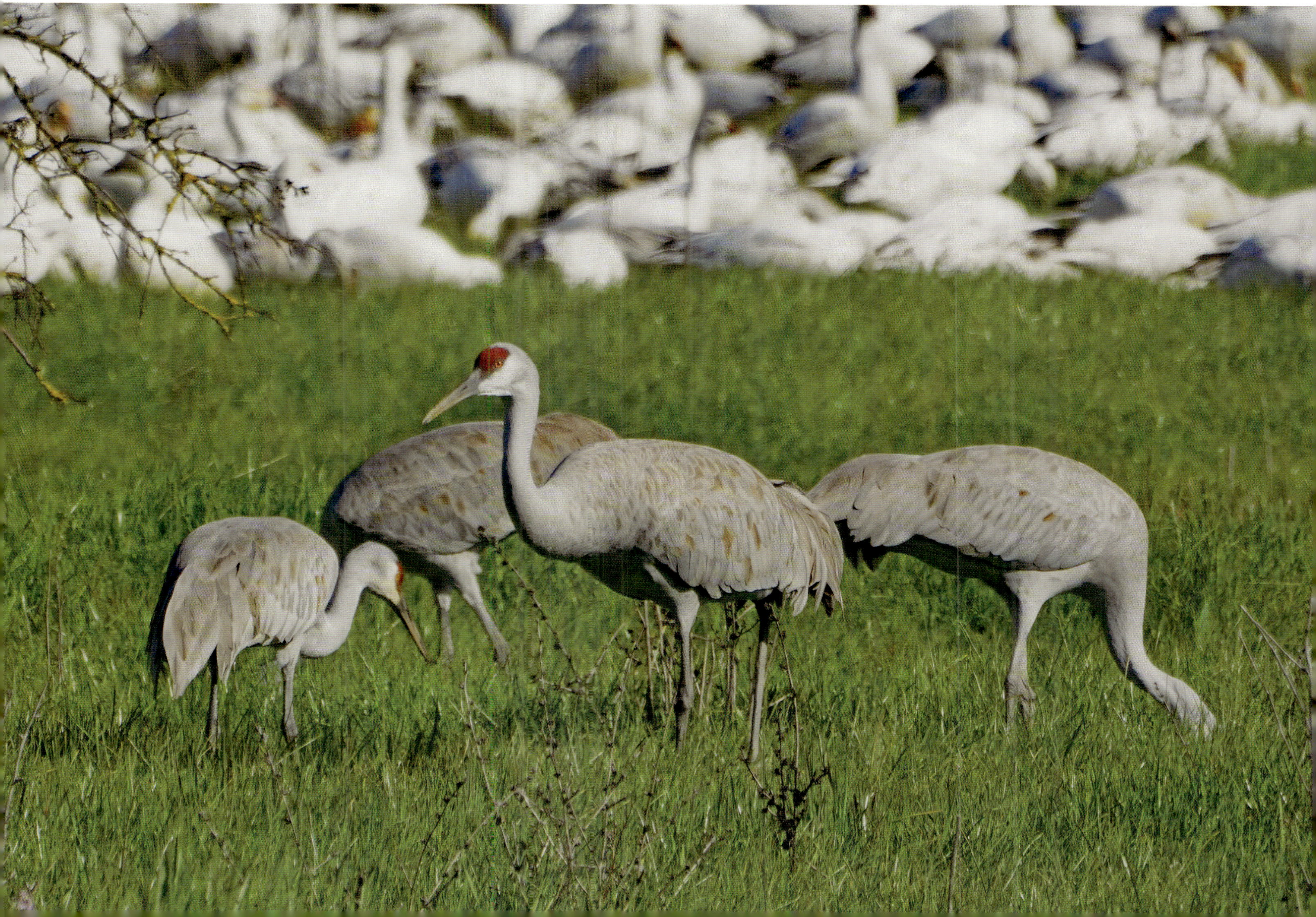

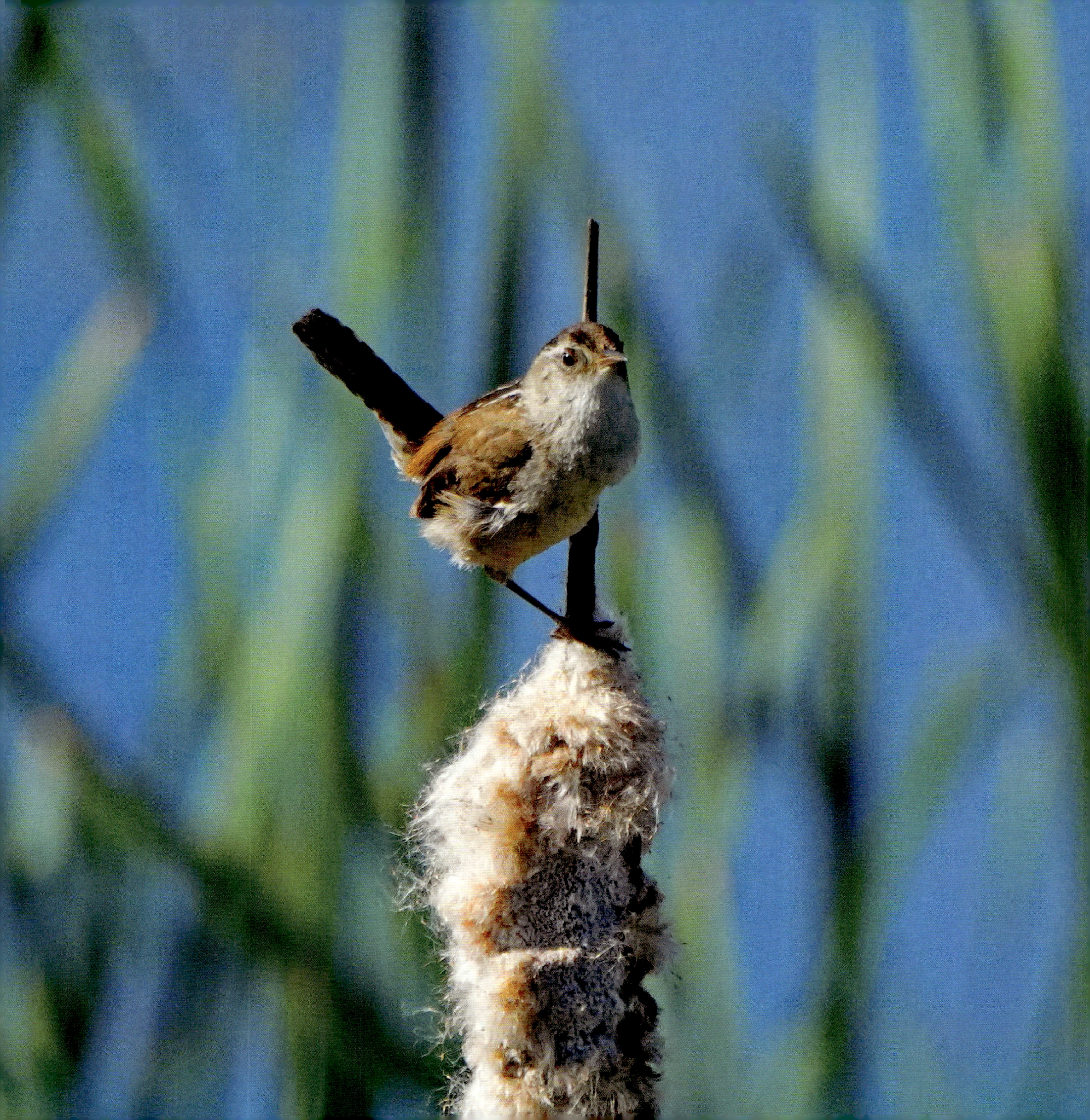

remaining grain scattered from the fall harvest and nibble on bugs in the sticky mud, but they can be displaced if the fields are drained before the birds' northbound migration begins.

Audubon California, Point Blue, and the Nature Conservancy have combined their expertise sunder the Migratory Bird Conservation Partnership, with the goal of making California more livable for birds. They've partnered with local rice farmers who have followed the scientists' advice to keep water on the fields throughout the spring, so that migratory birds aren't edged out of these wet refuges. In the fall, some rice farmers have begun to leave their land unplowed after harvest, so that the scattered grains aren't driven too deep underground for hungry birds to find them. "We give the birds the grain, and they provide us with fertilizer," says one farmer.

Now, rice fields in the Sacramento Valley are collectively designated as Shorebird Habitat of International Significance—one of the largest of these ecological sites on the continent. Park along a flooded field in the winter and you'll see thousands of birds take flight, swirling upward like scattered rice. Listen, and you can hear the dinosauric calls of Sandhill Cranes as they wade in the long shadow of the Sierras, as they have done for more than 2 million years.

Partnerships like this are increasingly putting conservation in the hands of people who work the land. In Marin County, two women—the dairy farmer Ellen Straus of Straus Family Creamery and the botanist Phyllis Faber—joined together to protect the agricultural richness and open space that make West Marin so special. In 1980, they founded the first farmland trust in the nation: Marin Agricultural Land Trust (MALT). MALT uses conservation easements to prevent development: They fundraise to buy the development rights for a property so that it can remain agricultural land—or open space—forever. As protected properties multiply, they become corridors that connect habitats and species. Point Blue partners with many of the land trusts in the greater Bay Area, helping guide climate change conversations, as well as advising and collaborating with farmers to arrive at data-supported land management practices that will benefit human and bird interests alike.

Vociferous and fearless, the fist-sized Marsh Wren (opposite) is a familiar trailside escort for those who spend time walking in the Bay Area's remaining marshes and wetlands.

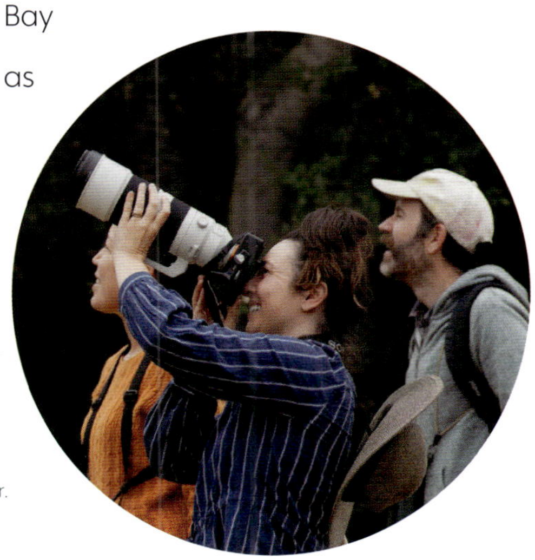

Skill swap! Photographer Dick Evans convinced the book's author Hannah Hindley (right) to trade her writing notebook for a big camera lens as they adventured through the Bay Area's bird sanctuaries together.

In her book *Braiding Sweetgrass*, Robin Wall Kimmerer describes how she once polled a class of environmental studies majors, asking them to name their knowledge of positive interactions between people and the land. Most of the students couldn't think of a single example. "How can we begin to move toward ecological and cultural sustainability," Kimmerer asks, "if we cannot even imagine what the path feels like?" In the Bay Area, that path forward is growing clearer.

To restore a wetland, to reconsider the shape of the rice-growing season, to rewild a creek flowing through ranchland—these are active solutions. But alongside restoration and management, conservation requires monitoring and understanding. Watching, and waiting. Collecting slow data is important work in a world that increasingly demands fast solutions and flashy stories. Often, a true story takes a long time to tell.

After 50 years of patient, steady data collection, only recently has Point Blue had the right technology to finally resolve what it first set out to understand: where Bay Area birds go when they migrate. Starting in 2010, biologists fitted Golden-crowned Sparrows with tiny backpacks with antenna that could measure light levels. As the little birds moved northward, light-level data from their backpacks helped researchers estimate their latitude and longitude to determine their location each day. Using this method, Point Blue pinpointed their migration routes: The same Golden-crowned Sparrows that sing for Bay Area residents every winter in the tall grasses and coastal sage spend their summers in breeding locations along the coast of Alaska, including Katmai National Park, Wrangell–St. Elias National Park, and the Chugach National Forest. Now, Motus (Latin for "movement") towers installed around the globe are using radio telemetry to identify migration routes in real time, anytime a bird equipped with a nanotag transmitter passes within a few miles of a tower.

Ours is a connective world. The migratory birds in our backyards depend on ecosystems half a globe away. The actions of a farmer can steer the course of a crane population that has existed four hundred times longer than the whole history of human agriculture. As the sun dips low at Arrowhead Marsh, a dozen Ridgway's Rails begin to call back and forth to each other, a ricocheting clatter that fills the air. The birds remain hidden in the dense cordgrass, speaking among themselves, but it has never been clearer: We're all part of the conversation too.

Golden-crowned Sparrows (opposite) are among our most beloved fall and winter visitors to our region, foraging under tangled thickets and filling the air with their melancholy minor whistles. Only recently did biologists discover where these familiar birds go on their northbound migration.

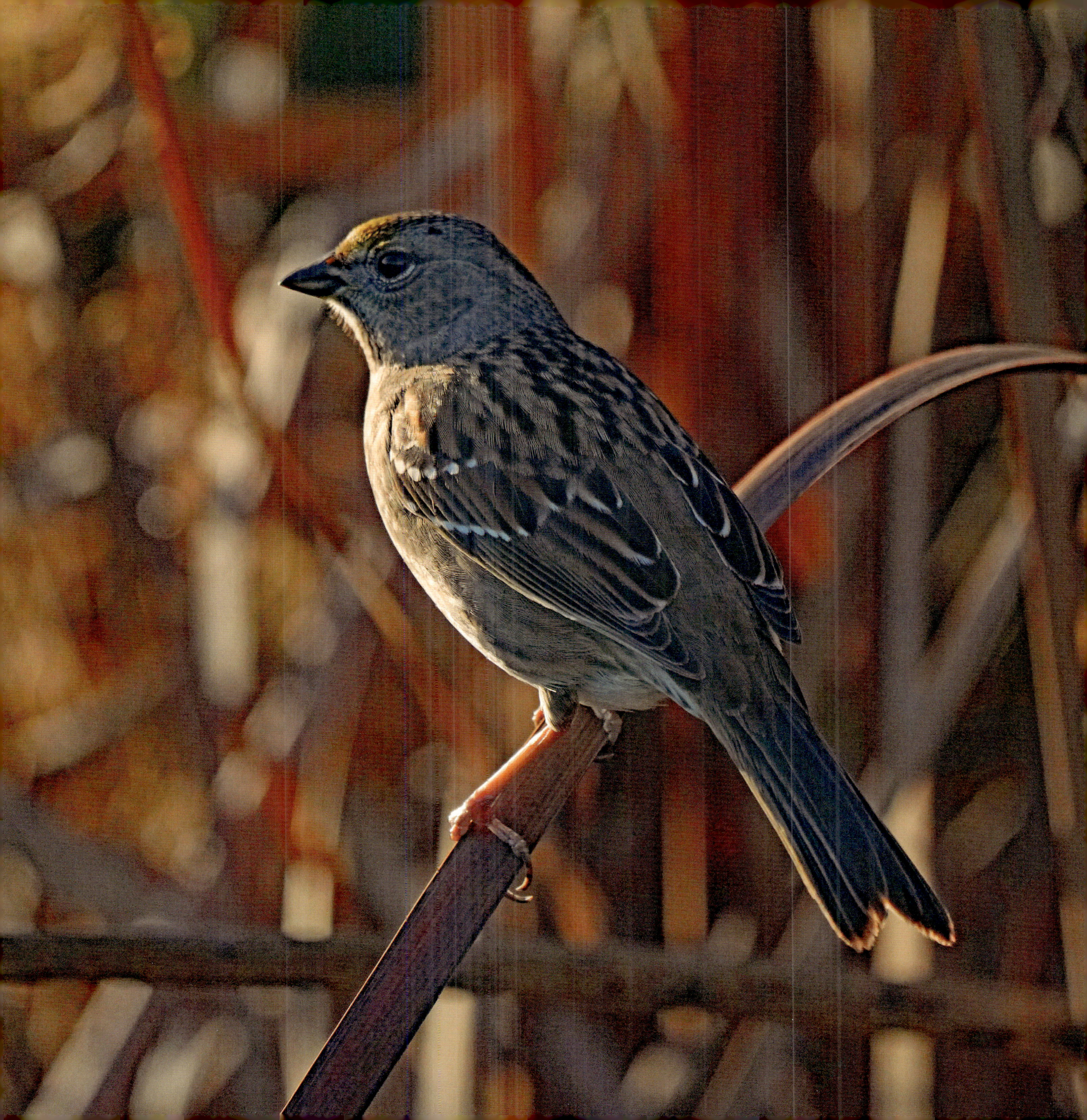

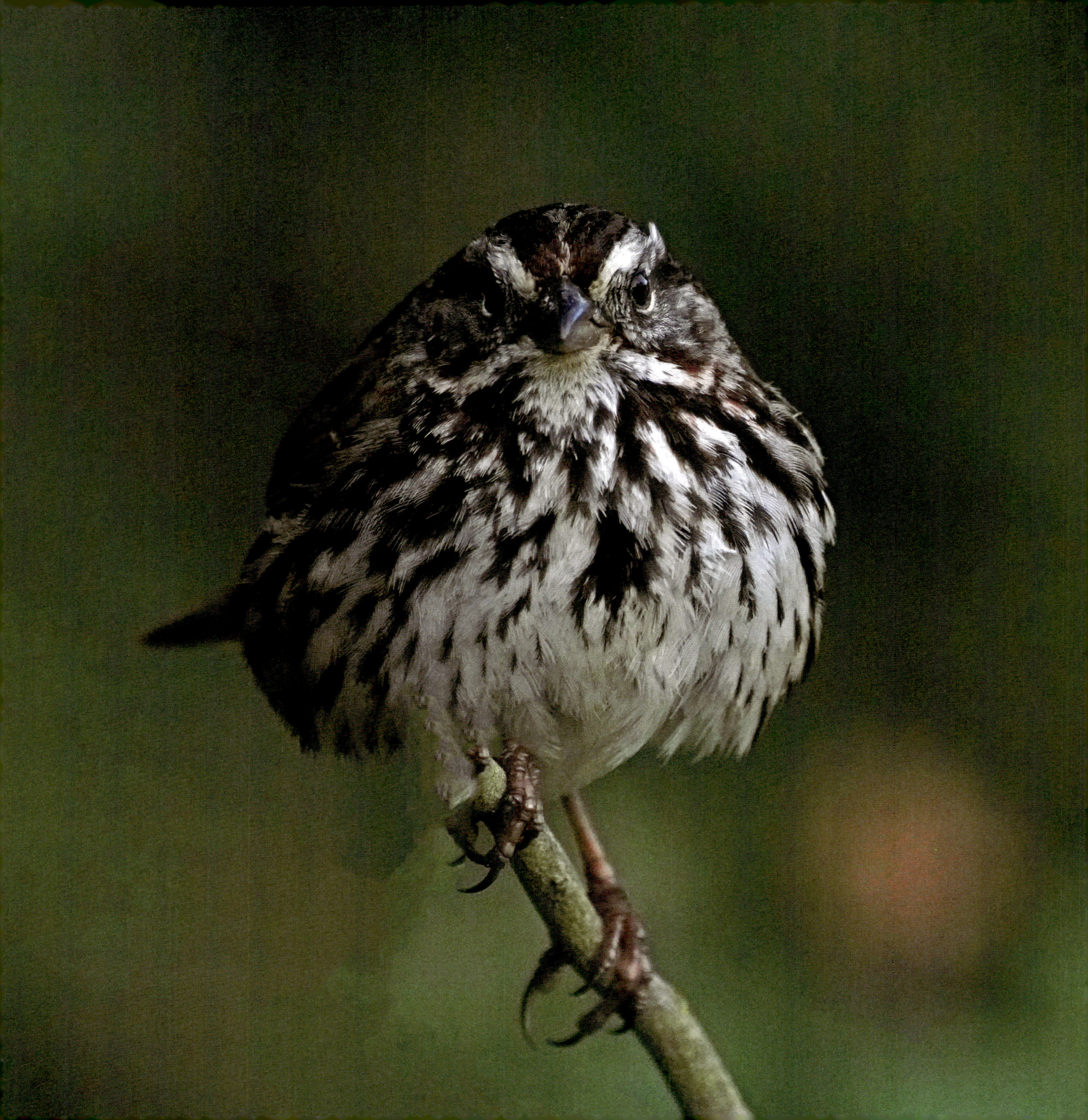

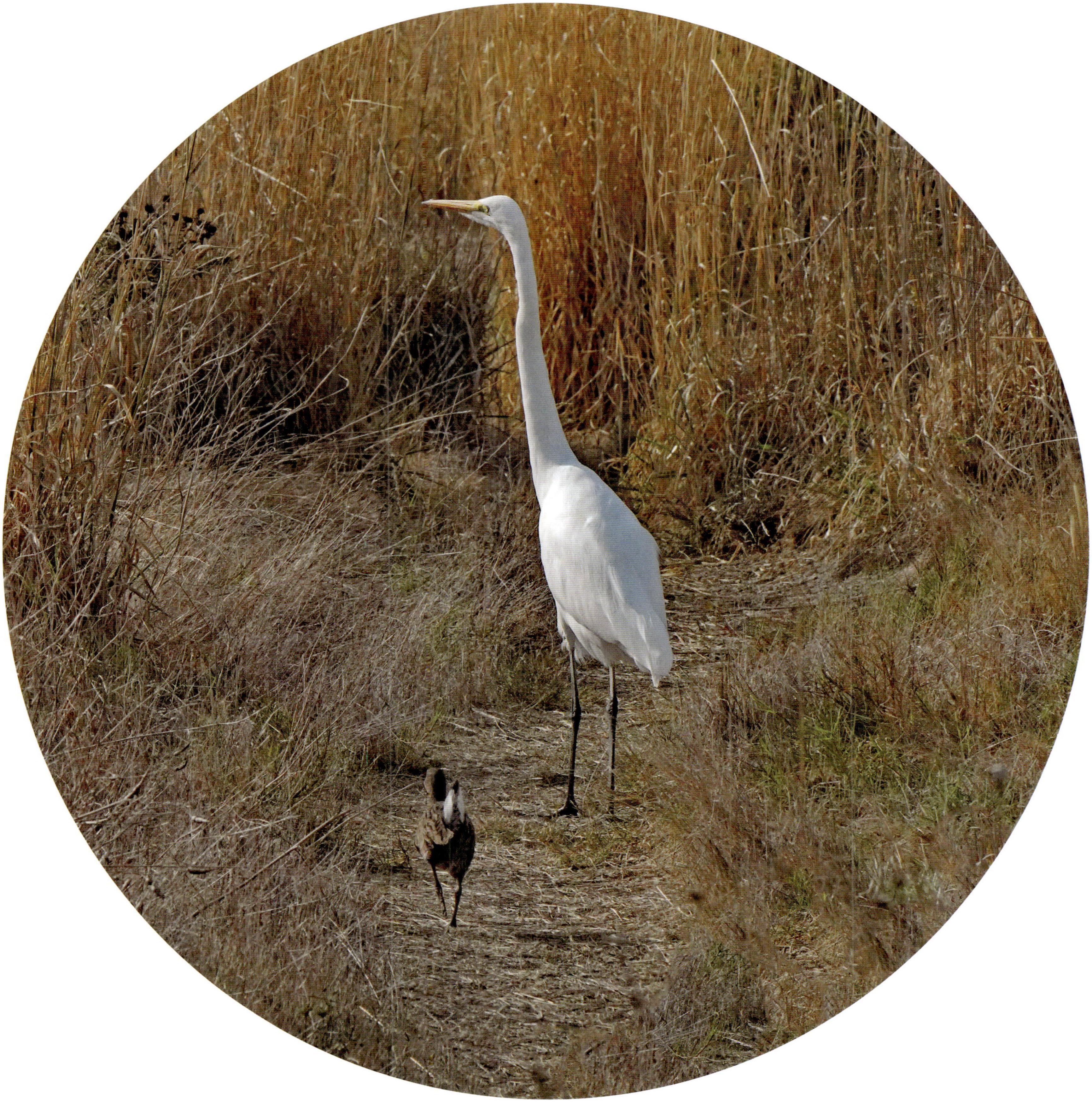

PRECEDING: A very round Song Sparrow (left) throws a piercing glance in the direction of the photographer. Song Sparrows, including various subspecies like the Alameda Salt Marsh Song Sparrow and Suisun Song Sparrow, are heavily reliant on salt marsh habitats in the San Francisco Bay area. A rare sight out in the open, a Ridgway's Rail (right) scampers behind a Great Egret in Muzzi Marsh. Secretive Ridgway's Rails are at the heart of the Bay Area's evolving story of wetland loss—and restoration.

RIGHT: Two Common Gallinules burst across the water in one of the Las Gallinas Valley Sanitary District's reclamation ponds. Gallinules are a species of rail; they can swim like ducks and walk on long splayed toes across floating marsh vegetation.

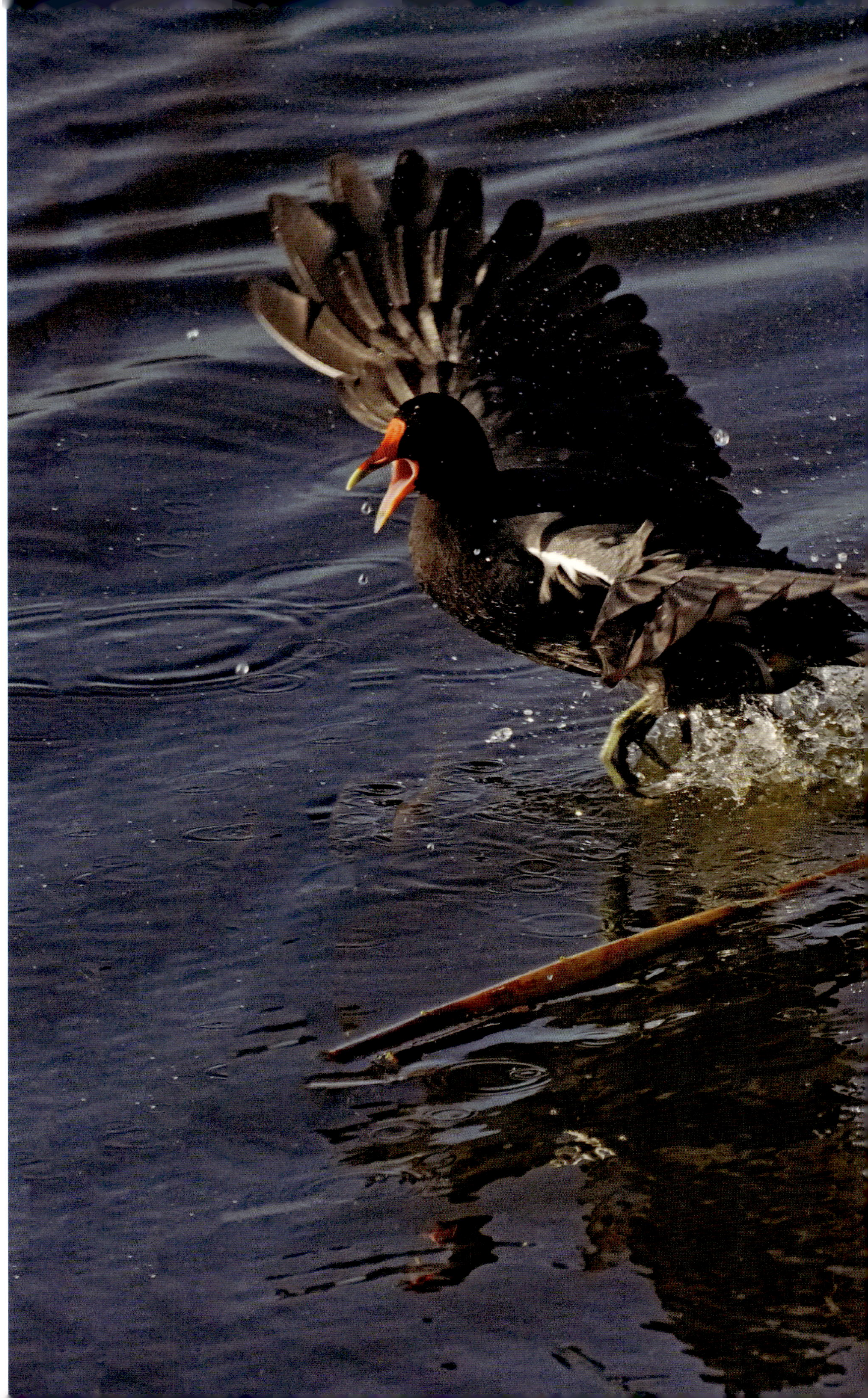

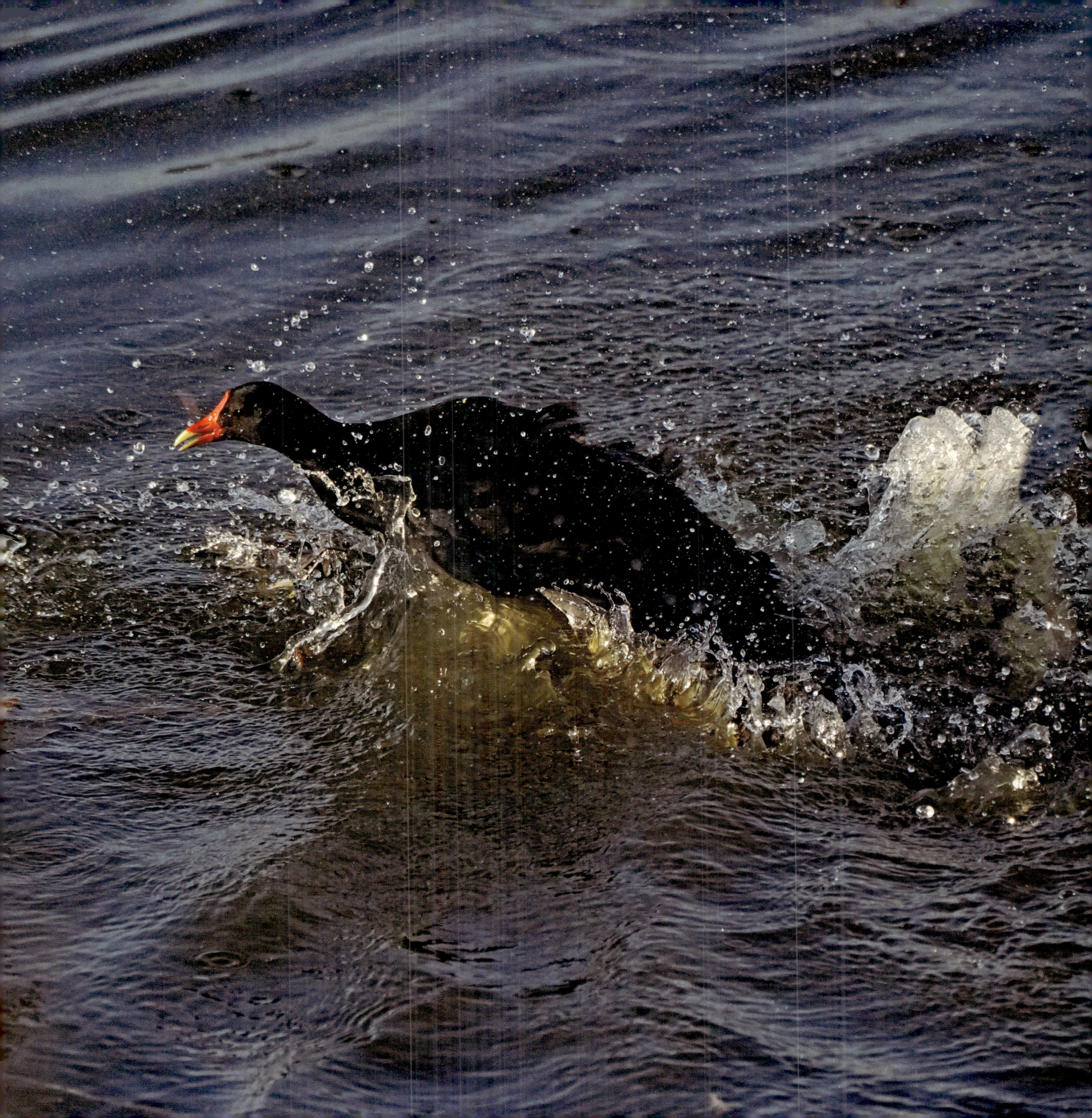

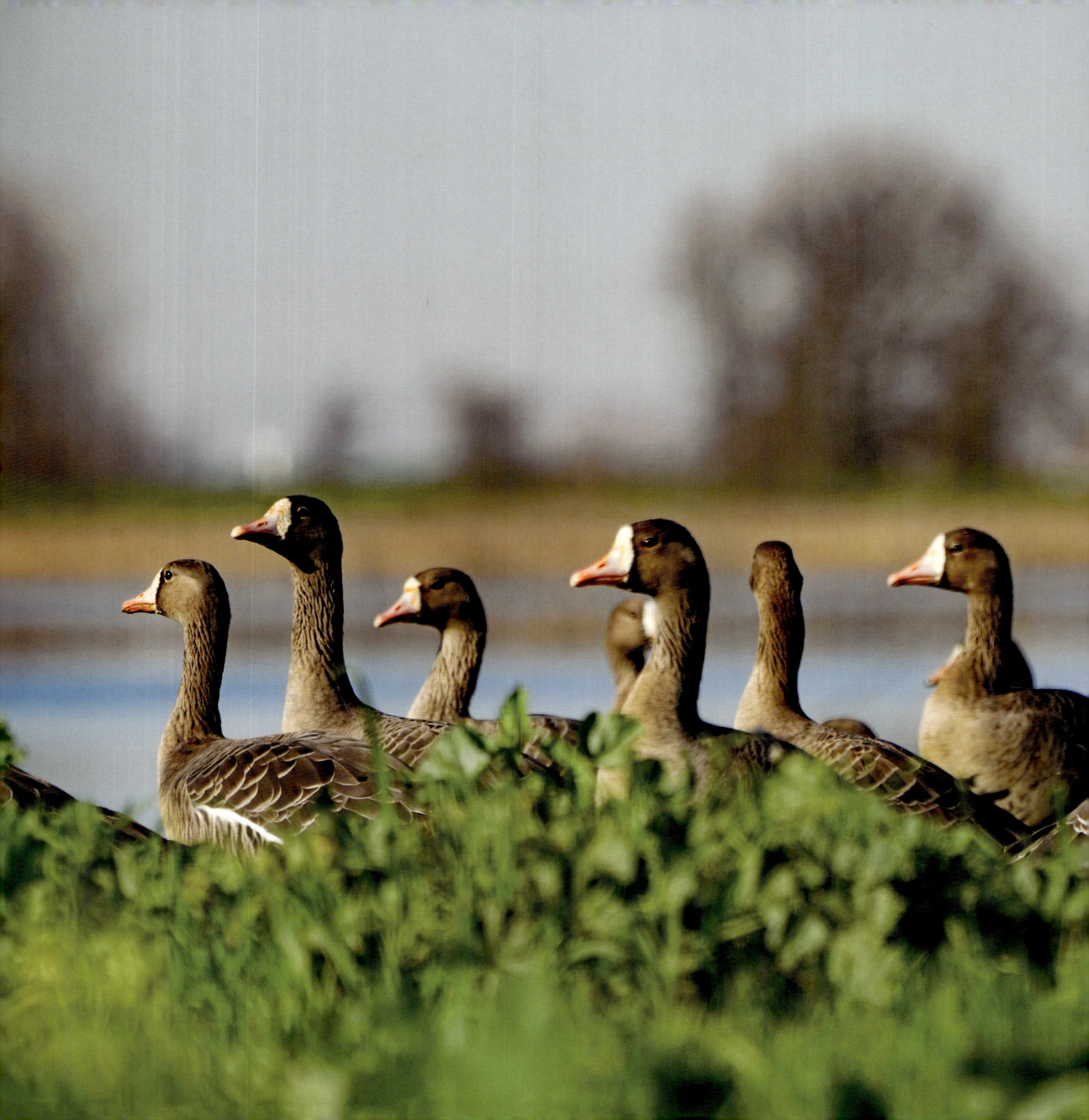

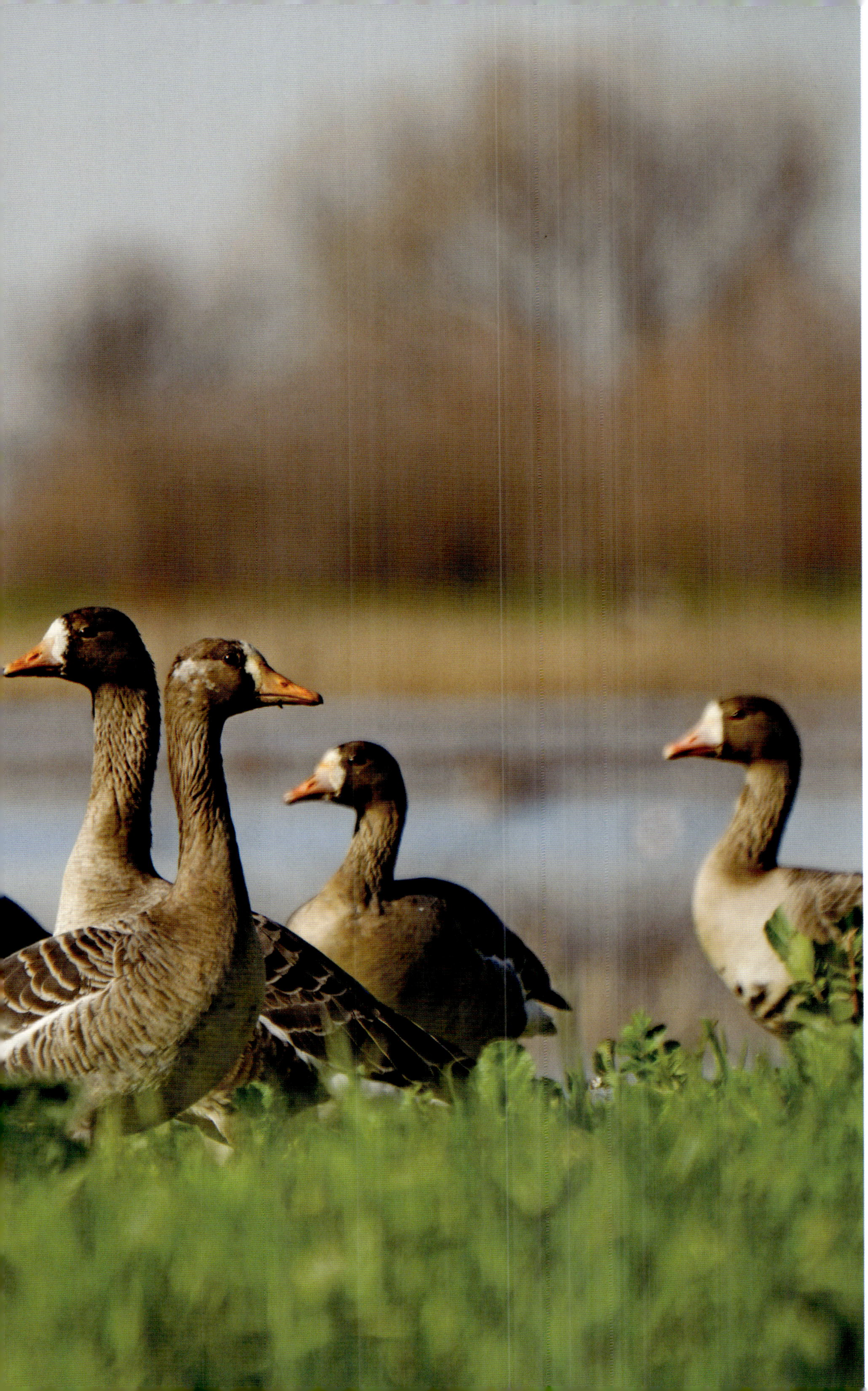

Greater White-fronted Geese follow the Pacific Flyway to the California Delta, where they flock together in wet agricultural fields and managed wetlands for winter food and shelter.

FOLLOWING: More than a million Snow Geese (left) return to California's wetlands to overwinter every year—these swirling flocks of birds are second only to Northern Pintails in terms of number of migratory waterfowl traveling the Pacific Flyway. While Canada Geese (right) have become a familiar year-round sighting in urban parks and golf courses around the Bay Area, their numbers swell in the winter as migratory populations arrive from their breeding grounds in Canada and the high Arctic. These geese form strong family bonds that include parents and siblings—when you look up and see their distinctive V-shaped formations passing overhead, those are likely to be family groups from the previous year.

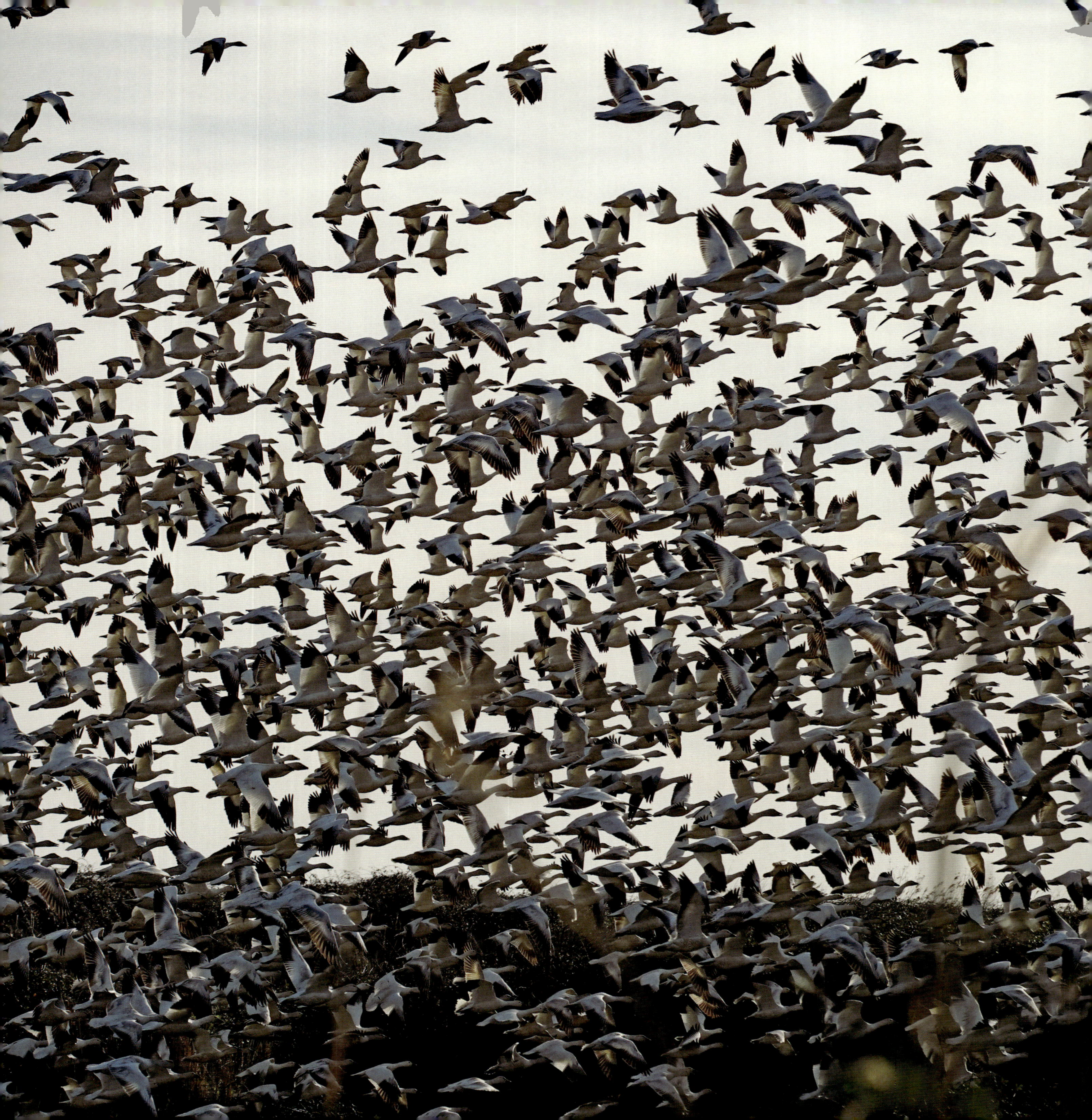

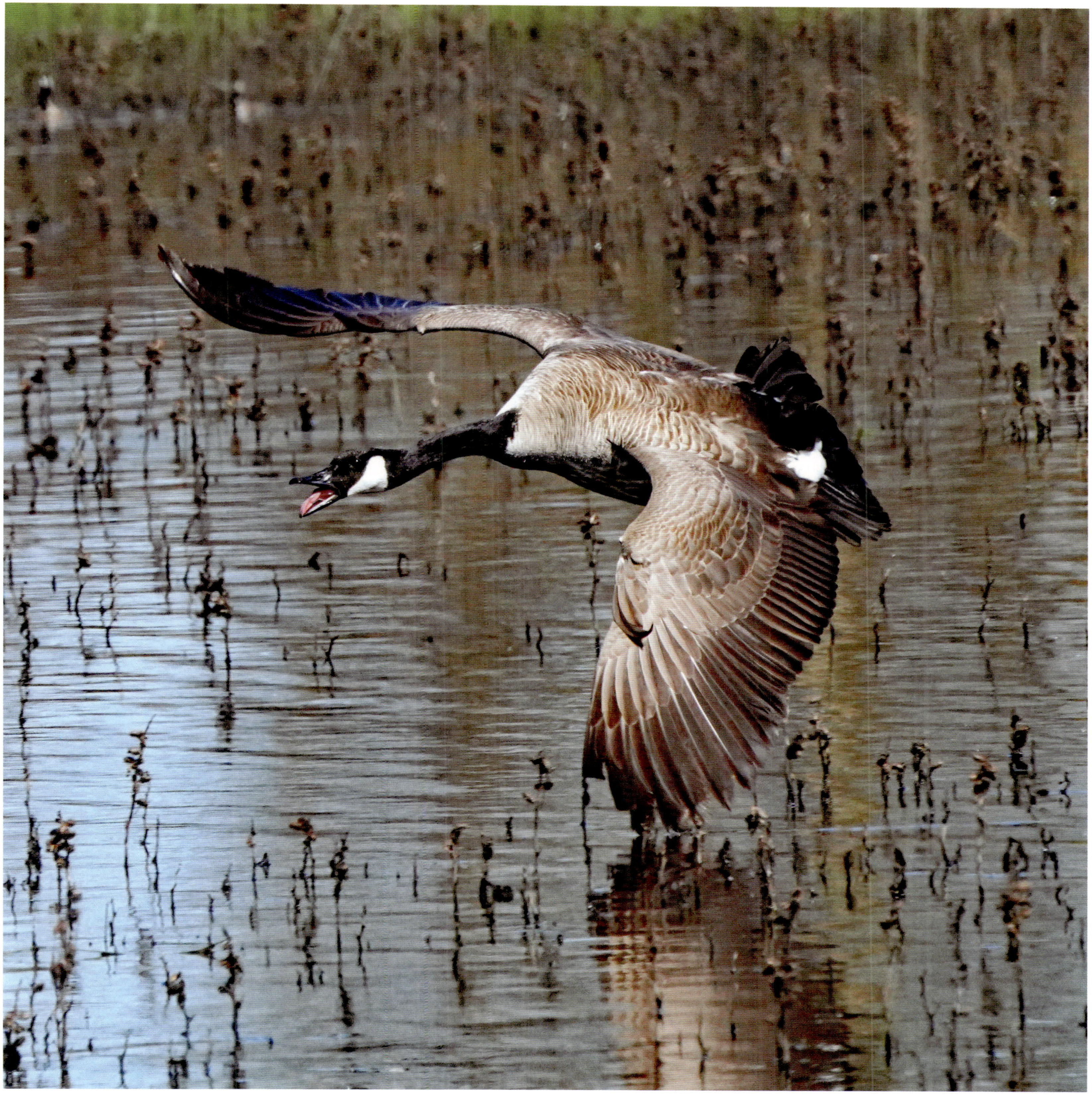

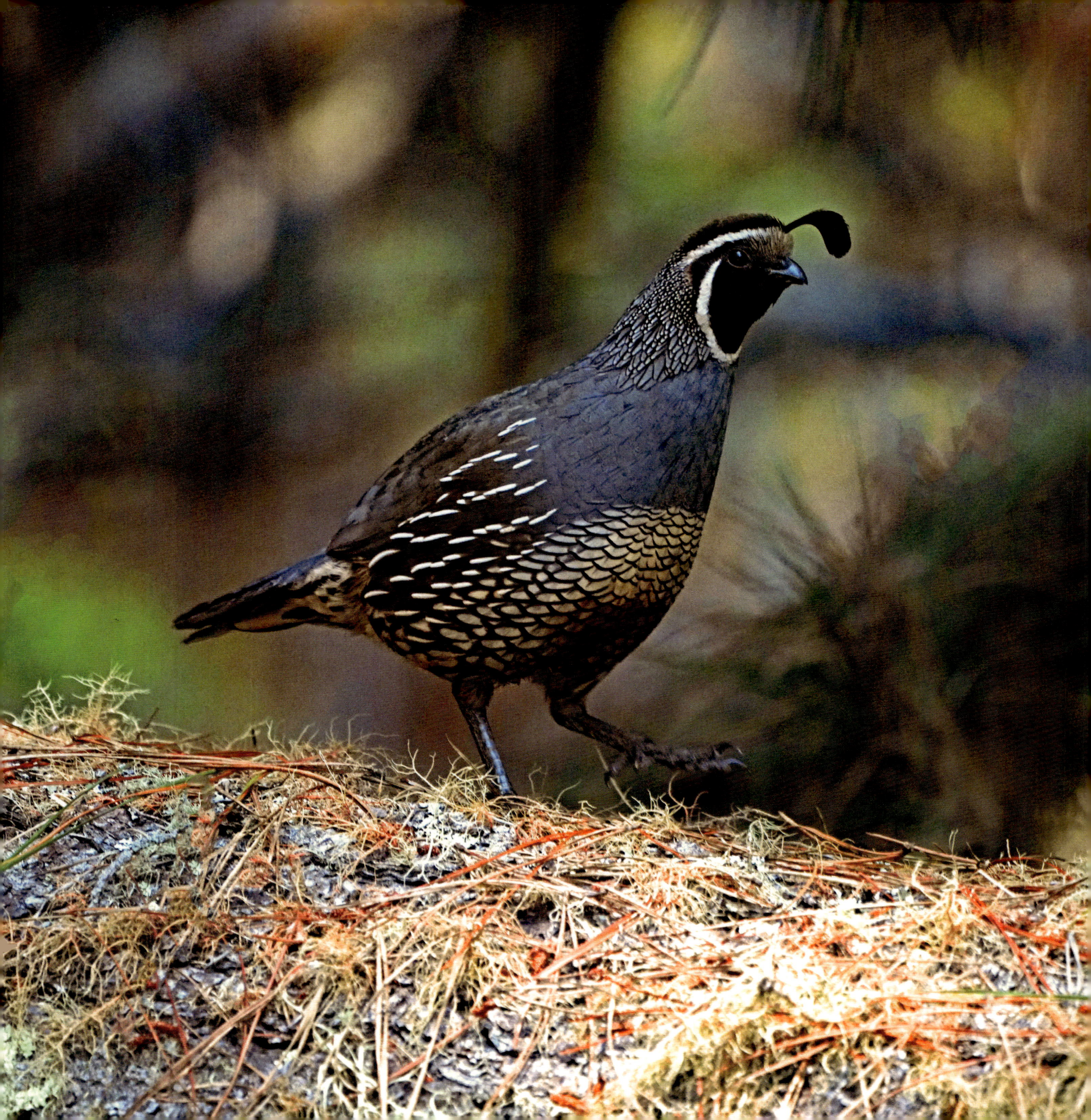

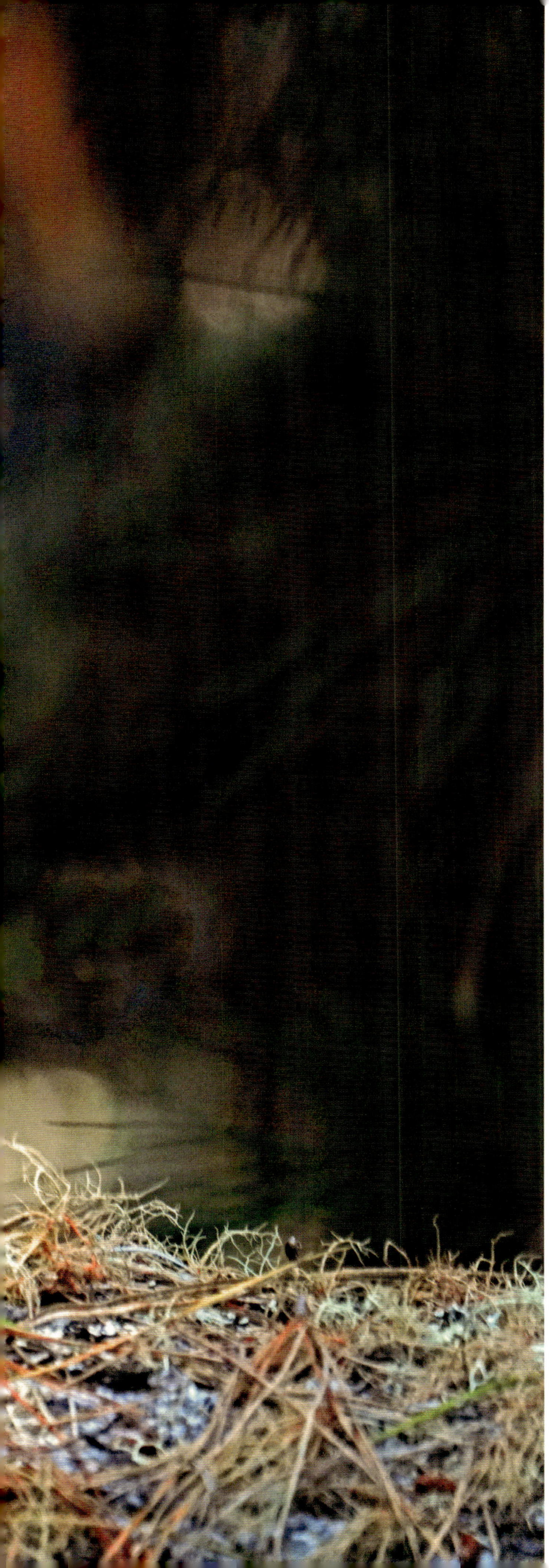

REWILDING RANCHLANDS

A long county road leads toward Stemple Creek Ranch, slicing west through the golden stubble of cattle pastures and dry inland hills. It's not yet noon on an August morning, but the ground already ripples with heat. At the turnoff, the road becomes gravel, and a tangle of wild roses folds across the entry gate, where a plaque reads "Protected Farmland Forever: Marin Agricultural Land Trust." California Quail call in the grass, and a songbird swoops over a cattle pond rimmed with lavish white blooms of Queen Anne's lace. Cool air pulses from a thick grove of riparian brush; the creekbed meanders like a dense green ribbon thick with vegetation, even in summer. A satellite map of the property shows that at the ranch boundary, the creek bed becomes abruptly treeless and dry. Even on Google Maps, it's clear that something special is going on here.

Today, we hike out to visit a tributary of Stemple Creek, where "beaver dam analogs" (BDAs) punctuate the path of the seasonal waterway. School groups from STRAW (Students and Teachers Restoring a Watershed) are collaborating with landowners to reshape and replant this tributary and countless others across the Bay Area. They've dug up nearby sediment and packed it between long interwoven branches. The string of dams—inspired by the ones that beavers build—slows the downhill flow and helps the land hold on to moisture. The BDAs cool the water, slow erosion, store carbon, and seed habitat complexity. It's an ecosystem service that beavers used to perform. Someday, a series of waterways stitched together by BDAs might create a greenway for beavers to follow back into their home range—bringing with them the birds and other life sustained by healthy watersheds.

PRECEDING: California Quail prefer a mix of open spaces for foraging and dense cover for protection. Ranchland restoration and the replanting of riparian areas help create safe habitat for these beloved native birds.

STRAW's work has deep roots. In 1992, a fourth-grade teacher named Laurette Rogers listened to her students' questions during a unit on endangered species: What could they do to help? The answer? Local, hands-on action. "This is about nature restoration," says the author Richard Louv, "but in a larger sense, this is about human restoration."

Today, Catie Clune—now the director of education at Audubon Canyon Ranch—recalls beginning the restoration work on this very tributary with a STRAW group when she was in fourth grade. "It looked really different. Just mud and grass," she says. Thirty years later, the banks are shaded with willow and brambles. Exuberant ranch dogs push between us to go racing through the shallow water that still pools in the low places, even after a summer of continuous heat waves. "Every year I see birds that I've never seen before," says Loren Poncia, the owner and ardent steward of Stemple Creek Ranch.

Stemple Creek Ranch is at the nexus of a web of partnerships engaged in restoring biodiversity to ranchlands. The ranch is one of three test cases for a 10-year study with the innovative Marin Carbon Project to explore how pastureland can help sequester carbon in the soil while restoring nutrients and moisture that nourish both livestock and a biodiverse web of life, from willows to warblers. Partners at the ranch also include the Marin Resource Conservation District and the Marin Agricultural Land Trust, the first land trust in the nation focused specifically on protecting farms and ranchland.

Later, I pull up a different kind of map—one that shows Stemple Creek Ranch as one block in a vast patchwork of lands protected by conservation trusts like MALT. Currently, over 25 percent of the entire Bay Area is conserved by land trusts, an area larger than San Francisco. That patchwork grows denser every year—and so do the flocks of birds that come in search of tangled creek banks and undeveloped fields.

TOP LEFT: Brown-headed Cowbirds historically followed wild herding animals like bison across open grasslands in America, and they are now often found close to ranchlands and livestock. They are parasitic, laying their eggs in other birds' nests, but studies show that riparian restoration can help restore balance.

TOP RIGHT: Spotted Towhees are among the first bird species to arrive when shrubs are replanted in riparian areas. They also respond well to oak woodland restoration.

BOTTOM LEFT: Townsend's Warblers stop in Bay Area forests on their spring and summer routes through the region. In making restoration decisions, it is important to consider the needs of our winter visitors as well as our summer breeders.

BOTTOM RIGHT: Song Sparrows, frequent hosts of cowbirds, can serve as indicators of successful habitat restoration. They readily colonize restored areas once there's enough vegetation for nesting and foraging, and monitoring Song Sparrow populations can help track the overall health of riparian ecosystems.

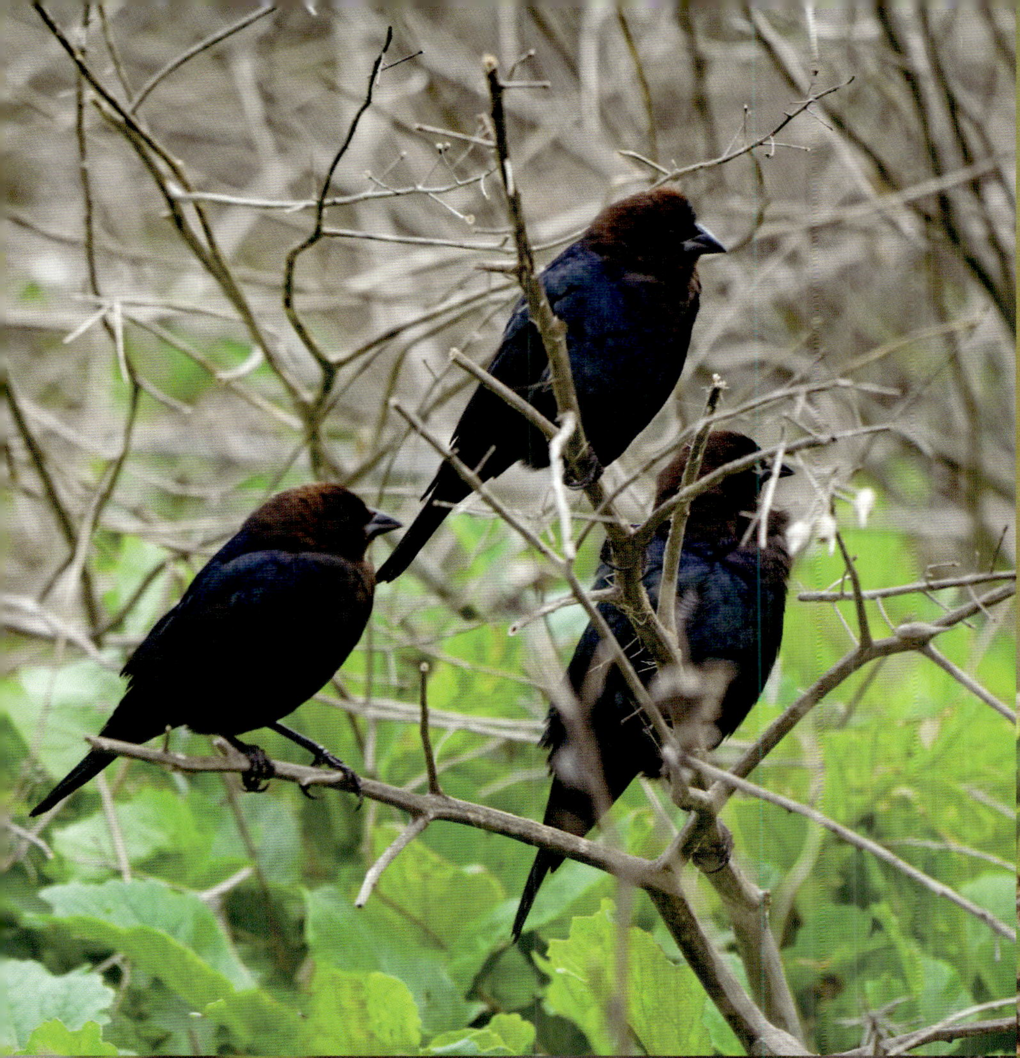

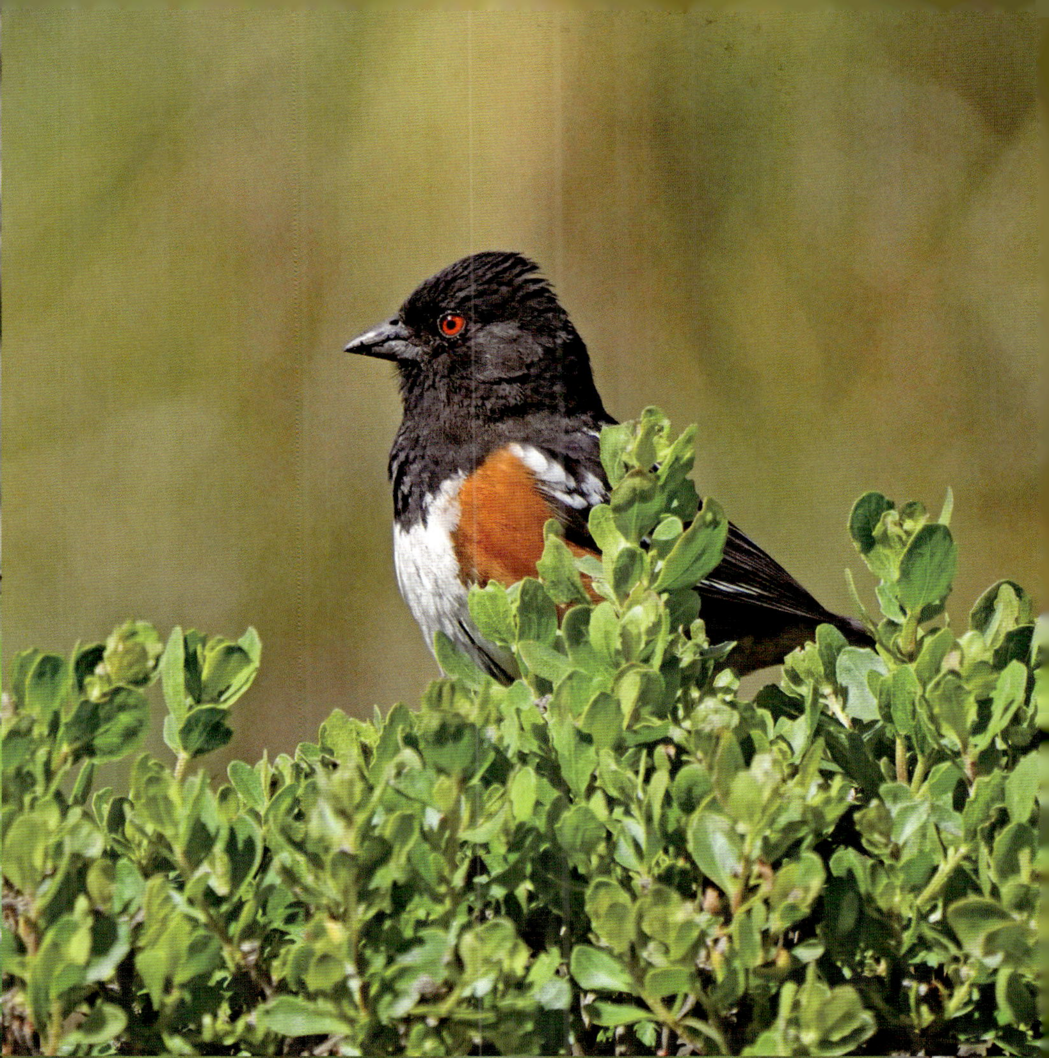

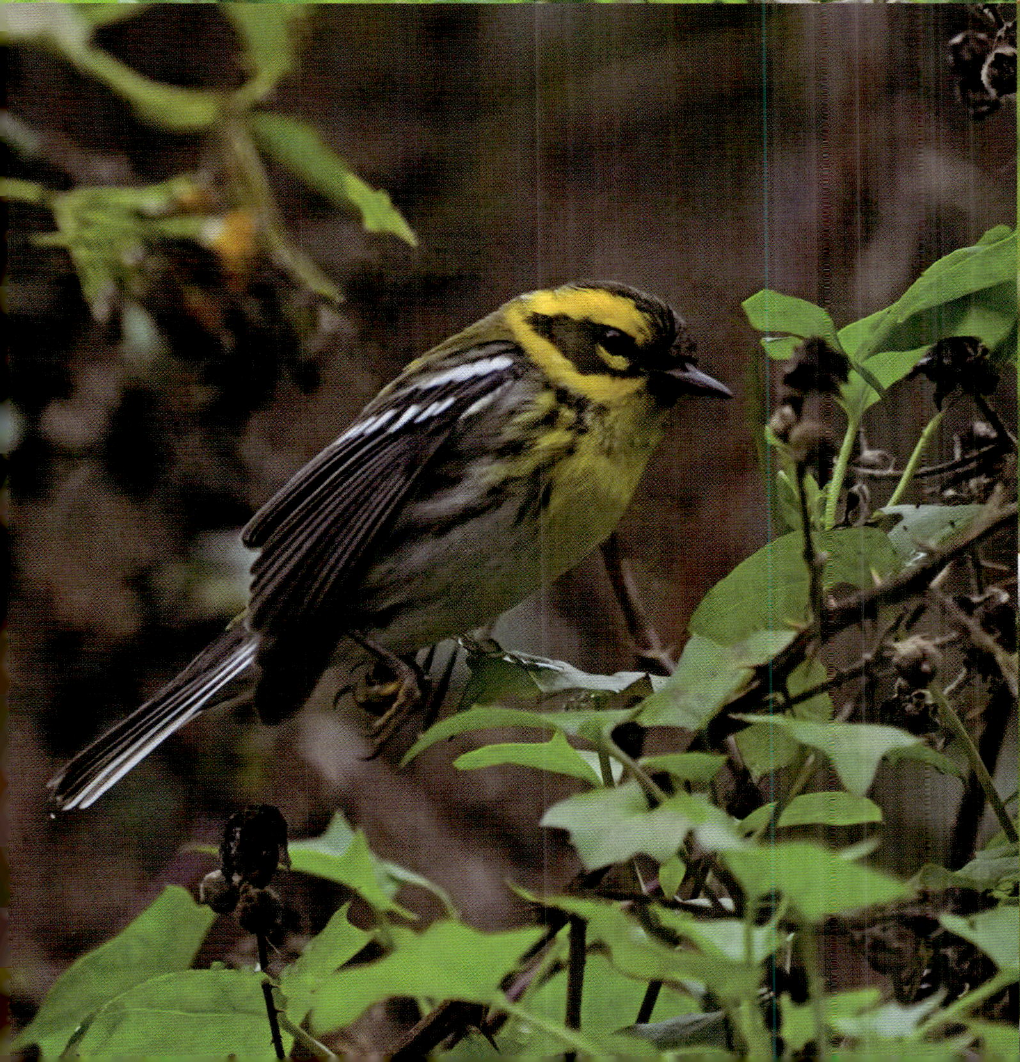

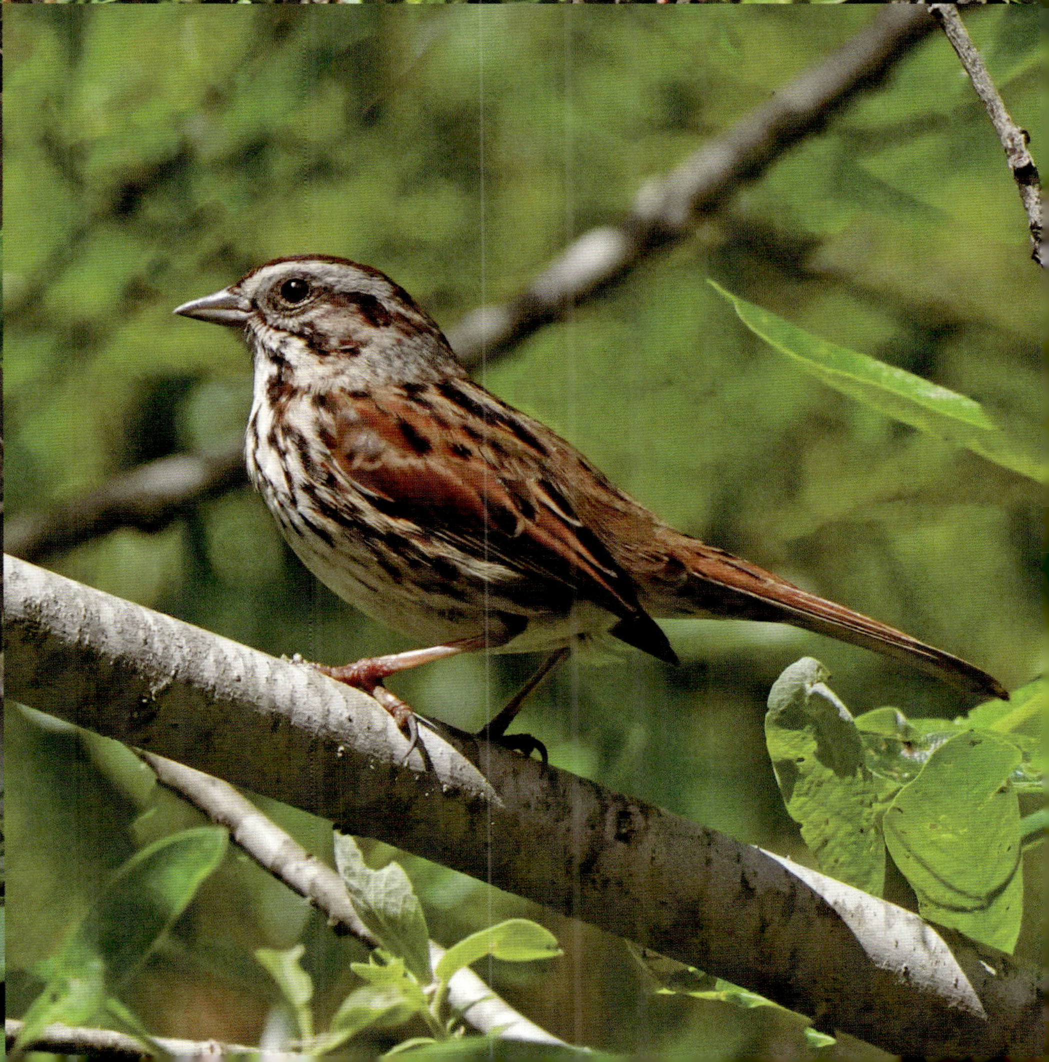

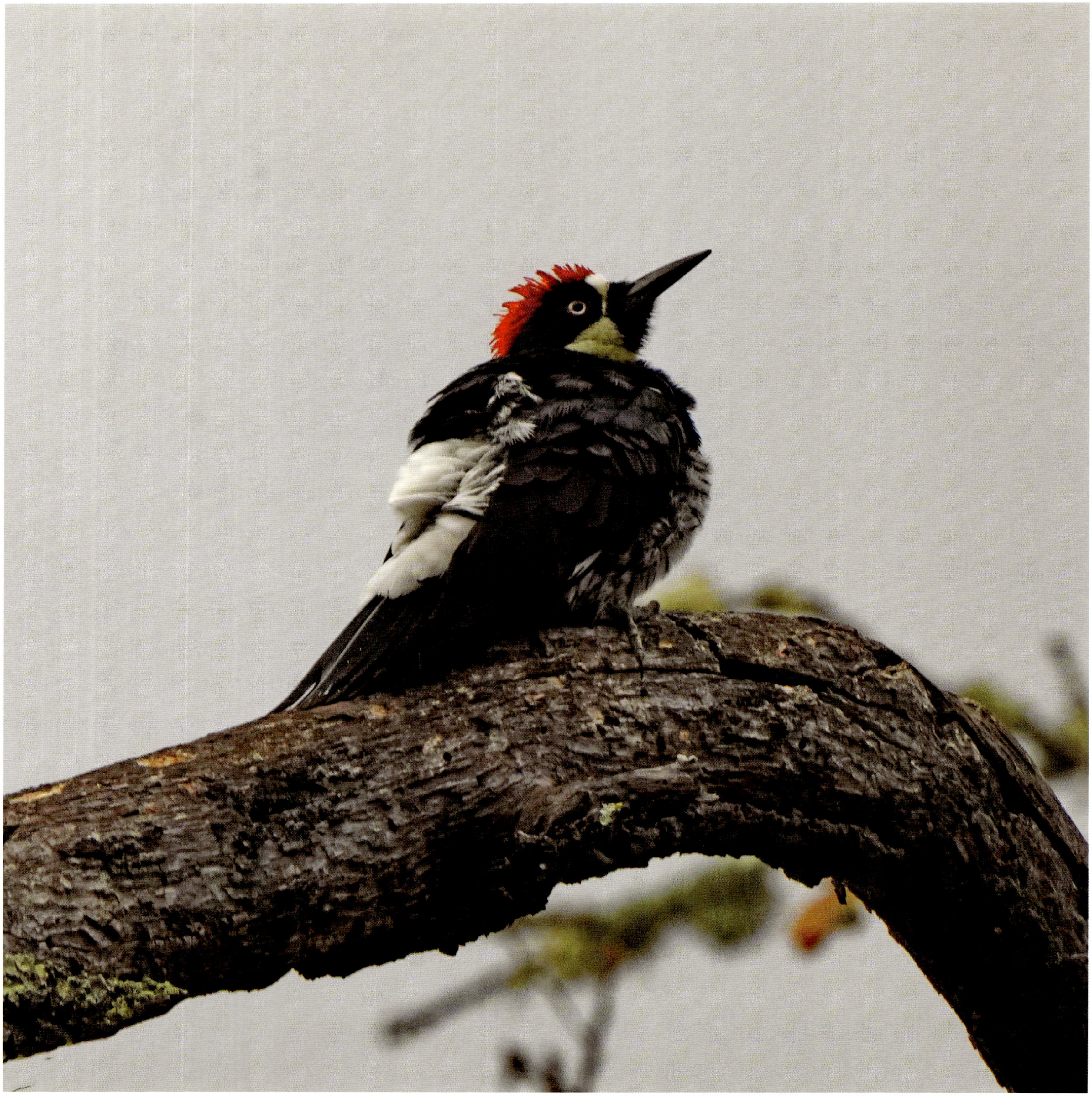

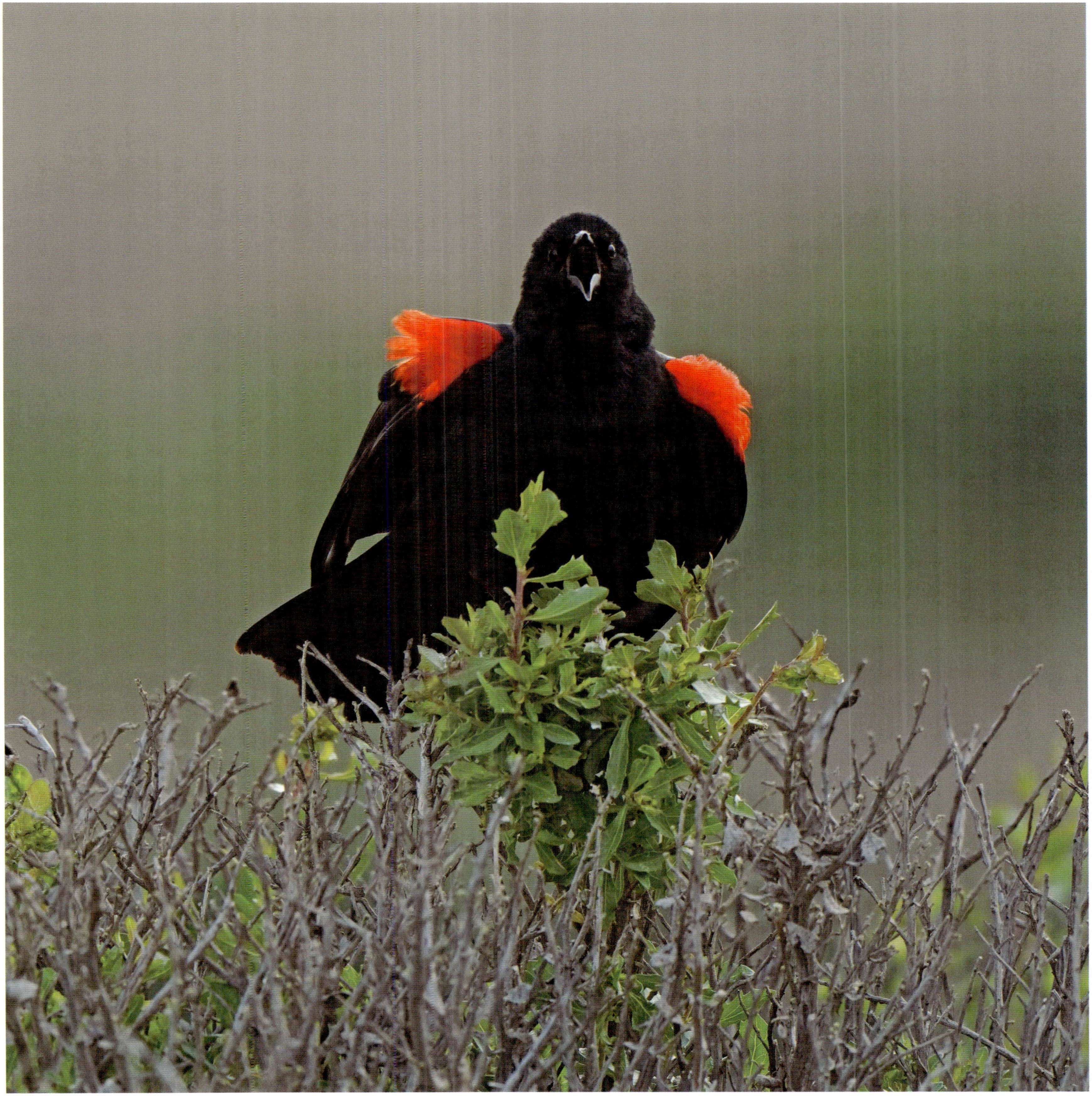

PRECEDING: Acorn Woodpeckers (left) are highly dependent on oak trees, which provide acorns as their primary food source, nesting cavities, and materials for building their communal granaries. Landowners can support these charismatic social birds by learning to manage existing oak ecosystems and by partnering to reforest areas with new saplings. Red-winged Blackbirds (right) are often found near water sources like cattle ponds, in marshes, and along the edges of fields. They will sometimes use agricultural areas for nesting and foraging.

RIGHT: A Great Horned Owl perches at the forested edge of a dairy farm on the Point Reyes Peninsula. For more than 150 years, ranchers have tended cattle on this windswept peninsula with leases that have allowed them to operate within the boundaries of Point Reyes National Seashore. In a historic deal in February 2025, all the dairy farmers and nearly all the ranchers chose a lease payout from the Nature Conservancy that will spark a new chapter of restoration work as 28,000 acres of pastureland are restored to native coastal prairie.

BELOW AND OPPOSITE: Other owls thrive in diverse landscapes around the Bay. A Barn Owl (below) hunkers in a snow-like field of fluffy seeds at the edge of an old military airfield outside of Novato. Where Malcolm Blanchard installed a nesting box in his backyard in Glen Ellen, a family of Screech Owls (opposite) moved in.

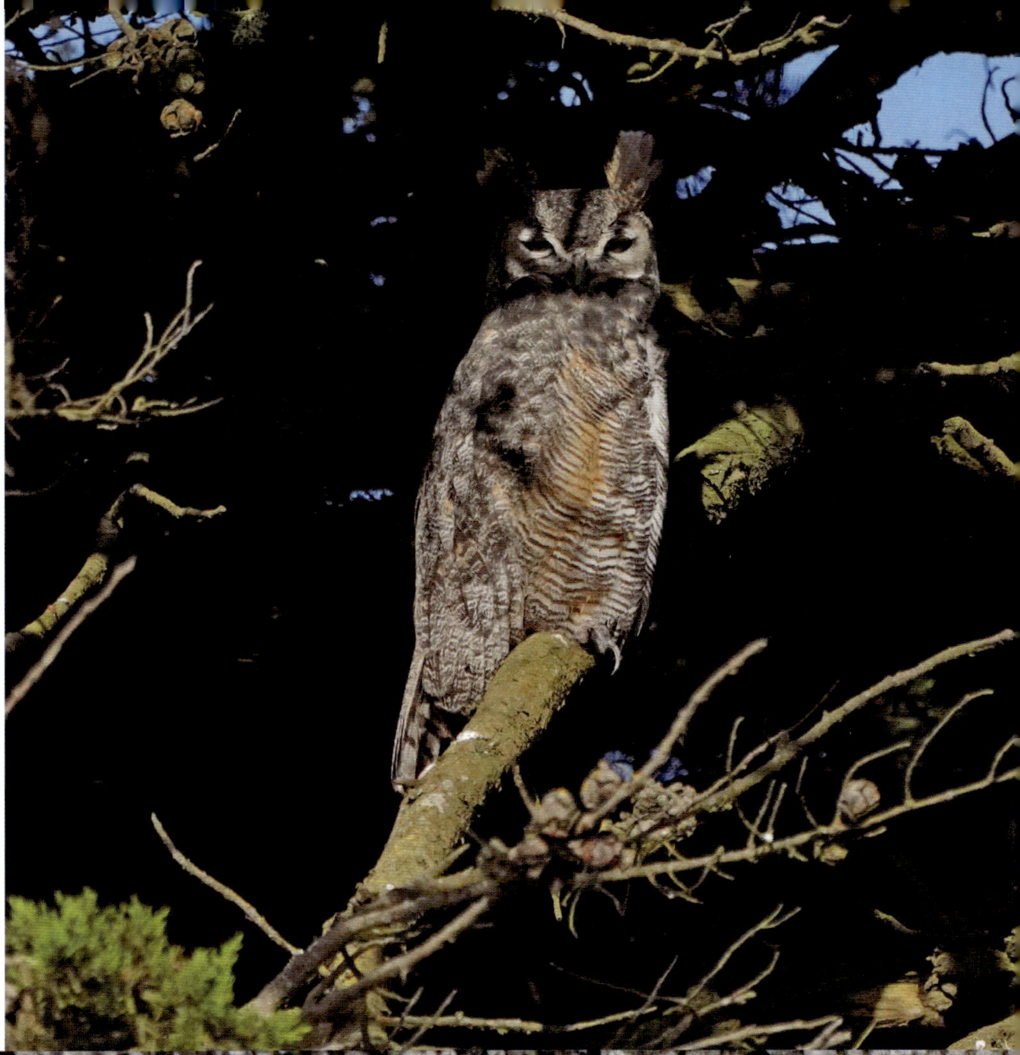

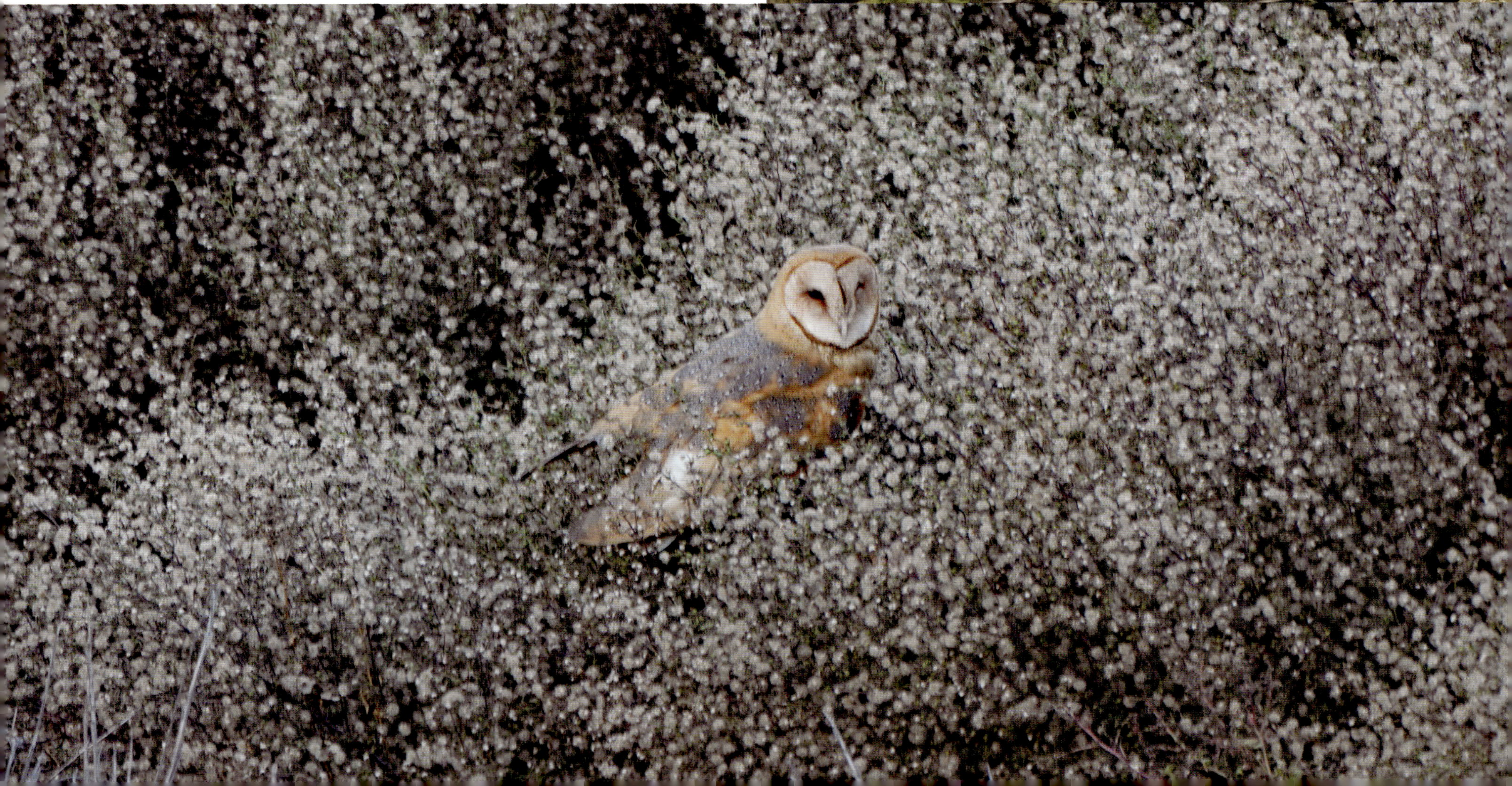

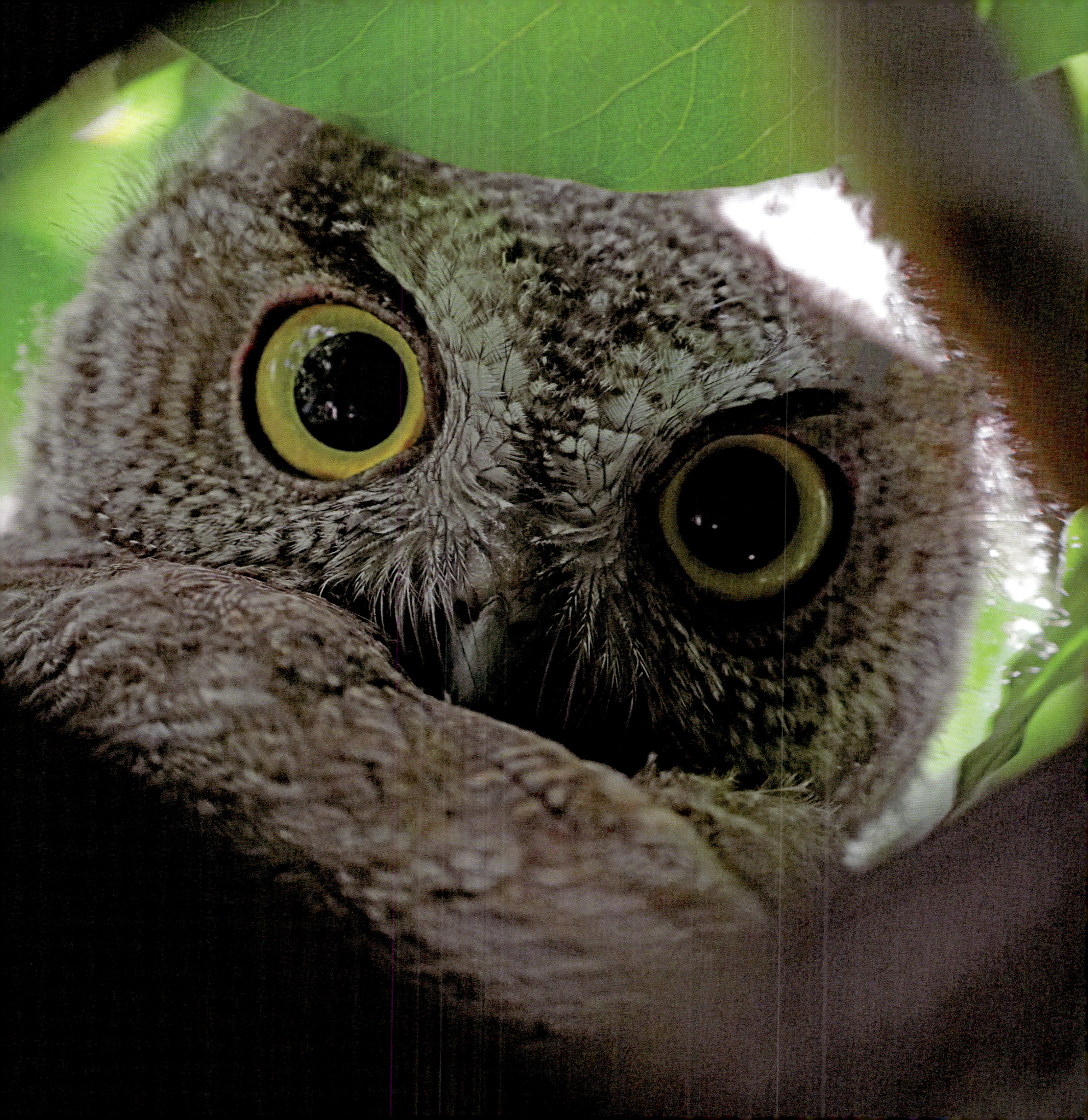

Western Bluebirds (left) are common in ranchlands throughout the Bay Area, particularly in areas with both open grasslands where they can hunt for insects and forests where they can build nests in tree cavities. They often use ranchland fences and utility lines as hunting perches, and landowners encourage their presence by building nest boxes in places where tree cavities aren't readily available.

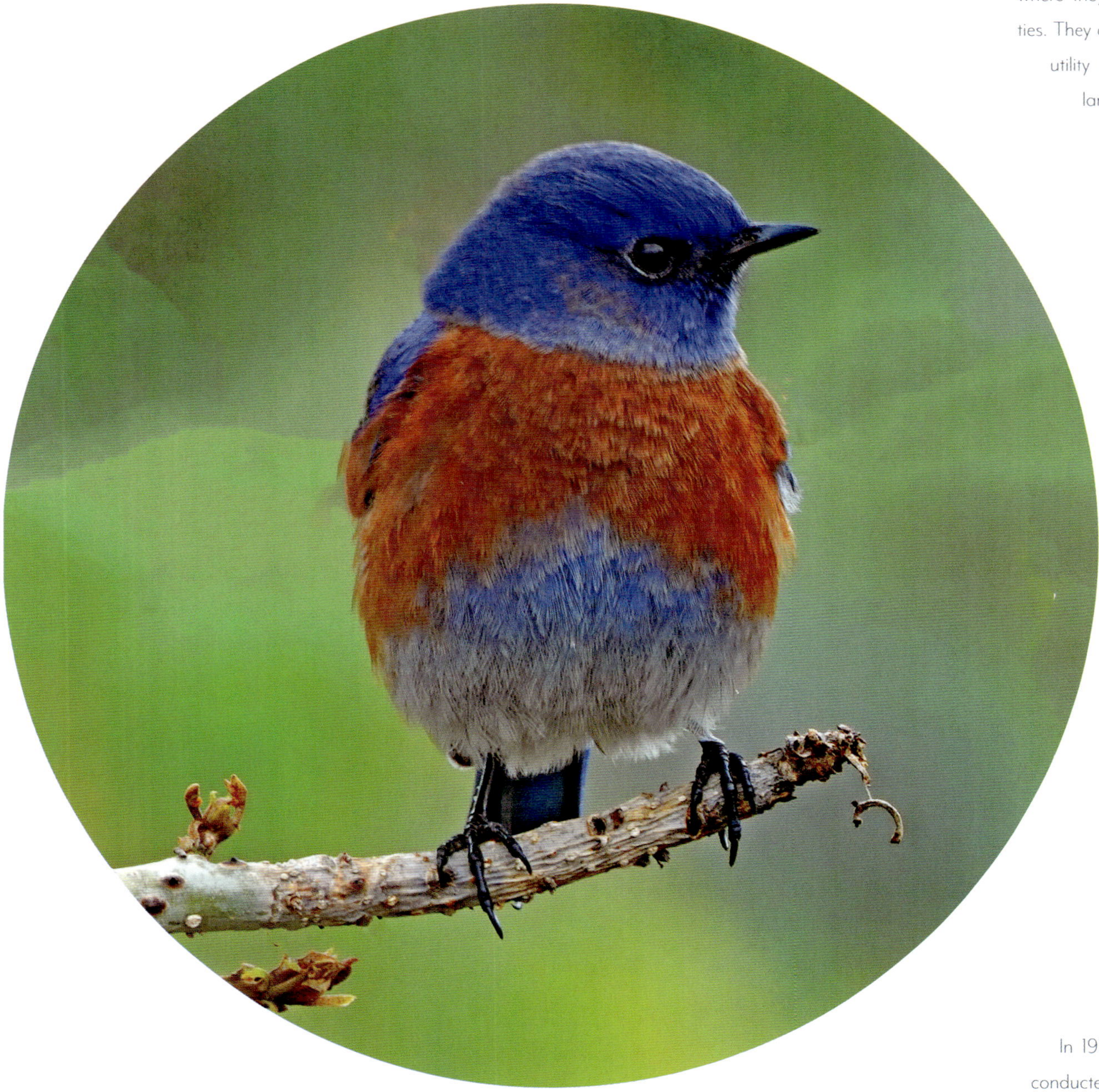

In 1914, grain growers in California conducted one of the earliest studies of the Western Meadowlark's (opposite) diet to understand whether or not the bird was a pest. The study showed that, lthough they do eat grain, Western Meadowlarks also help suppress populations of crop-damaging insects!

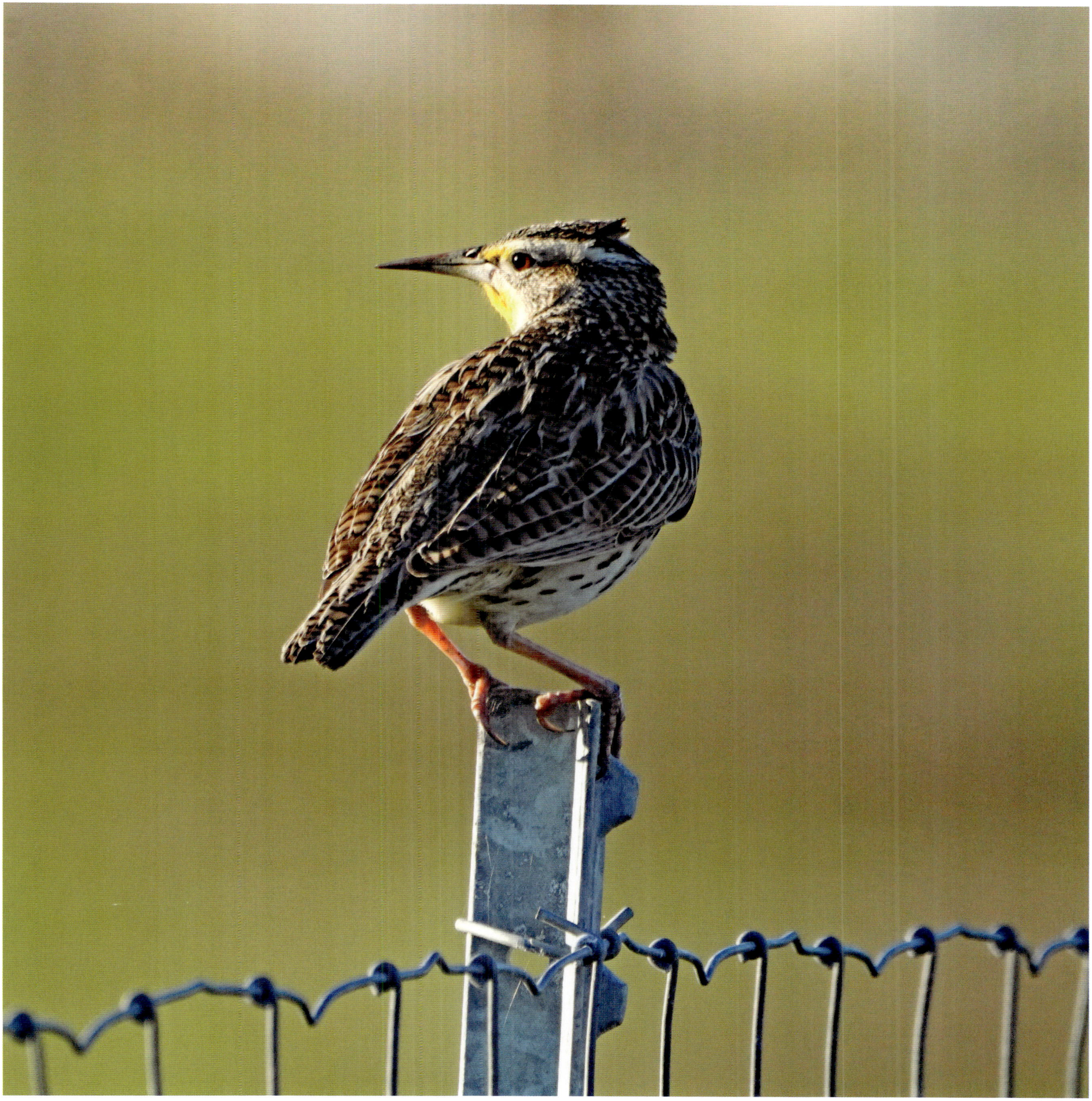

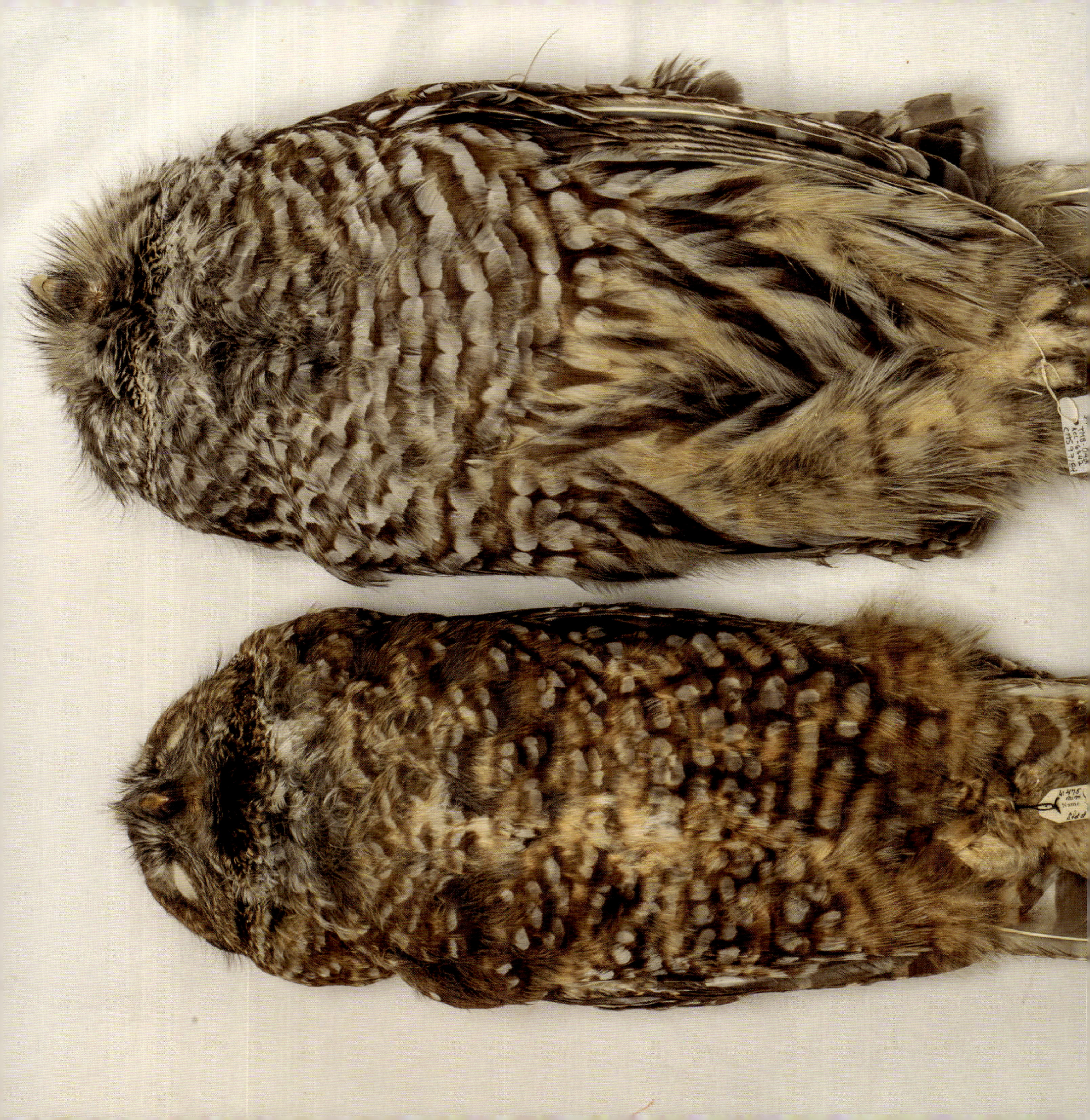

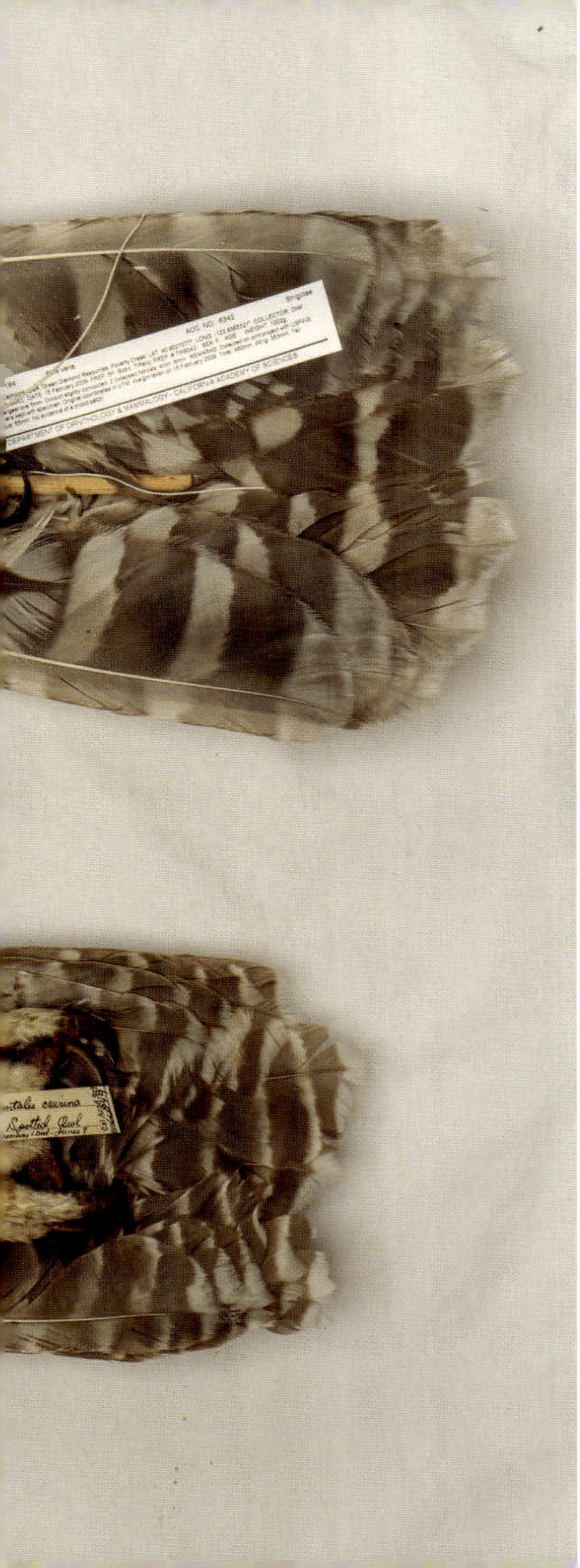

TO SACRIFICE AND TO SAVE

Go walking on a misty evening in the Doug-fir and bishop pine forests of Point Reyes National Seashore and you might hear it: two or three short, syncopated hoots followed by a longer, downward-lilting *whooooo*. Dark-eyed and elusive, Northern Spotted Owls tend to nest in mature, complex forests. Their presence is a biological indicator of an intact ecosystem. But these days, it is their absence that is noteworthy.

All down the west coast, as we logged their ancient homes to mill lumber and paper, we made it harder and harder for Spotted Owls to survive. Even after habitat protections were put in place following the Timber Wars of the late twentieth century, their numbers have plummeted—down by as much as 85 percent in some places. Scientists have determined that the cause of this most recent decline is a family feud of sorts. Their bigger cousins—Barred Owls, freshly arrived from Eastern North America sometime in the last hundred years—aggressively outcompete the smaller native species for space and food. One theory is that eastern Barred Owls managed to move west as homesteaders planted trees on the prairie and then in the wake of the Dust Bowl restored riparian woodlands, giving the interlopers a forested corridor to travel where otherwise the treeless Great Plains would have blocked passage. Another possibility is that anthropogenic climate change nudged Barred Owl populations north in latitude, high enough that they could use Canada's boreal forests to cross.

Compounding the issue, 200 years of logging old-growth forests on the West Coast has constricted what territory there is available for the two species to share. It is a chokepoint that prescribes loss, whether

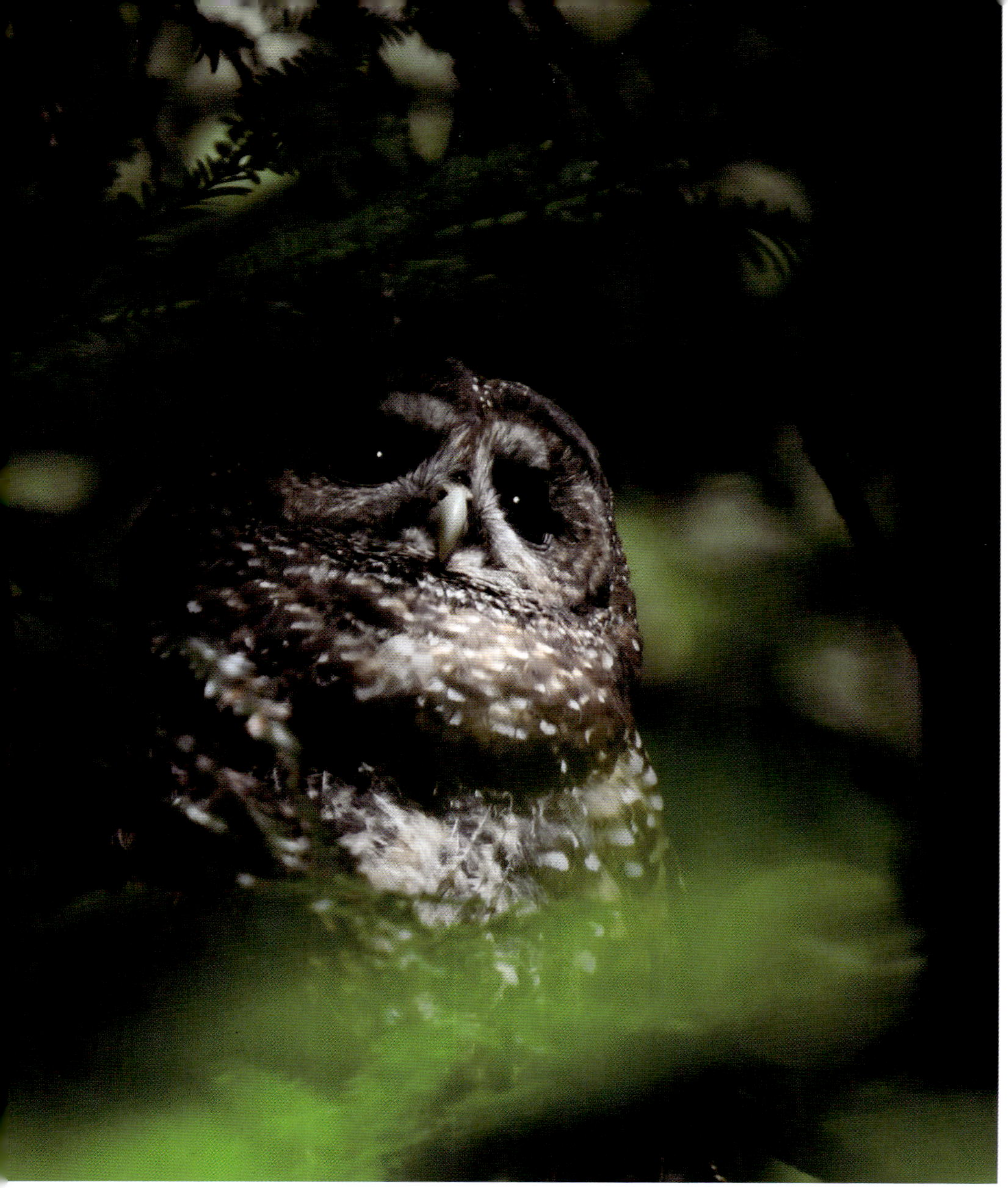

PRECEDING: Side by side, Northern Spotted Owl and Barred Owl specimens certainly look different from each other—native Spotted Owls are much smaller and more mottled. However, as the species hybridize, identification becomes trickier. Genomic sequencing and other lab resources at the California Academy of Sciences are helping tease apart the complex story of the decline of Northern Spotted Owls and the rise of Barred Owls in forests across the West. (Photo by Sharon Beals)

the management plan is martial or laissez-faire. Humans are implicated, inescapably. What do we owe our wild, native companions?

Recent genomic studies from the California Academy of Sciences show just how muddy the issue has gotten: Barred Owls have begun interbreeding with Spotted Owls, producing hybrids that further jeopardize the endangered owls' population. The results of a long-term study spanning Washington, Oregon, and California show a clear, if uncomfortable, solution: In places where the Barred Owls are culled, Spotted Owl populations stabilize. Now, the US Fish and Wildlife Service has proposed a controversial plan to eradicate 450,000 Barred Owls as a last-ditch effort to save a threatened species from extinction.

ABOVE: A secretive Northern Spotted Owl peers out through the branches of a redwood tree. While most Northern Spotted Owls prefer dense, shady old-growth conifer forests, Marin County populations will also nest in second-growth forests and in mature stands of bay laurel and ancient oaks. (Photo courtesy of Point Blue)

It's a tricky solution—humans don't have a great track record of removing apex predators without unforeseen cascading ecosystem impacts (consider the gray wolf). Is it better to let nature run its course—even if humans are the ones who tilted the balance? As with the other complex issues that managers are mitigating in the greater Bay Area—sea-level rise solutions that are workable for both marshland species and urban communities, negotiations between agriculture and conservation—it's hard to identify a path forward that is entirely without sacrifice.

But even as the reckoning between the two owls draws closer across the West, Marin County—which is the Northern Spotted Owl's southernmost range boundary—is a rare haven. While populations elsewhere continue to plummet every year, Marin's population of Northern Spotted Owls is currently stable, according to Point Blue scientists and national park biologists, who for years have been going out into the deep gulches and tall groves of our remaining old-growth redwood and fir forests looking for nests and listening for calls. The reason the owls aren't getting edged out by their big cousins? Geography. Just north of Marin County, the coniferous forests that stretch down from the Pacific Northwest through Sonoma County are interrupted by a break in the coastal mountain range. This windy, largely unforested expanse, like the Great Plains in miniature, is hard for interloping Barred Owls to cross from the north. Conservationists call this protective perimeter the "Petaluma Gap." This impasse makes Marin County the last stronghold for Northern Spotted Owls. Every year, a couple of Barred Owls creep across Marin's border—eventually, Bay Area wildlife managers may need to take up arms with the rest of the West against growing populations. But for now, the region is a refuge at complex crossroads.

The sun sets behind a windblown screen of Monterey Cypress near Land's End at the northwestern corner of San Francisco.

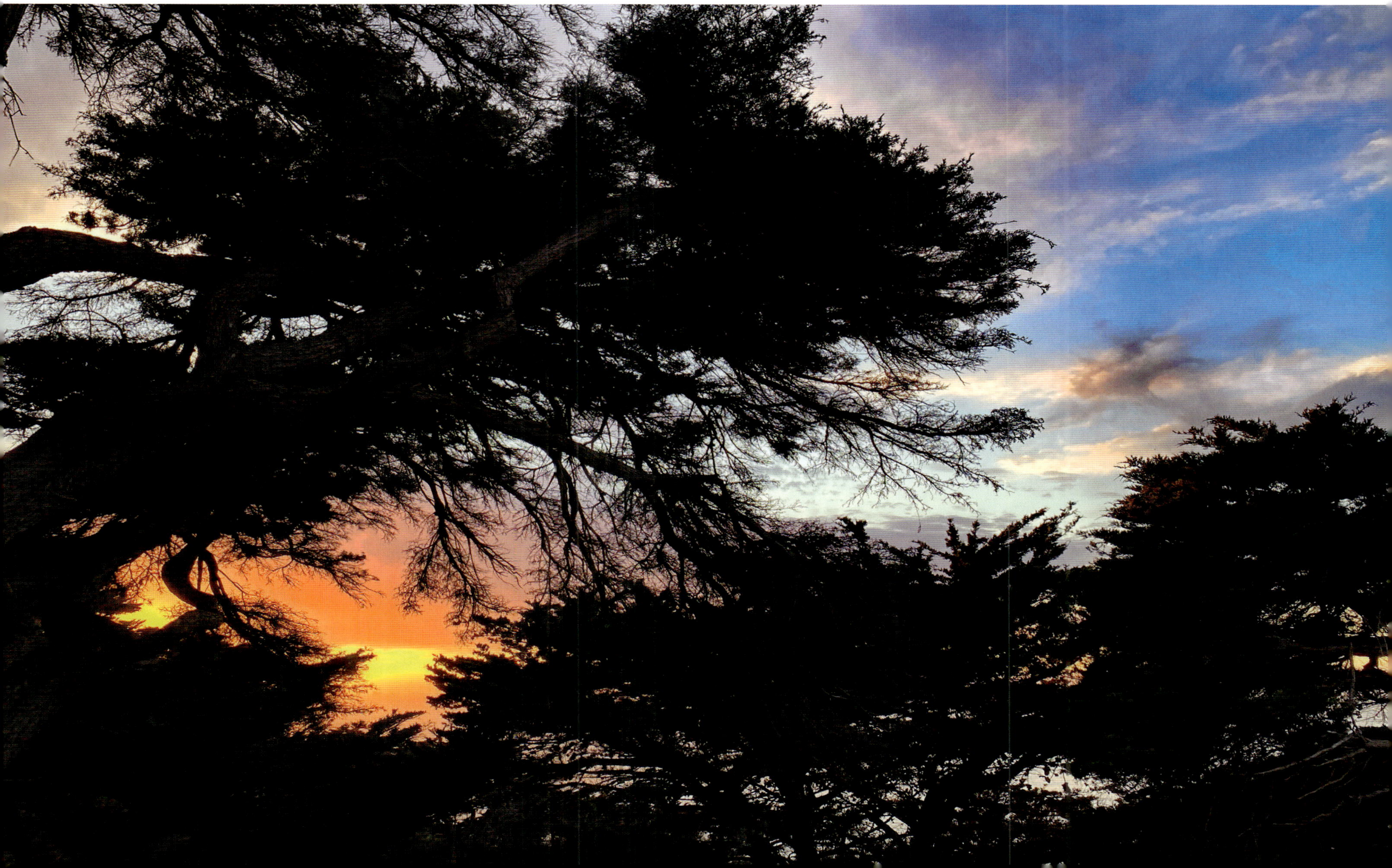

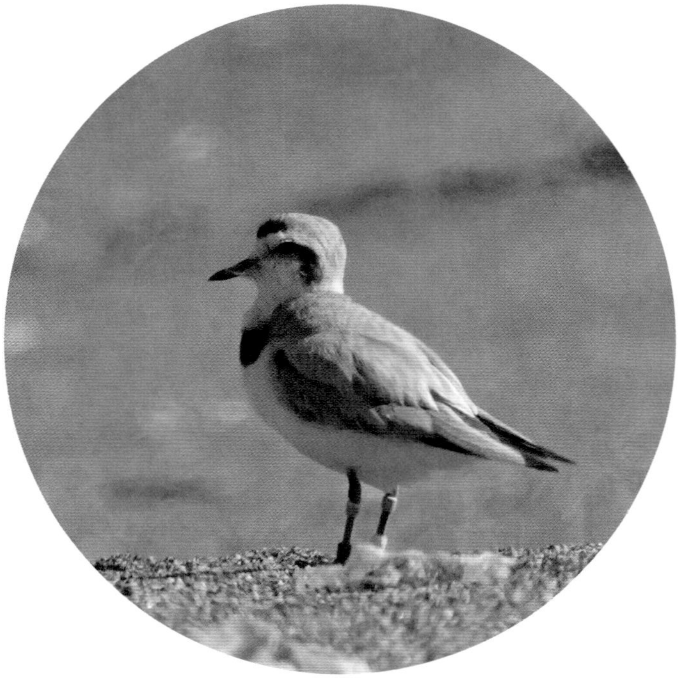

The number of Snowy Plovers on our beaches is an indicator of the health of our coastal ecosystem (adult plover pictured on the left, camouflaged eggs pictured opposite). These lovable, sparrow-sized shorebirds depend on dune habitat all along the California coast to nest and fledge their young between March and September. Their success is threatened every year by recreational beach use, coastal development, invasive dune plants, and predators like gulls and ravens that are drawn in by human activity. Snowy Plovers once numbered in the thousands along California's coast; Now no more than 2,500 breeding plovers remain in the state. Locally, Point Reyes National Seashore has partnered with Point Blue since 1995 to manage Snowy Plover recovery in the park's dunes. Everyday visitors can help too, by staying off dunes and leashing dogs during nesting season, or by volunteering to become docents on busy weekends.

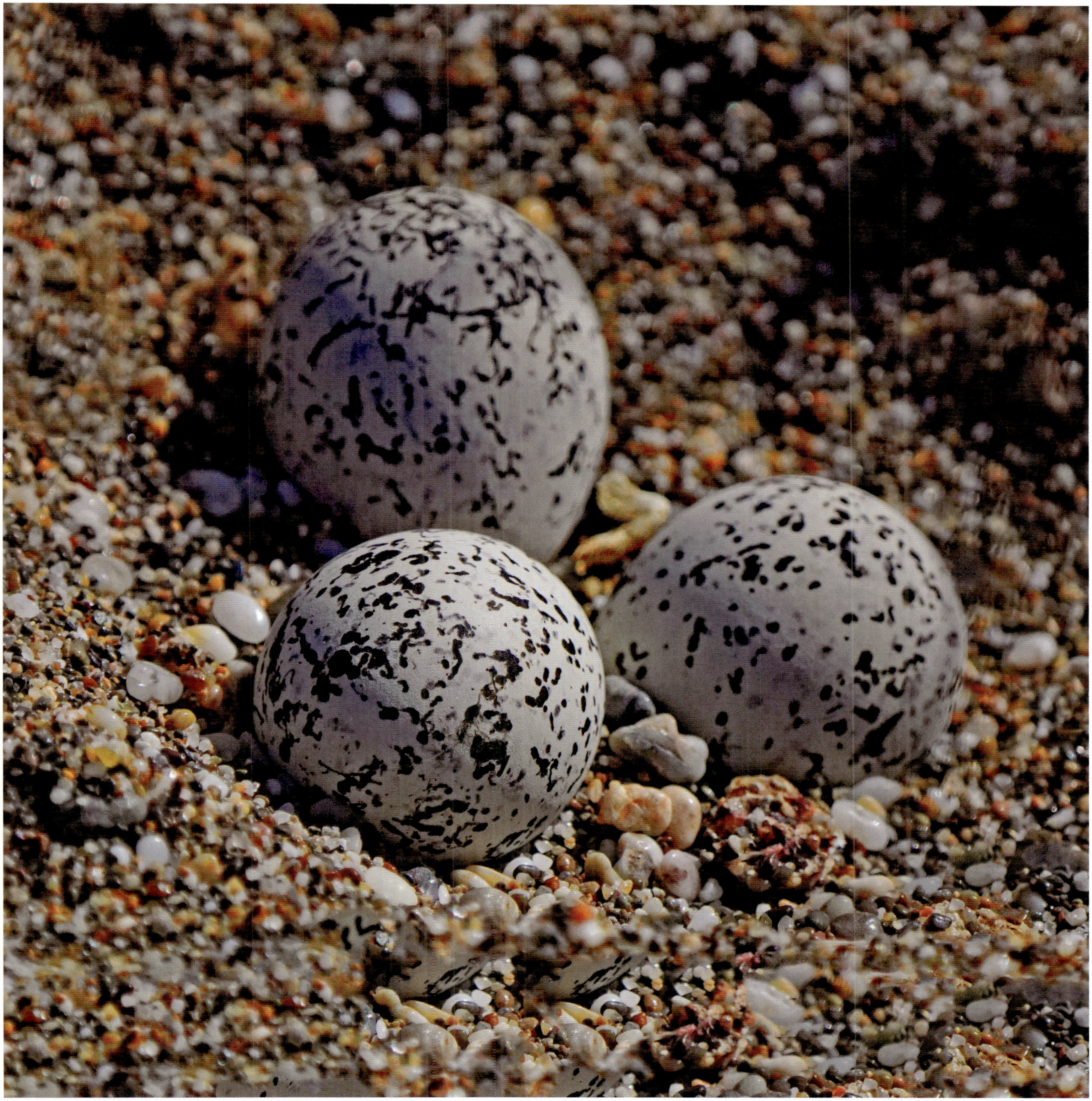

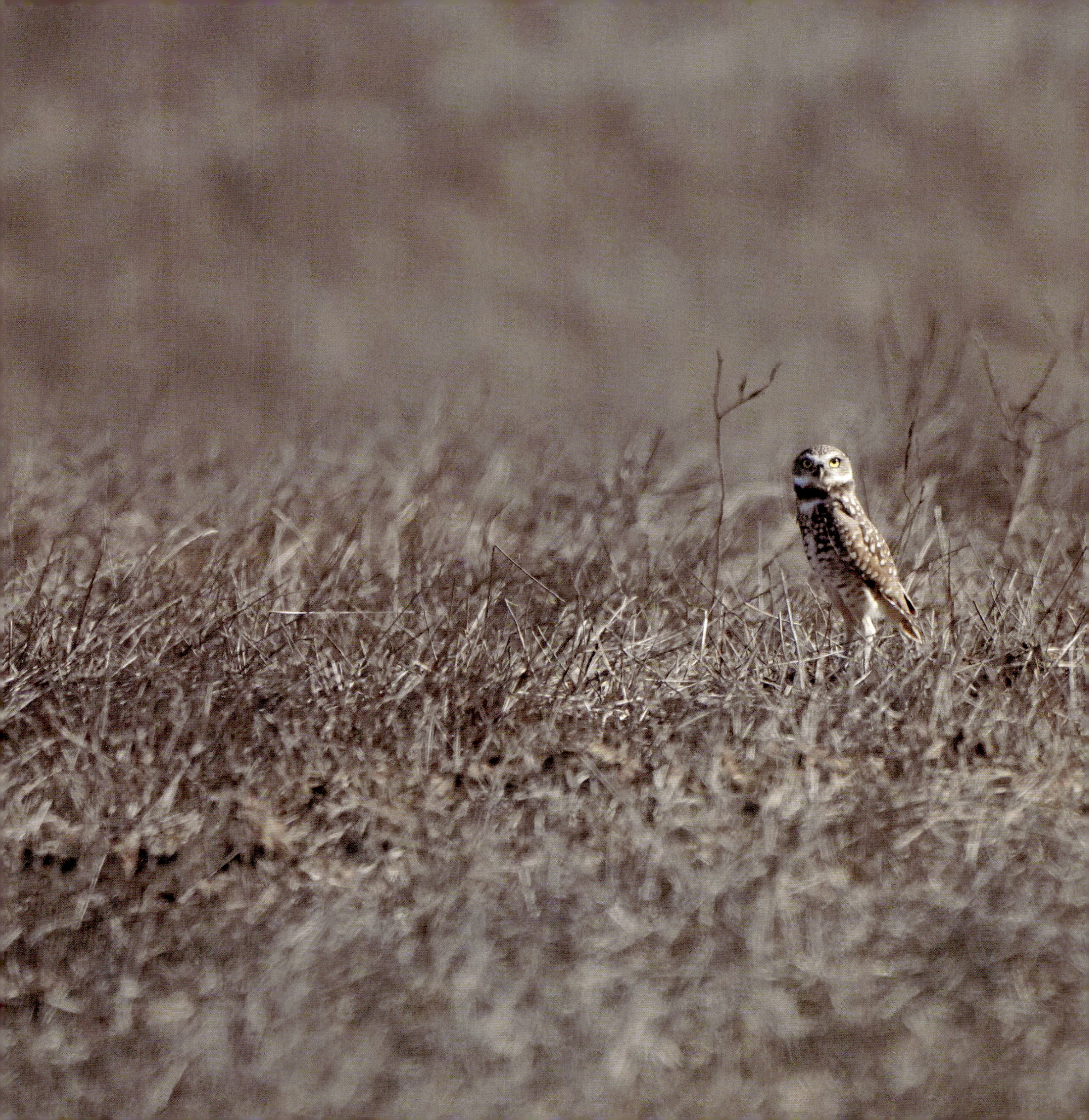

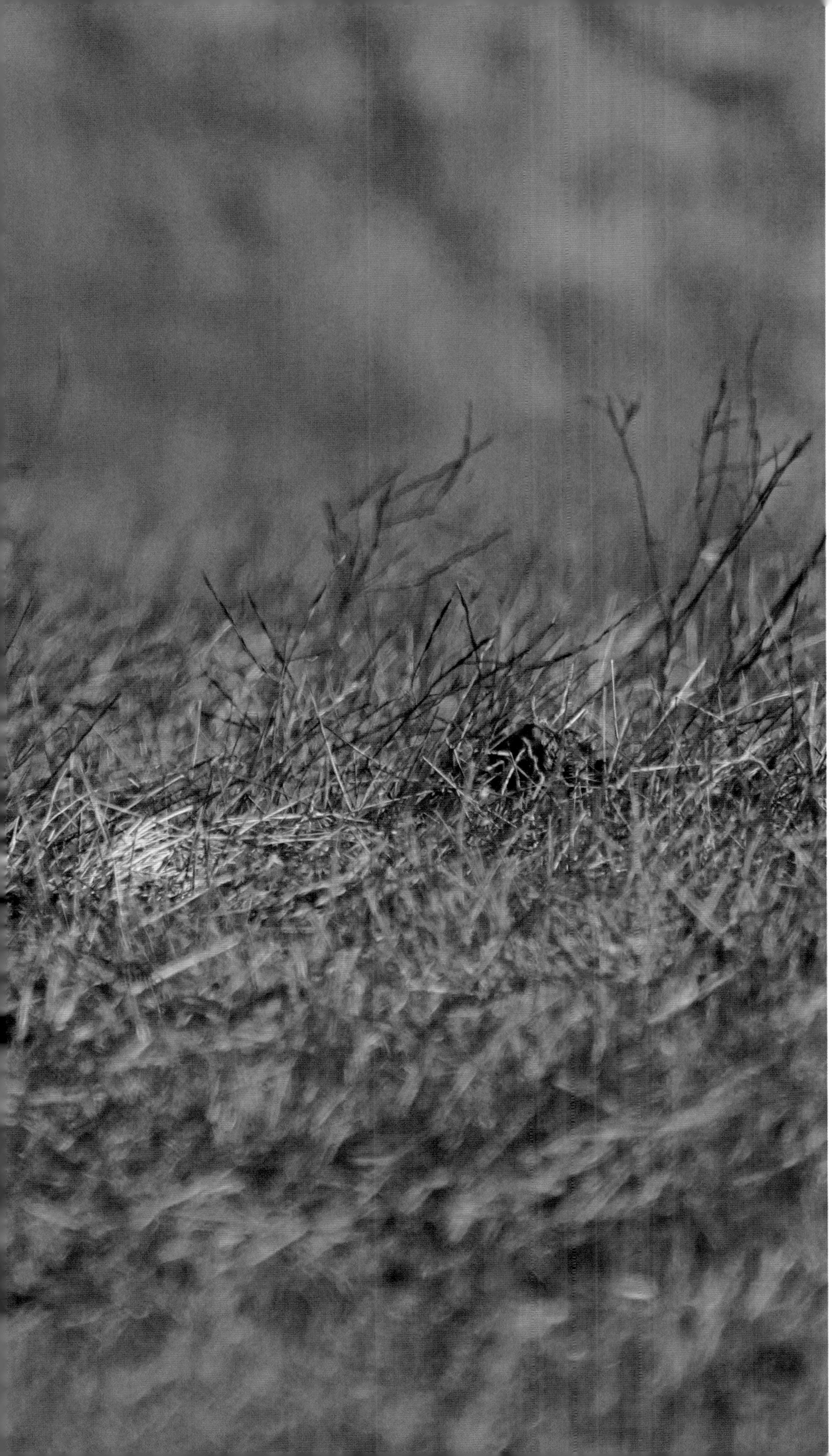

Tiny, charismatic Western Burrowing Owls once lived and nested at many locations throughout the Bay Area, but in recent years they have been listed by the State of California as a Species of Special Concern—threatened by urban development, intensive agriculture, and eradication of ground squirrels, whose holes the owls use as burrows. A complicated local conundrum has arisen on the Farallon Islands, where hyper-abundant populations of non-native House Mice attract Burrowing Owls in the fall. When winter mouse populations fall, the owls turn to preying on islands' Ashy Storm-Petrels. More than half the entire world's population of Ashy Storm-Petrels are thought to nest on the Farallones, each pair raising only one chick a year. The Burrowing Owls significantly impact the success of these slow-breeding birds. The US Fish and Wildlife Service (FWS) has proposed a plan to eradicate house mice from the South Farallon Islands using rodenticide, to encourage Burrowing Owls to move on before the Ashy Storm-Petrels arrive to nest. Yet conservation is complex: The plan must also mitigate the risk of owls temporarily increasing predation on petrels if they linger after the mice are gone.

FOLLOWING: The California Quail is both the State Bird of California and the official bird for the City of San Francisco. However, traffic, development, and feral cats make the city a tough habitat for these ground-dwelling birds. Once abundant, wild quail have completely disappeared from San Francisco, but a bold habitat restoration and reintroduction effort in partnership with the San Francisco Estuary Institute aims to bring them back to the city's Presidio. If they succeed, the benefits would include soil aeration and improved soil health, reduction of fire fuel loads, and better water penetration into the ground. Perhaps someday San Franciscans will once again hear the quail's syncopated *Chi-caaa-go* calls ringing out through the city fog.

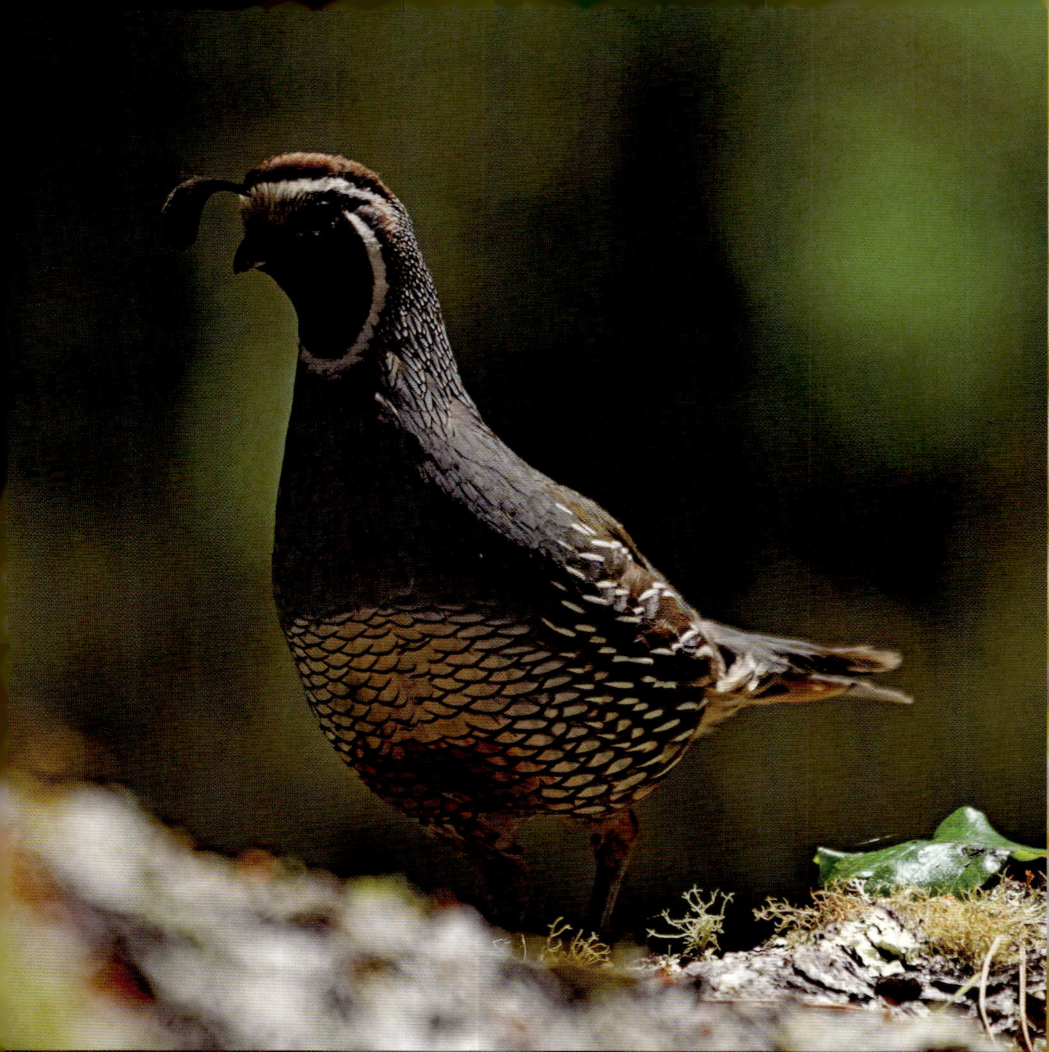
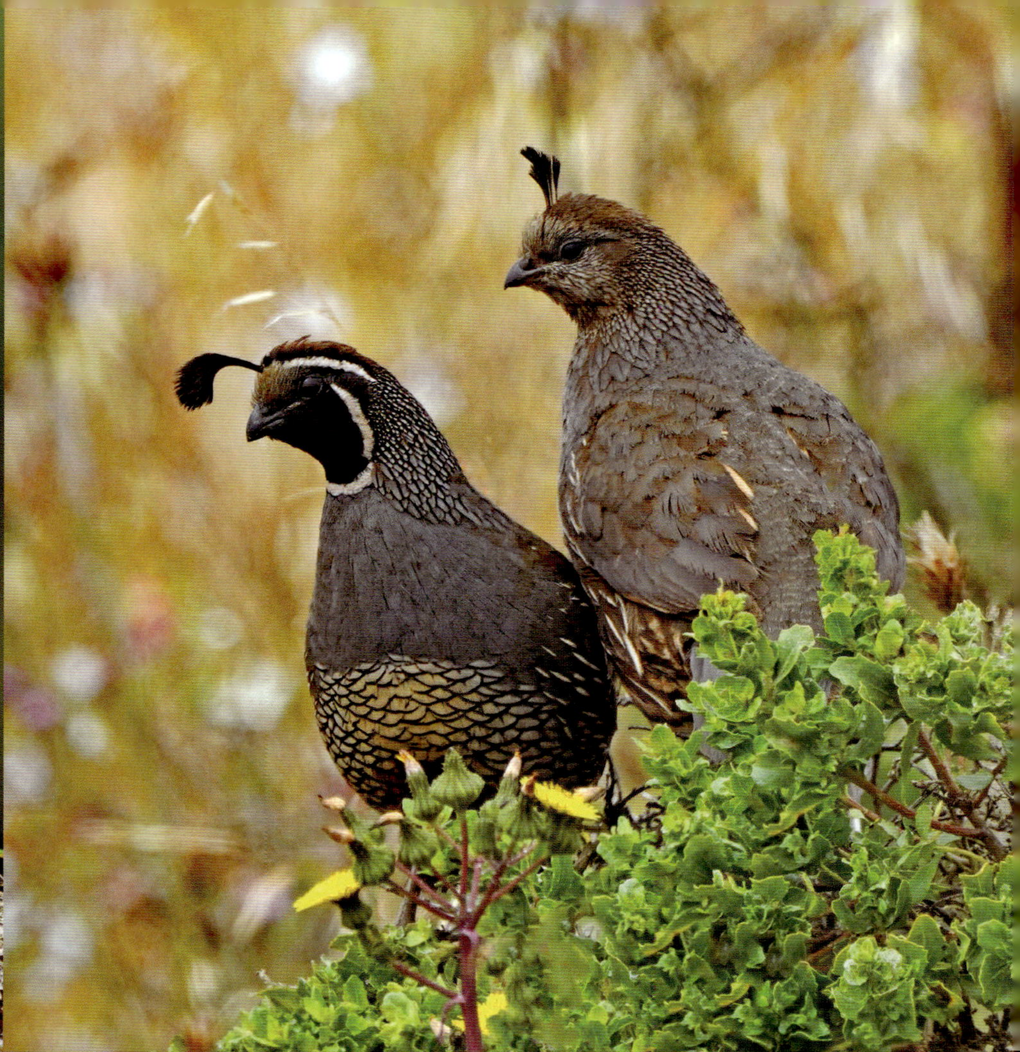
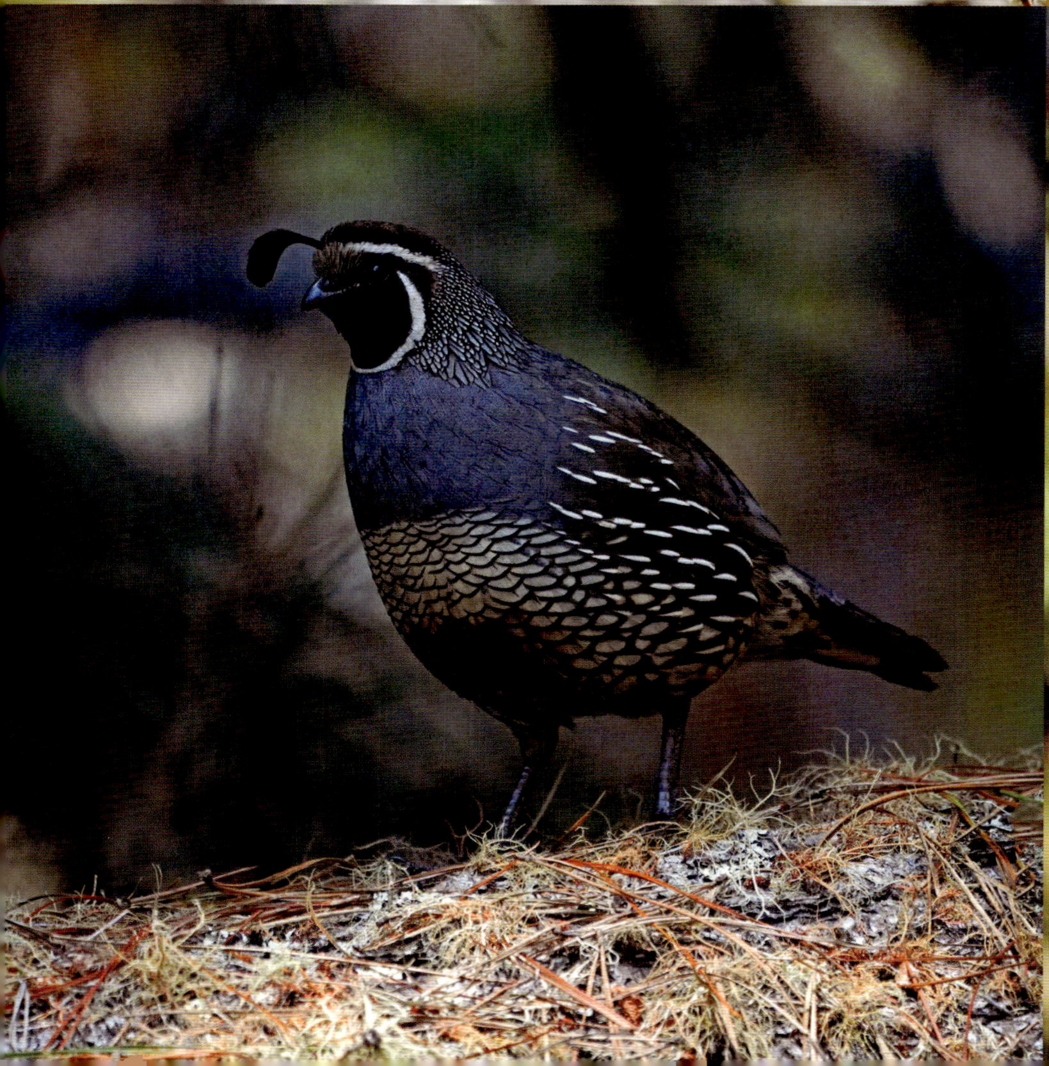
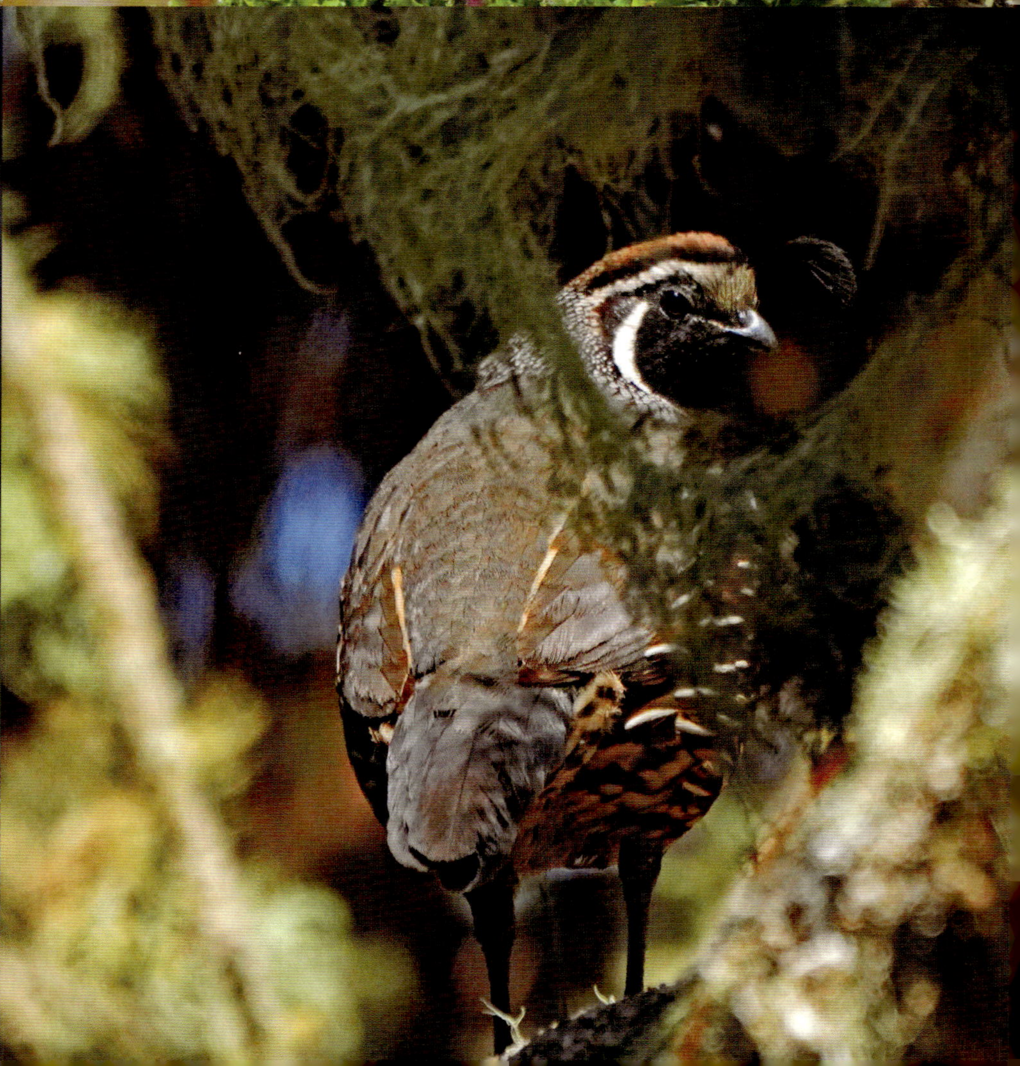

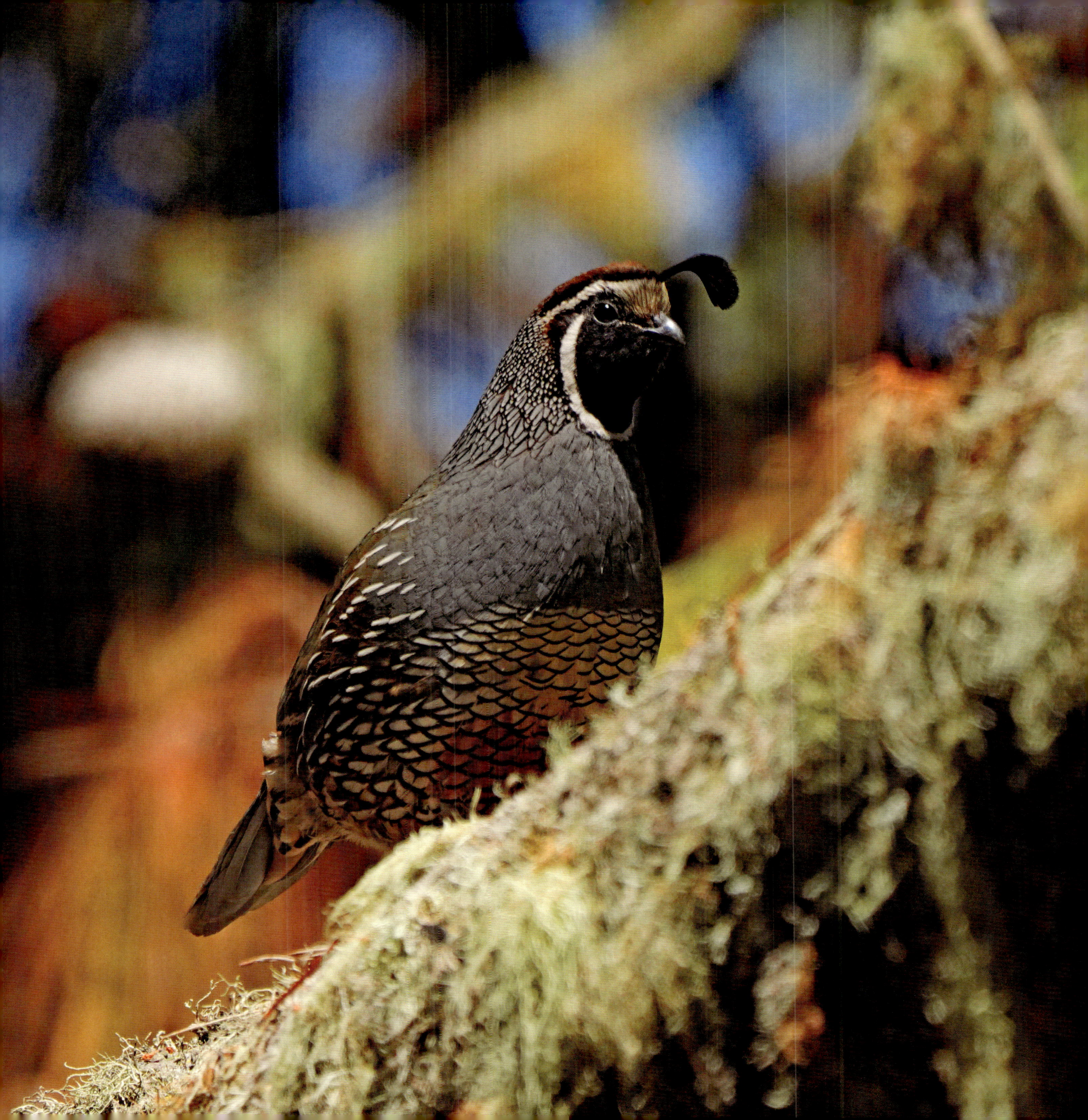

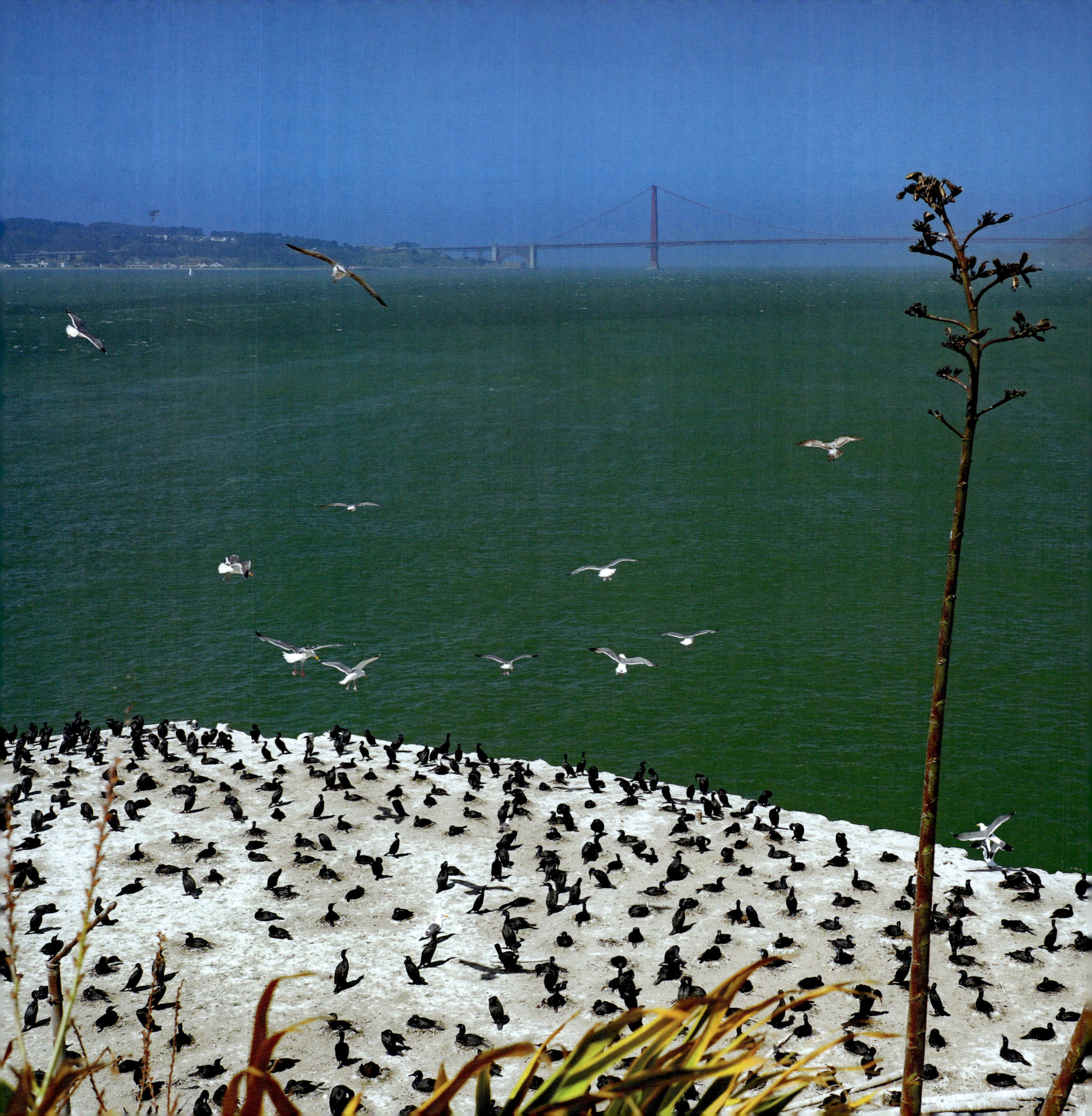

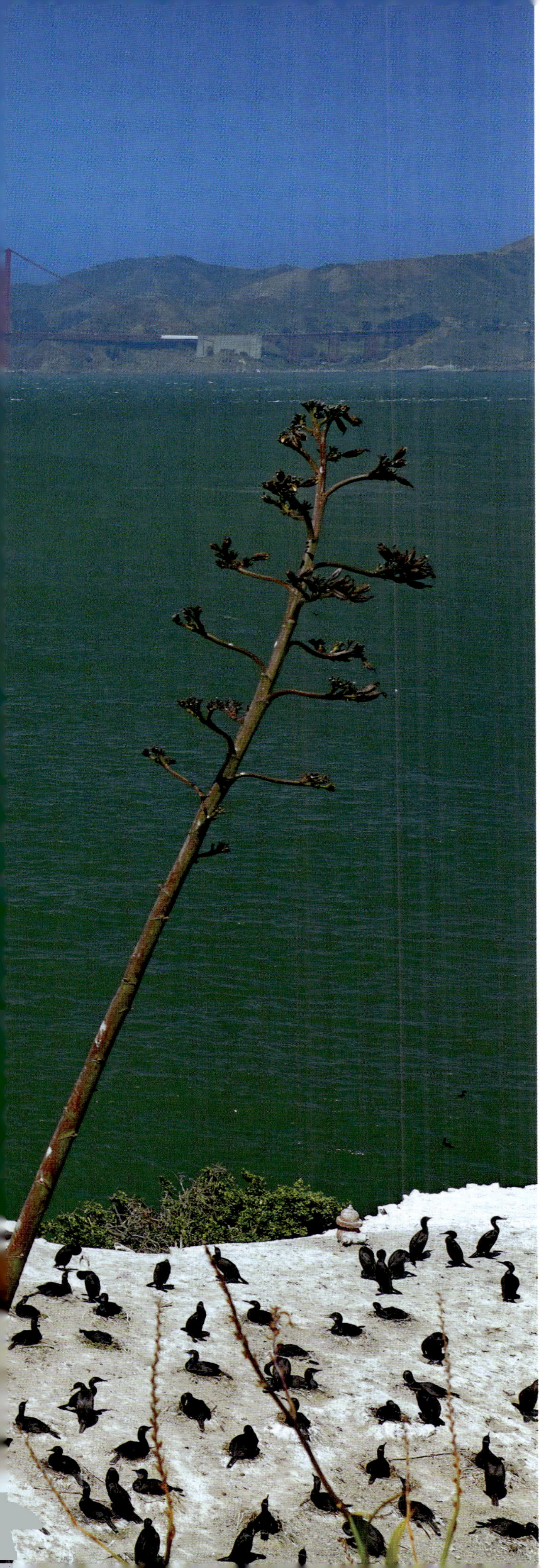

CANNONS AND CORMORANTS

On Alcatraz Island, a pair of Peregrine Falcons lifts off from the western ledge of the old penitentiary cell block. They float over views of the Golden Gate Bridge and the bay far below, then fold their wings into bullet-swift dives, faster than any bird on earth. Along the walkways traveled by visitors to the national park, Western Gulls nest on ledges, in flower beds, in the open-air fireplaces of the crumbling Officer's Club. They eye us coolly. On the windswept west side of the island, male Brandt's Cormorants flap in from the cold waves below, brown fronds of seaweed grasped in their beaks, to woo prospective mates. Their chins glint blue as they raise their faces to the sky. Thousands of webbed feet slap and shuffle across the old concrete courtyard—it is a dark expanse of movement, bird beyond bird.

Most visitors arrive by ferry to Alcatraz—"The Rock"—for a glimpse into its human history. The foundational structures on the island were built as a military fort offshore from San Francisco during the Civil War, with space to confine Confederate war prisoners, traitors, and, after the war, Indigenous leaders who resisted the American government's campaign to send their children to boarding schools. More than a hundred cannons, heavy as fire trucks, faced out to sea.

When the cell house closed in 1963, the 22-acre island was left to the birds. The steep cliffs and tide pools carved by army dynamite became havens for Black Oystercatchers and nesting gulls. Dense shrubs and trees took root in the unpeopled corners of the island, likely seeded by passing birds. Snowy Egrets came to roost, as did Black-crowned Night Herons with amber eyes. Now, Alcatraz is one of the largest nesting sites for Western Gulls on the West Coast. It is California's only publicly

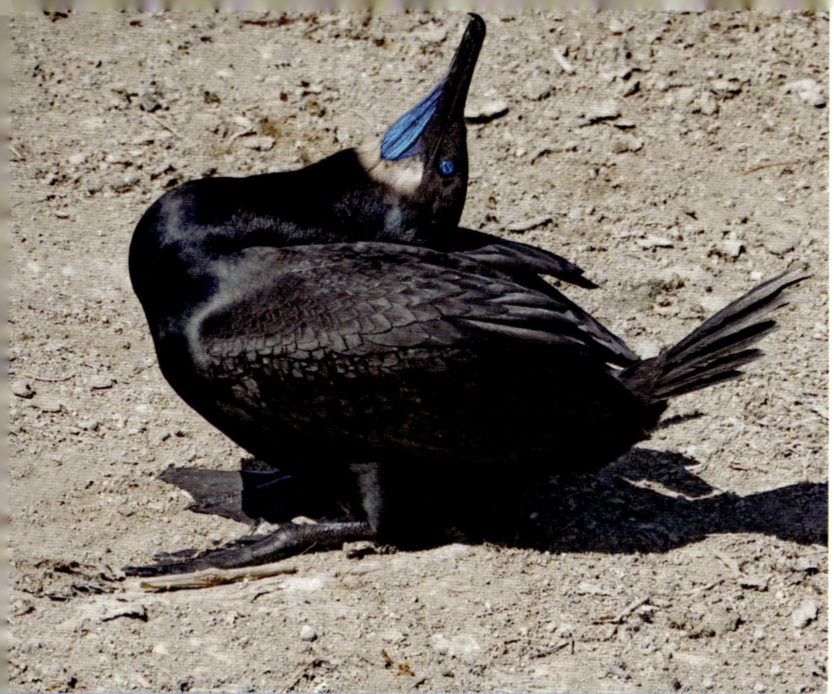

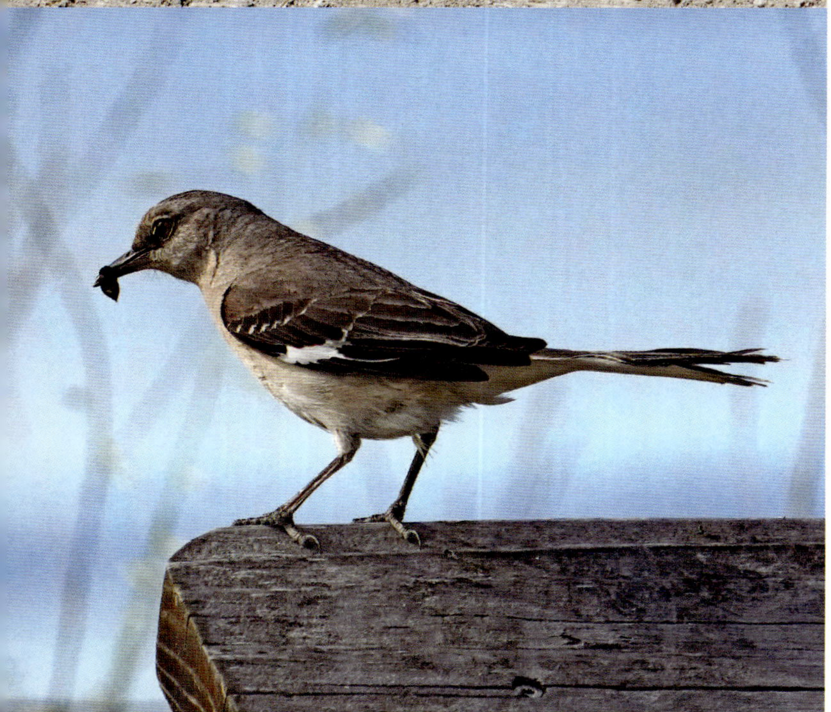

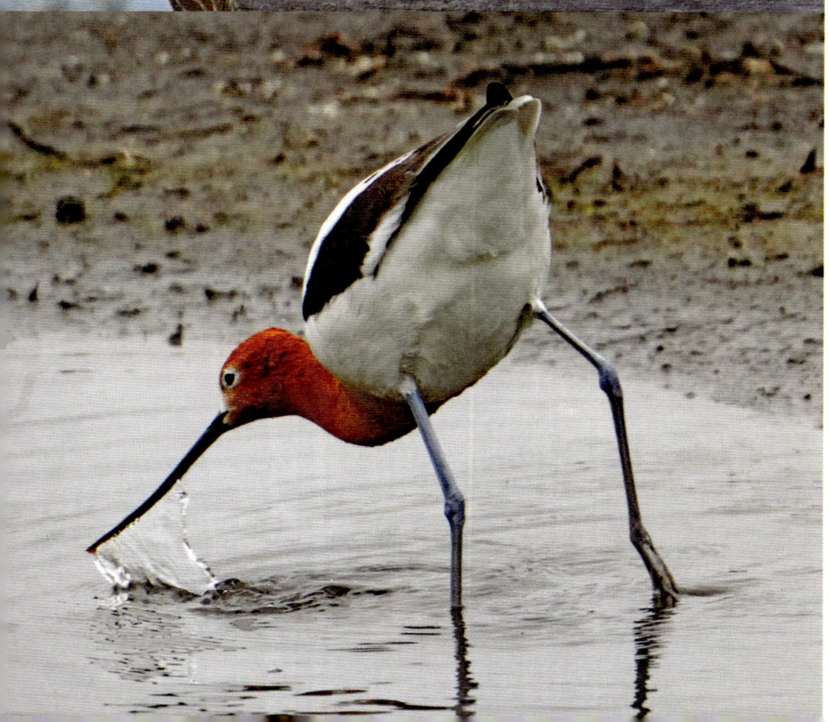

accessible breeding colony of Brandt's Cormorants, which arrive in the thousands from their oceanic fishing grounds in the spring.

Other military land in the greater Bay Area has also gone to the birds. White-crowned Sparrows forage in the decommissioned battery at Fort Mason, and rare Vesper Sparrows and Blackpoll Warblers flutter in the fort's foggy gardens. Annually, thousands of raptors ride warm air currents above historic Fort Cronkite in the Marin Headlands. Threatened Western Snowy Plovers nest where bombers once landed at the former Hamilton Air Force Base in Novato.

These unusual sites for birding are all now publicly accessible, but even active military sites offer powerful protection for birds: The US Department of Defense (DOD) has a legal mandate to steward the natural resources on the 27 million acres of land it manages. All across the nation, DOD lands are habitat stepping stones along avian migration routes, and important refuges for nesting birds. The department aims to "keep common birds common," and much of its conservation work also goes to ensuring that vulnerable bird species aren't pushed to endangerment. To achieve this, an understanding of whole-ecosystem health is necessary. Recognizing that birds need large swaths of unbroken habitat—not just preserved patches—to thrive, the military has shifted conservation from just focusing on saving individual animals to focusing on connecting lands to allow for safe movement. Collaboration between landowners such as ranchers, farmers, and the US military allows extensive open space to be put to birds' use, not just our own.

All DOD installations across the nation use the Avian Knowledge Network—a data infrastructure developed by Bay Area conservation scientists at Point Blue—to track and organize their bird data. The database has been adopted not only by the DOD but by the US Fish and Wildlife Service, the National Park Service, and thousands of global partners. "It might be the largest ecological database in the world," says Grant

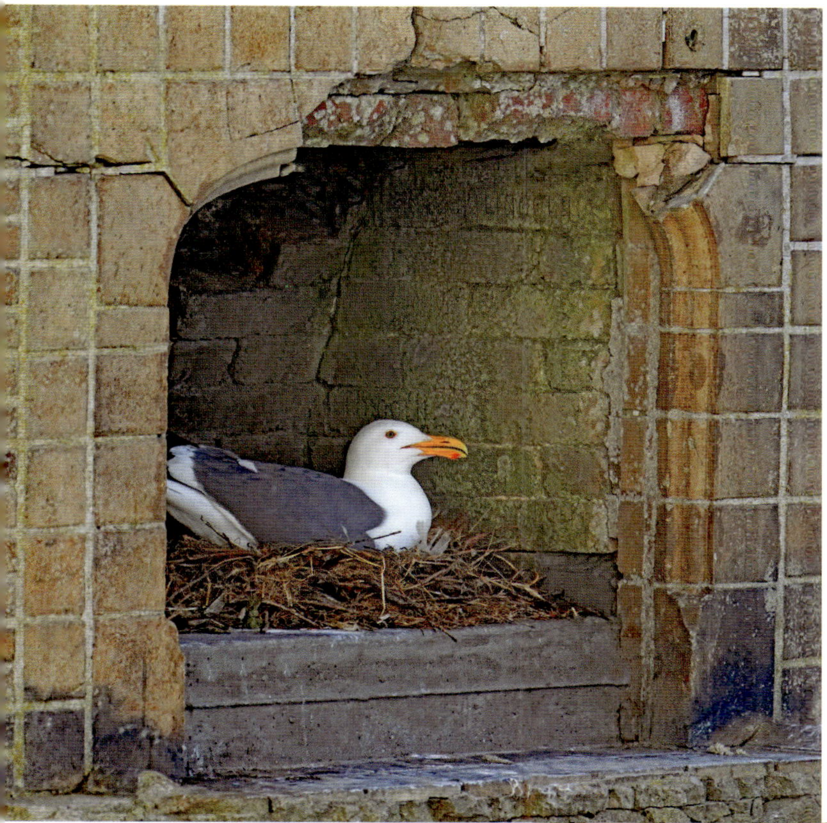

Ballard, chief science officer for Point Blue. With a partner as enormous as the US military, this locally developed network is uniting bird conservationists around a common language so that management plans can align with the stories the data tells.

Here on Alcatraz, we lean into the wind and watch birds, instead of military airplanes, loop above us. They cast shadows across the crumbling army parade ground, bridging the past with the future.

LEFT: A Western Gull repurposes a fireplace in old officers' quarters on Alcatraz Island.

BELOW: In 1969, Indians of All Tribes, a group of Native American activists, occupied Alcatraz Island for 19 months in protest of the US government's treatment of Native Americans and the breaking of treaties. Every year on Indigenous Peoples Day, Indigenous groups and their allies still gather at sunrise on Alcatraz—among the herons and Peregrine Falcons—affirming the resilience of Indigenous peoples in California and spotlighting ongoing struggles to protect sacred lands.

PRECEDING: In the shadow of the Golden Gate Bridge, Western Gulls soar over the densely packed nesting colony of Brandt's Cormorants on the western edge of Alcatraz Island.

OPPOSITE: Former military sites across the Bay Area are now home to birds of all feathers, including this Brandt's Cormorant captured mid-mating display on Alcatraz Island (top), this Northern Mockingbird with a beak full of bugs in at Hamilton Field (middle), and this American Avocet mid-hunt in Hamilton Wetlands (bottom).

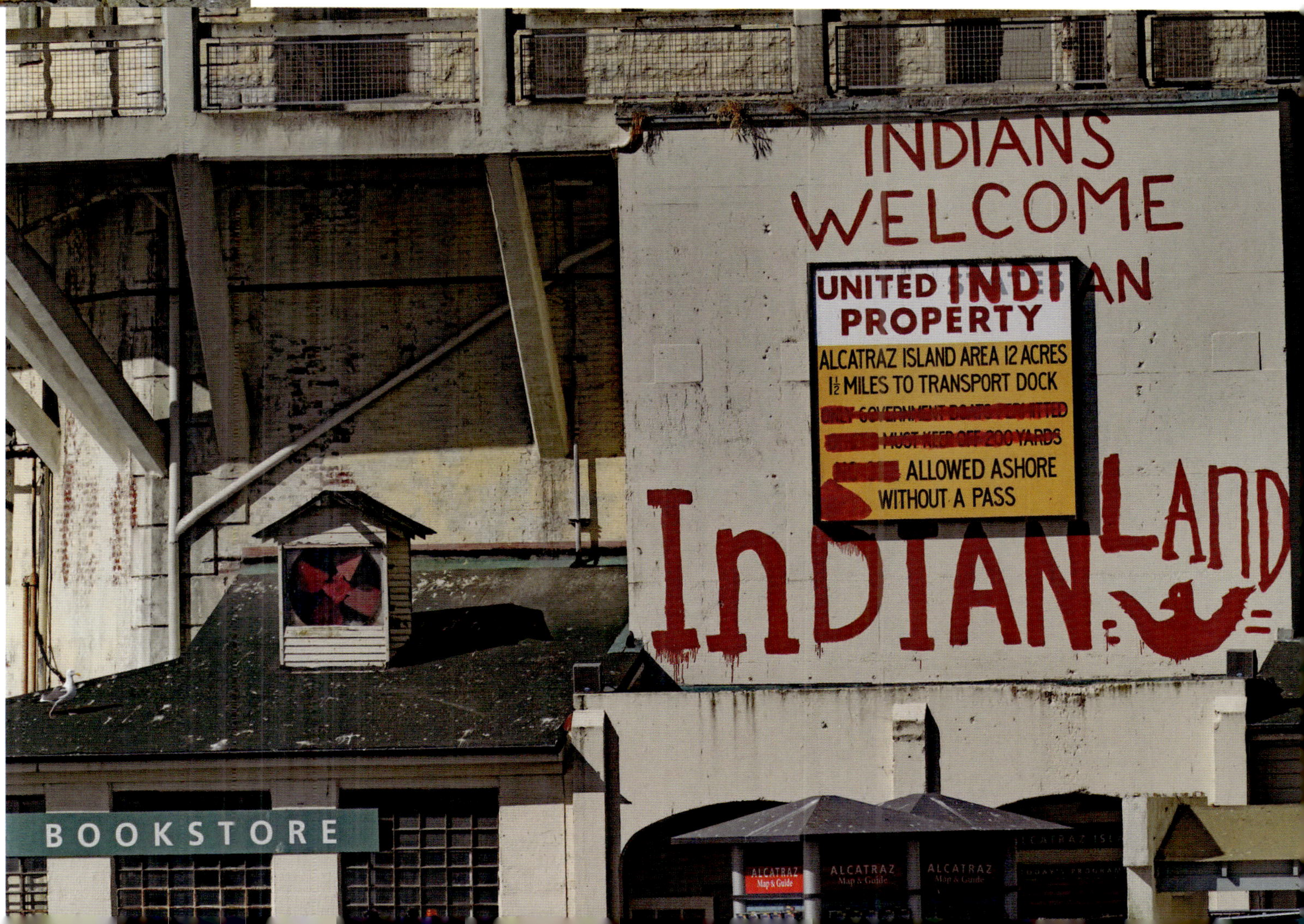

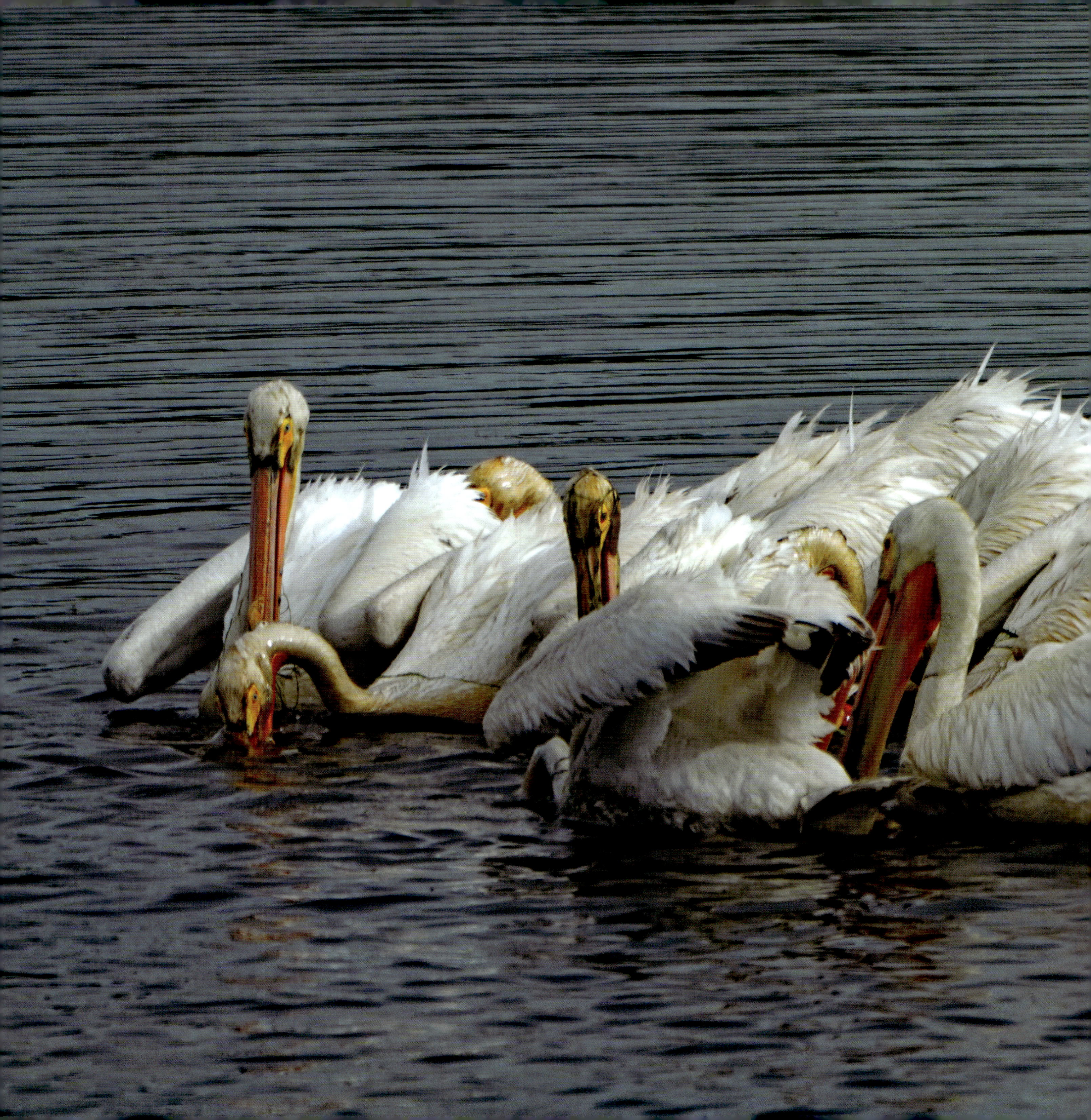

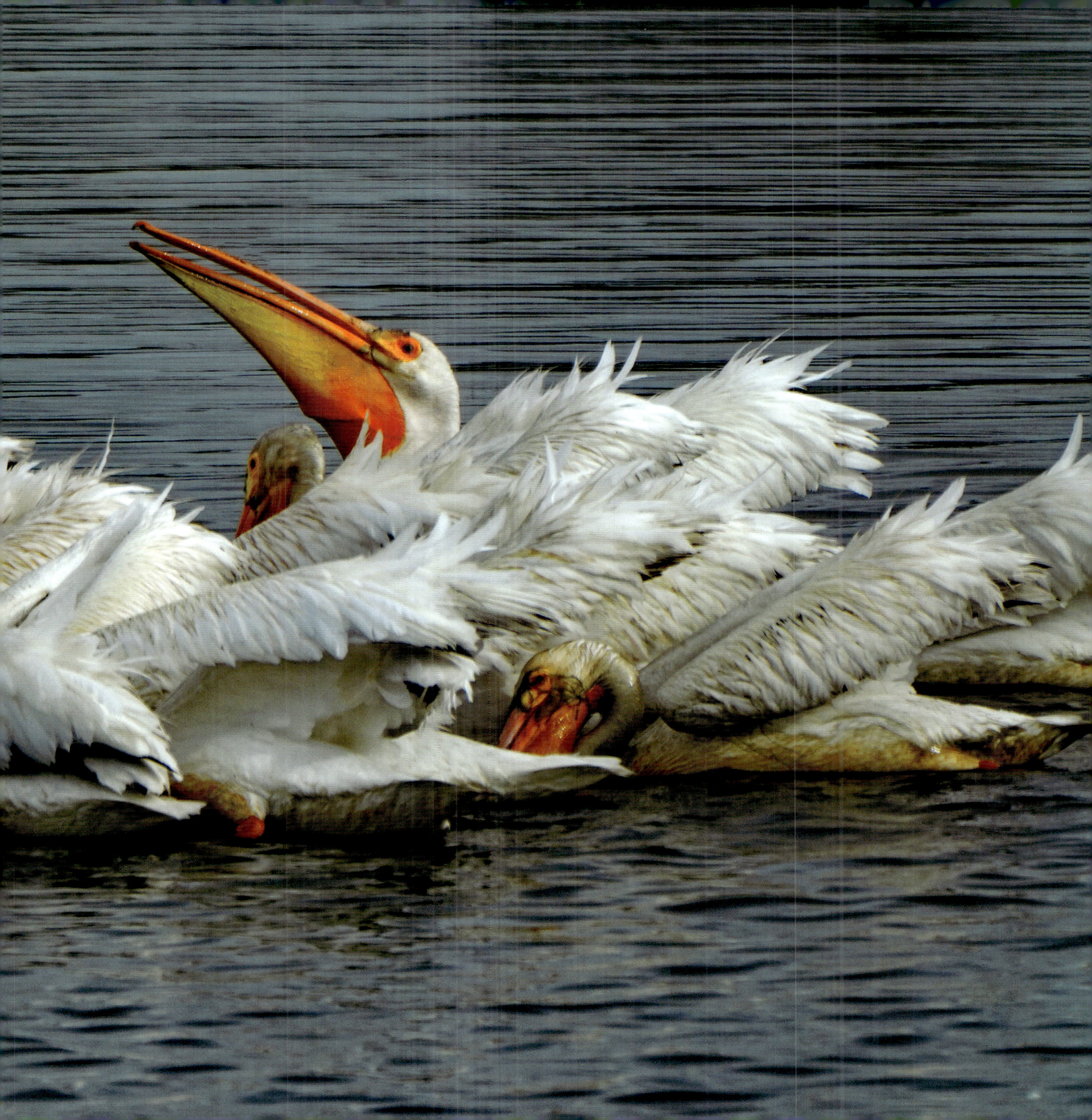

PRECEDING: With ruffled feathers and billowing bills, a fleet of White Pelicans parades across the water at Pacheco Pond in Hamilton Wetlands, a newly restored salt marsh recovered from the former Hamilton Air Force Base.

TOP RIGHT: A Black-necked Stilt balances in the shallows of Hamilton Wetlands, a thriving ecosystem restored from Novato's old military airport.

BOTTOM RIGHT: Rarely found more than a few miles inland, the Western Gull is a familiar sentinel of rocky outcroppings, wharves, jetties, and abandoned decommissioned military forts throughout the Bay Area.

OPPOSITE: Walk along the promenade at San Francisco's Crissy Field—the military's first Air Coast Defense Station on the Pacific coast—and you might be lucky enough to spot shining white flocks of Elegant Terns along their migration through the area in the summer and fall.

FOLLOWING: The Marin Headlands—the site of a number of historic military settlements and fortifications—are renowned for the throngs of hawks, eagles, and falcons that soar overhead on thermals during their annual fall migration. Using the Golden Gate as a narrow crossing point over the waters of San Francisco Bay, 19 different species of raptor pass through this region, including (left, clockwise from top) the Sharp-shinned Hawk, the Osprey, and the Red-shouldered Hawk. A White-tailed Kite (right) rests at the edge of the Corte Madera Marsh in between hunting flights. These raptors are exceptional at hovering in the wind—much like a human kite—in search of small prey like rodents, reptiles, and small birds. California is home to the largest concentration of White-tailed Kites in all of North America!

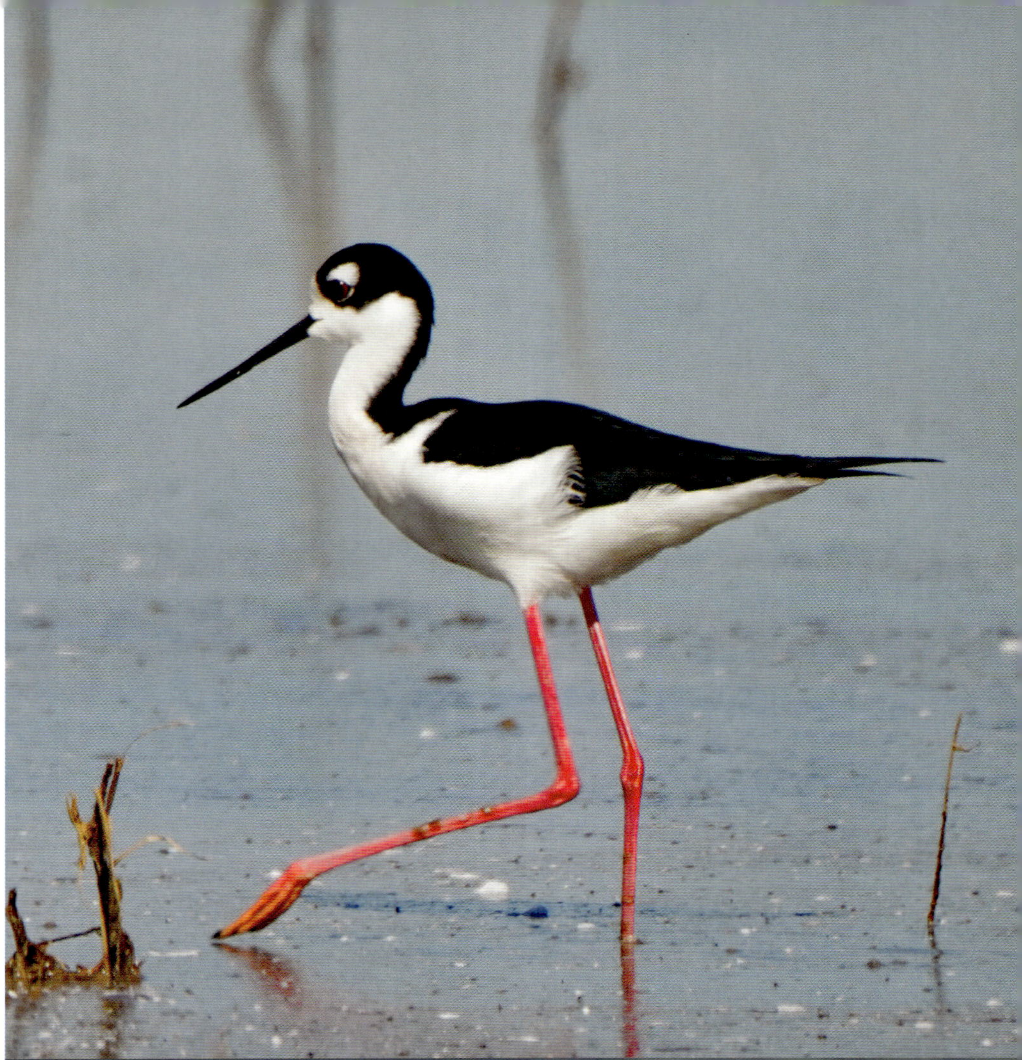

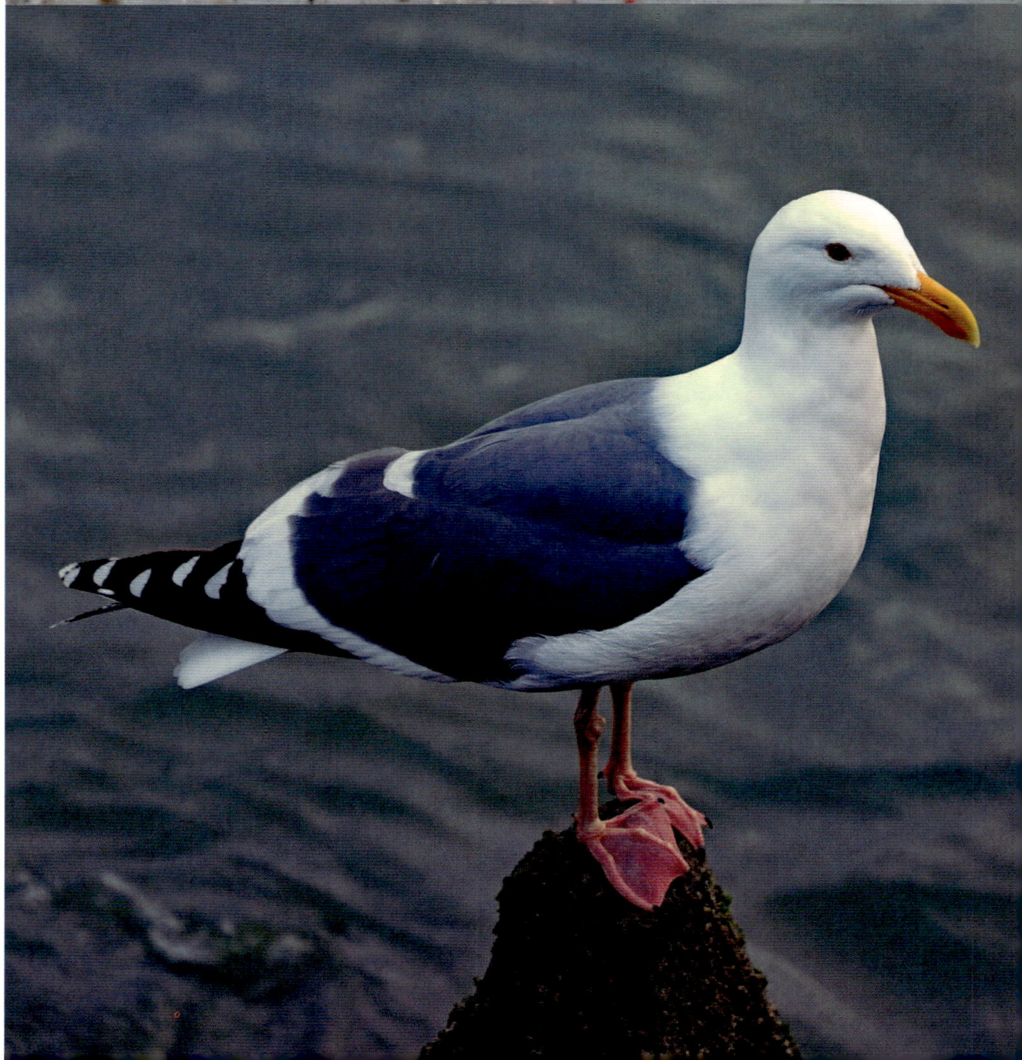

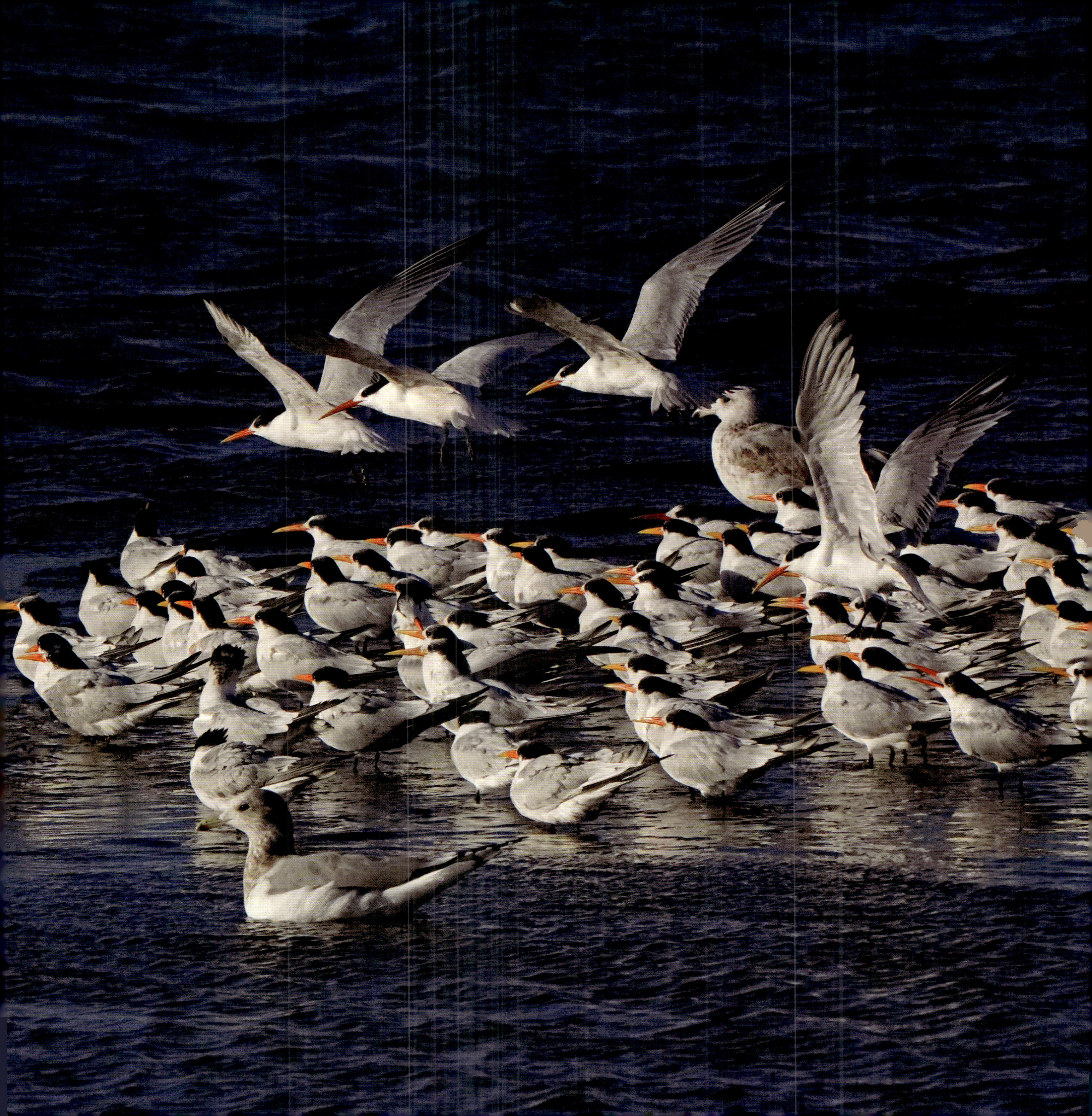

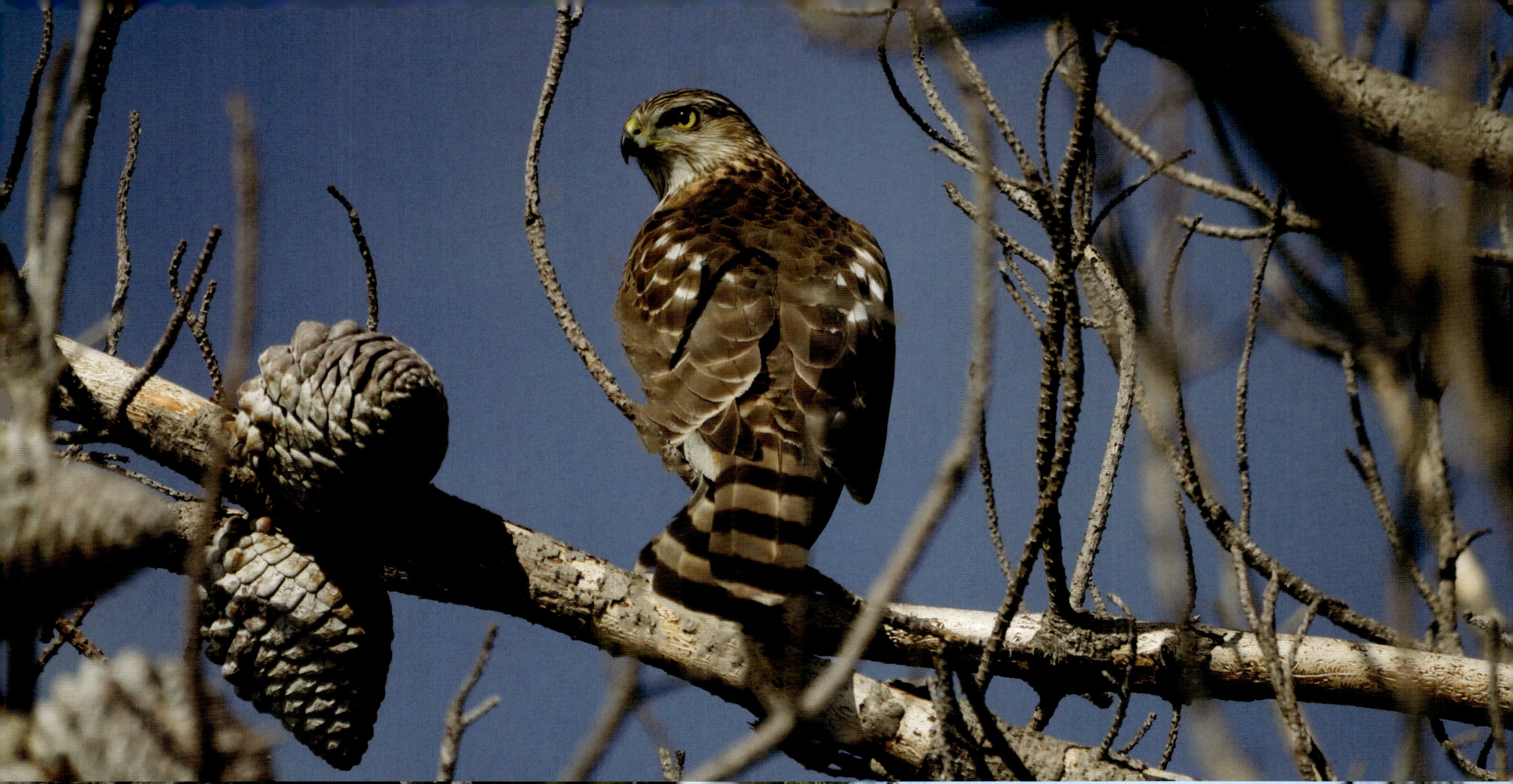
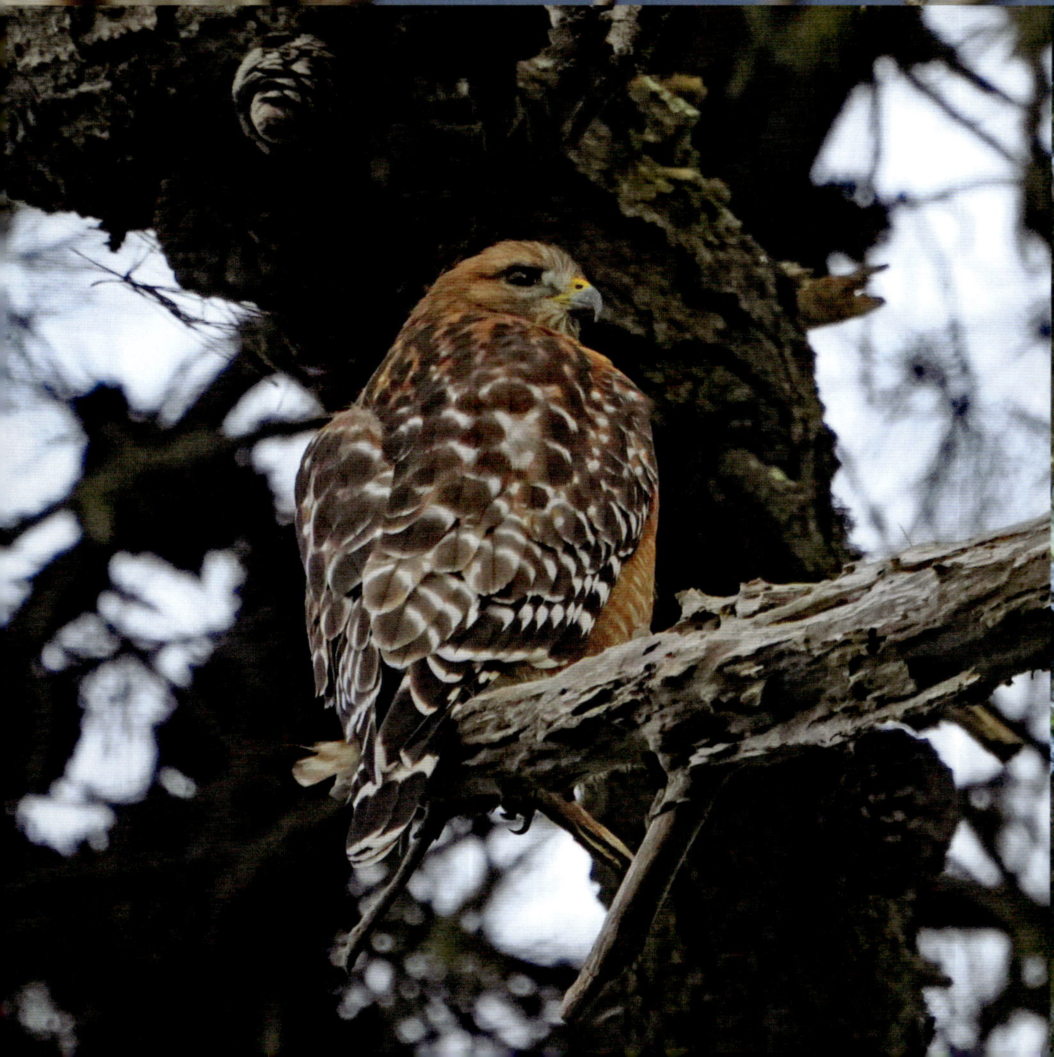
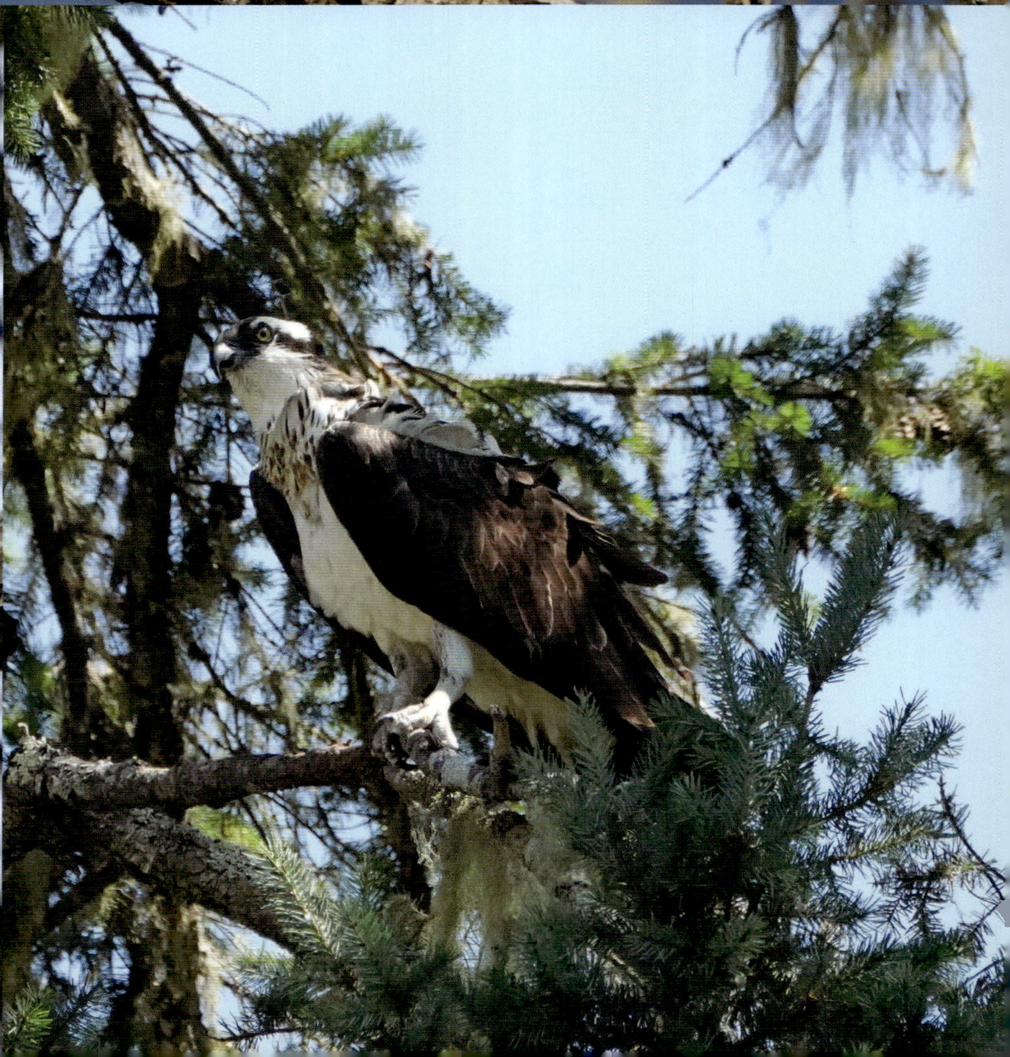

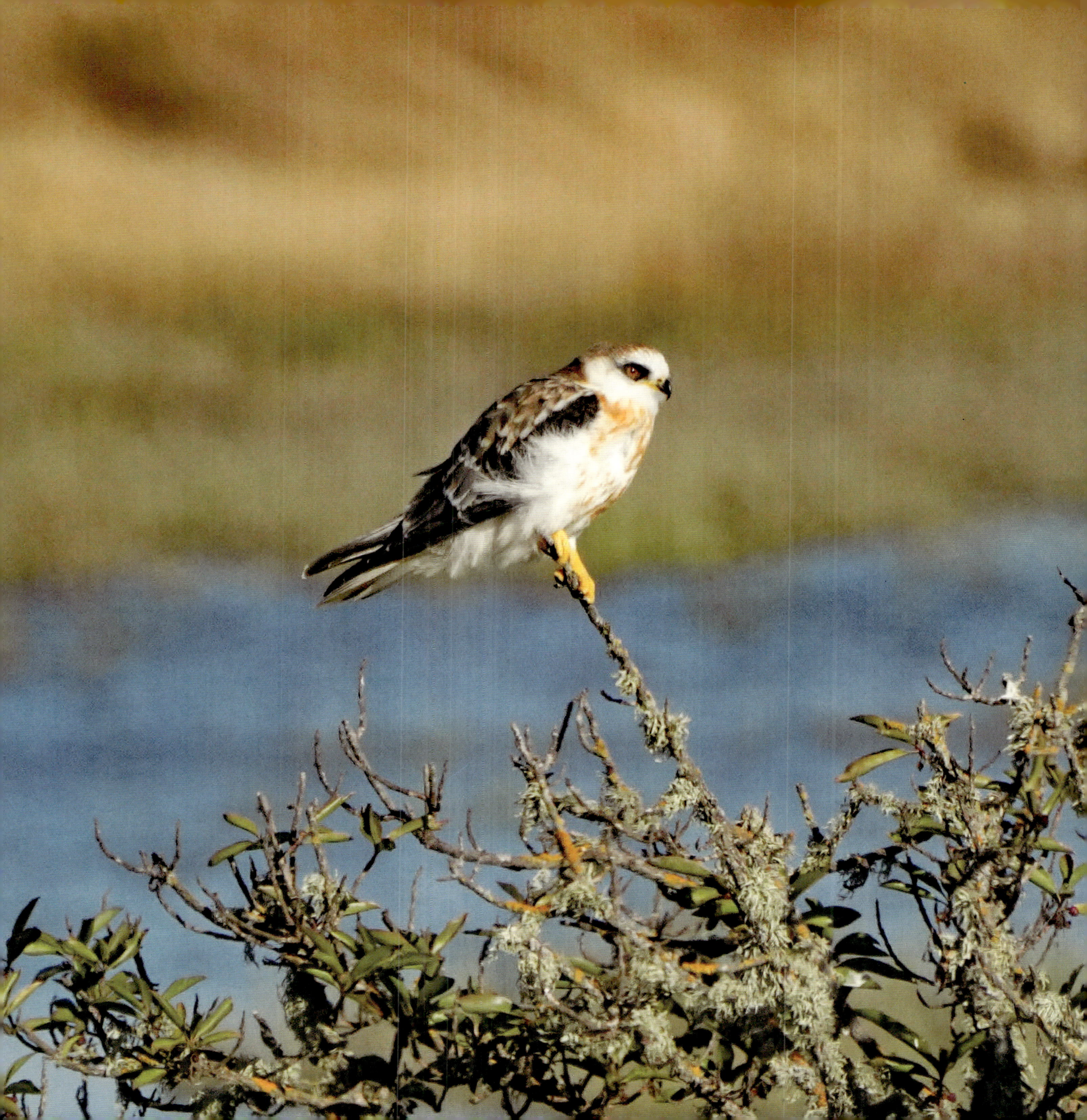

Over 51,000 acres of wetlands around the Bay Area provide low cover for cryptic, well-camouflaged birds like this mottled Wilson's Snipe (left).

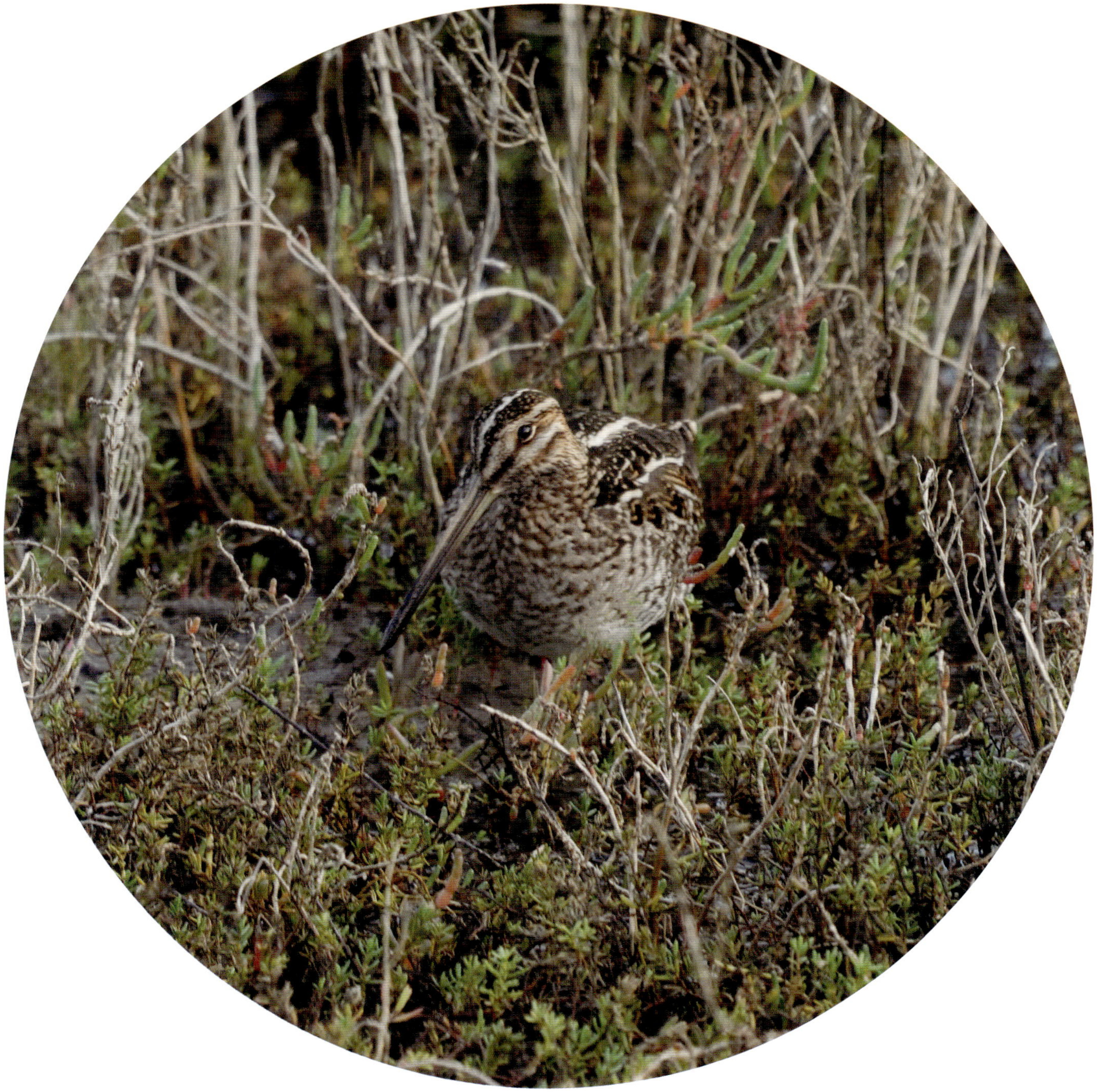

A Northern Pintail (opposite) weaves among tall stalks in the Cosumnes River Preserve, his choco-late-brown head and mottled plumage blending beautifully with the dry foliage. These ducks only flash their brightly-colored speculums—inner wing feathers—in flight.

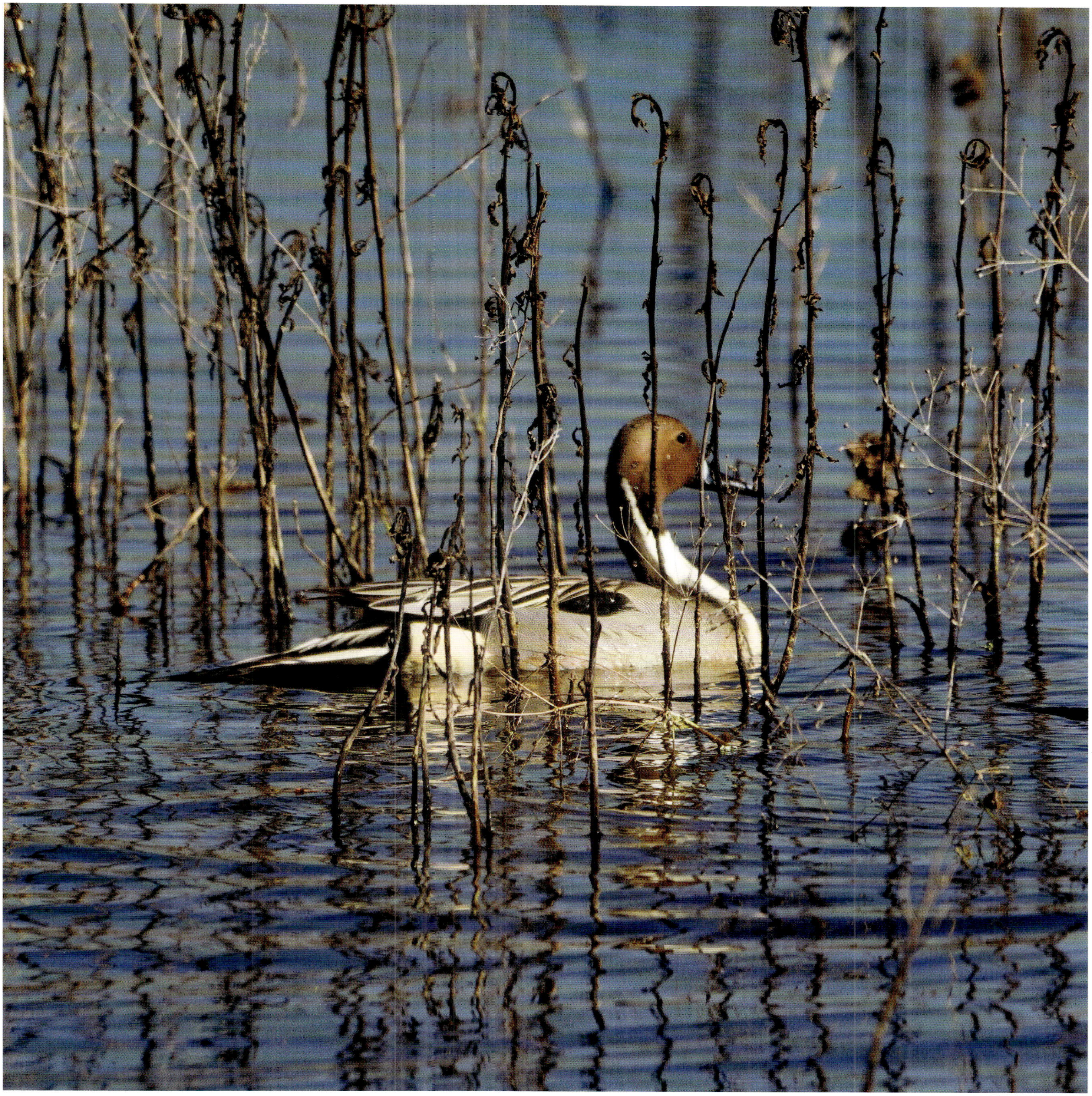

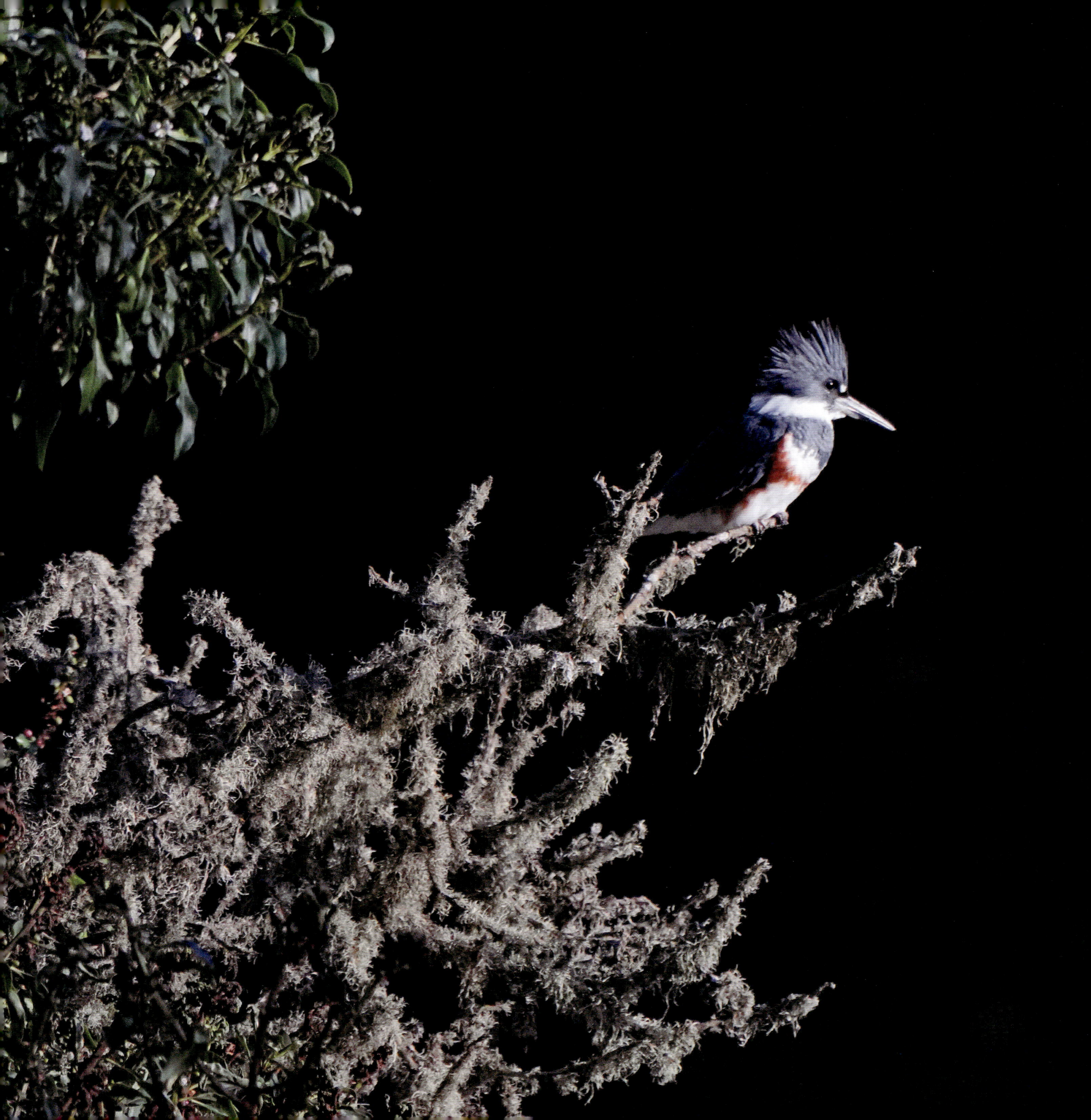

THE WAY IN

The rattling trill of a Belted Kingfisher ricochets off the calm surface of Spring Lake, and green shadows spill across the parking lot where we pass around binoculars and set off on our walk. Our two guides, Teresa and Miles Tuffli, follow their ears. We've only walked a few hundred feet when they pause, alerted by a metallic *tink* to the presence of a California Towhee. Soon we see the bird itself, hopping and scratching in the grass, bright orange and sphinx-brown. Miles describes how California Towhees use this short, plinking call to keep in touch with their mates from afar. "When a mated pair reunites," he says, "they burst into a duet squeal." Sure enough, the towhee we've been watching disappears into the bushes, and we hear the sound of a reunion: A loud, cascading *chirr* as the two birds—which mate for life—affirm their bond together.

Miles and Teresa are keen listeners and attentive storytellers. Their approach to birding isn't just to identify species but to understand the bigger narrative that birds are helping tell us through their interactions and behavior. It's an approach that changes the way the two educators listen, no matter where they are or what they're doing. I'm Birding Right Now is the name of their guiding company, but it's also their philosophy.

"What makes birding so incredible," says Miles, "is that anyone can learn, at any point in their life. You just have to be curious."

For Dick and me, still just coming to know our birds, we're learning from guides like Miles and Teresa to start by listening. I've been practicing in Sebastopol's Ragle Park. Here, the organization Weaving Earth holds free beginner-friendly workshops in "bird language" every month. "Bird language is already in our bodies," says David Hage, cofounder of Weaving Earth. "We're just helping people remember it." These are

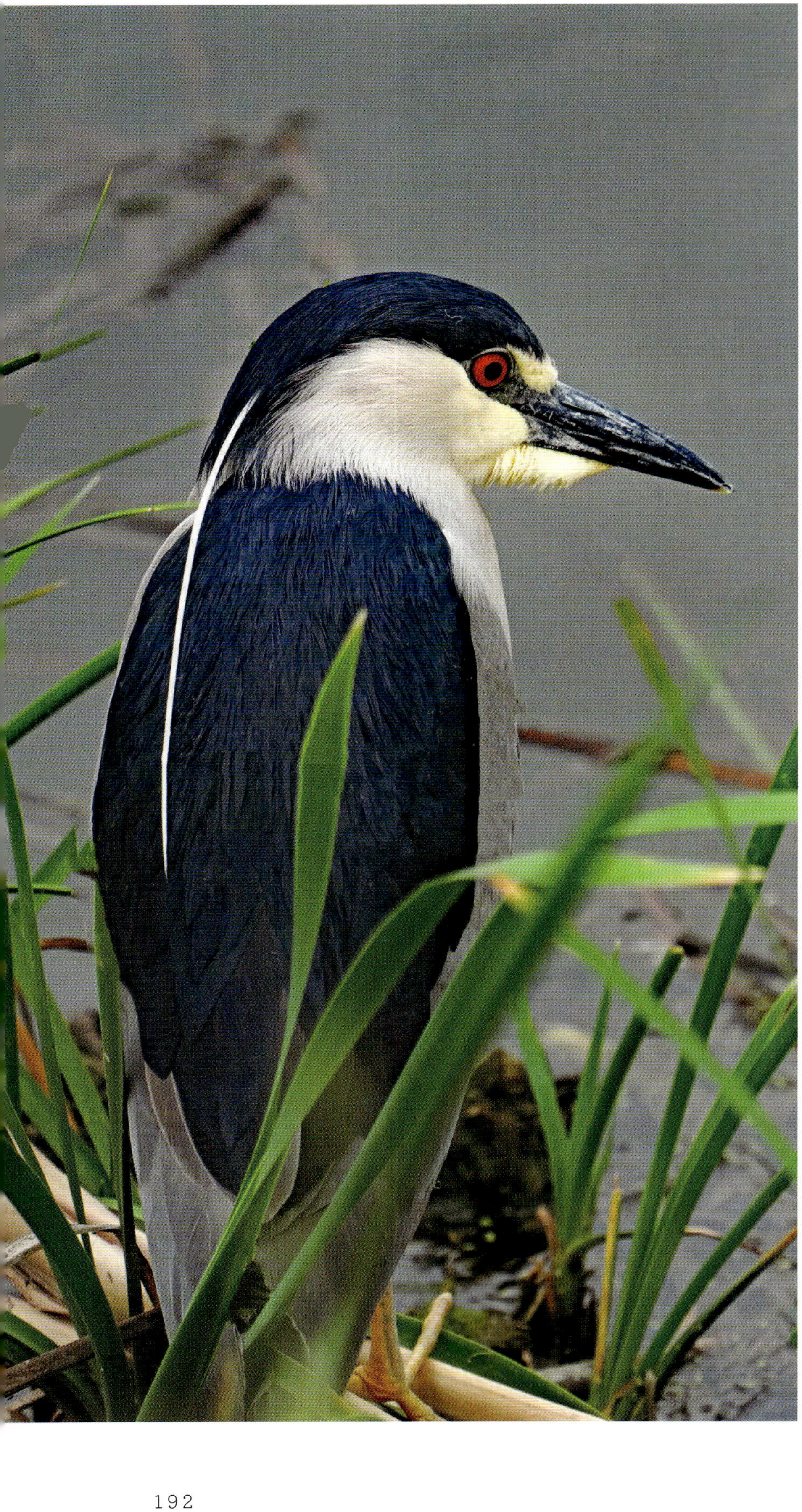

practice sessions for interpreting what birds are saying, certainly, but they are also acts of relationship building. By holding still and giving undivided attention to birds' choruses and silences, observers build intimacy with the earth and with their own bodies, and the boundaries between the two blur. Listen to the calls of geese swooping low across a meadow toward their evening roost; feel the cool dusk pooling around your own body, signaling day's end. In Weaving Earth's lessons, birds' actions and your own senses are no longer entirely separate. "How is my body impacting or being impacted in different spaces I move through?" the Weaving Earth teachers ask. In response, we tune in to ourselves as members of a community—with one another, with the birds, and with the land.

We begin by spreading out and finding a place to sit quietly, on our own. Then we come together to share what we observed. Each person's independent observations—rustling bird activity near where they were sitting, spikes in calls, pools of silence, the shadows of hawks passing overhead—weave together to form a clear view of the bigger story that unfolded around us while we sat. One person might describe hearing agitated cheeping from the nearby blackberry bushes, and another listener might add that where they sat, the birds grew silent. Together, we might note that near the end of our listening session, the call of a raptor pierced the air. The clues come together: Birds at the edge of a danger zone often call out to one another to warn of an approaching predator, but when a hawk flies directly overhead, all the birds go quiet—an audible event called an "accipiter ripple." And those little birds gathering and feeding near where another observer was sitting? They were using his body as a shield from more dangerous predators that might be nearby; as the clues come together, it becomes clear that when we enter an ecosystem, we're not just watchers but participants.

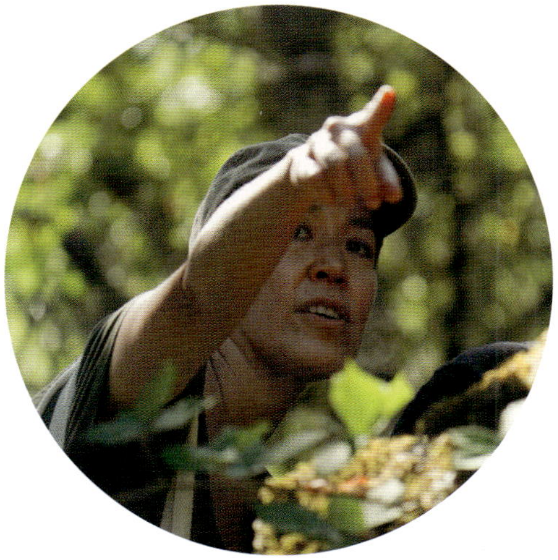

ABOVE: From the forest's edge along Spring Lake, Teresa Tuffli points out secretive birds.

Leslie Beach, who coordinates outdoor programs for the high school students at Marin Academy, recalls hearing the low background noise of California Quail all afternoon during one of these sessions, "but suddenly they got very loud—sharp indignant squeaks," and she knew something had shifted. Soon after, a long-tailed weasel emerged from the tall grass next to where the group was sitting, and the story became clear. "It was fascinating being in a group with mentors who helped me read the situation," she says. But the birds themselves are the real instructors, David says. "Birds teach us about connection across difference. How to recognize our interdependence. They show us how we're showing up in the world. They help us lift our heads and have a peaceful moment."

Back at Spring Lake, Teresa softly calls us over. Through a tangle of trees and brambles, she points out a Black-crowned Night Heron—only visible if we tilt our heads just right. We turn off the trail and pick our way closer to a marshy pond. For long minutes, we watch the heron slowly traverse the shore. As we linger, other birds emerge: a Great Egret, shy Virginia Rails, fat-bellied Soras with brindled feathers. Bird by bird, the secret life of the marsh is revealed: crisscrossing tracks in the squishy mud, the scuttle of a brown creeper in the oak branch above us, the glide and pause of feathered hunters at work. Through the veil of trees, hikers march past, unaware of the intricate scene playing out just off trail. Together, we linger and we listen. We let the birds be our guides.

BELOW: Point Blue Working Lands Science Director Ryan DiGaudio peers through a spotting scope in Point Reyes National Seashore, looking for vulnerable Snowy Plover nests—and teaching us how to identify them, too!

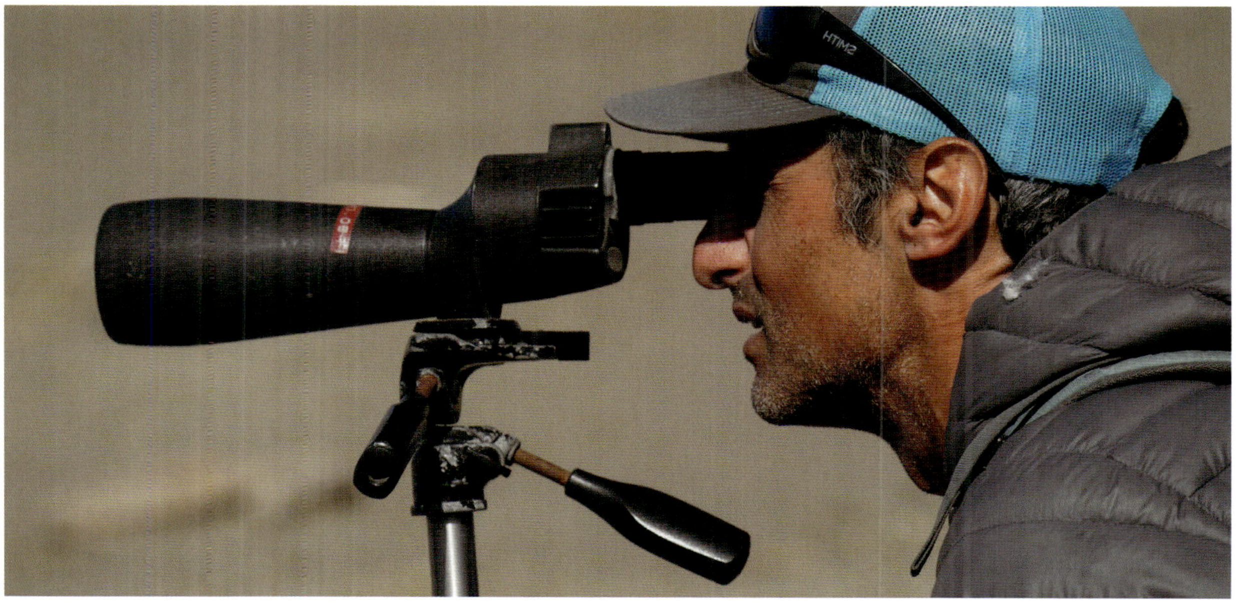

PRECEDING: The Belted Kingfisher is a familiar sight near water's edges throughout the Bay Area. They tunnel into muddy banks to build nests, and teach their young how to fish in dramatic, plummeting dives.

OPPOSITE: A Black-crowned Night Heron stands motionlessly in wait for prey. Aptly for their name, these birds are especially well suited for hunting at dusk and into the night: Their large, light-gathering red eyes give them sharp vision in low light.

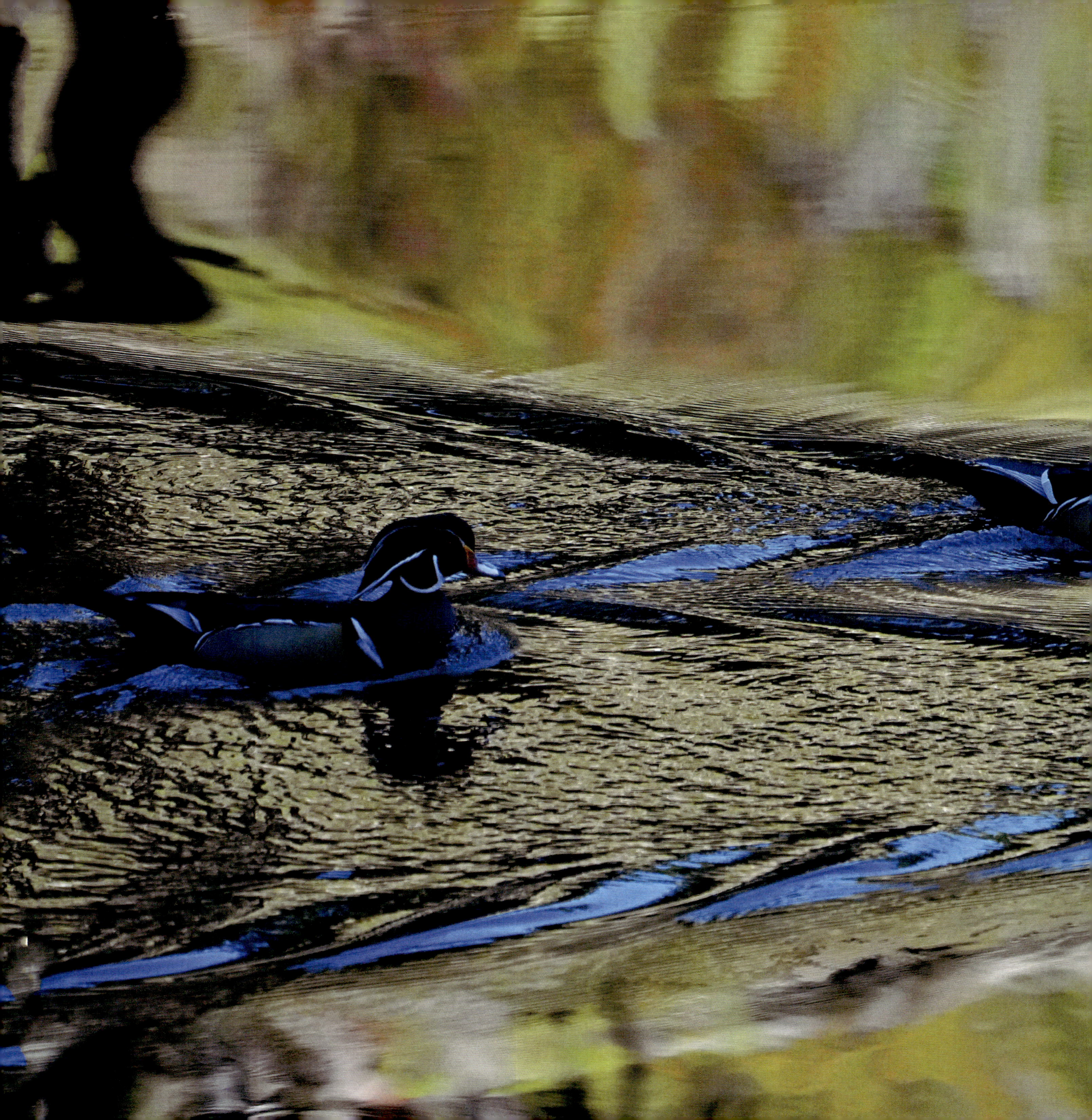

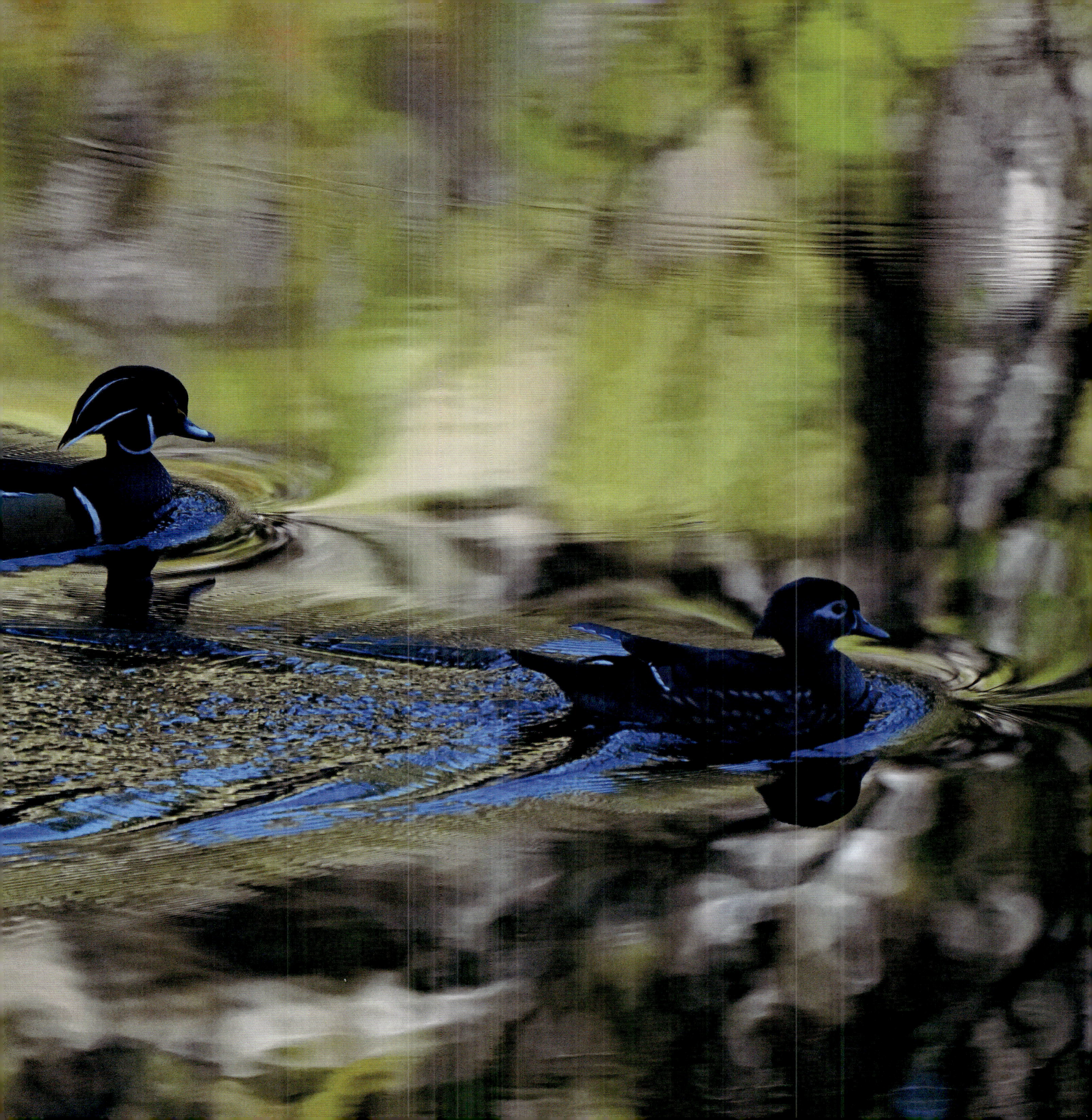

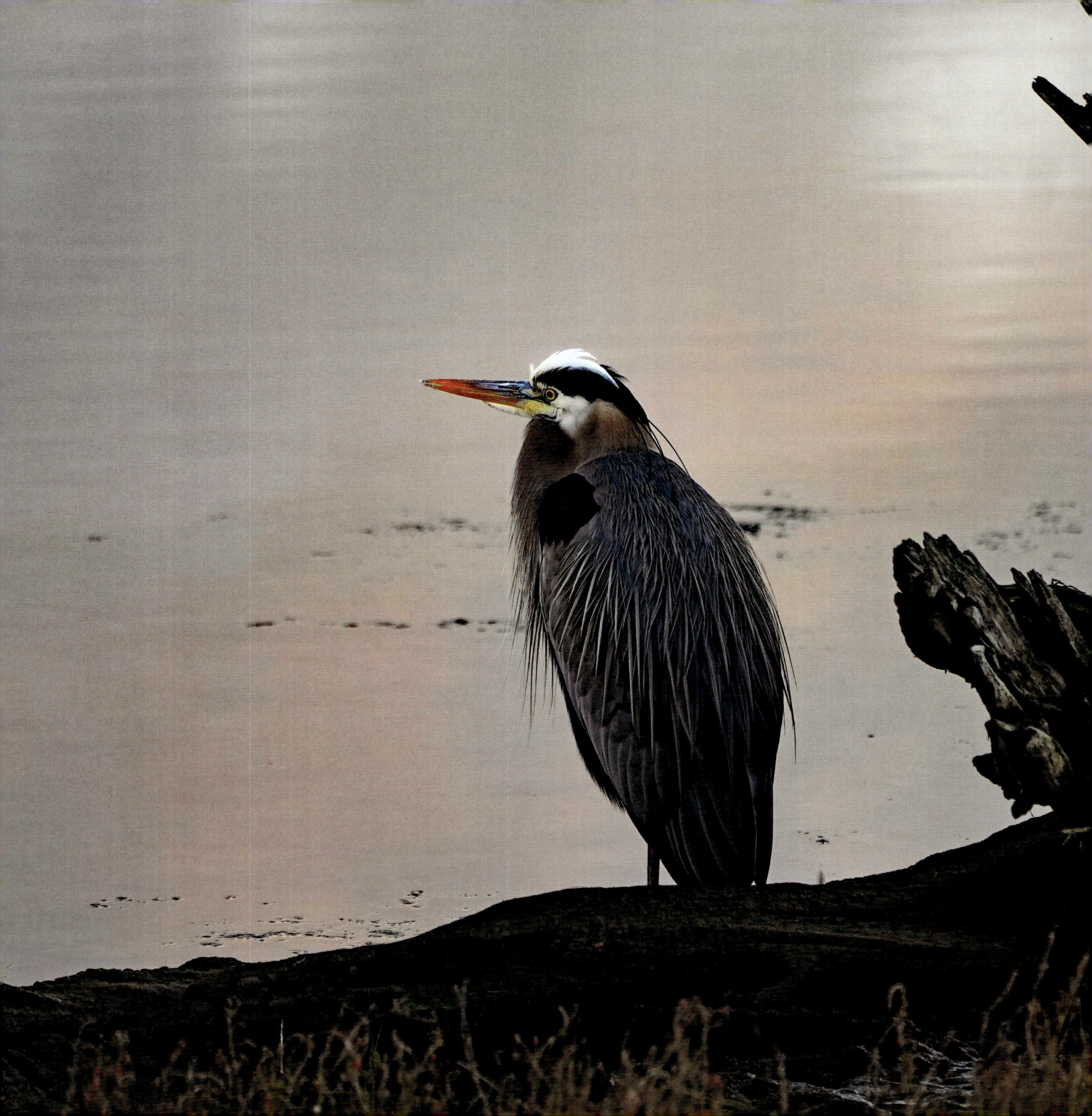

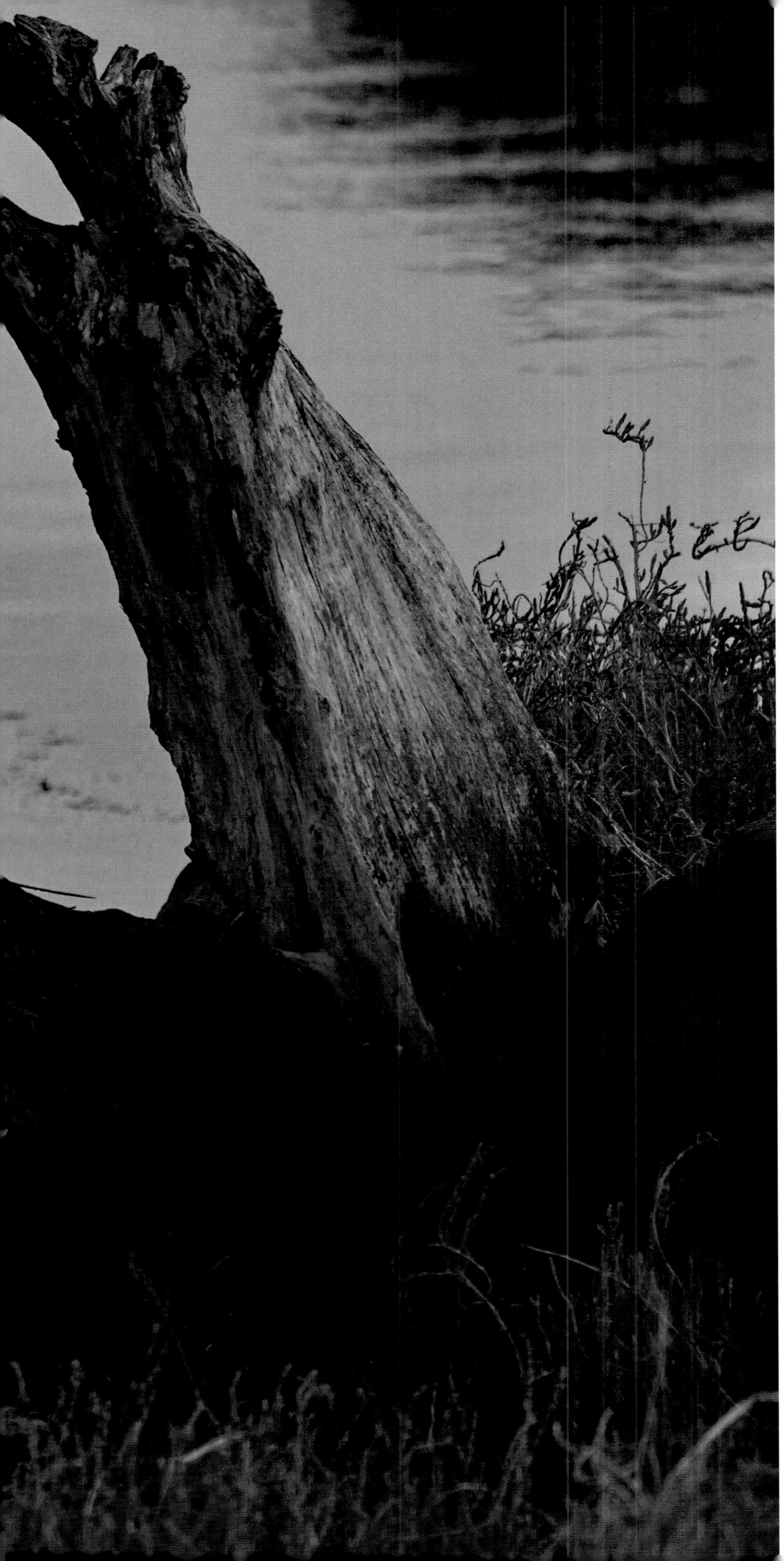

PRECEDING: Lisa Hug, a naturalist living in Sonoma County, has spent decades guiding trips across the North Bay in search of beautiful—and sometimes elusive—birds. She first led Dick Evans to Lake Lagunitas to discover Wood Ducks, and she returned later in the year with him to catch the males in their astonishingly colorful mating plumage. A good naturalist holds deep locational knowledge and understands seasonal changes—and the value of patience!

LEFT: Hidden below a Great Blue Heron's distinctive outer breast feathers is a special type of down that never molts, but continuously grows. The tips fray into a talcum-like dust called "powder down," which the heron rakes through its feathers with its toes, using the powder to clean off fish slime and pond oils from their feathers.

FOLLOWING: Bay Area White-crowned Sparrows (left) are well-known for having highly localized "dialects," with distinct groups of sparrows each singing distinct song patterns from one region to the next—even across short distances between parks or neighborhoods. These dialects are learned, with male sparrows copying songs from their parents and neighbors. Over time, variations emerge and become established within a particular area. We can train our ears to eavesdrop on these unique, local conversations! While Ridgway's Rails (right) are primarily walkers and runners, they are also capable swimmers. Sometimes they can be spotted paddling across narrow channels or flooded marsh areas.

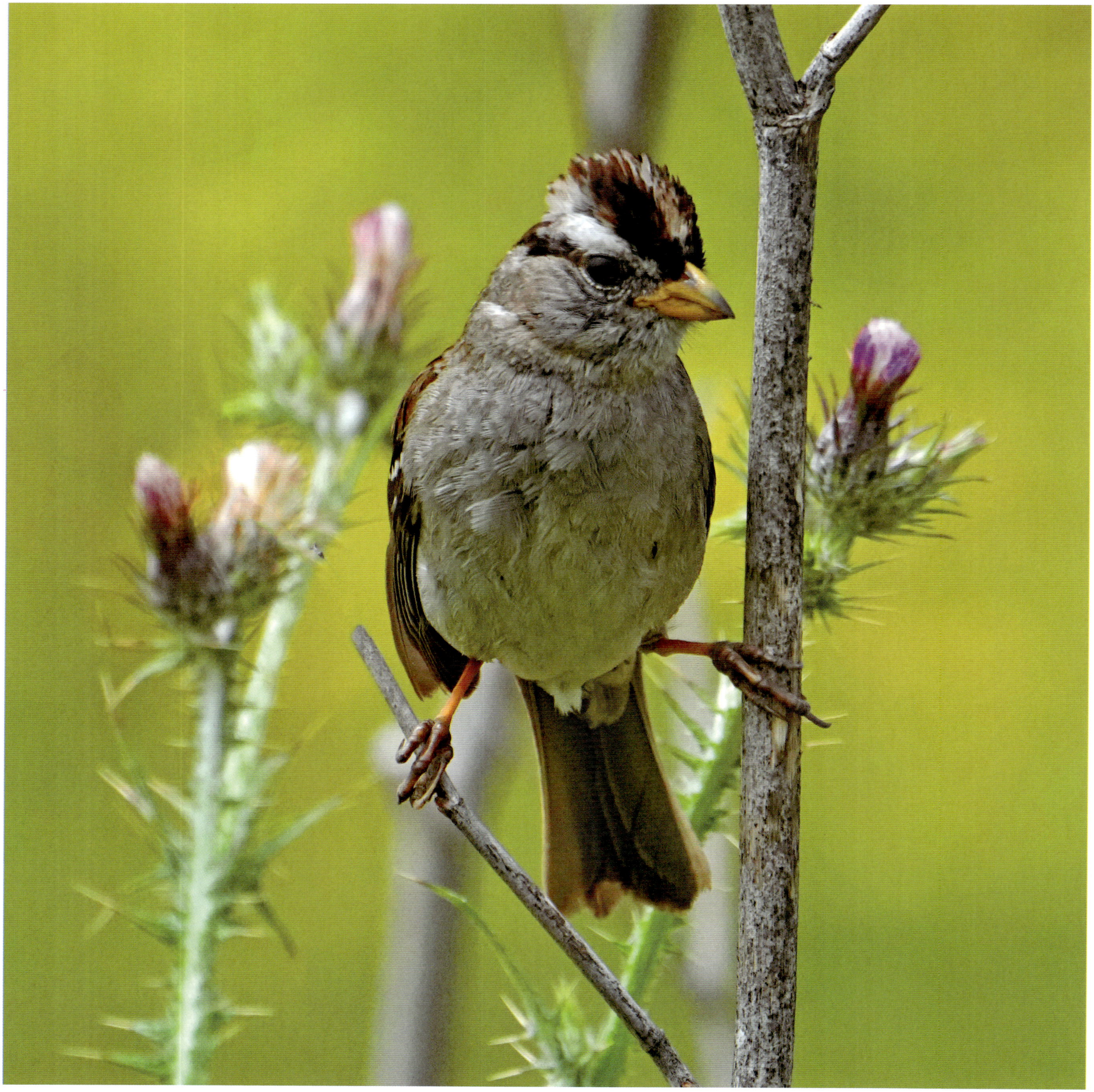

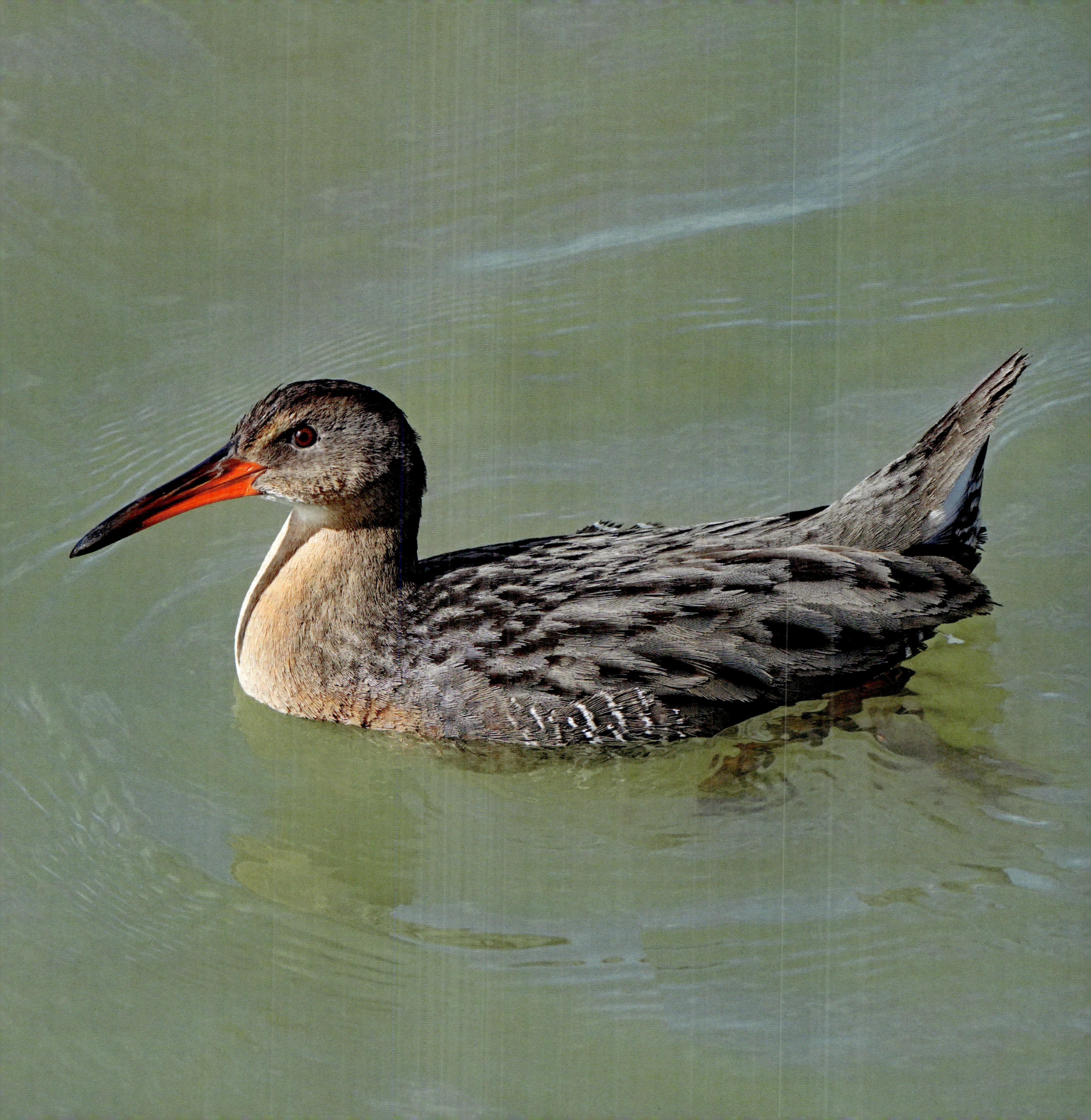

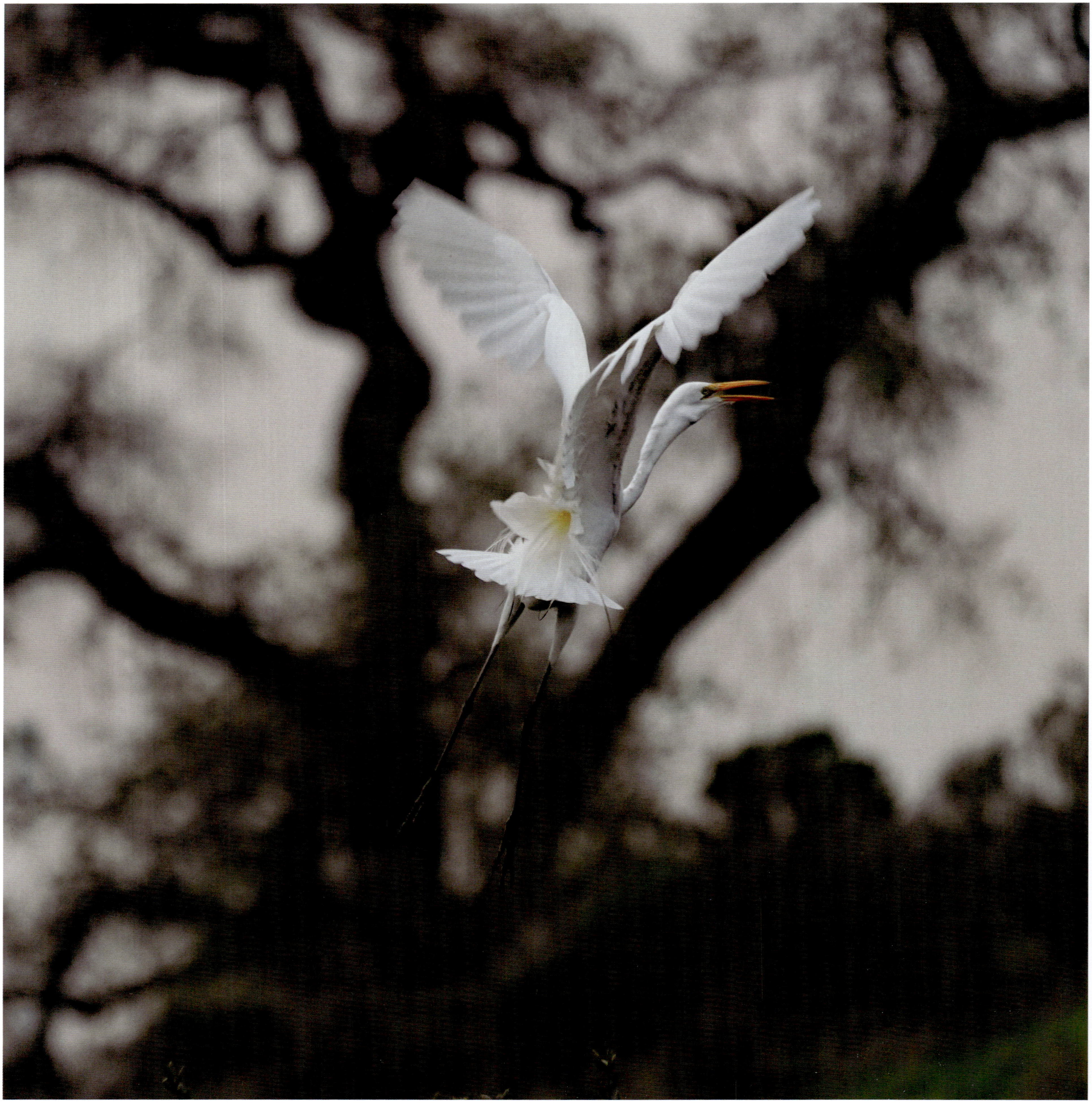

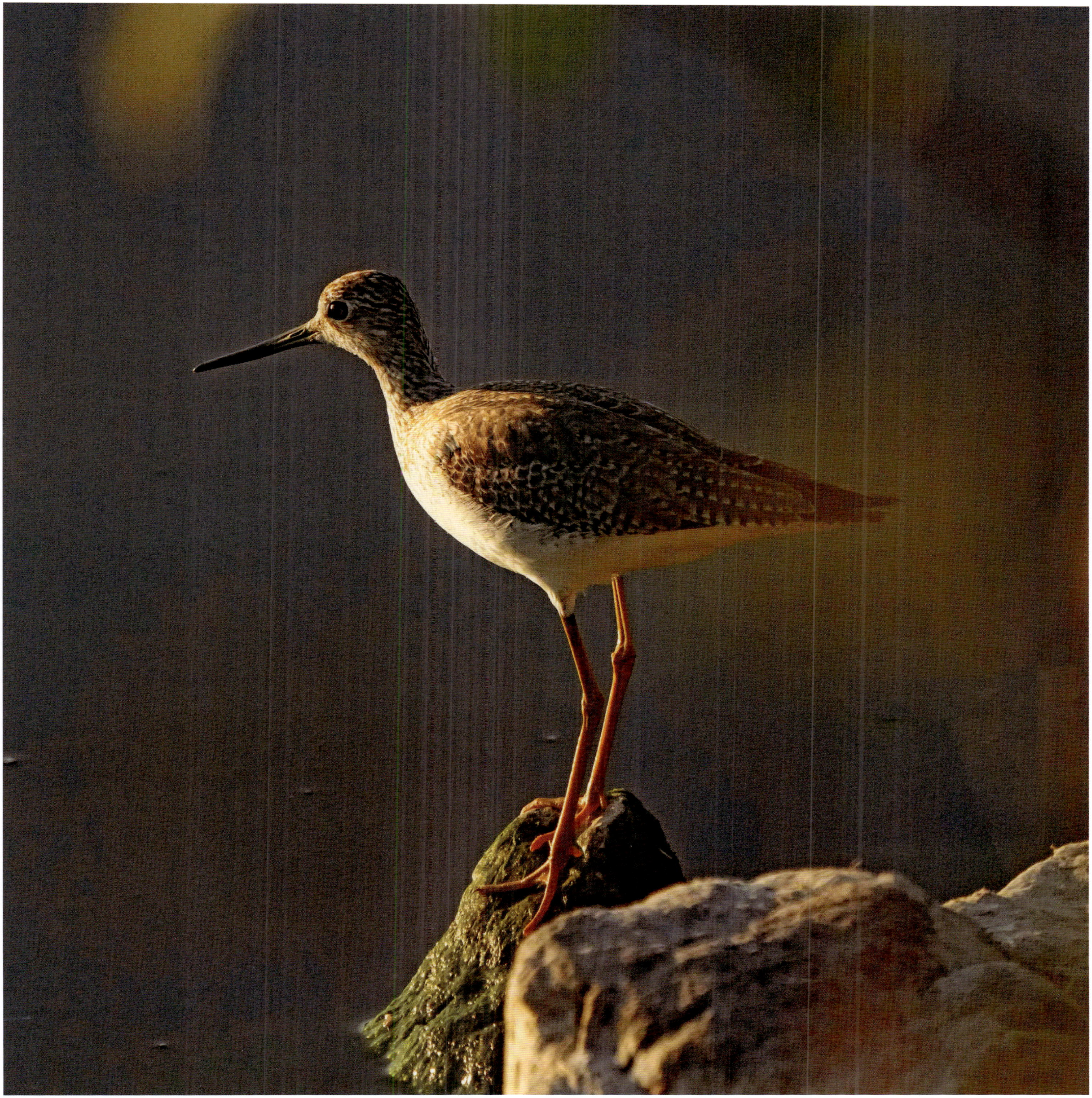

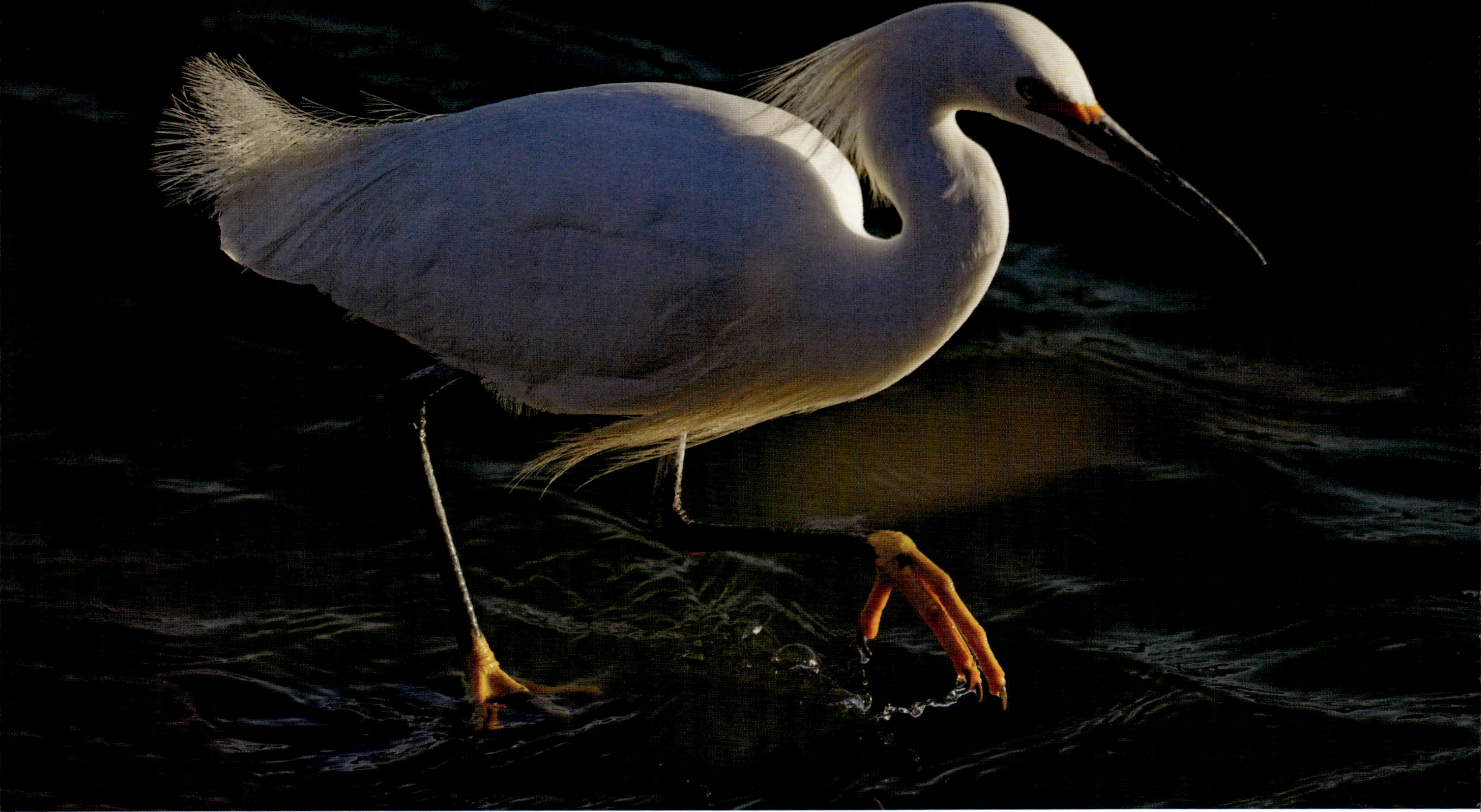

PRECEDING: A Great Egret (left) erupts into flight near Pacheco Pond, its lanky outstretched legs mirroring the tall oak trunks that rise behind it in the distance. From Southeast Alaska's rain-drenched peat bogs to South America's tidal flats, Greater Yellowlegs (right) stitch together the vast Pacific Coast, with watery Bay Area refuges at the heart of their incredible journey.

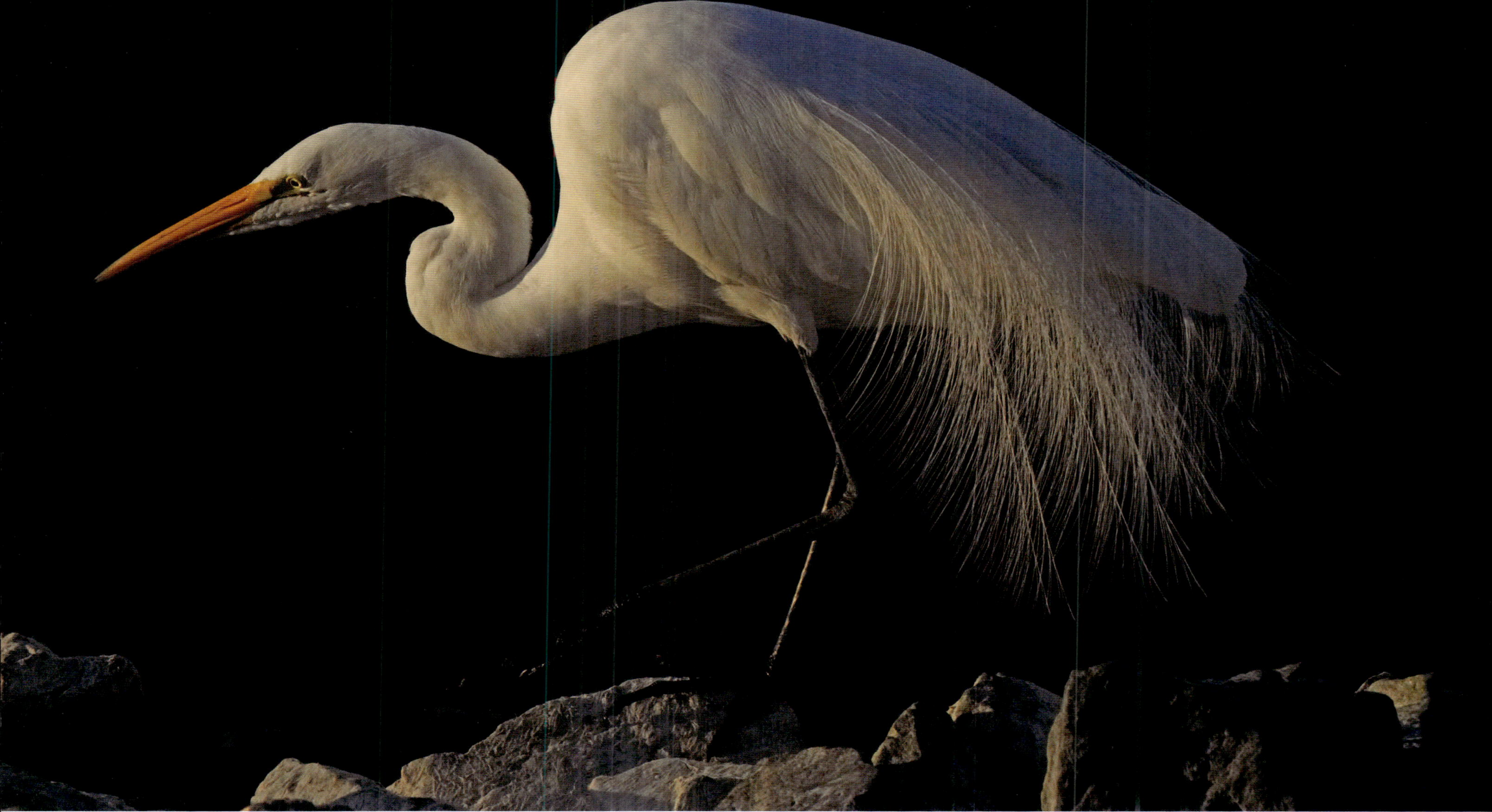

ABOVE: The more time we spend around birds, the more we learn to notice. Birding, after all, is an active practice of paying ever-closer attention. Though they appear similar from a distance, the Snowy Egret (left) has a dark bill and bright feet, as if it walked right through a spilled puddle of yellow paint. A Great Egret (right) has a yellow bill and black legs all the way down to the toes.

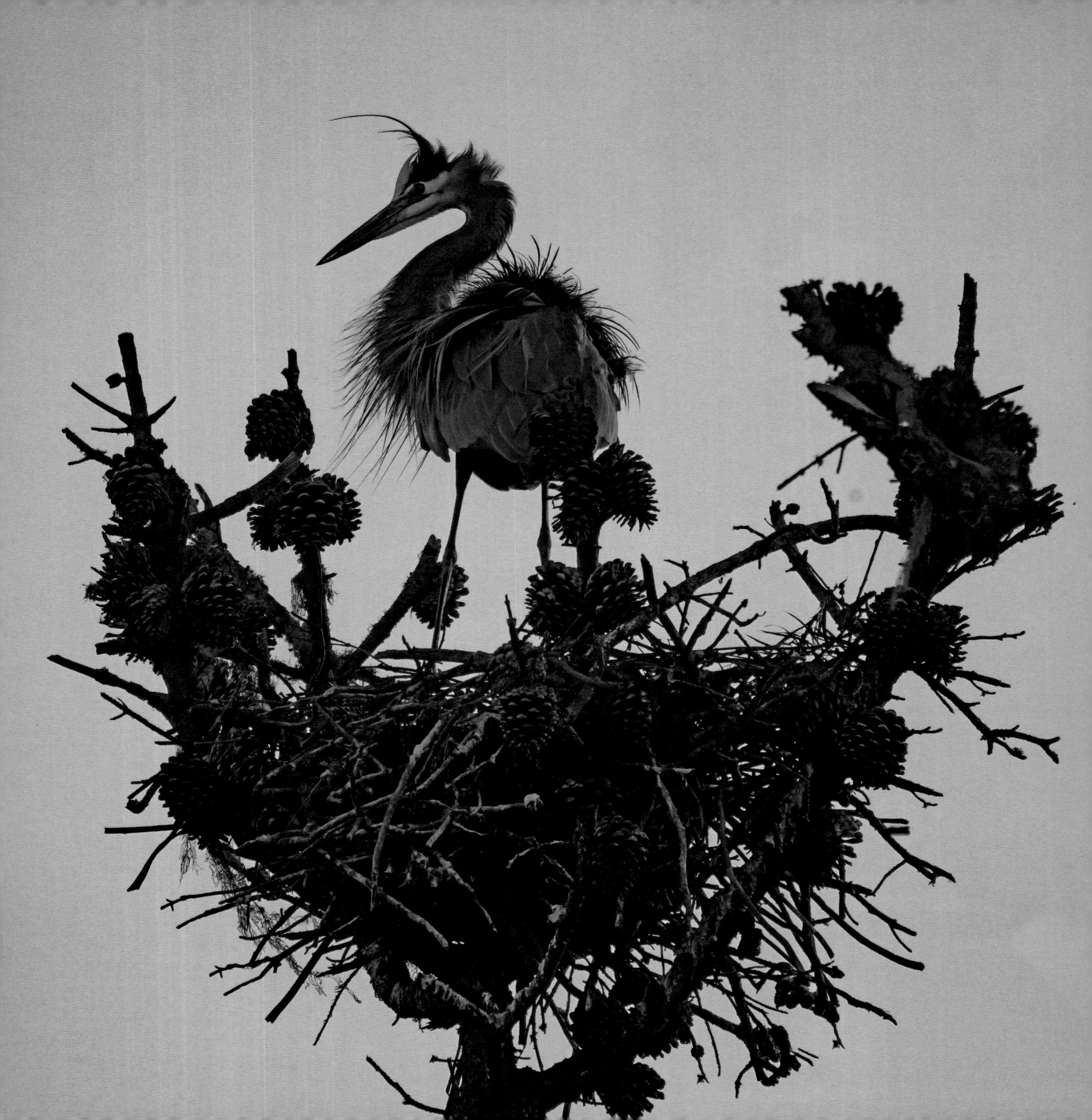

ACKNOWLEDGMENTS

Thank you to the dedicated team at Point Blue Conservation Science, without whom this book would never have grown wings: Grant Ballard, Manuel Olivia, Melissa Pitkin, Ryan DiGuadio, Julian Wood, Anne Chadwick, Diana Humple, Geoff Geupel, Meredith Elliot, Amanda Spears, Pete Warzybok, Mark Dettling, Laurette Rogers, Larissa Babicz, and all the current, past, and future apprentices at Point Blue, who care for our birds and continue to rise as conservation leaders in the shadow of the bridge and beyond. And a profound thank you to Christine Walker, for connecting us, accompanying us, and inspiring us every step of the way.

Beyond the extraordinary community at Point Blue, we are grateful for all the conservationists, educators, artists, and scientists who have guided us through this project. Thank you to Jack Dumbacher and Moe Flannery at the California Academy of Sciences for taking us behind the scenes into the bird collection, and to Sharon Beals for shooting dazzling photos of the specimens. Thank you to Susan Billy, Hopland Pomo basket maker. Thank you to masterful guides Teresa and Miles Tuffli, Lisa Hug, and Jean Lammers for taking us in search of birds and the stories that connect us to them. Thank you to Gerry McChesney, manager of the Farallon Islands National Wildlife Refuge—the longstanding partnership between USFWS and Point Blue has done more good than can ever be measured. Thank you to David Hage at Weaving Earth, for helping us deepen how we listen to birds. Thank you to Stemple Creek Ranch owner Loren Poncia for showing us how agriculture and conservation can powerfully interweave. Thank you to Keith Hansen for welcoming us into his art studio, and to Mary Ellen Hannibal for reading our book with an artist's eye and introducing it so beautifully in her foreword. And thank you to our Fault Line Feathers squad for the annual Rich Stallcup Bird-A-Thon—Malcom Blanchard, Rich Cimino, Grant Ballard, and Mary Ellen Hannibal—may our birding team continue to grow and triumph!

With gratitude, we acknowledge the birding community in the greater Bay Area—everyone we've met along this path has been bottomlessly open, supportive, curious, and kind. We are lucky to share this wild backyard with all of you.

A Great Blue Heron makes a striking silhouette atop its heavily built nest of branches and pinecones in Golden Gate Park.

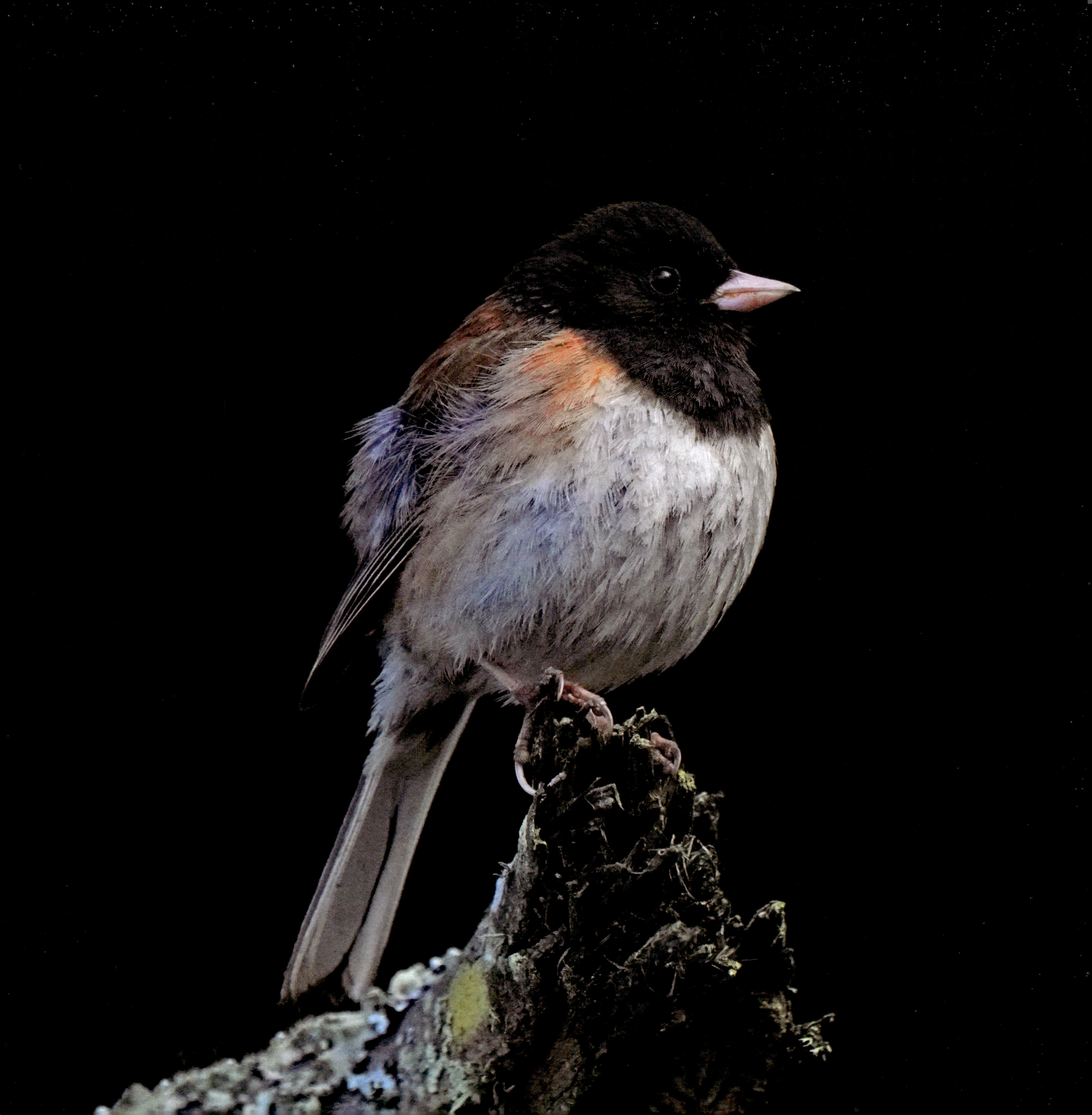

ABOUT THE AUTHORS

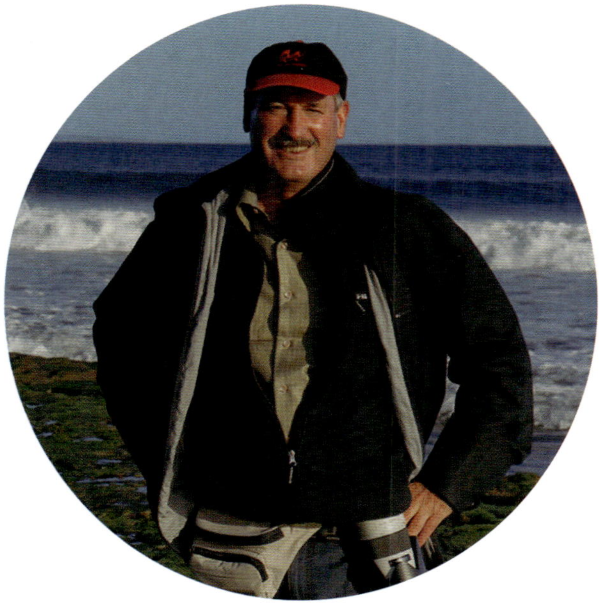

DICK EVANS BECAME INTERESTED IN PHOTOGRAPHY as a graduate student at Stanford University and continued his practice throughout a forty-seven-year career in the global metals industry that took him all over the world. San Francisco always remained home base, though, and he now lives in the city with his wife, Gretchen. Evans is the author or coauthor of the photography books *San Francisco and the Bay Area: The Haight-Ashbury Edition*, *The Mission*, and *San Francisco's Chinatown*.

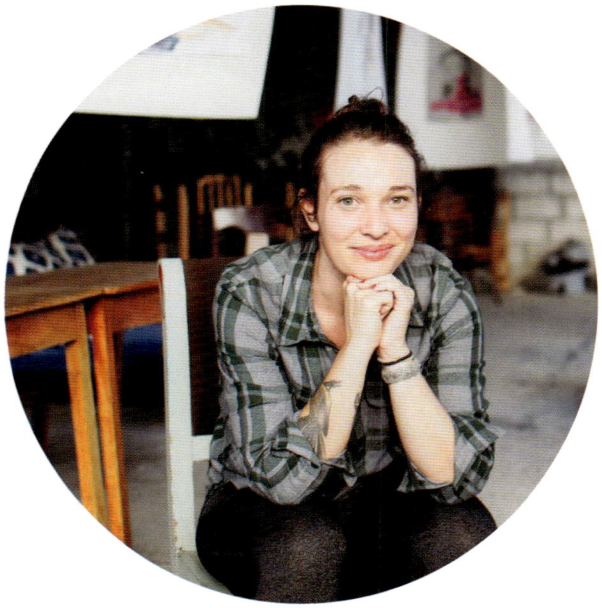

HANNAH HINDLEY IS A WILDERNESS GUIDE AND the recipient of the Thomas Wood Award in Journalism, the Ellen Meloy Desert Writers Award, and the Barry Lopez Prize in Nonfiction. She graduated from Harvard with degrees in English and evolutionary biology; she holds an MFA in creative nonfiction from University of Arizona. Her environmental essays can be found in *Bay Nature*, *The Sun*, *Hakai*, and more. Hannah writes about small creatures, big landscapes, and the scientists who love them.

Although some Dark-eyed Juncos migrate, many linger in the Bay Area year-round. Hikers might encounter these little ground-nesters in annual grasslands, coastal scrub, or forests, but you needn't venture far to find them—juncos are also frequent backyard visitors to bird feeders. They are familiar and endearing, filling the air with trilling *chips* as they hop and forage.

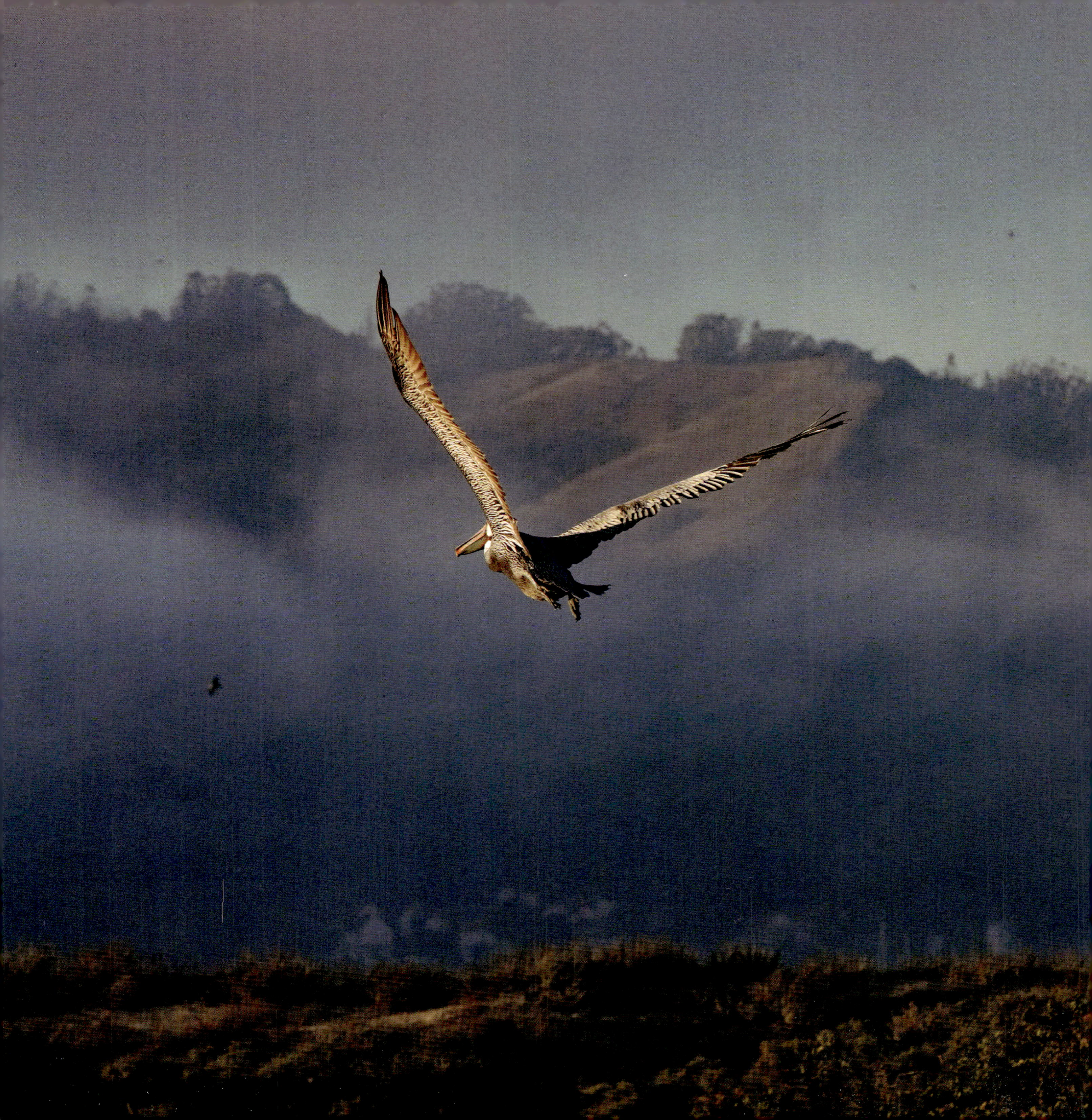

ABOUT POINT BLUE CONSERVATION SCIENCE

Point Blue Conservation Science is a nonprofit organization founded in 1965 as the Point Reyes Bird Observatory. Inspired by the deep connections between nature and communities in the Bay Area, Point Blue applies its expertise in science, partnerships, and community engagement to address the most pressing challenges facing wildlife and ecosystems—from the Pacific Ocean to the Sierra Nevada, and from Alaska to Antarctica.

Guided by the principles of connected conservation, Point Blue recognizes the interdependence of ecosystems, people, and global challenges. The organization's initiatives include stewarding migratory bird habitats, restoring wetlands, advancing climate-smart conservation, and safeguarding ocean ecosystems to support thriving marine wildlife and fisheries in the face of climate change.

Looking to the future, Point Blue is dedicated to harnessing cutting-edge science, integrating diverse perspectives, and empowering communities to implement nature-based solutions to the climate and biodiversity crises. With six decades of leadership, Point Blue is shaping a healthier, more sustainable future for wildlife and people alike.

LEFT: With a dinosauric wingspan of nearly 7 feet, a Brown Pelican evokes a sense of deep time as it takes flight, slow and powerful, in the fog above Crissy Field in San Francisco.

RIGHT: The Bald Eagle reminds us that conservation is connected. Its return to the Bay Area relies on healthy watersheds, protected open spaces, and thriving fish populations—all elements Point Blue works to understand and restore, from the Sierra to the sea, ensuring a wilder future for our region.

Point Blue Conservation Science

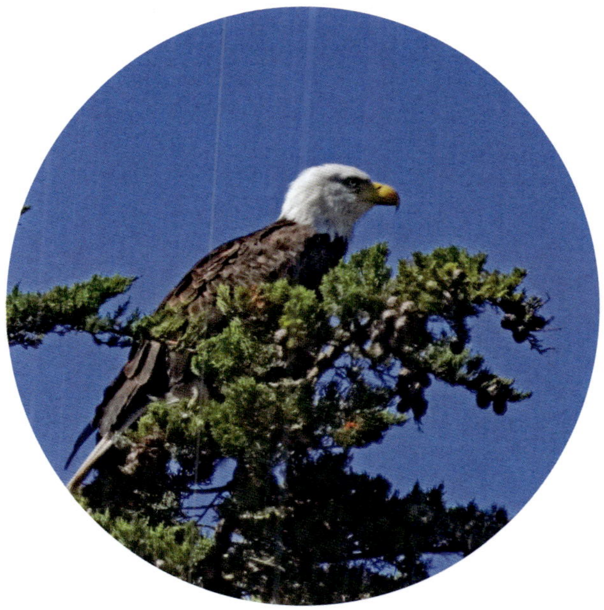